Coding Techniques

Coding Techniques

An Introduction to Compression and Error Control

Graham Wade

Principal Lecturer & Head of Communications Group
Department of Communication & Electronic Engineering
University of Plymouth

palgrave

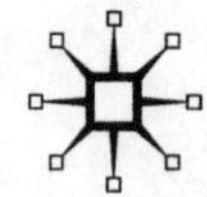

First published 2000 by
PALGRAVE
Houndmills, Basingstoke, Hampshire RG21 6XS and
175 Fifth Avenue, New York, N. Y. 10010
Companies and representatives throughout the world

PALGRAVE is the new global academic imprint of St. Martin's Press LLC Scholarly and Reference Division and Palgrave Publishers Ltd (formerly Macmillan Press Ltd).

ISBN 0–333–76011–5

A catalogue record for this book is available from the British Library.

This book is printed on paper suitable for recycling and made from fully managed and sustained forest sources.

Typeset by Ian Kingston Editorial Services, Nottingham, UK

10 9 8 7 6 5 4 3 2 1
09 08 07 06 05 04 03 02 01 00

Printed in Great Britain by
Creative Print and Design (Wales), Ebbw Vale

To Jenny
Dominic, Alistair, and Oliver

The fear of the Lord is the beginning of knowledge.
Prov 1 v 7

Contents

Preface

About the book

Coding techniques are fundamental to digital communications systems and have today matured into complex compression and error control algorithms. For example, the LZW algorithm is the basis of a number of lossless file compression techniques, CELP speech coding is used for personal communications, and MPEG-2 video coding can compress broadcast video by typically 30:1. In the field of error control, soft-decision Viterbi decoding is an industry standard in satellite systems, cyclic codes are the basis of protection in the Radio Data System (RDS), and Turbo decoders can approach the theoretical limit to within 1 dB. In addition, pseudo-random (PN) sequences are widely used in data modems, conditional access systems, CDMA, and more recently in image and video watermarking systems. The objective of this book is to examine the theoretical basis and practical implementation of these and similar coding techniques.

Every attempt has been made to include topical subjects and applications, such as coding for Teletext and RDS; GSM and CELP speech compression for voice communications; arithmetic coding (which can outperform Huffman coding); JPEG, JBIG and wavelet image compression; MPEG-2 compressed video; spread-spectrum based watermarking; frequency-domain Reed–Solomon decoding; soft-decision Viterbi decoding; and, of course, Turbo decoding for state-of-the-art error control.

Chapter 1 covers PCM fundamentals, since PCM is the first encoding step in a digital communications system. It covers theoretical and practical aspects of sampling and quantising. The chapter concludes with an overview of compression techniques and standards for speech, audio, data, image and video sources.

Chapter 2 covers lossless data compression techniques and commences with essential information theory. Much of the chapter is concerned with the efficient transmission of two-tone documents, as in the classical Huffman coding (variable-length coding) used in facsimile systems. The chapter also introduces the more advanced ideas of Arithmetic Coding (which can outperform Huffman coding), and the concepts of resolution reduction, dither and the advanced JBIG algorithm. The chapter concludes with the LZW technique, which is the basis of a number of lossless file compression formats.

Chapter 3 covers lossy compression techniques for speech, audio, image and video signals. After discussing the theory and practice of conventional DPCM, considerable emphasis is placed upon transform coding (the DCT), since this is fundamental to JPEG and MPEG systems. A separate section is devoted to the fundamentals of MPEG-2. The advanced concept of wavelet-based image compression is also introduced, and the chapter concludes with the theoretical basis of GSM and CELP speech codecs.

Chapter 4 covers aspects of channel coding other than forward error correction (FEC). It therefore covers the codes used for baseband digital transmission (interface and line codes), and some techniques used for magnetic recording (such as partial response signalling). Pseudonoise (PN) sequences are used extensively in digital communications systems for scrambling and spread spectrum applications, so PN sequence generation and related correlation aspects are considered in some detail. The application of PN sequences to video watermarking for copyright protection is also discussed.

Chapter 5 is the first of three chapters on error control coding and is concerned with the systems aspects of error control rather than the details of encoding. For instance, what are the bit rate and bandwidth implications of adding FEC, how good is a particular FEC system compared to the theoretically optimal system, and what is the decoded bit rate for a particular system? The chapter also discusses the concepts of Maximum Likelihood (ML) decoding, soft-decision decoding, and the important topic of MAP decoding (as used in Turbo decoders).

Chapter 6 commences with an introduction to the use of block codes for FEC and illustrates the theory with simple examples from Teletext and computer memory systems. It then progresses to the more general class of cyclic codes (binary BCH codes) and the methods for decoding them, and describes the use of such codes in the European Radio Data System (RDS). The algebraic theory of binary cyclic codes leads naturally to Reed–Solomon (RS) codes. These non-binary codes are used extensively in digital recording, terrestrial broadcasting, and satellite communications and so are discussed in some detail.

Chapter 7 covers convolutional codes, these being a practical alternative to block codes for FEC. It commences with the basic concepts of convolutional codes such as constraint length, code trellis, free distance and puncturing. The widely used Viterbi decoder is discussed in terms of both hard- and soft-decision decoding, and the latter is available as a C routine in Appendix C for an industry-standard code. The chapter then examines the advanced concept of iterative decoding (Turbo decoding and its variants) and includes a decoder analysis that can be used for the implementation of such a system. The theory is illustrated with a number of simulations carried out at the University of Plymouth, and the reader can gain experience of iterative decoding via the C routine in Appendix D. The chapter concludes with a look at some practical aspects of iterative decoding.

Readership

Students of electronic, communication or information engineering will find the book covers many of the signal and data coding topics that are usually studied to first degree level. The book actually goes beyond first degree level in some of these topics and so should also appeal to postgraduate students and engineers in industry. It contains many tables, so it should also be useful as a reference book.

Acknowledgements

Sections of the text have been reviewed and refined by experts in the field. I would like to acknowledge the kind help of:

A. Ambroze (University of Plymouth), S. Benedetto (Politecnico di Torino), M. Borda (University of Cluj-Napoca), P. Coulton (Lancaster University), M. Darnell (University of

Leeds), P. Fan (Southwest Jiaotong University), P. Farrell (Lancaster University), M. Ghanbari (Essex University), L. Hanzo (University of Southampton), P. James (University of Plymouth), N. Kingsbury (University of Cambridge), G. Nason (University of Bristol), L. Perez (University of Nebraska), S. Regunathan (University of California), I. Richardson (Robert Gordon University), R. Salmon (BBC Research & Development), R. Sewell (Cambridge Consultants Ltd.), P. Smithson (University of Plymouth), R. Terebes (University of Cluj-Napoca), M. Tomlinson (University of Plymouth), P. Van Eetvelt (University of Plymouth), M. Vetterli (Swiss Federal Institute of Technology), L. Wang (University of Texas).

Graham Wade
University of Plymouth
April 2000

Abbreviations

ADC	analogue-to-digital converter
ADPCM	adaptive differential pulse code modulation
AMI	alternate mark inversion
ARQ	automatic repeat request
ATM	asynchronous transfer mode
AWGN	additive white Gaussian noise
BCH	Bose–Chaudhuri–Hocquenghem
BER	bit-error rate
B*n*ZS	bipolar *n*-zero substitution
BSC	binary symmetric channel
CCIR	International Radio Consultative Committee
CCITT	International Telegraph & Telephone Consultative Committee
CELP	code-excited linear prediction
codec	coder–decoder
CR	compression ratio
CS-CELP	conjugate structure code-excited linear prediction
DAB	digital audio broadcasting
DAC	digital-to-analogue converter
DCT	discrete cosine transform
DED(C)	double error detecting (correcting)
DFT	discrete Fourier transform
DPCM	differential pulse code modulation
DWT	discrete wavelet transform
E	expectation
ETSI	European Telecommunications Standards Institute
FEC	forward error control
FFT	fast Fourier transform
GF	Galois field
GIF	graphics interchange format
GSM	Groupe Speciale Mobile (European global system for mobile communications).
HDB*n*	high-density bipolar *n*
HVS	human visual system
IDFT	inverse discrete Fourier transform
IEC	International Electrotechnical Commission
IFFT	inverse fast Fourier transform
IMT 2000	International Mobile Telecommunications 2000

ISI	intersymbol interference
ISO	International Organization for Standardization
ITU	International Telecommunications Union (formerly CCITT)
ITU-R	radio communications sector
ITU-T	telecommunications sector
JBIG	Joint Bilevel Image Group
JPEG	Joint Photographic Experts Group
LPC	linear predictive coding
LSB	least significant bit
LZW	Lempel–Ziv–Welch
MH	modified Huffman (coding)
MIPS	millions of instructions/second
MLSE	maximum likelihood sequence estimation
mmse	minimum mean square error
MPEG	Moving Picture Experts Group
MR	modified Read (coding)
NRZ	non-return to zero
NRZI	modified non-return to zero
NTSC	National Television Systems Committee
PAL	phase alternation line
PCM	pulse code modulation
pdf	probability density function
PSK	phase shift keying
PAM	pulse amplitude modulation
QPSK	quadrature phase shift keying
RDS	radio data system
RLC	runlength coding
RLL	runlength-limited (code)
RS	Reed–Solomon
RZ	return to zero
SEC(D)	single error correcting (detecting)
SNR	signal-to-noise ratio
SR	syndrome register
TIFF	tag image file format
UMTS	Universal Mobile Telecommunications System
UTRA	UMTS terrestrial radio access
VLC	variable-length coding
VOD	video-on-demand

CHAPTER 1
PCM fundamentals

The coding processes in a generalised digital communications system can be identified as analogue-to-digital conversion, followed by data compression (*source coding*), followed by coding for error control and transmission (*channel coding*). The majority of this book is concerned with source and channel coding, but first we examine the theoretical and practical aspects of *pulse code modulation* (PCM). Pulse code modulation is important because it is invariably the signal format at the source coder input. We start by outlining Shannon's ideal coding scheme, since this sets theoretical bounds for practical coding schemes, and then compare the performance of PCM to this ideal scheme. The chapter then examines two fundamental aspects of PCM (sampling and quantising) from theoretical and practical standpoints, and outlines how sampling rates and quantising noise can be 'adjusted' in order to efficiently convert analogue signals to PCM or vice versa. The chapter concludes with an overview of standard techniques for coding speech, data, images and video signals.

1.1 The ideal coding concept

Consider the receiver in Figure 1.1 (essentially a demodulator or decoder) where S_i and N_i are the average signal and noise powers for the input, S_o and N_o are the average signal and noise powers for the output, B_T is the transmission bandwidth and B_m is the message bandwidth. In general (but not always) a practical receiver improves the signal-to-noise ratio (SNR) in exchange for signal bandwidth, so that by choosing a modulation or coding scheme such that $B_T >> B_m$ it should be possible for the receiver to deliver a high-quality signal despite a low S_i/N_i ratio. Both coded and uncoded modulation schemes permit this type of exchange, and a well-known example of an uncoded scheme that does this is wideband frequency modulation (FM).

The most efficient exchange can only be achieved using a coded system. In fact, it is reasonable to seek a coded system that will give arbitrarily small probability of error at the receiver output, given a finite S_i/N_i. To this end it is helpful to have a broad view of coding which does not necessarily involve digital signals. Rather, we could view it as the transformation of a message signal or symbol into another signal or sequence of symbols for transmission. Using this broad concept of coding it is possible to regard a coded message as a vector in multidimensional signal space and to deduce from a noise sphere packing

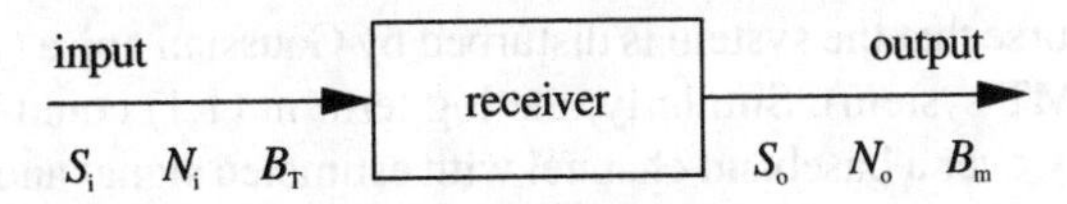

Figure 1.1 *Power-bandwidth exchange in a receiver.*

argument that it is theoretically possible to reduce the probability of error at the receiver output to zero, *despite finite channel noise*. A rigorous treatment of this problem was given by Shannon in 1949 and, broadly speaking, his *ideal coding scheme* involves restricting the source output to a finite set of M different messages and assigning a particular message to a randomly selected signal or code vector of duration T seconds. For a binary digital system a particular message would be assigned to an n-digit codeword selected at random from the 2^n possible code vectors. The optimum receiver would involve correlation or matched filter detection and a minimum-distance (Maximum-Likelihood) decision criterion.

Unfortunately, it turns out that to achieve arbitrarily small probability of error at the receiver output it is necessary for T (or n) to extend to infinity. In other words, the ideal coding scheme requires infinite coding/decoding delay! In turn, this means that to maintain a constant information rate of $(1/T)\log_2 M$ bits/s (see Section 2.2) then M must increase exponentially. These factors make such an ideal coding scheme impractical, although the concept is still extremely important since it defines the ultimate that can be achieved with respect to error-free transmission rate and bandwidth-SNR exchange. For his ideal coding scheme, Shannon showed that

$$C = B\log_2\left(1+\frac{S}{N}\right) \text{ bits/s} \tag{1.1}$$

where C is the *channel capacity*, B is the channel bandwidth (in Hz), S and N are the average received signal and noise powers respectively, and the noise is additive white Gaussian noise. Equation (1.1) is referred to as the *Shannon–Hartley* law and it acts as a benchmark for various practical modulation/demodulation schemes since it defines the absolute maximum information rate, R, which can be *reliably* (without error) sent over the channel. We will use it in Chapter 5 to establish a theoretical limit for error control systems.

EXAMPLE 1.1

Equation (1.1) can be used to obtain an upper performance bound for a practical system. For example, since (1.1) applies to bandpass channels as well as to lowpass channels, it can be used to estimate an upper limit to the achievable bit rate in a *discrete multitone* (DMT) system. In a DMT system the usable bandwidth is divided into N subchannels (carriers) and a high data rate is assigned to subchannels with high SNR, and vice versa. Assuming independent subchannels, an upper bound to the capacity of such a system could be estimated as

$$C = \sum_{j=1}^{N} C_j = \sum_{j=1}^{N} B_j \log_2(1+SNR_j) \tag{1.2}$$

This assumes of course that the system is disturbed by Gaussian noise (which is not necessarily true for a DMT system). Similarly, the log term in (1.1) could be integrated with respect to frequency over a baseband channel with estimated signal and noise characteristics in order to obtain an upper limit to its capacity. For example, this type of calculation reveals that very high rate digital subscriber line (VDSL) copper access systems have a target capacity of the order of 50 Mbit/s over short metallic cables.

As just stated, Shannon's noisy coding theory also defines the optimal bandwidth-SNR exchange in a communications system. Suppose for example that the receiver in Figure 1.1 is part of Shannon's ideal coding/decoding scheme. Since there will be no loss of information in the receiver we have

$$B_T \log_2\left[1+\frac{S_i}{N_i}\right] = B_m \log_2\left[1+\frac{S_o}{N_o}\right] \quad (1.3)$$

If $S_o/N_o >> 1$ and $S_i/N_i >> 1$ then

$$\frac{S_o}{N_o} \approx \left(\frac{S_i}{N_i}\right)^{B_T/B_m} \quad (1.4)$$

The ideal coding scheme therefore permits an exponential increase in SNR with transmission bandwidth, that is, a low input SNR can give a high output SNR provided $B_T >> B_m$. Similar (but less efficient) exchange of bandwidth for SNR is achieved in PCM and spread spectrum systems.

It should now be clear that every practical communications channel has finite information capacity owing to finite bandwidth and finite noise and signal powers. In order to transmit information reliably, the information rate, R, must be less than the channel capacity, and this is underscored by Shannon's *noisy coding theorem*:

Theorem 1.1 It is theoretically possible to transmit information through a noisy channel with arbitrarily small probability of error provided that the information or source rate, R, is less than the channel capacity, that is, $R < C$ for reliable transmission.

Simply sampling a signal and transmitting the analogue samples cannot lead to an arbitrarily small error probability since the corresponding information rate will be infinite (see (2.3))! Rather, according the Theorem 1.1 we must restrict the source information rate if we are to have a small probability of error at the decoder output. Adopting the concept behind Shannon's ideal coding scheme, this can be achieved by restricting the source to a finite alphabet of M discrete messages or symbols transmitted every T seconds. This *discrete source* is usually generated via *quantisation* of the samples using an analogue-to-digital converter (ADC). We will see in Section 2.2 that a discrete source of M symbols has a maximum information rate or *entropy rate* of $R = (1/T)\log_2$ bits/s and clearly this could be made less than the available channel capacity, C. If this is the case, we can conjecture that virtually all the transmitted information will be received uncorrupted or, put another way, *noise in the channel will have negligible effect upon the received signal quality.* The major impairment in such a system will then be the errors due to quantising, and this is the fundamental concept behind PCM. Looked at another way, if the channel does not have the required capacity – that is, it generates a high decoded bit error rate (BER) – then (1.1) implies that we could increase the signal power (specifically E_b/N_o, as in Fig. 5.11(a)).

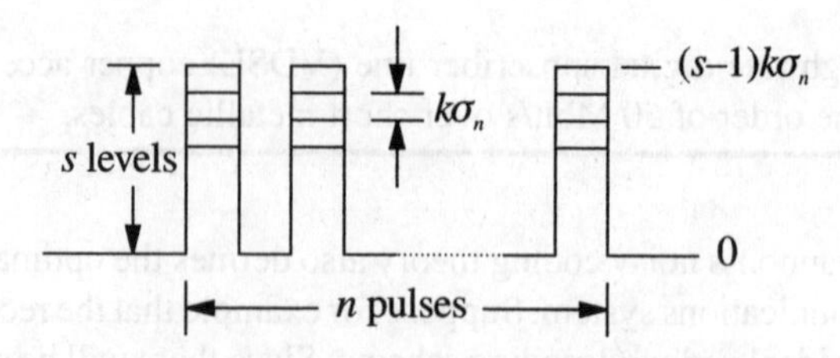

Figure 1.2 *PCM code group for s-ary pulses.*

Alternatively, it is probably more economic to add *forward error control* (FEC). It is shown later that this usually increases the channel bandwidth, so (1.1) again implies that the channel capacity will tend to increase.

1.2 PCM bandwidth

A PCM coder simply applies waveform sampling, quantising and coding (usually binary number, offset binary or Gray coding is used, although others are possible). We shall see in Section 1.4 that if the analogue source is bandlimited to B_m then the signal can be sampled at the *Nyquist rate* of $2B_m$ (or greater), and all the information will be retained in the samples. To create a discrete source, we next quantise each sample to one of M levels and assume that the analogue signal forces each level to be equiprobable. As shown later, this assumption gives the maximum possible average information per sample ($\log_2 M$ bits), so the discrete source will be generating information at the maximum rate of

$$R_{max} = 2B_m \log_2 M \quad \text{bits/s} \tag{1.5}$$

Each quantised sample is now coded by assigning it to a code group as indicated in Figure 1.2 and s-ary PCM is assumed where $s < M$ for a coded system. Since n s-ary pulses have s^n combinations, we require that $s^n = M$, corresponding to $\log_s M$ pulses per sample, and the serial transmission rate becomes $2B_m \log_s M$ pulses/s. At this point we call upon Nyquist's bandwidth–signalling speed relationship for a transmission channel:

Theorem 1.2 An ideal lowpass channel of bandwidth B can transmit independent symbols at a rate $R_s \leq 2B$ without intersymbol interference (ISI).

From this theorem it follows that the minimum possible baseband transmission bandwidth for PCM is

$$B_T = B_m \log_s M \quad \text{Hz} \tag{1.6}$$

Example 1.2
For industrial closed-circuit television (CCTV) satisfactory monochrome images are obtained using a 6-bit ADC. This corresponds to $s = 2$, $M = 64$, so at least six times the video bandwidth would be required for baseband serial transmission of the digitised signal.

1.3 PCM channel capacity

If no information is to be lost during transmission (no decoding errors) then, according to Theorem 1.1 and (1.5) the PCM channel capacity must be

$$\begin{aligned} C_{\mathrm{PCM}} &= 2B_{\mathrm{m}} \log_2 M \\ &= 2B_{\mathrm{m}} \log_s M \log_2 s \\ &= B_{\mathrm{T}} \log_2 s^2 \end{aligned} \tag{1.7}$$

For binary PCM ($s = 2$) this means that the maximum transmission rate is simply $2B_{\mathrm{T}}$ bits/s or 2 bits/s/Hz. In order to compare the power requirements of PCM with that of Shannon's ideal coding scheme, it is necessary to express s in terms of $S_{\mathrm{i}}/N_{\mathrm{i}}$ and to assume that the channel is disturbed by additive white Gaussian noise. Let the s-ary levels be separated by $k\sigma_n$, where k is a constant, σ_n is the rms noise in the channel, and the normalised noise power is $N_i = \sigma_n^2$. Since the analogue signal is assumed to have a flat probability density function (pdf), it follows that the s levels are equiprobable, in which case the average normalised signal power at the demodulator input is

$$S_i = \frac{1}{s} \sum_{r=0}^{s-1} (rk\sigma_n)^2 \tag{1.8}$$

$$= \frac{(k\sigma_n)^2}{s} \sum_{r=0}^{s-1} r^2 \tag{1.9}$$

and this reduces to

$$S_i = \frac{(k\sigma_n)^2 (s-1)(2s-1)}{6} \tag{1.10}$$

Note that the s-ary pulses are unipolar and that the corresponding DC component has an average power

$$S_{\mathrm{DC}} = \frac{(k\sigma_n)^2 (s-1)^2}{4} \tag{1.11}$$

Since this component does not convey useful information, we can resort to bipolar transmission of the s-ary pulses and so reduce S_{i}. In this case

$$S_{\mathrm{i}} = \frac{(k\sigma_n)^2 (s-1)}{24} [4(2s-1) - 6(s-1)] \tag{1.12}$$

$$= \frac{(k\sigma_n)^2 (s^2 - 1)}{12} \tag{1.13}$$

so

$$\frac{S_{\mathrm{i}}}{N_{\mathrm{i}}} = \frac{k^2 (s^2 - 1)}{12} \tag{1.14}$$

Finally, substituting for s^2 in (1.7) gives

$$C_{\text{PCM}} = B_{\text{T}} \log_2 \left(1 + \frac{12}{k^2} \frac{S_{\text{i}}}{N_{\text{i}}} \right) \tag{1.15}$$

Note that by making this substitution we have introduced a practical coding scheme which, in turn, implies that the decoded error probability will not be arbitrarily small. In other words, while (1.1) assumes a complex coding technique yielding *arbitrarily small* error probability, (1.15) specifies the maximum rate at which information can be transmitted with *small* error. For example, considering a binary PCM system with $k = 10$, the bit-error probability $P(\varepsilon)$ for a Gaussian noise channel is approximately 3×10^{-7} (which turns out to be a reasonable objective for a PCM system handling broadcast-quality video signals, for example). Bearing this in mind, we can compare (1.1) and (1.15) for the same C/B ratios to give

$$\left(\frac{S_{\text{i}}}{N_{\text{i}}} \right)_{\text{PCM}} = \frac{k^2}{12} \left(\frac{S_{\text{i}}}{N_{\text{i}}} \right)_{\text{ideal}} \tag{1.16}$$

This means that the PCM system needs to increase the signal power by a factor $k^2/12$ above that for Shannon's ideal coding scheme. For $k = 10$ this is an increase of 9.2 dB and even then the error probability for a binary system will be about 3×10^{-7} rather than zero. Increasing k will however reduce $P(\varepsilon)$ towards zero.

It is also important to note what happens as k (or equivalently $S_{\text{i}}/N_{\text{i}}$) is reduced. Typically, a decrease of several dB in the SNR can increase $P(\varepsilon)$ by several orders of magnitude and $P(\varepsilon)$ can rapidly become significant! Clearly, in contrast to uncoded systems, if the SNR can be kept above this effective threshold then the PCM channel can be regarded as virtually transparent.

To summarise, PCM is important because of its simple implementation, widespread use and versatility (it is not *signal-specific*). It is also used as a practical benchmark for comparison of data compression schemes (as for example in Table 1.4). Finally, we have highlighted the possibility of exchanging signal power for bandwidth in coded systems, and a common way of doing this is to incorporate FEC.

1.4 Ideal sampling and reconstruction of signals

Conversion of a signal to PCM simply involves sampling, quantising, and the coding of the quantising levels. The ADC carries out all three processes, and the reverse operations are carried out by the DAC and *reconstruction filter*. The essential elements of a PCM system are shown in Figure 1.3, the bandlimiting and reconstruction filters being crucial components.

Sampling theory is neatly and powerfully handled using the Fourier transform and various convolutional relationships, and we shall adopt that approach here. In particular, *ideal sampling* can be neatly explained using the Fourier transform pairs in Figure 1.4. Suppose that an arbitrary continuous time signal $x(t)$ is bandlimited to the frequency $\omega_{\text{m}} = 2\pi f_{\text{m}}$ and is sampled at the rate $\omega_{\text{s}} = 2\pi f_{\text{s}}$ (we assume that the signal has *finite* energy so that its Fourier transform exists). If this signal is real, then the magnitude $|X(\omega)|$ of its Fourier transform is an even function, as illustrated in Figure 1.4(a). For ideal or *impulse* sampling, $x(t)$ is multiplied by the impulse sequence $\delta_{\text{T}}(t)$ to give the sampled

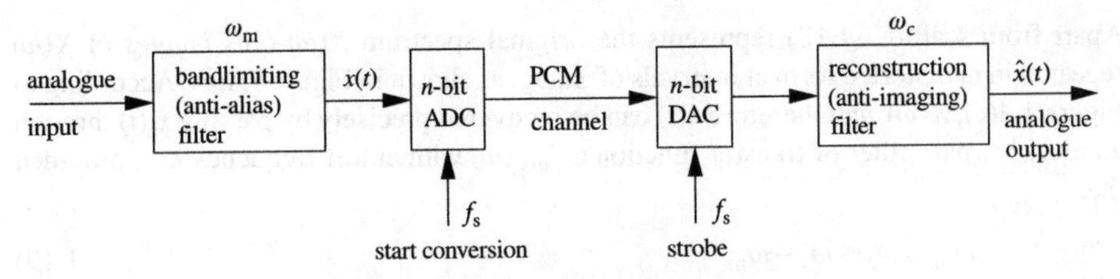

Figure 1.3 *Basic elements of a PCM system.*

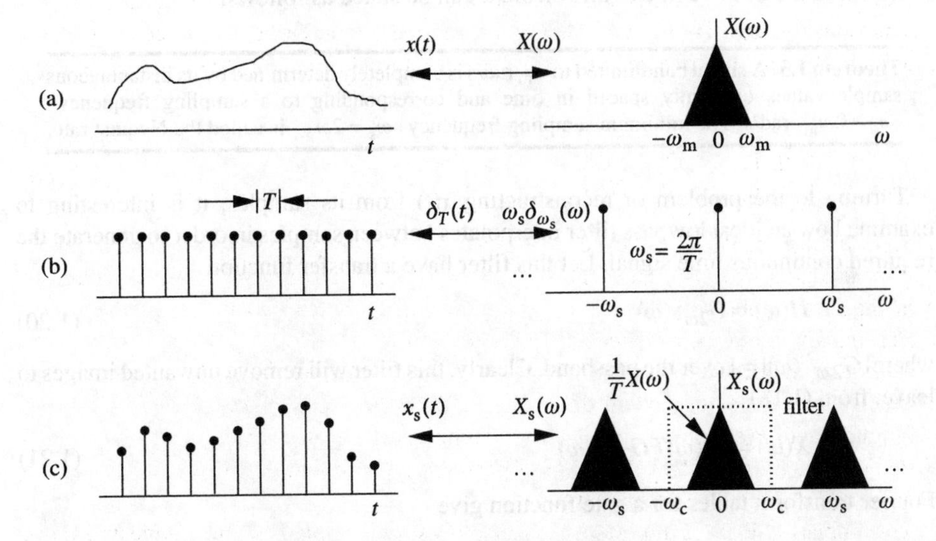

Figure 1.4 *Fourier transform pairs for ideal sampling (transform phase is omitted for clarity).*

signal $x_s(t)$, and the Fourier transform of $x_s(t)$ follows from the frequency convolution theorem:

$$
\begin{aligned}
X_s(\omega) &= \frac{1}{2\pi}\left[X(\omega) * \omega_s\delta_{\omega_s}(\omega)\right] \\
&= \frac{1}{T}\left[X(\omega) * \sum_{n=-\infty}^{\infty}\delta(\omega - n\omega_s)\right]
\end{aligned}
\tag{1.17}
$$

where n is an integer. Using the distributive convolution law

$$
\begin{aligned}
X_s(\omega) &= \frac{1}{T}\sum_{n=-\infty}^{\infty} X(\omega) * \delta(\omega - n\omega_s) \\
&= \frac{1}{T}\sum_{n=-\infty}^{\infty} X(\omega - n\omega_s)
\end{aligned}
\tag{1.18}
$$

Apart from scaling, (1.18) represents the original spectrum $X(\omega)$ plus *images* of $X(\omega)$ repeated in undistorted form at intervals of $\pm n\omega_s$, as shown in Figure 1.4(c). According to Figure 1.4(c), $X(\omega)$ and therefore $x(t)$ can be recovered precisely by passing $x_s(t)$ through an ideal lowpass filter of transfer function $G_{2\omega_c}(\omega)$ with cutoff frequency ω_c, provided that

$$\omega_m \le \omega_c \le \omega_s - \omega_m \tag{1.19}$$

If this is true, then spectral overlap or *aliasing* is avoided and the sampling theorem is satisfied. In the above context, this theorem can be stated as follows:

Theorem 1.3 A signal bandlimited to ω_m rad/s is completely determined by its instantaneous sample values uniformly spaced in time and corresponding to a sampling frequency $\omega_s \ge 2\omega_m$ rad/s. The minimum sampling frequency $\omega_s = 2\omega_m$ is called the Nyquist rate.

Turning to the problem of reconstructing $x(t)$ from its samples, it is interesting to examine how an ideal lowpass filter interpolates between samples in order to generate the required continuous time signal. Let this filter have a transfer function

$$H(\omega) = G_{2\omega_c}(\omega) \tag{1.20}$$

where $|G_{2\omega_c}(\omega)| = 1$ over the passband. Clearly, this filter will remove unwanted images to leave, from (1.18)

$$X(\omega) = X_s(\omega) T G_{2\omega_c}(\omega) \tag{1.21}$$

Fourier transform tables for a gate function give

$$\frac{\omega_c}{\pi}\mathrm{Sa}(\omega_c t) \Leftrightarrow G_{2\omega_c}(\omega) \tag{1.22}$$

where

$$\mathrm{Sa}(x) = \frac{\sin x}{x} \tag{1.23}$$

so that, using the time convolution theorem,

$$\begin{aligned} x(t) &= T x_s(t) * \frac{\omega_c}{\pi}\mathrm{Sa}(\omega_c t) \\ &= \frac{2\omega_c}{\omega_s} x_s(t) * \mathrm{Sa}(\omega_c t) \\ &= \frac{2\omega_c}{\omega_s} \sum_{n=-\infty}^{\infty} x(nT)\delta(t-nT) * \mathrm{Sa}(\omega_c t) \\ &= \frac{2\omega_c}{\omega_s} \sum_{n=-\infty}^{\infty} x(nT)\mathrm{Sa}[\omega_c(t-nT)] \end{aligned} \tag{1.24}$$

For Nyquist rate sampling this reduces to

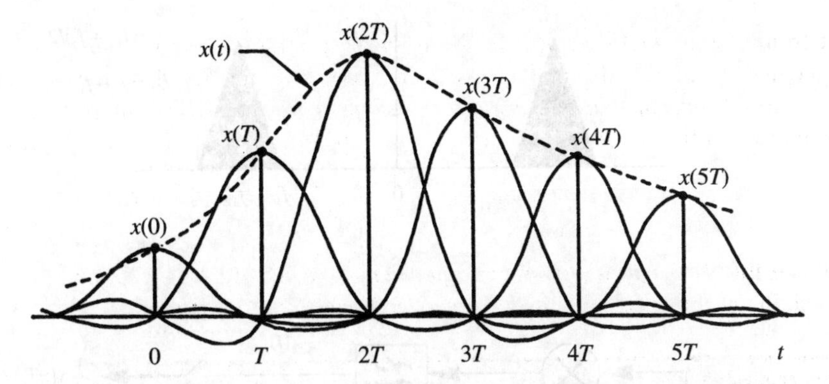

Figure 1.5 *Reconstruction of a signal x(t) from its Nyquist rate samples.*

$$x(t) = \sum_{n=-\infty}^{\infty} x(nT)\mathrm{Sa}\left[\frac{\pi}{T}(t-nT)\right] \qquad (1.25)$$

The sampling function $\mathrm{Sa}[\pi(t-nT)/T]$ has zeros for all $t = kT \neq nT$ and is weighted by the value of $x(t)$ at $t = nT$. A graphical interpretation of (1.25) is shown in Figure 1.5. Effectively, the ideal lowpass filter is replacing each sample $x(nT)$ with an interpolating function $x(nT)\mathrm{Sa}[\pi(t-nT)/T]$ and the value of $x(t)$ at any point in time is the sum of contributions from all interpolating functions. Note that the cutoff frequency of this filter is $1/2T$ and effectively it is transmitting $(\sin x)/x$-shaped pulses at a rate of $1/T$. This illustrates a limiting form of data transmission and we shall return to this concept in Section 4.1.1. Also, it should now be clear why the bandlimiting filter in Figure 1.3 is sometimes referred to as an *anti-alias* filter and why the reconstruction filter is sometimes referred to as an *anti-imaging* filter.

1.4.1 Bandpass sampling

Sometimes it is necessary to apply digital signal processing (DSP) to a bandpass signal $v(t)$ with a spectrum centred on $f_c = (f_1 + f_2)/2$: Figure 1.6(a). According to Theorem 1.3 we could treat this as a lowpass signal and sample at a rate $f_s \geq 2f_2$ although, clearly f_s will often be ridiculously high! A better approach is to use a narrowband signal model and represent $v(t)$ as

$$v(t) = x(t)\cos(\omega_c t) - y(t)\sin(\omega_c t) \qquad (1.26)$$

Here, $x(t)$ and $y(t)$ are independent, low-frequency (baseband) signals each bandlimited to $B_T/2 = (f_2 - f_1)/2$. Given $v(t)$ in this form it is apparent that $x(t)$ and $y(t)$ can be recovered using the quadrature carrier system in Figure 1.6(b), which, in turn, means that each low-frequency signal can be sampled at a rate B_T. The net effective sample rate for $v(t)$ is therefore

$$f_s = 2B_T \qquad (1.27)$$

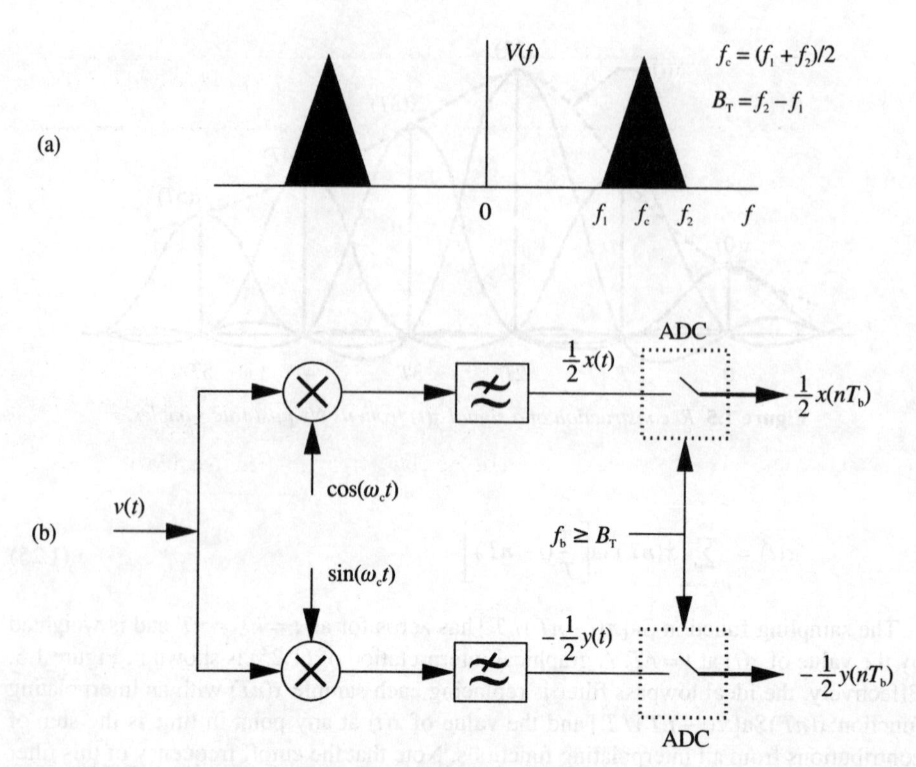

Figure 1.6 *Bandpass sampling: (a) spectrum of a bandpass signal v(t); (b) quadrature sampling of v(t).*

It is also possible to sample the bandpass signal directly at a similar sample rate provided f_s is carefully selected to avoid spectral aliasing. By considering image spectra arising from sampling it can be shown that aliasing is avoided if

$$2B_T \frac{k}{N} \le f_s \le 2B_T \frac{k-1}{N-1} \tag{1.28}$$

where $k = f_2/B_T$, $k \ge N$ and N is a positive integer. Equation (1.28) defines a set of frequency bands suitable for direct bandpass sampling and the *minimum possible* sampling frequency can be written

$$f_s = 2B_T \frac{f_2 / B_T}{\lfloor f_2 / B_T \rfloor} \tag{1.29}$$

where $\lfloor x \rfloor$ denotes the greatest integer in x. Clearly, if f_2 is an integer multiple of B_T, or if $f_2 >> B_T$, then the minimum sample rate is $2B_T$, whilst the worst case occurs when $f_2 \approx 2B_T$ corresponding to a minimum rate of $4B_T$. We can therefore state a bandpass sampling theorem as follows:

Theorem 1.4 The minimum sampling frequency for a bandpass signal of bandwidth B_T lies in the range $2B_T \leq f_s \leq 4B_T$.

A practical example of bandpass sampling is found in sub-band coding (Example 3.5).

1.4.2 Multidimensional sampling

So far we have considered one-dimensional (1D) forms of the sampling theorem. It is, however, important to realise that signals can exist in several dimensions and that we need to obey similar rules for such signals. Consider the two-dimensional (2D) spatial sampling of a static image $f(x,y)$. In this case a form of the 2D sampling theorem could be expressed as

$$\frac{1}{\Delta x} \geq 2W_u \qquad \frac{1}{\Delta y} \geq 2W_v \tag{1.30}$$

where Δx and Δy are the spatial sampling periods horizontally and vertically, and w_u and w_v are the maximum horizontal and vertical *spatial* frequencies (in cycles/picture width(height)) respectively of the image transform.

Example 1.3
A video signal is a 3D function $g(x,y,t)$ with a corresponding 3D spectrum, and the scanning device samples the scene in the vertical (y) direction and in the temporal (t) direction. For an interlaced video signal with line-locked sampling the sampling structure and the corresponding spectrum are shown in Figure 1.7. Note that temporal sampling occurs at both the field rate ($1/T$) and the frame rate ($1/2T$), and horizontal sampling (not shown) occurs during A/D conversion.

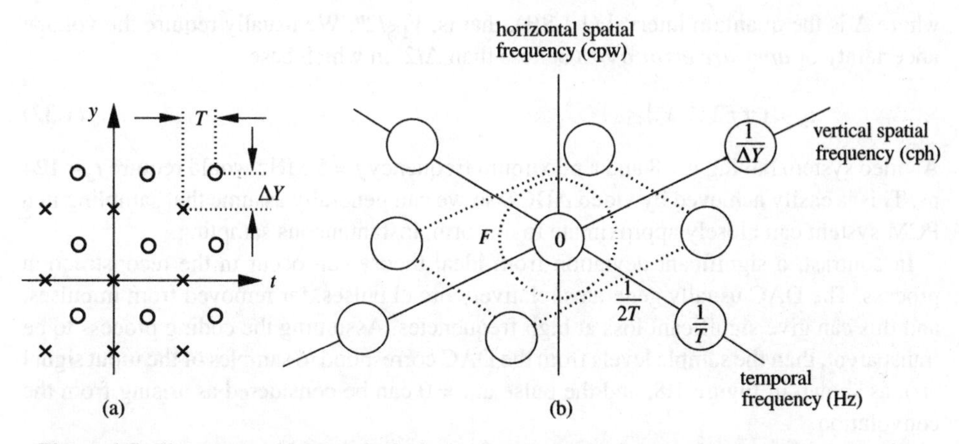

Figure 1.7 *Illustration of 2D line-locked sampling in an interlaced video system: (a) sampling structure in vertical/temporal plane; (b) repeat spectra for vertical/temporal sampling.*

The spectra are bandlimited because of the inherent spatial and temporal filtering before scanning, although aliasing can sometimes still occur. Occasionally the sampling rates are insufficient for the picture source being transmitted and the spectra overlap, giving spatial or temporal aliasing. Rotation reversal is a well-known example of temporal aliasing. The construction process is similarly non-ideal. In practice, the images tend to be filtered out by the receiver display limitations and by the 3D spatial–temporal lowpass filter provided by the human visual system (HVS). This is denoted by filter F in Figure 1.7.

1.5 Practical sampling and reconstruction

In this section we discuss the sampling and reconstruction aspects of practical systems, noting at the outset that both impulse sampling and ideal lowpass filtering are physically unrealisable. In practice the anti-alias filter in Figure 1.3 should not generate significant distortion of the input signal, and it should reduce alias-generating components to an acceptable level. For example, the filter cutoff frequency could be selected at, say, 20 times the fundamental frequency of a complex periodic signal, or at roughly the reciprocal of the pulse width for a single event. Filter stopband attenuation is determined by the basic need to keep alias-generating components below the quantisation threshold of the ADC, that is, below 1/2 LSB. For an n-bit ADC this is $V_{FS}/2^{n+1}$ where V_{FS} is the ADC full-scale input, so that, for example, a 10-bit ADC would need a filter attenuation $>2^{11}$ or at least 66 dB near $f_s/2$.

Sometimes practical systems can come very close to the theoretical ideal, as for example in the sampling process of an ADC. Ideally we wish to define a precise sampling instant. However, even if the sampling pulse applied to the ADC is jitter-free, there is still uncertainty in the sample instant arising from the ADC itself and this time uncertainty is called the ADC *aperture time*, t_a. Over time t_a a sinusoidal signal of frequency f Hz occupying the full quantising range of an n-bit ADC can change by as much as

$$\delta V = \pi f \Delta 2^n t_a \tag{1.31}$$

where Δ is the quantum interval (1 LSB), that is, $V_{FS}/2^n$. We usually require the voltage uncertainty or *aperture error* δV to be less than $\Delta/2$, in which case

$$t_a < (\pi f 2^{n+1})^{-1} \tag{1.32}$$

A video system having $n = 8$ and a maximum frequency $f = 5$ MHz would require $t_a < 124$ ps. This is easily achieved by video ADCs, so we can generally assume that sampling in a PCM system can closely approximate to uniform, instantaneous sampling.

In contrast, a significant deviation from ideal theory can occur in the reconstruction process. The DAC usually generates relatively broad pulses, far removed from impulses, and this can give significant loss at high frequencies. Assuming the coding process to be transparent, then the sample levels from the DAC correspond to samples of the input signal $x(t)$, as shown in Figure 1.8, and the pulse at $t = 0$ can be considered as arising from the convolution

$$x(t)\delta(t) * G_\tau(t)$$

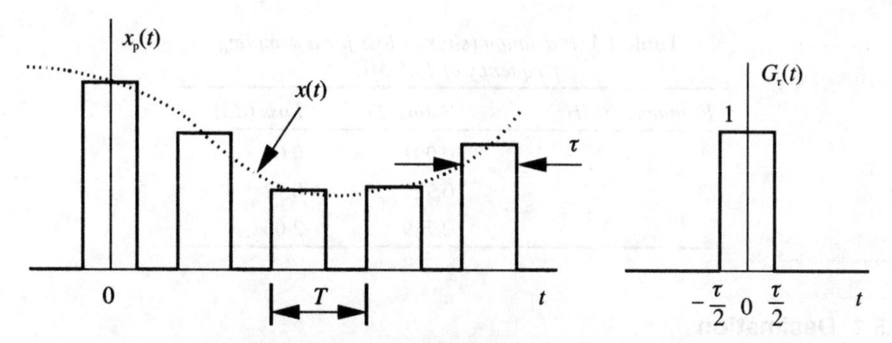

Figure 1.8 *PAM signal generated by a DAC.*

where $G_\tau(t)$ is a gate function of duration τ. Similarly, the PAM signal generated by the DAC can be expressed as

$$x_p(t) = [x(t)\delta_T(t)] * G_\tau(t) \tag{1.33}$$

where $\delta_T(t)$ is the impulse train in Figure 1.4(b). Since

$$G_\tau(t) \Leftrightarrow \tau \mathrm{Sa}\left(\frac{\omega\tau}{2}\right) \tag{1.34}$$

and

$$x(t)\delta_T(t) \Leftrightarrow \frac{1}{T}\sum_{n=-\infty}^{\infty} X(\omega - n\omega_s) \tag{1.35}$$

then

$$x_p(t) \Leftrightarrow \frac{\tau}{T}\mathrm{Sa}\left(\frac{\omega\tau}{2}\right)\sum_{n=-\infty}^{\infty} X(\omega - n\omega_s) \tag{1.36}$$

Usually the DAC output approximates to NRZ form ($\tau \approx T$) in which case

$$x_p(t) \Leftrightarrow \mathrm{Sa}\left(\frac{\omega\tau}{2}\right)\sum_{n=-\infty}^{\infty} X(\omega - n\omega_s) \tag{1.37}$$

Finally, passing $x_p(t)$ through the reconstruction filter to remove all image spectra gives the continuous-time signal

$$\hat{x}(t) \Leftrightarrow \mathrm{Sa}\left(\frac{\omega T}{2}\right)X(\omega) \tag{1.38}$$

Equation (1.38) shows that the ideal spectrum $X(\omega)$ is weighted by a $(\sin x)/x$ function, so high baseband frequencies may suffer significant attenuation. For instance, Table 1.1 shows the $(\sin x)/x$ loss for several video frequencies when the sampling frequency is 13.5 MHz. For broadcast systems this sort of loss is sufficient to warrant the use of an anti-imaging filter with a $(\sin x)/x$-corrected passband response.

Table 1.1 *Maximum (sin x)/x loss for a sampling frequency of 13.5 MHz.*

Frequency (MHz)	Sa($\omega T/2$)	*Loss (dB)*
1	0.991	0.08
3	0.921	0.72
5	0.789	2.06

1.5.1 Decimation

Many signal processing applications require a change of sample rate, and this invariably requires the use of a *multirate* filter. This class of filters converts a set of input samples to another set of samples that represent the same analogue signal sampled at a different rate. Suppose we wish to reduce the sample rate by an integer factor M, from $\omega_s = 2\pi$ to $\omega'_s = 2\pi / M$. Ideal sampling theory tells us that this can be achieved without danger of aliasing provided the sampled signal is first bandlimited to a maximum frequency

$$\omega_m \le \pi / M \tag{1.39}$$

In other words, to satisfy the sampling theorem the input signal must be bandlimited to one-half the *final* sample rate. Typical magnitude Fourier transforms illustrating the limiting case of a sample rate reduction process are shown in Figure 1.9(a). The $\downarrow M$ symbol denotes sample rate compression by an integer factor M, which means that the sequence $y(m)$ is formed by extracting every Mth sample from the sequence $w(n)$, that is

$$y(m) = w(Mm) \tag{1.40}$$

Since bandlimitation is essential for sample rate reduction it is usual to associate an anti-alias lowpass filter with the compressor and to refer to the combination as a *decimation filter*; Figure 1.9(b). As indicated, the role of the lowpass filter is to ensure that components of $w(n)$ above $\omega = \pi/M$ are negligible. Note that the decimator is a time-varying system, so it cannot be described in terms of an impulse response or transfer function.

The anti-alias lowpass filter is usually implemented using a *transversal finite impulse response* (FIR) structure as shown in Figure 1.10(a). Here, z^{-1} corresponds to a one (input) sample delay, and $h(0)$ to $h(N-1)$ denote N digital multipliers, the outputs of which are summed before being applied to the compressor. The significant point about Figure 1.10(a) is that, for every M inputs to the compressor, $M-1$ are discarded – a clear waste of computational effort! Fortunately, it can be shown using signal-flowgraph theory that the two branch operations of compression and multiplication (gain) *commute*, and this leads to the more computation-efficient structure in Figure 1.10(b). This structure reduces the computation rate by a factor M, since the multipliers and adders are now working at the lower sample rate of $2\pi/M$.

Example 1.4

Decimation filters are widely used in professional audio systems in order to eliminate high-precision analogue components. Typically we require a 16-bit (preferably 20-bit)

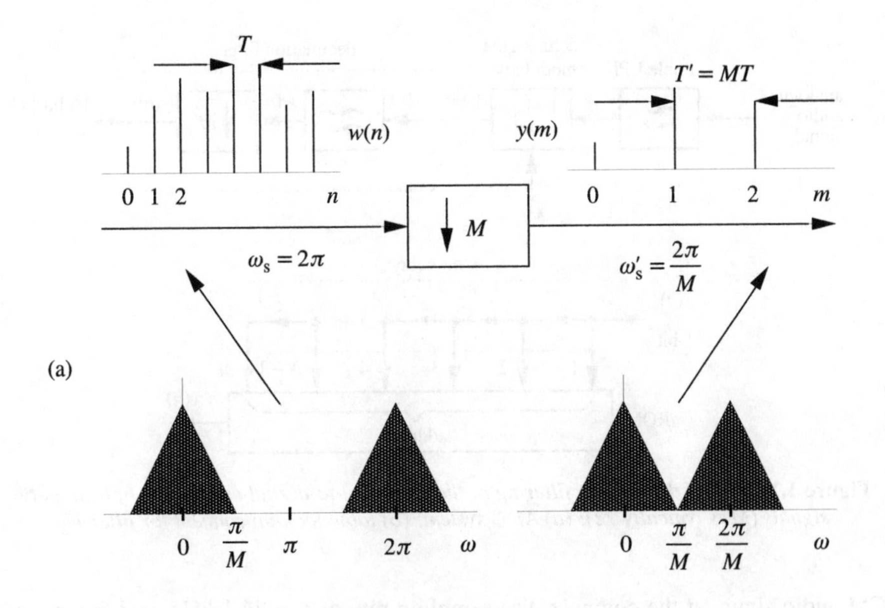

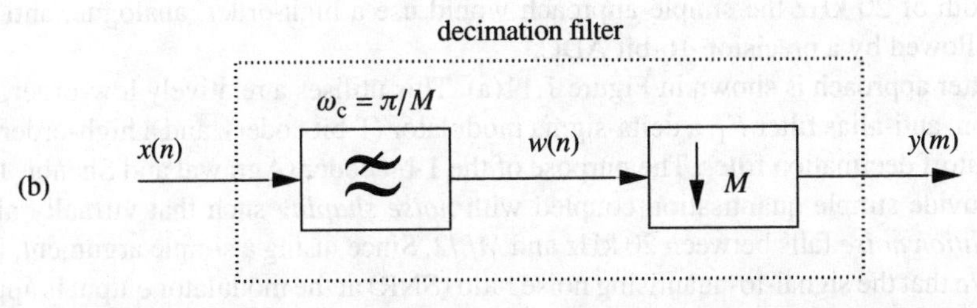

Figure 1.9 *Decimation filtering: (a) sampling rate compressor; (b) model for a decimation filter.*

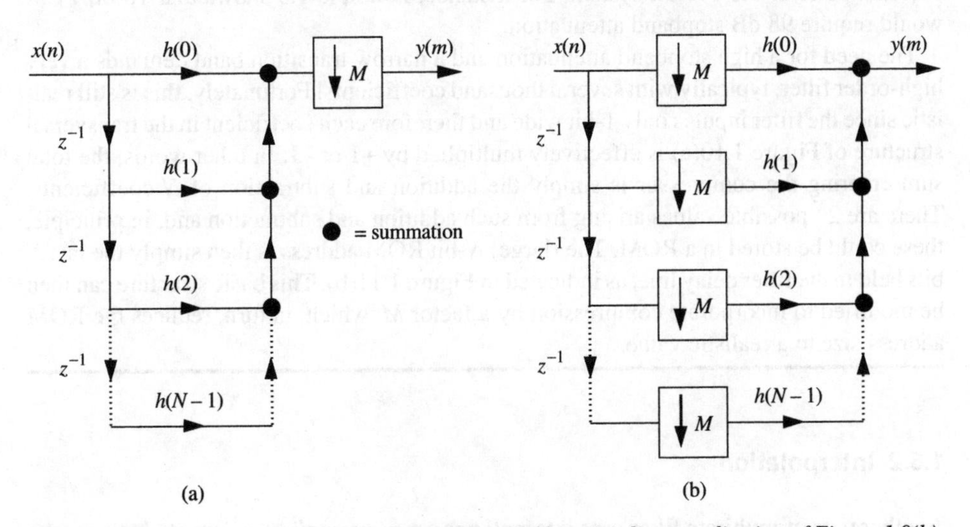

Figure 1.10 *Transversal realisation of a decimation filter: (a) direct realisation of Figure 1.9(b); (b) more efficient realisation.*

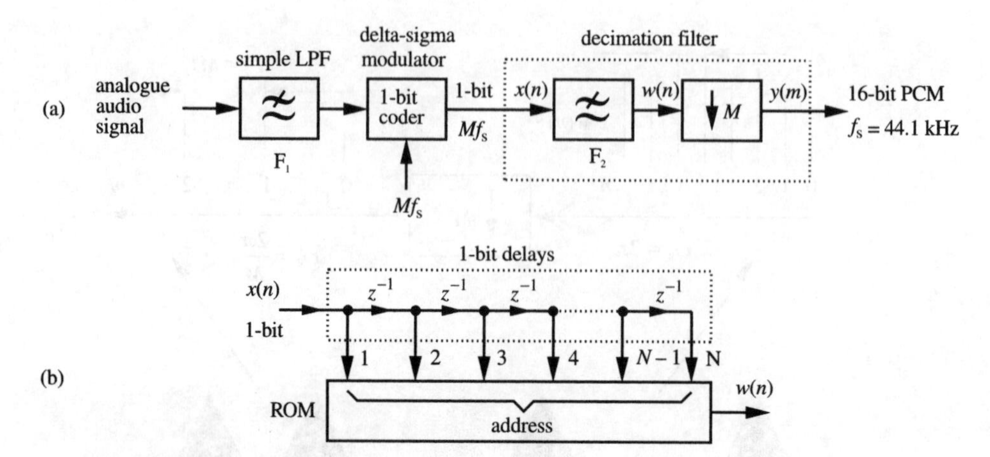

Figure 1.11 *Use of decimation filtering in the analogue-to-digital conversion of hi-fi audio signals (M is typically 72): (a) ADC system; (b) table look-up concept for filter F_2.*

PCM audio signal at the compact disc sampling rate of f_s = 44.1 kHz, and for an audio bandwidth of 20 kHz the simple approach would use a high-order, analogue, anti-alias filter followed by a precision 16-bit ADC.

A better approach is shown in Figure 1.11(a). This utilises a relatively low-order, low-precision, anti-alias filter F_1, a delta-sigma modulator (1-bit coder), and a high-order very sharp cutoff decimation filter. The purpose of the 1-bit coder (Agrawal and Shenoi, 1983) is to provide simple quantisation coupled with *noise shaping* such that virtually all the *quantisation noise* falls between 20 kHz and $Mf_s/2$. Since, using a simple argument, it can be shown that the signal-to-quantising noise ratio (SNR) at the modulator output is approximately 0 dB, this means that the stopband attenuation of filter F_2 essentially defines the required SNR of the overall system. For instance, Example 1.6 shows that 16-bit PCM would require 98 dB stopband attenuation.

The need for a high stopband attenuation and a narrow transition band demands a very high-order filter, typically with several thousand coefficients! Fortunately, this is still realistic since the filter input is only 1-bit wide and therefore each coefficient in the transversal structure of Figure 1.10(a) is effectively multiplied by +1 or –1. In other words, the total sum entering the compressor is simply the addition and subtraction of N coefficients. There are 2^N possible values arising from such addition and subtraction and, in principle, these could be stored in a ROM. The (large) N-bit ROM address is then simply the last N bits held in the filter delay line, as indicated in Figure 1.11(b). This basic structure can then be modified to incorporate compression by a factor M, which, in turn, reduces the ROM address size to a realistic value.

1.5.2 Interpolation

Another type of multirate filter uses interpolation or *oversampling* in order to increase the sample rate. Suppose we wish to increase the sample rate by an integer factor L. A model

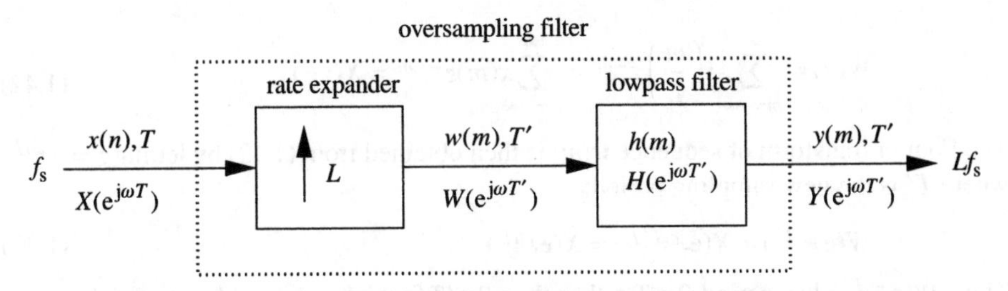

Figure 1.12 *Model for increasing the sample rate by an integer factor L.*

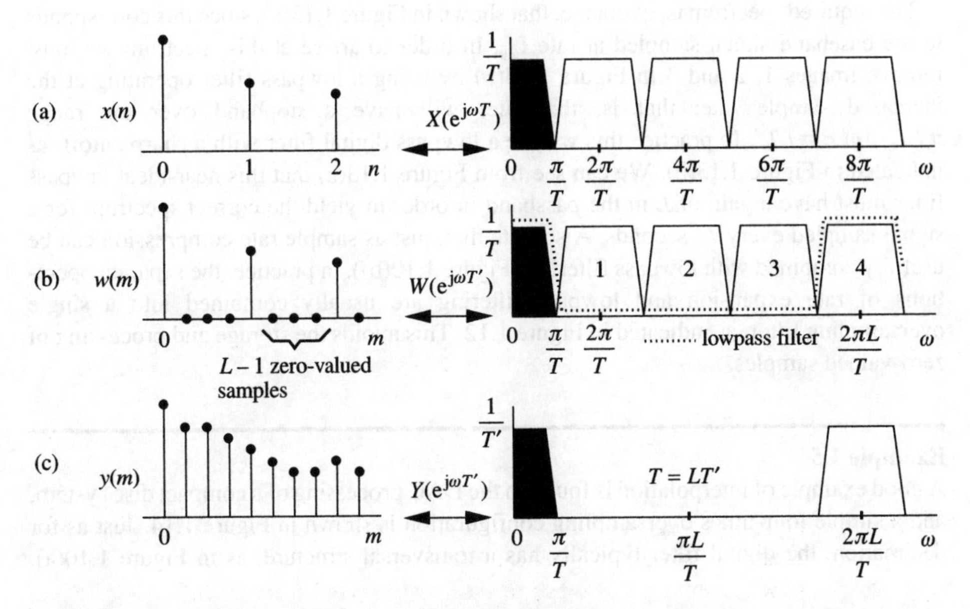

Figure 1.13 *Sequences and their transforms for interpolation by L = 4 (diagram shows transform magnitudes).*

of the required process is shown in Figure 1.12 and the corresponding signals for $L = 4$ are illustrated in Figure 1.13. For a bandlimited signal $x(n)$ sampled at the rate $f_s = 1/T$, ideal sampling theory gives the spectrum $X(e^{j\omega T})$ shown in Figure 1.13(a). Here, $X(e^{j\omega T})$ is the Fourier transform of sequence $x(n)$. The *rate expander* now inserts $L - 1$ zero-valued samples after each input sample to give sequence $w(m)$ (Figure 1.13(b)), but intuitively we might expect the spectrum to remain the same, since essentially nothing has been added to $x(n)$. Mathematically,

$$w(m) = \begin{cases} x\left(\dfrac{m}{L}\right) & m = 0, \pm L, \pm 2L, \ldots \\ 0 & \text{otherwise} \end{cases} \tag{1.41}$$

Taking the z-transform of sequence $w(m)$ gives

$$W(z) = \sum_{m=-\infty}^{\infty} x\left(\frac{m}{L}\right) z^{-m} = \sum_{m=-\infty}^{\infty} x(m) z^{-Lm} = X(z^L) \qquad (1.42)$$

The Fourier transform of sequence $w(m)$ is then obtained from (1.42) by letting $z = \mathrm{e}^{\mathrm{j}\omega T'}$ where T' is the new sampling period:

$$W(\mathrm{e}^{\mathrm{j}\omega T'}) = X(\mathrm{e}^{\mathrm{j}\omega T' L}) = X(\mathrm{e}^{\mathrm{j}\omega T}) \qquad (1.43)$$

Thus, $W(\mathrm{e}^{\mathrm{j}\omega T'})$ has period $2\pi/T$ rather than $2\pi/T'$ which we would expect for a more general sequence of sampling period T'.

The required spectrum is, of course, that shown in Figure 1.13(c), since this corresponds to the baseband signal sampled at rate Lf_s. In order to arrive at this spectrum we must remove images 1, 2 and 3 in Figure 1.13(b) by using a lowpass filter operating at the increased sample rate; that is, the filter will have a stopband over the range $\pi/T < |\omega| < \pi/T'$. In practice this will be a lowpass digital filter with a sharp cutoff, as indicated in Figure 1.13(b). We can see from Figure 1.13(c) that this near-ideal lowpass filter must have a gain of L in the passband in order to yield the correct spectrum for a signal sampled every T' seconds. Also note that, just as sample rate compression can be usefully combined with lowpass filtering (Figure 1.10(b)), in practice, the separate operations of rate expansion and lowpass filtering are usually combined into a single oversampling filter, as indicated in Figure 1.12. This avoids the storage and processing of zero-valued samples.

Example 1.5

A good example of interpolation is found in the DAC processing of a compact disc system, and a simple four-times oversampling configuration is shown in Figure 1.14. Just as for decimation, the digital filter typically has a transversal structure, as in Figure 1.10(a).

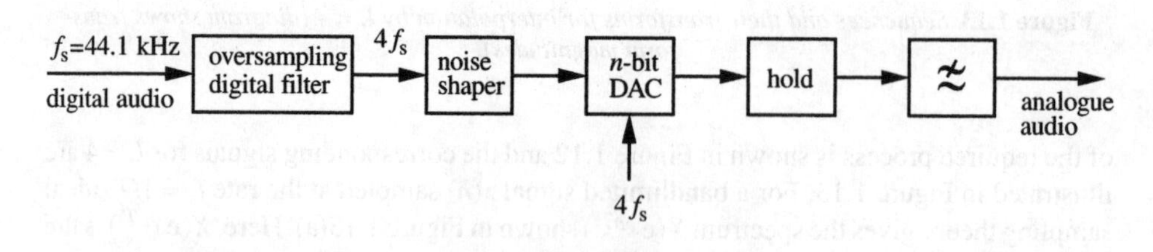

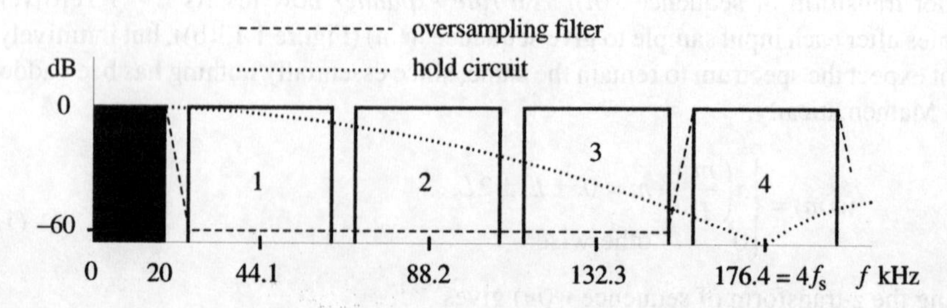

Figure 1.14 *A simple four-times oversampling system for digital-to-analogue conversion in compact disc systems.*

However, in order to perform interpolation, each sample shifted into the filter is now multiplied by four different coefficients in turn, rather than by a single coefficient. The products are also summed four times during each sample period, *T*, and passed to the output, thereby giving an output rate of $4f_s$. As indicated in Figure 1.14, the filter is of high order (typically 100 or more coefficients), cutting off sharply at the audio limit of 20 kHz so that images 1, 2 and 3 are rejected. The hold circuit following the DAC deliberately generates $(\sin x)/x$ distortion and, according to (1.38), gives a first zero at the output sampling rate, that is, at $4f_s = 176.4$ kHz. The $(\sin x)/x$ response, together with that of a simple lowpass analogue filter is sufficient to remove image 4.

The significant point about Figure 1.14 is that it avoids the need for a sharp cutoff analogue lowpass filter. This filter would also need a linear phase response in the passband to avoid impairment of pulsed sounds. In contrast, linear phase, sharp cutoff and high stability are relatively easy to achieve using the oversampling filter.

Noise shaping

Configurations similar to Figure 1.14 are also used to simplify the DAC considerably. First note that the wordlength of the oversampling filter output is typically 28 bits (corresponding to a 16-bit input and 12-bit coefficients) and that this must be rounded off to n bits at the DAC input. In practice this is achieved by a *noise shaper* which, in its first-order form, is essentially a quantiser with the rounding error fed back and subtracted from its input.

Clearly, since the quantising error will tend to be similar for successive samples of a low-frequency signal, subtraction of the error from the next sample will tend to reduce the average quantising noise at low frequencies. More specifically, since the signal bandwidth is considerably less than half the sampling frequency due to oversampling, this process enables the noise density to be shaped such that it is low at low frequencies and high outside the signal bandwidth. The low density over the signal bandwidth gives a SNR gain, and this means that the resolution, n, of the DAC can be reduced. In fact, in more complex oversampling systems, n can be reduced to just 1 bit, whilst the overall system performance still corresponds to a 16-bit DAC working at a sample rate of 44.1 kHz (Naus *et al.*, 1987).

1.6 Quantisation

This section first illustrates typical quantising and coding characteristics. It then takes a general theoretical look at quantising noise in terms of its power and spectral content, and concludes by computing the SNR for simple PCM systems. Quantisation in the context of MPEG coding is discussed in Section 3.3.3.

1.6.1 Quantising characteristics and coding

We saw in Section 1.2 that amplitude quantisation is fundamental to the success of PCM, and that in a well-designed PCM system quantising error or quantising noise should be the dominant impairment. There are essentially two ways of approximating an input sample x

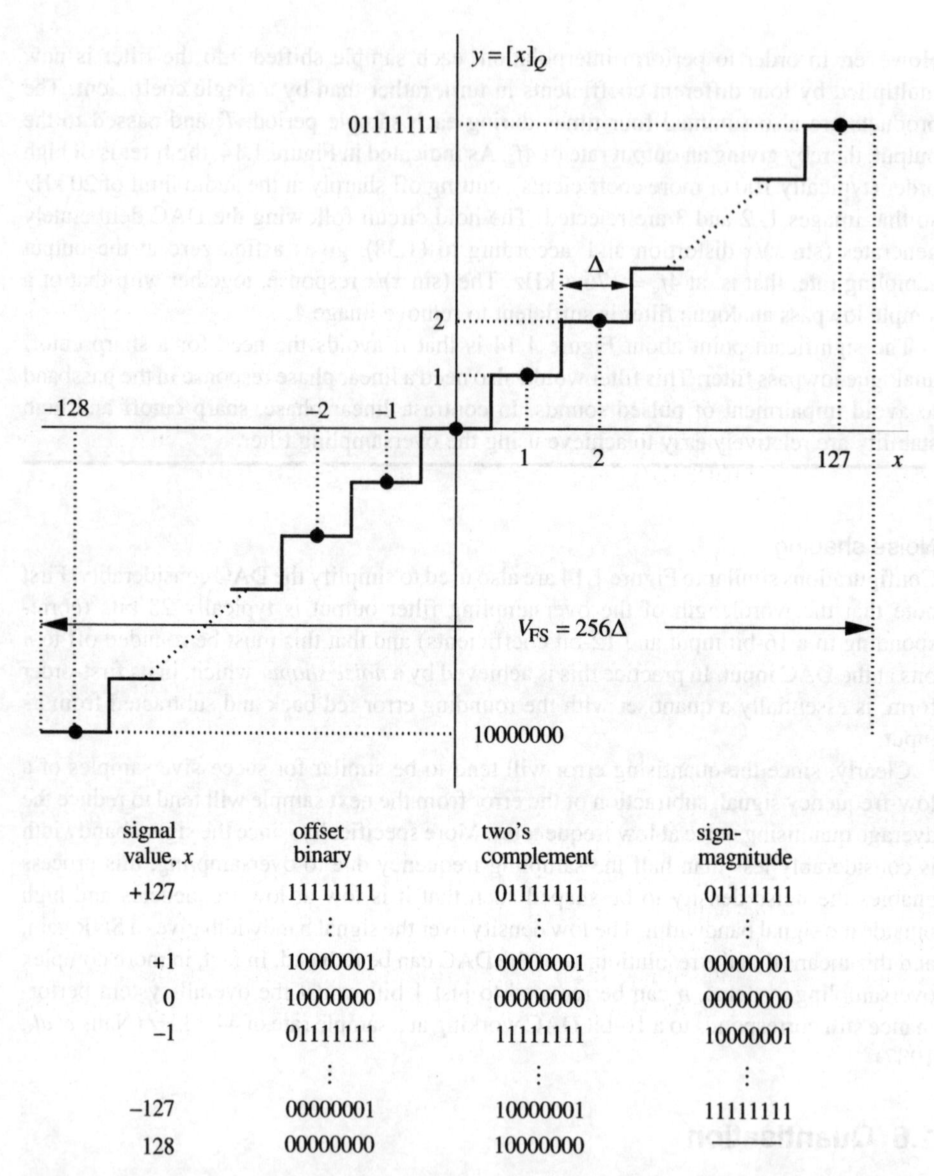

signal value, *x*	offset binary	two's complement	sign-magnitude
+127	11111111	01111111	01111111
⋮	⋮	⋮	⋮
+1	10000001	00000001	00000001
0	10000000	00000000	00000000
–1	01111111	11111111	10000001
⋮	⋮	⋮	⋮
–127	00000001	10000001	11111111
128	00000000	10000000	

Figure 1.15 *Typical uniform, bipolar quantising characteristic and coding of an 8-bit ADC.*

to a quantised value or *representative level* y_k, $k = 1, 2, ..., M$. The usual method is to select the quantising characteristic such that, if x lies between $(m - 0.5)\Delta$ and $(m + 0.5)\Delta$, where Δ is the quantum interval, then x is *rounded* to $m\Delta$. This approach minimises the quantising noise power and is illustrated in Figure 1.15: for example, $x = 1.8$ is rounded to 2. If the characteristic were to be shifted $\Delta/2$ to the right then an input lying between $m\Delta$ and $(m + 1)\Delta$ would be *truncated* to $m\Delta$, and the quantisation noise would increase. In Figure 1.15 the full-scale input is defined as $V_{FS} = 256\Delta$ or $2^n\Delta$ for an n-bit ADC, so if V_{FS}

= 10 V we have $\Delta \approx 39$ mV for $n = 8$ and $\Delta \approx 153\ \mu$V for $n = 16$. Therefore, despite this rather large input range, a 16-bit ADC still requires an exceptionally stable circuit and very high-precision voltage dividers, and in this case there is good reason to resort to alternative techniques to achieve analogue-to-digital conversion (see Example 1.4).

The quantiser output y is a quantised version of x, that is, $y = [x]_Q$, and it can use one of a number of codes. For example, if we wish to use straight binary coding for a bipolar input it is necessary to 'offset' it, as shown in Figure 1.15. Whilst this is a popular coding technique for ADCs, difficulties can arise when performing arithmetic operations. In particular, if the offset binary values are multiplied by a factor, the offset (10000000) is also multiplied, which causes an unwanted shift of the signal zero position. Most arithmetic operations are in fact performed using two's complement representation, where a negative number is formed by bit-by-bit inversion of the magnitude, plus 1 LSB (in practice this is often obtained by simply inverting the MSB of the offset binary representation). For the characteristic in Figure 1.16 a two's complement number can be written

$$y = [x]_Q = 128\left(-b_0 + \sum_{i=1}^{7} b_i 2^{-i}\right) \tag{1.44}$$

where bit b_i, $i = 0, 1, \ldots, 7$ is 1 or 0, which also means that the zero signal is conveniently represented by the all-zero code. Finally, note that sign-magnitude coding can sometimes be advantageous in applications involving symmetrical positive and negative signals since only the sign bit (the MSB) changes between positive and negative values of a number.

The foregoing discussion and Figure 1.15 illustrate the common case of *uniform* quantisation. This is appropriate for digitizing video signals for example, where the signal pdf $p(x)$ tends to be 'flat'. In this case we usually assume that the quantising noise power is independent of the input signal, which in turn means that the SNR will decrease with decreasing signal level. On the other hand, when $p(x)$ is non-uniform it is possible to quantise the signal with a smaller error variance by using a *non-uniform* quantising characteristic. Typically the signal pdf peaks at $x = 0$ (as in the case of speech signals) and the corresponding optimised quantising characteristic will be of the form shown in Figure 1.16. The idea is to use more quantising levels where $p(x)$ is relatively high and vice versa. Clearly, this type of *pdf-optimised quantiser* is susceptible to mismatch with respect to

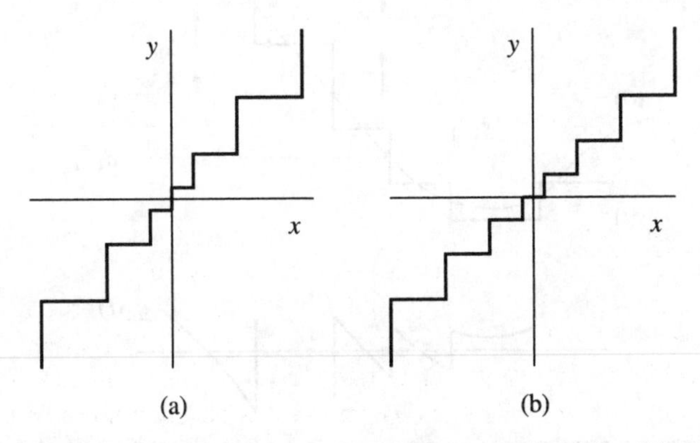

Figure 1.16 *Non-uniform quantising characteristics: (a) midrise; (b) midtread.*

both pdf-shape and input variance and this implies that the SNR is only optimal for one specific variance. A good example of pdf-optimised quantisers and non-uniform quantising is discussed in Section 3.1.2 under DPCM.

1.6.2 Quantising noise

The quantising process for an arbitrary waveform $x(t)$ is illustrated in Figure 1.17. In this diagram an input in the range

$$d_k \leq x < d_{k+1} \tag{1.45}$$

(where d_k is a *decision level*) is quantised to the representative level y_k and the quantum interval corresponding to y_k is Δ_k. Over this quantum interval the instantaneous quantising error or quantising noise is

$$q_\mathrm{n} = x - y_k \tag{1.46}$$

and we are interested in computing its mean-square value for a particular quantising problem. Note that, whilst q_n is obviously a function of time, it is also a function of the instantaneous signal amplitude x. Put more formally, the random variable Q_n corresponding to quantising noise is a function of the random variable X since

$$Q_\mathrm{n} = X - Y \tag{1.47}$$

where Y is the discrete-valued random variable representing the quantiser output levels. Having recognised this, we can now apply simple statistics to evaluate the mean-square value of q_n. For a continuous random variable X with a pdf $p(x)$, the expected value of a function $f(X)$ is

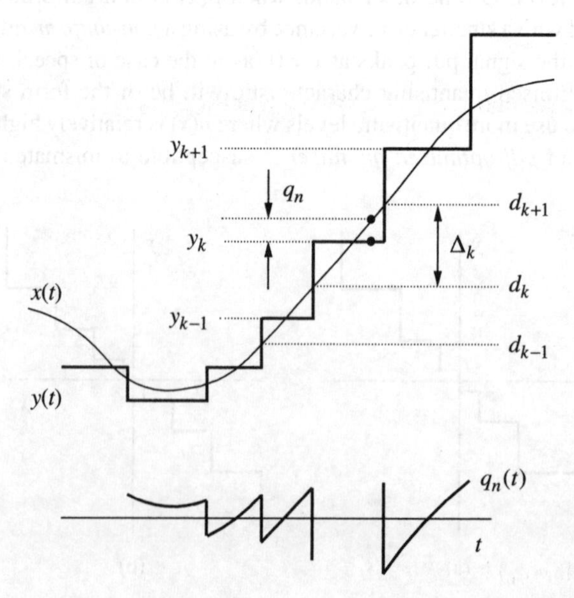

Figure 1.17 *Quantising error for a non-uniform characteristic.*

$$E[f(X)] = \int_{-\infty}^{\infty} f(x)p(x)\mathrm{d}x \tag{1.48}$$

Using this, we can write the mean-square quantising noise for L quantising levels as

$$\overline{q}_n^2 = E[Q_n^2] = \sum_{k=1}^{L} \int_{d_k}^{d_{k+1}} (x - y_k)^2 p(x)\mathrm{d}x \tag{1.49}$$

This can be simplified by making several assumptions. Firstly, provided that quantising is not too coarse, then $p(x)$ can be approximated by a constant $p(y_k)$ over the kth quantum interval, that is, over this interval,

$$p(x) = p(y_k) = P_k / \Delta_k \tag{1.50}$$

where P_k is the probability that x falls within the kth quantum. Secondly, using the approximation in (1.50) it can be shown that a necessary condition for minimising $\overline{q}_n^2$ is that the representative level y_k be placed midway between the corresponding decision levels. In other words, it is reasonable to assume that y_k is the quantised level for

$$y_k - \Delta_k / 2 \le x < y_k + \Delta_k / 2 \tag{1.51}$$

This is an important assumption which greatly simplifies analysis. Combining equations (1.49)–(1.51),

$$\overline{q}_n^2 = \sum_{k=1}^{L} \frac{P_k}{\Delta_k} \int_{y_k-\Delta_k/2}^{y_k+\Delta_k/2} (x - y_k)^2 \mathrm{d}x \tag{1.52}$$

$$= \frac{1}{12} \sum_{k=1}^{L} P_k \Delta_k^2 \tag{1.53}$$

For the special but common case of uniform quantising,

$$\overline{q}_n^2 = \Delta^2 / 12 \tag{1.54}$$

where Δ is the quantum interval, and we conclude that, if quantising is uniform and reasonably fine, *then $\overline{q}_n^2$ is approximately independent of the pdf of the input signal*! Note that in the case of fine uniform quantising we could assume that

$$p(q_n) = \begin{cases} 1/\Delta & -\Delta/2 \le q_n \le \Delta/2 \\ 0 & |q_n| > \Delta/2 \end{cases} \tag{1.55}$$

and use a more direct calculation for $\overline{q}_n^2$ from (1.48):

$$\overline{q}_n^2 = \int_{-\Delta/2}^{\Delta/2} q_n^2 \, p(q_n) \, \mathrm{d}q_n = \Delta^2 / 12 \tag{1.56}$$

Also, since the pdf $p(q_n)$ has zero mean, then

$$\overline{q}_n^2 = E[Q_N^2] = \sigma_q^2 = \Delta^2 / 12 \tag{1.57}$$

where σ_n^2 is the quantising noise variance or normalised noise power.

1.6.3 System considerations

At this point it is useful to model the principal operations of a PCM system in order to determine how the quantising noise propagates to the system output, and a suitable model (ignoring coding) is shown in Figure 1.18(a). We shall assume fine uniform quantising and for mathematical convenience this is placed before (ideal) sampling. In practice, sampling precedes quantising, but these two operations commute and so they can be represented as shown. Also, the lowpass filters are assumed to be ideal with cutoff frequency $\omega_c = \omega_s/2$. Intuitively, it appears from Figure 1.18 that quantising noise is a relatively wideband signal with a *power spectral density* (PSD) $P_n(\omega)$ which extends well above the sampling frequency ω_s. Assuming this to be the case, then the PSD $P_{ns}(\omega)$ of the sampled quantising noise will be a highly aliased spectrum (Figure 1.18(b)) and the noise power P_{no} at the output of the reconstruction filter will be

$$
\begin{aligned}
P_{no} &= \frac{1}{2\pi}\int_{-\omega_s/2}^{\omega_s/2} P_{ns}(\omega)\,d\omega \\
&= \frac{1}{2\pi}\sum_{k=-\infty}^{\infty}\int_{-\omega_s/2}^{\omega_s/2} P_n(\omega - k\omega_s)\,d\omega \\
&= \frac{1}{2\pi}\sum_{k=-\infty}^{\infty}\int_{k\omega_s-\omega_s/2}^{k\omega_s+\omega_s/2} P_n(\omega)\,d\omega
\end{aligned}
\tag{1.58}
$$

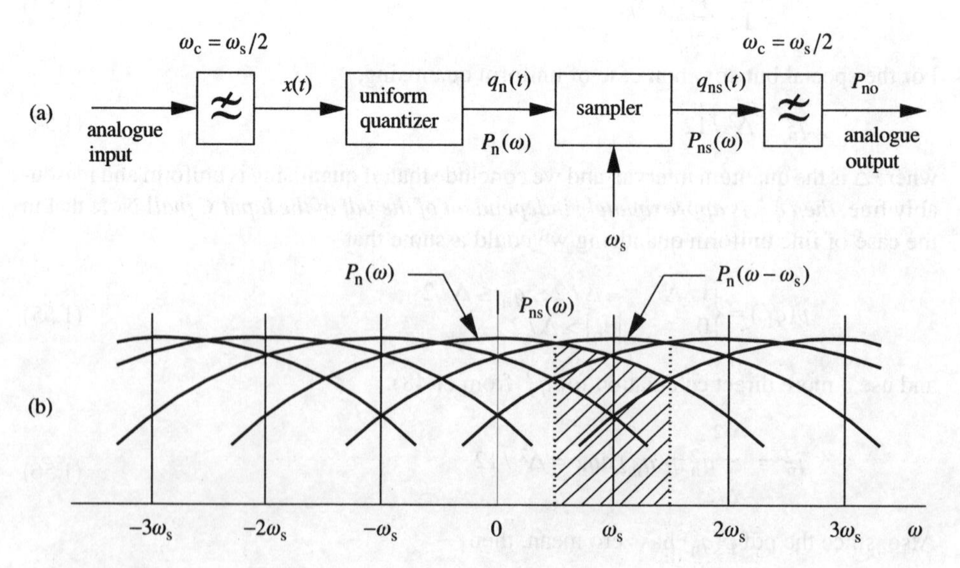

Figure 1.18 *Quantising noise: (a) model of a PCM system; (b) quantising noise spectra.*

For $k = 1$ the integral corresponds to the shaded region in Figure 1.18(b), so summation over all k must give the area under $P_n(\omega)$. Therefore,

$$P_{no} = \frac{1}{2\pi}\int_{-\infty}^{\infty} P_n(\omega)d\omega = \frac{\Delta^2}{12} \tag{1.59}$$

by definition of the PSD. *This interesting result states that all the quantising noise power generated by the quantiser appears in the band $0 - \omega_s/2$ at the analogue output.* In practice, $\omega_s > 2\omega_m$ and $\omega_c < \omega_s/2$ (see Figure 1.4(c)), so the output noise power will be somewhat less than $\Delta^2/12$.

The actual output noise power P_{no} for $\omega_c < \omega_s/2$, and the characteristics of the output quantising noise, can be estimated by considering the sampled noise signal $q_{ns}(t)$ (a train of impulses in our model). Except for special cases (periodic or slowly changing inputs), it is reasonable to assume that the samples of quantising noise are uncorrelated and this means that the *autocorrelation function* (Section 4.5.1) of $q_{ns}(t)$ over a large time window tends to an impulse function. According to the *Wiener–Khinchine theorem*, the PSD $P_{ns}(\omega)$ is the Fourier transform of this autocorrelation function and so $P_{ns}(\omega)$ (and therefore the output noise spectrum) tends to be flat. This is readily confirmed for finely quantised practical systems, say 10 bits/sample or more for uniformly quantised sound signals and 6 bits/sample or more for uniformly quantised video signals. In these cases, quantising noise is similar to white Gaussian noise, although the noise spectra become progressively more peaked as the digital resolution is reduced. Subjectively, these deviations from white noise characteristics are referred to as *granular noise* for small sound signals and *contouring* (as in a map) for slowly varying video signals.

Assuming a flat PSD and a reconstruction filter of cutoff frequency $\omega_c < \omega_s/2$, it follows from the definition of PSD in (1.58) that

$$P_{no} = 2\sigma_q^2\frac{\omega_c}{\omega_s}; \quad \sigma_q^2 = \frac{\Delta^2}{12} \tag{1.60}$$

1.6.4 SNR in PCM systems

Knowing the quantising noise power we can now compute a signal-to-quantising noise ration (SNR) at the output of a PCM system (after the reconstruction filter). In doing so, we must be careful to state how the SNR is defined, together with the exact measurement conditions. For instance, the overall SNR depends upon ω_c and ω_s (equation (1.60)), and upon any increase in noise level due, say, to multiple coding–decoding operations in tandem or to the addition of a *dither* signal at the ADC input. Also, in practice a *weighting factor* (in dB) is often added to the basic computed SNR to account for the subjective effect of different noise spectra.

Example 1.6

The SNR in digital audio systems is often defined as the ratio of full-scale signal power to the noise power. Taking the full-scale signal at the output of the DAC as $2^n\Delta$, then the power of a digitised sinusoidal signal is simply $(2^{n-1}\Delta/\sqrt{2})^2$. Assuming fine uniform quantising and that all the quantising noise power ($\Delta^2/12$) is available at the reconstruction filter output, we have

$$\text{SNR} = 10\log_{10}\left[\frac{\text{signal power}}{\text{noise power}}\right] \text{ dB} \tag{1.61}$$

$$= 20\log_{10}[2^{n-1}\sqrt{6}] \text{ dB}$$

$$= 6n + 1.8 \text{ dB} \tag{1.62}$$

With the above assumptions, this means, for example that the SNR in a uniformly quantised 16-bit PCM audio system is limited to about 98 dB.

Example 1.6 highlights several important points. First, the SNR changes by 6 dB/bit, and this is generally true for other definitions of SNR. Secondly, in general, SNR calculations using (1.62) or a similar expression cannot be directly compared with signal-to-random-noise measurements in an analogue system. This is because the subjective effect of quantising noise in audio and video systems differs from that of truly random noise, particularly for coarse quantisation.

Cascaded codecs

A practical system may consist of N identical ADC–DAC pairs (codecs) interspaced with analogue signal processing, as in Figure 1.19. We shall ignore the analogue processing and also assume that each codec incorporates an input and output lowpass filter with cutoff frequency $\omega_c < \omega_s/2$. We require the overall SNR at the output of N codecs in tandem and this raises the question of how the noise variance at the input to a codec affects the noise variance at its output. To answer this, consider the random variables X and Y in Figure 1.19(b); from the discussion in Section 1.6.2 we can write

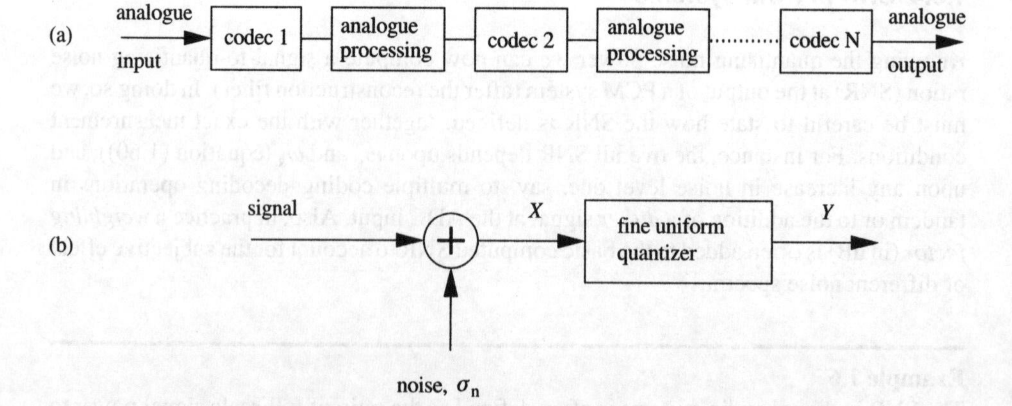

Figure 1.19 *Combining noise sources in a PCM system: (a) multiple codecs in an analogue/digital environment; (b) a model for computing total output noise variance.*

$$Y = X - Q_n \tag{1.63}$$

where Q_n is the random variable representing the quantising noise. The variance of the quantiser output is then

$$\sigma_y^2 = E[(Y - \mu_y)^2] \tag{1.64}$$

where $\mu_y = E[Y]$. Noting that $E[Y] = E[X] = \mu x$, since $E[Q_n] = 0$, this reduces to

$$\sigma_y^2 = \sigma_x^2 + \sigma_q^2 - 2E[(X - \mu_x)Q_n] \tag{1.65}$$

If we now make the reasonable assumption that Q_n is independent of X (as implied by (1.54)) we have

$$\sigma_y^2 = \sigma_x^2 + \sigma_q^2 \tag{1.66}$$

(Note that this assumption is not true and equation (1.66) does not hold when the quantiser is optimised to the input pdf). Finally, if the noise component of X is uncorrelated with the signal component of X, then the total output noise power from the quantiser is

$$P_n = \sigma_n^2 + \sigma_q^2 \tag{1.67}$$

where σ_n^2 is the input noise variance and $\sigma_q^2 = \Delta^2/12$. In other words, *to a good approximation we can add noise powers at the input and output of a fine uniform quantiser*. It therefore follows from (1.60) that the quantising noise power at the output of N codecs in tandem is

$$P_N = \frac{2N\omega_c}{\omega_s}\sigma_q^2 \tag{1.68}$$

In video systems the reference signal range used in the SNR definition usually occupies a fraction k of the total quantising range of $2^n\Delta$ (assuming n bits/sample) and so the overall SNR at the output of N video codecs is

$$\text{SNR} = 20\log_{10}\left(\frac{\text{reference signal range}}{\text{total rms noise}}\right) \text{ dB} \tag{1.69}$$

$$= 20\log_{10}\left(\frac{k2^n\Delta}{\sqrt{P_N}}\right) \text{ dB} \tag{1.70}$$

$$= 6n + 10.8 + 10\log_{10}\left(\frac{k^2\omega_s}{2N\omega_c}\right) \text{ dB} \tag{1.71}$$

Example 1.7

CCIR-601 (1982) recommends the following digital encoding parameters for the luminance signal in 625-line video systems (4:2:2 coding standard):

Digital resolution n	8-bits/sample, uniform quantisation
Sampling frequency	13.5 MHz (864×line frequency)
Reference signal range	220Δ

For video signals the reference range is the luminance signal range (black level to peak white level), and in the 4:2:2 standard this is 220/256 of the total quantising range. Table 1.2 shows SNR estimates using the above parameters in (1.71) for several values of N.

Table 1.2 *Overall SNR for N video codecs in tandem (coding is to the 4:2:2 standard).*

f_c (MHz)	SNR (dB)		
	$N = 1$	$N = 2$	$N = 3$
5	58.8	55.8	54
5.5	58.4	55.4	53.6

1.7 Overview of coding techniques and standards

So far we have concentrated on 'straight PCM', that is, waveform sampling, quantising and simple binary coding as performed by an ADC. In communications systems this is usually only the first step in a sequence of complex signal processing and coding operations prior to transmission or data storage, and it is usual to split this processing into *source coding* and *channel coding,* as shown in Figure 1.20. In this model, the ADC is simply the 'front end' of the source encoder. Source and channel coding techniques are discussed in depth in later chapters, so here we simply outline the basic ideas.

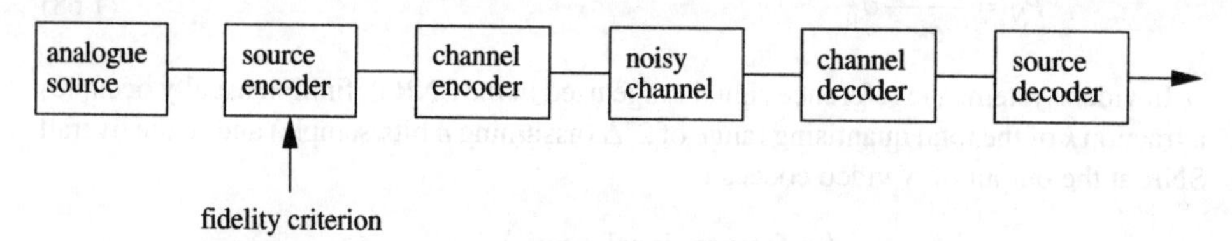

Figure 1.20 *A coding model for digital communications systems.*

1.7.1 Source coding

The broad objective of source coding is to exploit or remove 'inefficient' redundancy in the PCM source in order to reduce the transmission bit rate. Source coding therefore achieves data compression. This is beneficial since a reduced transmission rate can lower the cost of the link, or enable more channels to be used on the same link, or enable more video information to be stored on disk. For example, it would be unnecessary and uneconomic to directly transmit a broadcast quality video signal for video conferencing, and for this application source coding techniques result in more cost effective bit rates of the order of 1 Mbit/s. From an information theory point of view, we could regard source coding as processing the PCM source into a form that has an inherently lower information rate R (this is discussed in Section 2.2). Since R must be less than the channel capacity C for reliable transmission (Theorem 1.1), then source coding could be regarded as a way of minimising the required channel capacity C.

Sometimes we require data compression without information loss, as in the compression of computer data files, and in this case we use some form of completely reversible or *lossless* technique. On the other hand, non-reversible or *lossy* compression is commonly used for speech and video signals, and, by definition, this results in some distortion or loss of quality. The loss may be imperceptible, as in broadcast quality MPEG-2 video signals, or quite apparent, as in video telephone systems. In general, therefore, it is usual to associate a quality or *fidelity* criterion with source coding, as indicated in Figure 1.20. Besides fidelity, other factors to be considered when compressing data are the required *compression ratio* (CR) and the computational complexity and cost. Source coding standards are often set by such bodies as the ITU (formerly CCITT), ETSI and ISO.

A consequence of the efficient coding of the PCM source will be that the source encoder output tends to be a sequence of *independent* (and therefore *uncorrelated*) symbols. In fact, the statistical independence of source encoder outputs is a *necessary* condition for optimal coding. Looked at another way, we might regard the source encoder as attempting to generate a random signal at its output. Unfortunately, since much of the redundancy (sample-to-sample correlation) has been removed, the source encoder output generally tends to be less *robust* and more susceptible to transmission errors. The situation is similar to listening to speech in a noisy environment – it may be intelligible only because of the high redundancy in the English language. The usual solution is to protect the source output against transmission errors using channel coding.

1.7.2 Channel coding

Channel coding is the process of matching the source encoder output to the transmission channel, and in general it involves several distinct processes:

- error control coding
- transmission coding
- scrambling

Consider first the concept of error control coding. In general the noisy channel in Figure 1.20 represents a digital modulator, a physical analogue channel and a digital demodulator. Taking a simple example, the demodulator could make a definite (hard) decision on the data, but due to noise in the analogue channel it is inevitable that some decisions will be incorrect. It is then the role of the error control decoder to correct these decision errors as far as possible using the *parity* or *check* bits inserted by the channel encoder. This process is called *forward error control* (FEC), since correction is achieved by forward transmission only. Paradoxically, whereas the source encoder removes 'inefficient' redundancy, the channel encoder inserts 'efficient' redundancy in the form of parity bits, and one might ask if anything is gained! Fortunately, in practice the added redundancy (coding overhead) is usually only a small fraction of that which is removed by the source encoder. For example, an MPEG-2 video (source) encoder can reduce the bit rate by typically 30:1, while a (convolutional) error control encoder used to protect the video transmission over a satellite channel will typically only increase the bit rate by a factor 2.

There are several more ways in which a channel encoder can match a signal to a transmission channel. Sometimes the basic binary code is translated to a code with a more favourable spectrum, either through a code translation process or via a simple coding process. These techniques could be described as *transmission* coding. Also, very often, the basic binary code is 'scrambled' using a *pseudonoise* (PN) sequence in order to provide one or a number of very useful features (Section 4.4).

A case study

At this point it is helpful to illustrate an actual channel coding scheme in order to introduce some commonly used terms and to see where they appear in the context of a communications system. A good example is found in the area of digital video transmission. The Moving Picture Experts Group (MPEG) has defined a number of standards for video source coding, such as MPEG-1, MPEG-2 and MPEG-4. A look at the MPEG-2 specification reveals that it defines the MPEG transmission *bitstream*, but it leaves the design of channel coding to the system designer since the medium could be satellite, cable or terrestrial. As a case study we could consider the channel coding used in the European satellite-based Digital Video Broadcasting (DVB) system, shown in Figure 1.21 (the same coding and interleaving is used for terrestrial-based DVB). Like other compressed data, the MPEG programme stream is non-robust and requires FEC for protection. In fact, for DVB, *two* FEC codes are used in a *concatenated* (inner/outer) configuration. Typically, the raw bit error rate (BER) at the input to the FEC decoder in the receiver could be 10^{-4} and the FEC decoder should give a decoded error rate of at least 10^{-10} if the sensitive compressed video is to be decoded with acceptable quality.

For satellite channels the predominant impairment is noise which leads to *randomly distributed* single-bit errors. This can be handled by making the inner code in Figure 1.21 the industry standard (133,171) constraint-length 7 *convolutional* code, with *Viterbi decoding* at the receiver. A process of *puncturing* can give various inner *code rates*, ranging from 1/2 to 7/8. Unfortunately, like all FEC decoders, the inner code will sometimes be overloaded (fail) and generate a *burst* of errors (impulsive noise will also generate error bursts). The outer code should therefore have good burst error correction capability,

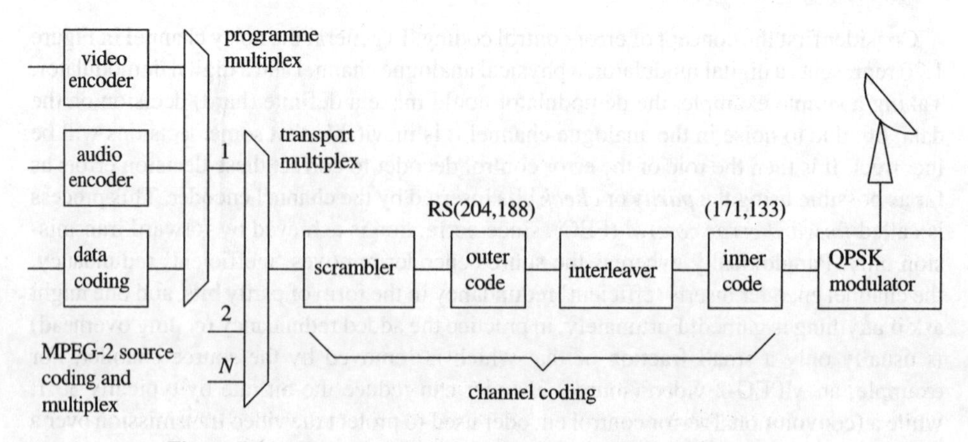

Figure 1.21 *Channel coding in the European DVB satellite system.*

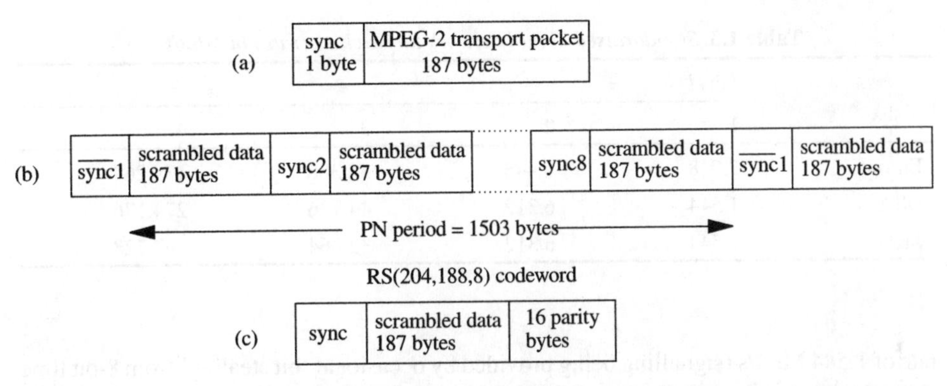

Figure 1.22 *DVB transport packet processing.*

as found for example in *Reed–Solomon* (RS) codes. Consider the MPEG-2 packet structure in Figure 1.22(a). Each MPEG-2 packet is 187 bytes plus a sync byte, and these are scrambled to assist clock extraction at the receiver and to smooth the modulated spectrum. Scrambling is achieved by modulo-2 addition of a PN sequence to the transport stream, the PN sequence being reset every eight 204-byte packets, as indicated in Figure 1.22(b). Each scrambled transport packet is then protected using an (n,k,t) RS code – Figure 1.22(c). Here n is the *block length* in bytes (204), k is the number of data bytes to be protected (188), and t is the number of byte errors that can be corrected in n bytes. For RS codes it is shown later that

$$n = 2^m - 1, \quad n - k = 2t \tag{1.72}$$

where m is a positive integer. Since we require $k = 188$, then clearly a minimal value of n is 255. For DVB $t = 8$, so the basic code is the RS (255,239,8) code. This can then be *shortened* to the RS (204,188,8) code in order to give the required data block size of 188 bytes. The coding overhead is therefore only 16/188 = 8.5% and is acceptably low. Note that between the inner and outer encoders the DVB scheme includes a process of *interleaving* (Figure 1.21). At the receiver a *de-interleaver* restores the original data, but in the process this breaks down (disperses) any lengthy bursts of errors reaching the outer decoder. The net effect is to reduce the required power of the outer code.

1.7.3 Speech coding

In practice the target bit rate for a source encoder is often determined by the need to meet a specific CCITT bit rate. Consider the standardised gross (total package) rates for digital networks in Table 1.3. For Europe, level 1 corresponds to CCITT 30-channel PCM, that is, 30 voice channels and 2 framing/signalling channels. The 32 8-bit time slots span 125 μs, giving a gross rate of 2.048 Mbit/s and a voice channel rate of 64 kbit/s (8 kHz sampling and 8 bits/sample). A speech rate of 64 kbit/s achieves *toll quality* or commercial telephony quality (which is higher than *communication* quality) by using non-uniform quantising or digital companding. In Europe this is an approximation to *A*-law companding. For the USA (and generally for Japan), level 1 corresponds to CCITT 24-channel PCM, that is, 24 8-bit time slots plus 1 framing bit span 125 μs and give a gross

Table 1.3 *Standardised digital network hierarchies (rates in Mbit/s).*

	Level 1	2	3	4
Europe	2.048	8.448	34.368	139.264
USA	1.544	6.312	44.736	274.176
Japan	1.544	6.312	32.064	97.728

rate of 1.544 Mbit/s (signalling being provided by occasional 'bit stealing' from 8-bit time slots). In this system digital companding is an approximation to μ-law companding and the gross channel rate of 64 kbit/s includes signalling. Also note that 64 kbit/s is the capacity of an *Integrated Services Digital Networks* (ISDN) B channel.

We might regard conversion to 64 kbit/s *A*-law or μ-law PCM as the 'front end' processing of the source encoder in Figure 1.20. For example, the 64 kbit/s PCM data stream could then be applied (after first conversion to uniform PCM) to an *adaptive differential PCM* (ADPCM) encoder in order to reduce the bit rate to 32 or 16 kbit/s under ITU Rec. G.721 and G.726 respectively; see Table 1.4. The ADPCM encoder acts as a *transcoder* between the uniform PCM system and the channel encoder.

Table 1.4 lists some speech coding techniques' generating rates from 64 kbit/s down to 2.4 kbit/s (at 2.4 kbit/s, nearly one hour of continuous speech can be stored on a 1 Mbyte DRAM). RPE-LPC, VSELP and PDC VSELP are cellular standards, while LPC-10E, CELP and MELP are US DoD secure standards. CELP coding is also used by MPEG-4 for speech rates in the range 4–24 kbit/s (see Figure 1.23). G.729 was created primarily for use

Table 1.4 *Some speech coding techniques (* denotes a digital cellular standard)*

Coding technique	*Bit rate, kbit/s*	*Quality*
PCM (G.711)	64	Toll
ADPCM (G.721)	32	Toll
ADPCM (G.726)	16, 24, 32, 40	Toll
LD-CELP (G.728)	16	Toll
RPE-LPC* (GSM) – Europe	13	Communications
CS-ACELP (G.729)	8	Toll
VSELP* (IS-54) – USA	7.95	Communications
PDC VSELP* - Japan	6.7	Communications
MPC-MLQ (G.723.1)	5.3, 6.3	Toll
CELP (FS-1016)	4.8	Synthetic/comms.
IMBE (INMARSAT std.-M)	4.15	Synthetic/comms.
AMBE (DVSI)	4	Toll
MELP	2.4	Synthetic/comms.
LPC10E (FS-1015)	2.4	Synthetic

on noisy radio channels. The G.723.1 standard merits some comment. It was developed as the speech standard for very low bit rate (PSTN) video telephony and as such it was incumbent to use as few bits as possible. The result was a trade-off between bit rate, quality, complexity and delay. It has a lower bit rate than G.729, but has virtually the same quality under clean speech and error-free conditions. In fact, under such conditions its quality at 6.3 kbit/s is equivalent to 32 kbit/s G.726! On the other hand, it has a 30 ms frame size compared with 10 ms for G.729, resulting in a one-way total codec delay of 67.5 ms, compared to 25 ms for G.729. The interesting point is that, besides videophones, G.723.1 has also been adopted as a standard for *Internet telephony* or *Voice over Internet Protocol* (VoIP). In other words, G.723.1 specifies the speech codec used in many PSTN-IP gateways.

Generally speaking, as the bit rate is reduced, the speech quality reduces from *toll* quality to *communications* quality to *synthetic* quality. Roughly speaking, toll quality corresponds to commercial telephony quality, whilst communications quality corresponds to detectable distortion but still very little degradation of intelligibility. A more quantitative assessment of speech quality is done by using a *Mean Opinion Score* (MOS). Traditionally this is obtained via real-time *subjective* assessment of speech on a scale of 1 to 5 (1 is 'bad', 5 is 'excellent'), although the exact MOS is somewhat subject to the exact test conditions. As an example, under certain test conditions, G.711 scored 4.1 compared with 3.9 for G.729 and 3.5 for the European cellular standard (GSM). In a separate test, G.726 scored 3.9 compared with 3.8 for G.723.1. Rather than perform formal subjective testing, it is often more convenient to obtain an *objective* measure. One such comparison method is the *Perceptual Speech Quality Measure* (PSQM), which uses a model of human sound perception to compare codec input and output.

Finally, if the low-delay PCM standard G.711 has a complexity normalised to 1, then the 4.8 kbit/s CELP standard would have a relative complexity of about 400, and the military LPC10E standard a relative complexity of about 100. Section 3.4 discusses the second-generation cellular mobile system (GSM), and the CELP system in some detail.

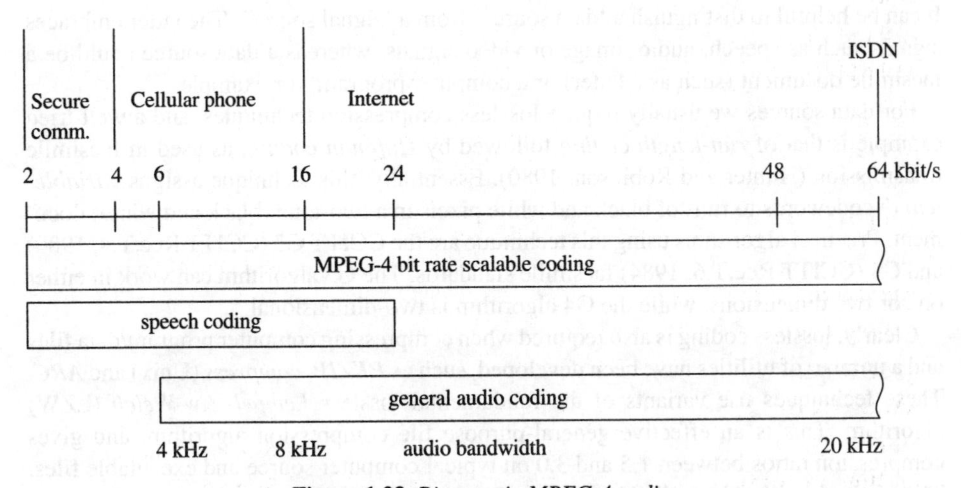

Figure 1.23 *Bit rates in MPEG-4 audio.*

1.7.4 Audio coding

High-quality digital audio signals (mono or stereo) are usually sampled at 32, 44.1 or 48 kHz. The first sampling rate is applicable to broadcasting, the second to compact disc (CDs), and 48 kHz is used for professional studios and DAT digital tape recorders. For stereo signals encoded at 16 bits/sample, the raw audio rate is therefore nominally 1.5 Mbit/s and significant compression is often required. Compression is usually to a standard bit rate, such as 384, 256, 128 or 64 kbit/s for stereo signals (ISO/IEC, 1992). These rates would be required for example in stereo digital audio broadcasting (DAB), stereo cable radio, the Internet and mono-audio for ISDN.

A combination of two compression mechanisms is usually used. The first removes redundant information, such as the perceptual redundancy that exists between the right- and left-hand channels of a stereo signal. The second mechanism takes advantage of psychoacoustic phenomena like spectral and temporal masking. For example, a sound preceded by another, higher level, sound will be perceived only if it lies above a *masking curve*, which is itself dependent upon the frequency and time intervals separating the two sounds.

The MPEG audio standards [ISO/IEC 1992,1994] essentially analyse the signal using a filter bank (subband filtering), exploit perceptual redundancy (joint stereo coding), and apply bit allocation and quantising between subbands based upon a computed masking curve. The MPEG standard can be implemented as one of three increasingly complex compression schemes known as layers I, II and III. This permits a universal coding scheme to have various applications, such as a professional recording and consumer recording. MPEG layer III for example is used for transmitting music on the Internet (under the name MP3) and can give CD quality at a compression ratio of 12. MPEG-4 audio uses 'bit rate scalability' in order to achieve bit rates from 2 kbit/s to beyond 64 kbit/s (Fig. 1.23).

1.7.5 Data coding

It can be helpful to distinguish a 'data source' from a 'signal source'. The latter embraces signals such as speech, audio, image or video signals, whereas a data source could be a facsimile document (such as a letter) or a computer program, for example.

For data sources we usually require lossless compression techniques, and a well-tried example is that of *run-length coding* followed by *Huffman coding*, as used in facsimile transmission (Hunter and Robinson, 1980). Essentially, this technique assigns *variable-length* codewords to runs of black and white pixels in a two-tone (black and white) document. Practical algorithms using this technique are the CCITT G3 (CCITT Rec.T.4, 1980) and G4 (CCITT Rec.T.6, 1984) facsimile standards. The G3 algorithm can work in either one or two dimensions, while the G4 algorithm is two-dimensional.

Clearly, lossless coding is also required when compressing computer program/data files and a number of utilities have been developed, such as *PKZIP*, *compress* (Unix) and *ARC*. These techniques use variants of the fundamental lossless *Lempel–Ziv–Welch* (LZW) algorithm. This is an effective general-purpose file compression algorithm, and gives compression ratios between 1.5 and 3.0 on typical computer source and executable files. The LZW algorithm is studied in detail in Section 2.5.

1.7.6 Still image coding

A *still image* is a fixed frame or picture as distinct from a time-sequence of frames (video), and it can be two-tone or continuous tone, monochrome or colour. Computer images are often formatted as GIF or TIFF files and, as in computer data file compression, GIF and TIFF support the use of the LZW compression algorithm. Since LZW is a general-purpose algorithm it is not 'tuned' to image compression and compression ratios can vary greatly, depending upon the image; while a factor 2 is common, it could be as high as 10. It can give acceptable results for bi-level (two-tone) images, although its compression is somewhat inferior to that achieved by the G4 algorithm.

Continuous-tone digital images can also be compressed using the JPEG image compression standard (ISO/IEC, 1991; Wallace, 1991). The *baseline* JPEG algorithm is a lossy technique based upon the DCT (although the JPEG standard also has a lossless compression mode based upon DPCM). At a compression of 1.5–2 bits/pixel, JPEG coded images are usually indistinguishable from the original, while 0.5–0.75 bits/pixel gives good quality images sufficient for most applications. At 1 bit/pixel, a 720 × 576 pixel image would take about 6.5 seconds to transmit over a 64 kbit/s link.

Whilst JPEG is a standard for coding greyscale or colour images, the JBIG algorithm is primarily for coding bi-level images (ISO/IEC, 1993). The algorithm is lossless, and also has a limited greyscale capability. Its *progressive* coding mode captures images as a compression of a low-resolution or *bottom layer* image plus a sequence of *delta files* or *differential layers* that each provide another level of resolution enhancement. This flexibility is useful, for example, where output devices such as PCs and laser printers have varying resolution, and for image browsing over low rate communications links. It is clearly applicable to facsimile images and generally outperforms the G4 algorithm (the most efficient of the G3/G4 algorithms). This is particularly true for bi-level images rendering greyscale using halftoning, since the fixed Huffman coding table is not well matched to the short runs present in a halftoned image. Finally, a key point of the JBIG algorithm is that it uses an efficient coding technique called *Arithmetic Coding* (AC) within each layer. This advanced lossless technique is discussed in Section 2.4.

1.7.7 Video coding

Figure 1.24 summarises the main standards for uncompressed analogue and digital video. It also shows that transcoding is possible between the analogue domains, and between the digital domains. For example, conventional analogue processing takes analogue component signals, *YUV* (essentially luminance, *Y*, and *B–Y*, *R–Y*), and encodes them to a composite PAL colour signal. Translation from analogue to digital format only adds quantisation impairment, for example, translation from analogue *YUV* to 4:2:2 *component digital* format by appropriate sampling, or translation from an NTSC composite analogue signal to a *composite digital* signal by sampling at 4 times colour subcarrier frequency $4f_{sc}$. In this section we are primarily concerned with component digital video formats and their relevance to MPEG-2 coding.

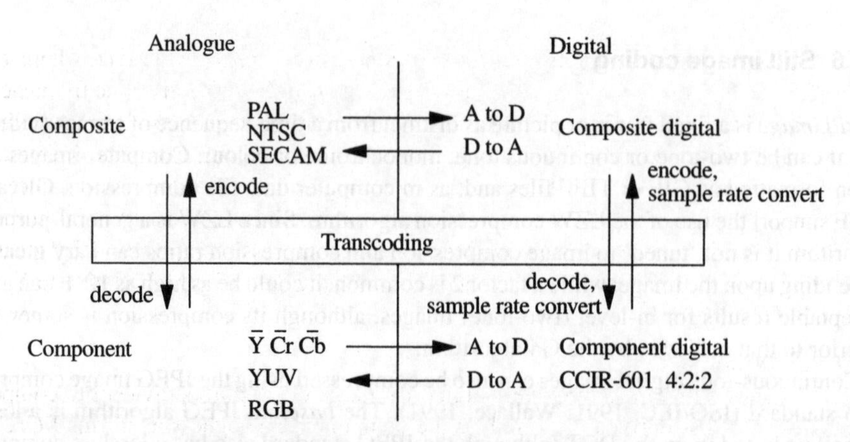

Figure 1.24 *Coding standards in video systems.*

The A:B:B format

A colour signal from a camera is usually a *component analogue* signal in RGB format (Figure 1.24). Since each of these components contains both luminance (greyscale) information and chrominance (colour) information, it is beneficial to convert the signal to the more useful $\overline{Y}\overline{Cr}\overline{Cb}$ format. This is a form of *color space conversion*. The exact conversion depends upon the transmission standard and for the ITU-R standard the matrix equations are (CCIR, 1982)

$$\begin{aligned}\overline{Y} &= +0.587G' + 0.299R' + 0.114B' \\ \overline{Cr} &= \frac{0.5}{0.701}(R' - Y') = -0.419G' + 0.500R' - 0.081B' \\ \overline{Cb} &= \frac{0.5}{0.886}(B' - Y') = -0.331G' - 0.169R' + 0.500B'\end{aligned} \quad (1.73)$$

Here, $\overline{Y}$ represents the analogue luminance (essentially greyscale) component ranging between 0 and 1, $\overline{Cr}$, $\overline{Cb}$ are the analogue chrominance (essentially colour) components ranging between –0.5 and 0.5, and R', G' and B' are gamma-corrected analogue values in the range 0 to 1. The significant point here is that CCIR-601 (1982) defines digitisation parameters for video signals in this particular component format. The general sampling format is A:B:B (see ITU-R BT.624), where A reflects the international standard sampling rate for Y signals and is normalised to 4. This gives flexibility to generate lower and higher standards. B then reflects the relative colour difference or chrominance sampling frequency. Possible sampling formats for a $\overline{Y}\overline{Cr}\overline{Cb}$ component analogue input, are:

- 8:4:4 a possible format for higher definition television (HDTV)
- 4:4:4 a useful format for computer graphics
- 4:2:2 a standard interface format for studio equipment (contribution quality)
- 4:2:0 a common format for distribution applications
- 4:1:0 a special format with significantly reduced chrominance resolution
- 2:1:1 a possible low bit rate format suitable for use in electronic news-gathering

CCIR-601 prescribes a main sampling rate of the form $f_s = nf_h = 13.5$ MHz, where $n = 864$ in 625 line systems, $n = 858$ in 525 line systems and f_h is the respective line frequency. Clearly, the sampling frequency is *line locked* (corresponding to an *orthogonal* sampling pattern), and is independent of the scanning standard. The number 4 in the above formats therefore corresponds to 13.5 MHz, and the number 2 denotes that the chrominance is sampled at 6.75 MHz, regardless of the scanning standard. More precisely, 4:2:2 means that chrominance is *horizontally* subsampled by a factor 2 relative to *Y*, whilst 4:2:0 means that chrominance is horizontally *and vertically* subsampled by a factor 2 relative to *Y*.

The 4:2:2 format

Figure 1.24 shows that this component digital format is the digital equivalent to component analogue and is derived by sampling the component analogue waveforms $\overline{Y}\overline{Cr}\overline{Cb}$. The (8-bit) component digital signal is an offset and scaled version of (1.73):

$$\begin{aligned} Y &= 219\overline{Y} + 16 \\ Cr &= 224\overline{Cr} + 128 \\ Cb &= 224\overline{Cb} + 128 \end{aligned} \tag{1.74}$$

(quantisation forcing the nearest integer value in each case). In (1.74), *Y* will be in the range 16–235, and *Cr*,*Cb* will be in the range 16–240. CCIR-601 specifies the sampling as four *Y* samples for two *Cb* samples and two *Cr* samples; see Figure 1.25(a). Note that ⊕ corresponds to a luminance sample and *two* chrominance samples per pixel. Since both *Cr* and *Cb* are simultaneously available each line, the *vertical* chrominance resolution is the same as for luminance. On the other hand, the chrominance resolution is half that of luminance in the *horizontal* direction, but this is of little consequence since the colour acuity of the eye is less than that for luminance. Each sample has 8-bit precision with a provision for

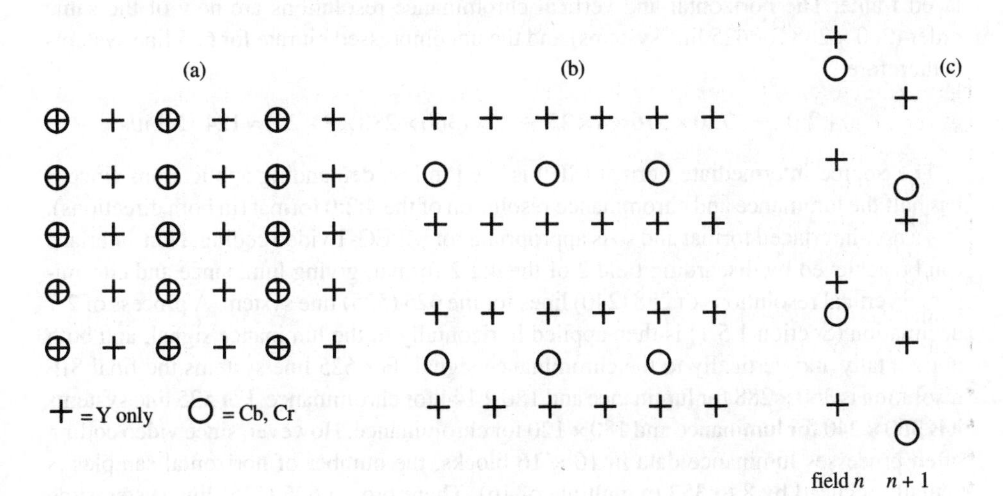

Figure 1.25 *Luminance-chrominance sampling patterns (a) 4:2:2; (b) 4:2:0; (c) for interlaced 4:2:0 frame.*

extension to 10 bits/sample. For 8-bit coding the raw bit rate for 4:2:2 sampling is therefore $13.5 \times 8 + 2 \times (6.75) \times 8) = 216$ Mbit/s, and this corresponds to an effective sample rate of 27 MHz when the samples are time multiplexed.

In practice this extremely high bit rate is reduced since samples in line and field blanking intervals are actually redundant. The *active* line is actually only 720 samples in both 625 line and 525 line systems, compared, say, with 864 samples/full scan line in a 625 line system. The corresponding number of active samples/line for each chrominance component is 360 in both 625 and 525 line systems. Similarly, there are 576 *active* lines for 625 line systems (and 480 active lines for 525 line systems). Allowing for blanking, the uncompressed bit rate for 625 line systems sampled according to the 4:2:2 standard is therefore

$$4:2:2 \qquad 720\times576\times8\times25\ +\ 2\times(360\times576\times8\times25)\approx166\ \ \text{Mbit/s}$$

A similar calculation for 525 line systems gives the same figure. Note that 166 Mbit/s arises from the video alone, and there will be an additional bit rate due to ancillary signalling in blanking (such as Teletext and subtitling data). Clearly, there is still a need for compression. For instance, a compression ratio of 4.9 is required to reach the 34 Mbit/s level in Table 1.3 (again, 34 Mbit/s is the 'total package' rate embracing video error correction, sound and ancillary services, so the actual compressed rate will be around 30 Mbit/s).

The 4:2:0, SIF and CIF formats

This format is obtained from the 4:2:2 format by using the same chrominance samples for two adjacent lines (the luminance sampling is the same as for 4:2:2). The 4:2:0 sampling format is shown in Figure 1.25(b), which indicates that the *Cb* and *Cr* samples are generated by interpolating 4:2:2 samples of the two adjacent lines that they will 'colour'. Figure 1.25(c) shows the position of the 4:2:0 chrominance samples within each field of an interlaced frame. The horizontal and vertical chrominance resolutions are now of the same order (360×288 for 625 line systems) and the uncompressed bit rate for 625 line systems is therefore

$$4:2:0 \qquad 720\times576\times8\times25\ +\ 2\times(360\times288\times8\times25)\approx124\ \ \text{Mbit/s}$$

The Source Intermediate Format (SIF) is used in less demanding applications since it has half the luminance and chrominance resolution of the 4:2:0 format (in both directions). It is a non-interlaced format and so is appropriate for MPEG-1 video coding. Non-interlace can be achieved by discarding field 2 of the 4:2:2 format, giving luminance and chrominance vertical resolutions of 288 (240) lines for the 625 (525) line system. A process of 2:1 decimation (Section 1.5.1) is then applied horizontally to the luminance signal, and both horizontally and vertically to the chrominance signal. For 625 line systems the final SIF resolution is 360×288 for luminance and 180×144 for chrominance. For 525 line systems it is 360×240 for luminance and 180×120 for chrominance. However, since video coding often processes luminance data in 16×16 blocks, the number of horizontal samples is actually reduced by 8 to 352 (a multiple of 16). Therefore, in 625 (525) line systems the actual SIF resolution is 352×288 (240) for luminance and 176×144 (120) for chrominance.

To enable codecs to interwork between regions of the world with 625 line and 525 line TV standards, CCITT Rec. H.261 specified a non-interlaced 'Common Intermediate Format' (CIF). This is a compromise between the American and European SIF formats. The CIF spatial resolution is 352 × 288 for luminance and 176 × 144 for chrominance, using an orthogonal sampling pattern. The maximum CIF temporal resolution is 30 frames/s (strictly 29.97 frames/s), this being the 525 line system frame rate. There are several extensions to CIF. QCIF (1/4 CIF) has a resolution of 176 × 144 for luminance and 88 × 72 for chrominance, and this is the main input format used for ISDN H.261 videophone. SQCIF has a luminance resolution of 128 × 96 and is supported by the H.263 standard. In addition, H.263 has optional support for 4CIF (with a luminance resolution of 704 × 576) and 16CIF (with a luminance resolution of 1408 × 1152).

1.7.8 MPEG video coding

The MPEG-2 international standard for video coding (ISO/IEC, 1994) (also referred to as ITU-T Rec.H.262) supports efficient coding of *interlaced* video, whereas MPEG-1 was specifically designed for *sequential* images. Its importance is emphasised by the numerous applications, as for example in cable TV, DBS, DVD, VOD and HDTV.

The basic role of an MPEG encoder is to take analogue or digital video signals and compress them into packets of information that are suitable for disk storage, as on a video server, and hence for transmission. It is important to note, however, that the standard only defines the *bitstream syntax* and the *decoding process*; it does not define how an encoder should be implemented, only that it should be compliant with the bitstream. *This means that all encoders will not necessarily generate the same quality video at a given bit rate.*

The CCIR-601 4:2:2 and 4:2:0 formats are important here because they are specified as the *input formats* for MPEG-2 coding, see Table 1.5 (note that main level in 525 line systems corresponds to 720 × 480, and so on). Essentially they define the input signal quality to the encoder. For example, MPEG-2 is capable of compressing standard

Table 1.5 *MPEG-2 profile/level combinations.*

	Simple profile No B frames 4:2:0	Main profile Not scalable 4:2:0	SNR scalable profile SNR scalable 4:2:0	Spatially scalable profile SNR scalable Spatial scalable 4:2:0	High profile SNR scalable Spatial scalable 4:2:0 or 4:2:2
High level ≤ 1920 × 1152	×	≤ 80 Mbit/s	×	×	≤ 100 Mbit/s
High level ≤ 1440 × 1152	×	≤ 60 Mbit/s	×	≤ 60 Mbit/s	≤ 80 Mbit/s
Main level ≤ 720 × 576	≤ 15 Mbit/s	≤ 15 Mbit/s	≤ 15 Mbit/s	×	≤ 20 Mbit/s
Low level ≤ 352 × 288	×	≤ 4 Mbit/s	≤ 4 Mbit/s	×	×

definition 4:2:0 video down to the range 3–15 Mbit/s, but around 3 Mbit/s the coding and decoding artefacts can be objectionable. For digital terrestrial television broadcasting of standard-definition 4:2:0 video, a bit rate of about 6 Mbit/s is therefore a good compromise, the corresponding compression ratio being about 20.

Table 1.5 shows that MPEG-2 encoding can be to one of a range of resolutions formally defined by *profile/level* combinations. As a rule, each profile defines a set of algorithms added as a superset to the algorithms in the profile below. A level specifies the range of the parameters that are supported by the implementation, such as frame size, frame rate and bit rates. The primary *distribution* standard is MPEG-2 main profile (MP) and corresponds to a 4:2:0 input format. Note that main profile/main level MP@ML coding corresponds to a maximum resolution of 720 × 576 and a bit rate ≤ 15 Mbit/s. In contrast, HDTV is specified by MP@H1440L, or higher. Whilst MPEG-2 MP@ML coding is appropriate for video distribution, it is inadequate for studio production, such as editing. To this end, MPEG-2 also specifies the 4:2:2 input format for high-end applications (Table 1.5). This is sometimes referred to as *contribution* standard or MPEG-2 4:2:2 profile.

MPEG-4 video coding

MPEG-4 has the formal designation ISO/IEC 14496 and became an International Standard in 1999. The video coding uses a DCT-based algorithm with advanced intraframe coding, optimised VLC, and advanced prediction. MPEG-4 is not meant to replace MPEG-2; rather, it provides for a wide range of alternative applications and new types of content. Typical applications include multimedia delivery via the Internet, mobile communications, low bit rate (often wireless-based) communications and broadcasting. A key feature of MPEG-4 is that it includes *object-based* coding in the sense that a scene can comprise various objects. Typical objects could be the computer-generated walls and floor of a room, Web site information, an arbitrarily shaped image and a wall picture derived from a HDTV video. Moreover, users can modify scenes in a truly interactive way by, for example, adding, deleting, repositioning or animating objects.

The basic idea is shown in Figure 1.26. Since no specific transport mechanism is defined in MPEG-4 a broad choice of transport protocols can be used, such as MPEG-2, ATM and DAB. At the sync layer, *packetised elementary streams* (PESs) are reassembled based on their timing information. The information in a PES includes a primitive object, how it is to be decoded (in terms of *object descriptors*, ODs), how it is organised within the scene and any interactive (upchannel) information from the terminal. Object descriptors can be added or deleted dynamically as the scene changes using Binary Format for Scenes (BIFS) commands.

A key feature of MPEG-4 video coding is the use of 'sprites' to achieve bit-saving. A sprite is a still image depicting a largely unchanging background, for example, a computer-generated wall of a room, or a crowd at a tennis match. Taking the latter, the sprite could be a still image describing the background content of a tennis match over all frames of the sequence (it is assumed that it is extracted from the video sequence prior to coding). The sprite would be transmitted to the receiver only once as the first frame of the sequence, and stored in a sprite buffer. In each consecutive frame only the camera parameters relevant for the background are transmitted, thus allowing the receiver to reconstruct the background image for each frame. A 'foreground object' could be the tennis player and this is segmented from the background. The texture of the foreground object is compressed and is

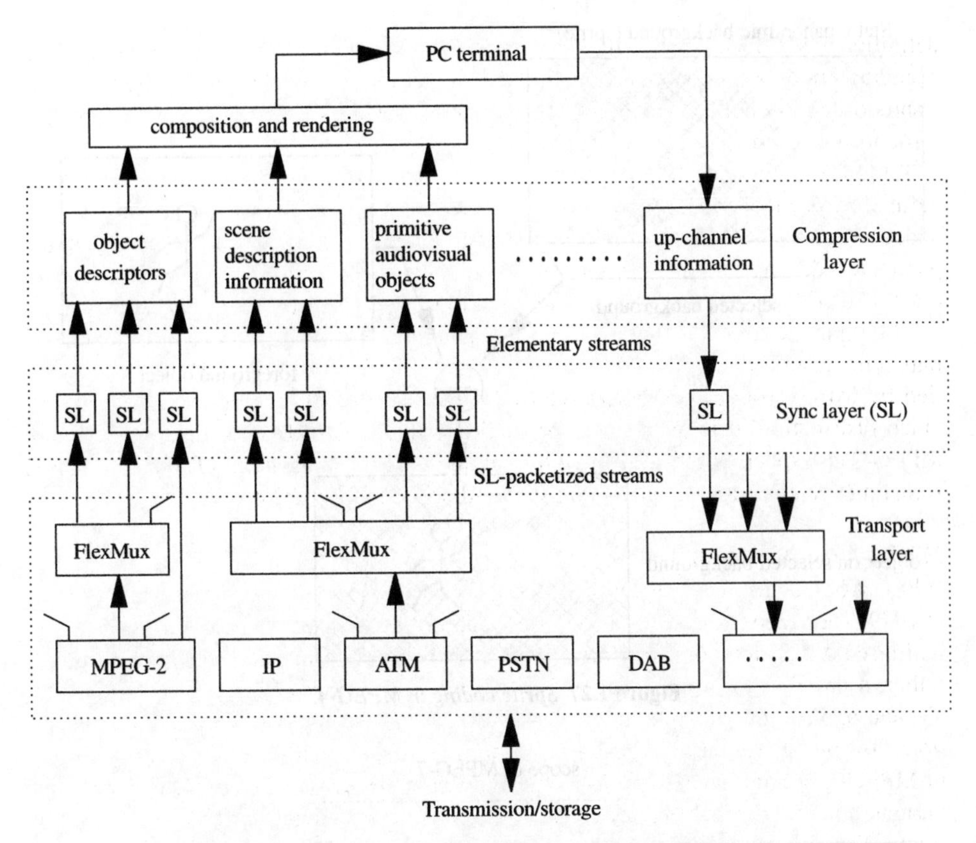

Figure 1.26 *The interactive and 'object' basis of MPEG-4.*

transmitted along with its arbitrary shape. The receiver then composes foreground and background images to reconstruct each frame (Figure 1.27).

MPEG-7 video coding

Standard Internet search engines allow searching for textual information but do not permit searching for audio-visual content. For instance, textual-based search engines do not permit a Web search for a particular song melody, or a video sequence of a rocket launch. Similarly, they do not permit a simple graphic to be drawn on a screen in order to search for a similar graphic. This is because there is no generally recognised description of such material. On the other hand, given suitable audio-visual descriptors, the search engine could be applicable to digital libraries, multimedia directory services ('yellow pages'), and broadcast media selection (channel selection), for example.

The objective of MPEG-7 is therefore to expand textual-based Internet search methods by standardising the description of multimedia material regardless of whether it is locally stored, in a remote database or broadcast. In other words, MPEG-7 specifies a *standard set of descriptors* that can be used to describe various types of multimedia information (Figure 1.28). The descriptors are associated with the content itself to permit fast and efficient

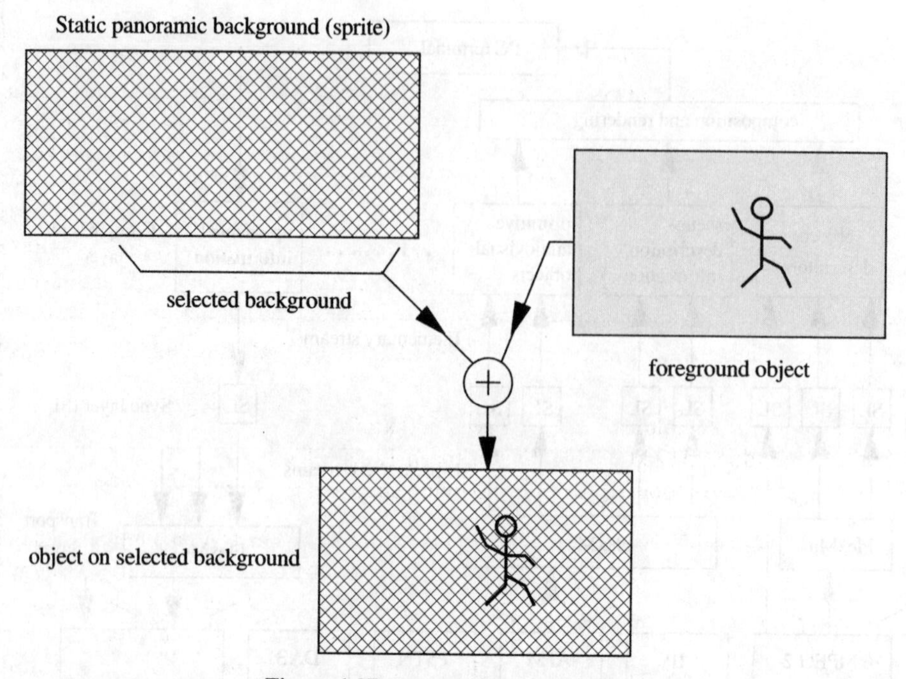

Figure 1.27 *Sprite coding in MPEG-4.*

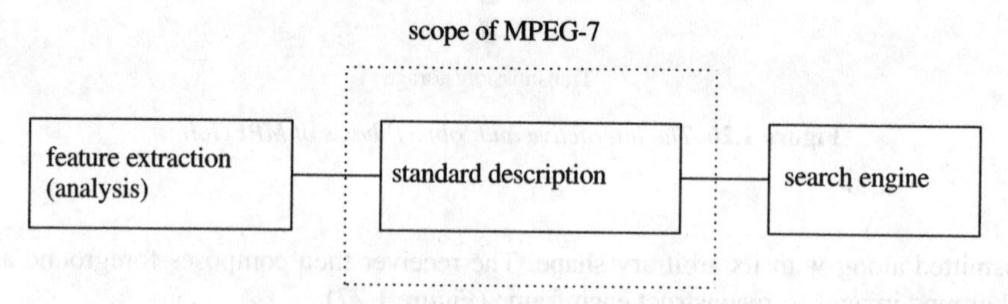

Figure 1.28 *Scope of MPEG-7.*

searching for material of user interest. In particular, using MPEG-4 coding, it is possible to attach descriptions to objects *within* a scene.

Descriptors can be high-level or low-level. For example, a user searching for pictures could input a sketch as a 'wildcard' image, or input a general verbal description of the required picture, such as 'a sunset over the sea'. These are high-level descriptors. On the other hand, shape, texture, colour, movement and position are classed as low-level descriptors.

1.7.9 Universal Mobile Telecommunications System (UMTS)

An overview of telecommunications standards would be incomplete without mention of third-generation mobile standards. The second generation mobile telecommunications system based on GSM handles voice telephony, facsimile and electronic mail, but is not well suited to wideband applications like video, multimedia or high-speed Internet access. This created a need for a third-generation system, which has the broad objective of bringing the capability of mobile networks significantly closer to that of fixed networks. In other words, in addition to the usual voice, facsimile and data services, provision is made for high-speed Internet access, video telephony and conferencing, entertainment services, and online banking and shopping. UMTS is a European contribution to the IMT 2000 family of global third-generation mobile standards. In contrast to GSM, UMTS embraces a truly global concept, enabling roaming on a worldwide basis; that is, compact lightweight terminals enable data files to be sent or received anywhere in the world. Maximum target bit rates in UMTS depend upon the user's environment:

- Rural outdoor: 384 kbit/s, maximum speed 500 km/h
- Suburban outdoor: 512 kbit/s, maximum speed 120 km/h
- Indoor/low range outdoor: 2 Mbit/s, maximum speed 10 km/h

Access is provided to cordless, cellular and satellite networks from a single handheld terminal, and all this is achieved using advanced source and channel coding technology. The choice of channel coding depends upon the propagation environment and the spectrum efficiency and quality requirements of the various services. Applications of large cells, especially in the case of satellite transmission, usually require more powerful channel coding, while microcellular systems used in a pedestrian environment may allow less complex channel coding. Typical UMTS transmission using CDMA access technology uses convolutional error control coding and interleaving, both of which are discussed in depth in later chapters.

Bibliography

Agrawal, R. and Shenoi, K. (1983) Design methodology for $\Sigma\Delta M$, *IEEE Trans.*, **COM-31**, 360–369.

CCIR Rec. 601 (1982) Encoding parameters of digital television for studios, *XVth Plenary Assembly*, Geneva, 1982, Vol. XI, part 1.

CCITT Rec. T.4 (1980) *Standardisation of Group 3 facsimile apparatus for document transmission.*

CCITT Rec. T.6 (1984) *Facsimile coding schemes and coding control functions for Group 4 facsimile apparatus.*

CCITT Rec. H.261 (1990) *Video codec for audiovisual services at p × 64 kbit/s.*

Hunter, R. and Robinson, A. (1980) International digital facsimile coding standards, *Proc. IEEE*, **68**(7), 854–867.

ISO/IEC 10918 (1991) Digital compression and coding of continuous-tone still images, *JTC1/SC2/WG10 (JPEG)* ISO CD.

ISO/IEC 11172-3 (1992) *Coding of moving pictures and associated audio for digital storage media at up to 1.5 Mbit/s, Audio Part.*

ISO/IEC 11544 (1993) *Coded representation of picture and audio information – Progressive bi-level image compression.*

ISO/IEC 13818-2 (1994) Generic coding of moving pictures and associated audio information, *JTC1/SC29/WG11 Part 2: Video.*

ISO/IEC (1998) Coding of moving pictures and audio, MPEG-7: Context and Objectives, *JTC1/SC29/WG11 N2460.*

ISO/IEC (1999) Coding of moving pictures and audio, MPEG-4 Overview, *JTC1/SC29/WG11 N2725.*

IEEE (1997) MPEG digital video-coding standards, *IEEE Signal Processing Magazine*, September, pp. 82–100.
Naus, P. *et al.* (1987) A CMOS stereo 16-bit D/A converter for digital audio, *IEEE J. Solid-State Circuits*, **SC-22**(3), 390–394.
Wallace, G. (1991) The JPEG still picture compression standard, *Communications of the ACM*, **34**(4), 30–44.

CHAPTER 2

Lossless source coding

Figure 2.1 provides a classification of many common data compression techniques and at the outset we see that compression can either be 'lossless' or 'lossy' in the information sense. In this chapter we examine the lossless (or completely reversible) techniques whereby no information is lost in the compression process. The decoder output data is identical to the encoder input data. Lossless techniques are particularly (but not exclusively) applicable to 'text' type data, such as English text or a computer program. A notable exception is the lossless *JBIG* algorithm for image compression. Conversely, lossy techniques are particularly applicable to 'signal' type data, such as sampled speech or video signals.

Figure 2.1 shows that lossless techniques could be classified into three basic types:

- *Ad hoc* methods, such as *run-length coding* (RLC) of image or text data, are often heuristic and not mathematically rigorous, although they are still used extensively. Blanking suppression could be regarded as a form of RLC.
- Entropy encoders achieve compression by exploiting non-uniformity in the probabilities of source symbols. In other words, compression is achieved by exploiting the *statistical* redundancy in the signal, in contrast to, say, ADPCM which exploits *correlation* redundancy. In practice, the probabilities could be drawn from a fixed table or estimated on the fly due to insufficient or time-varying statistics. Formally, we consider the distribution of probabilities for the next input symbol to be computed by a *probabilistic model*, and assume that an *entropy mapper* uses the input symbol string and the model output to generate a (usually) compressed string of symbols.
- A dictionary technique looks for repeated substrings of data (as often found in text files) and effectively factorises them out in the sense that no code string is transmitted upon recognising a substring. The dictionary therefore comprises common substrings and is usually constructed on the fly at both encoder and decoder. A good practical example here is the *LZW* (Lempel–Ziv–Welch) coding of text or computer files. We will see that LZW coding does not separate modelling from coding as distinctly as the entropy encoder. On the other hand, since it still exploits the statistical redundancy in the source, it could also be classified as an entropy coding technique.

2.1 Run-length coding

This simple technique converts a string of repeated characters into a compressed string of fewer characters. A general format for RLC is shown in Figure 2.2(a), together with several examples. A special character *S* is sent first to indicate that compression follows and *X* and *C* denote the repeated character and character count respectively. Clearly, *S*

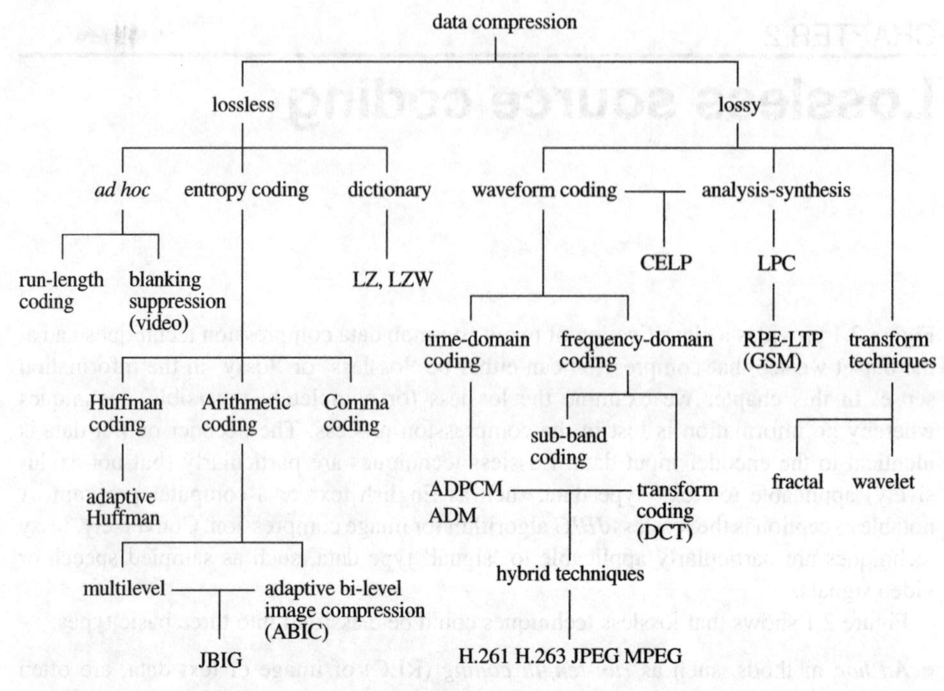

Figure 2.1 *Classification of some data compression techniques.*

should not naturally occur in the data string, and the format in Figure 2.2(a) is only useful for strings of four or more repeated characters. Alternatively, if *S* denotes a following run of *specific* characters, such as a run of 'space' (or null) characters, the format reduces to *S* and the character count, *C*. In other words, null suppression in text requires just two characters.

RLC is also used extensively for compressing two-tone (black and white) images. Typical images are pages of text and/or drawing (as found in facsimile) and cartoon-type

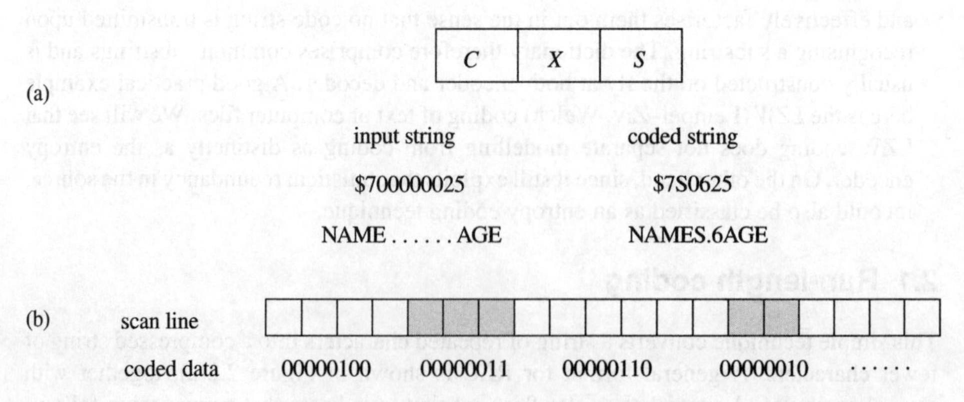

Figure 2.2 *Run-length coding: (a) typical three-character format for text; (b) simple fixed wordlength coding for two-tone graphics.*

images for low bit rate visual communication. For this application the basic RLC scheme operates in one dimension (horizontally) and exploits the correlation between pixels along a line. The simplest scheme uses fixed wordlength coding and Figure 2.2(b) illustrates this for a single scan line and 8-bit words. The transmitted sequence is simply the run-lengths 4, 3, 6, 2, ..., and each new symbol denotes a tone change. A facsimile system scanning at the CCITT rate of 1728 samples/line is quite capable of generating (usually white) runs of the order of 100 or larger, so the use of 8-bit words could achieve significant suppression; that is, a run of up to 255 would be represented by just 8 bits. In practice, for a facsimile document short runs are usually more probable than long ones and this statistical redundancy is often exploited by applying *variable-length coding* (VLC) to the run-lengths. Run-length/variable-length coding for two-tone graphics is discussed in Example 2.7.

Example 2.1
Multispectral (visible and infrared) image data from weather satellites can be clustered into three homogeneous regions representing a surface class (sea plus land) and two cloud classes. This format is suitable for image dissemination on twisted-pair telephone lines (PSTN) (Wade and Li, 1991) and the homogeneous regions lend themselves to RLC prior to transmission (Figure 2.3). If the run-length $r < 64$ then 8-bit words are selected, the first two bits denoting the class. Occasionally, $r \geq 64$ and 16-bit words are used, with the first two bits indicating the extended wordlength.

For this application typically 94% of runs were found to have $r < 64$, and so 8-bit coding is used for most of the time (corresponding to an average wordlength of 8.5 bits/run). This simple byte-orientated run-length/variable-length coding scheme is computer-compatible and typically achieves 4:1 compression for weather images.

Example 2.2
An $N \times N$ pixel block of a digitised image can be transformed to an $N \times N$ block of 'frequency domain' coefficients using the technique of *transform coding* (TC). This technique is discussed in detail in Section 3.2. After coefficient quantisation the non-zero coefficients tend to be clustered around the origin or DC component as illustrated in Figure 2.4 for $N = 8$. Clearly, much of the quantised block comprises zero-valued coefficients and it is common to exploit this redundancy in order to increase transmission efficiency. In particular, if the coefficients are transmitted by addressing them in a zigzag way, as shown, then the non-zero coefficients are concentrated at the start of the transmitted sequence and the efficiency of RLC is improved. Each non-zero coefficient can be regarded as a run terminator, and the run-length followed by the value (level) of the non-zero coefficient can be treated as a source symbol, $a_j, j = 1, 2, 3, ...$. For example, the zigzag scanning in Figure 2.4 yields the symbols

$$a_1 = (0,9) \quad a_2 = (0,6) \quad a_3 = (0,4) \quad a_4 = (2,2) \quad a_5 = (4,1) \quad a_6 = (7,1)$$

and these can be encoded using some agreed VLC, whereby the most probable short run and relatively low-level symbols are assigned the shortest codewords. The remaining zero-valued coefficients can then be denoted by simply transmitting an end-of-block character,

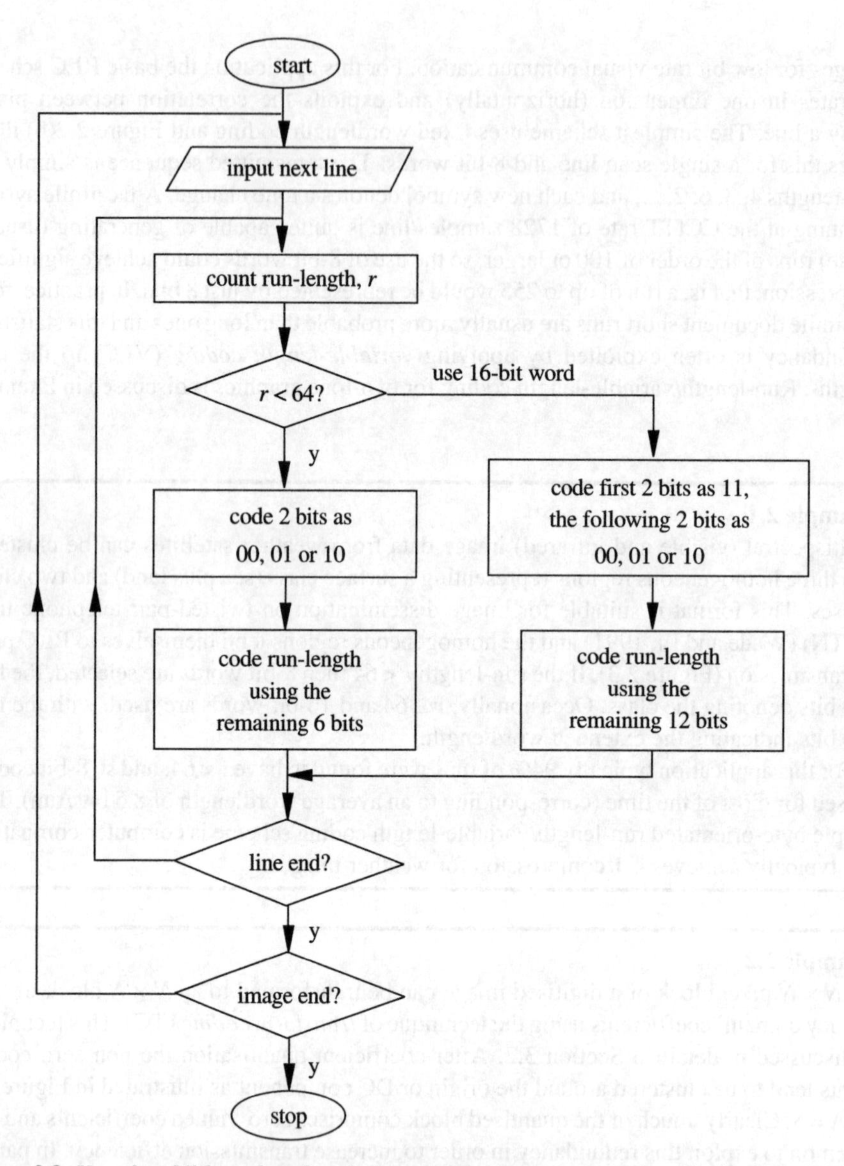

Figure 2.3 *Use of an 8/16 run-length code for weather image compression (00, 01, 10 code three different classes in the image).*

that is, the coefficients after the last non-zero one are not transmitted. This zigzag VLC technique is used for example in the CCITT H.261 video codec recommendation and in the MPEG-2 ISO/IEC 13818-2 standard. Table 2.1 shows part of a typical MPEG-2 VLC table and highlights the point that short run, relatively low-level symbols are assigned the shortest codewords.

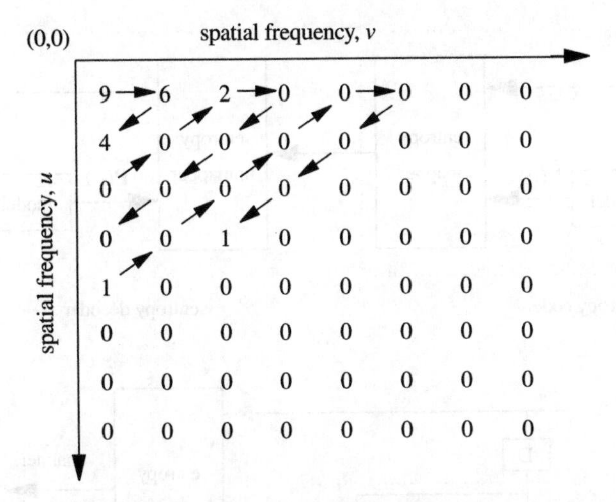

Figure 2.4 *Zigzag addressing of quantised coefficients values generated by a transform coder (the DC coefficient is quantised to a value of 9).*

Table 2.1 *Part of an MPEG-2 VLC table (the last bit, s, denotes the sign of the level).*

VLC	*Run*	*Level*
1s	0	1
0100s	0	2
00101s	0	3
011s	1	1
000110s	1	2
00100101s	1	3
0101s	2	1
0000100s	2	2
0000001011s	2	3

2.2 Information measure and source entropy

Before discussing VLC we need a quantitative description of the discrete source in terms of its *information rate*, *R*. Consider a *discrete source* having a finite alphabet *A* of *M* different symbols $a_j, j = 1, 2, ..., M$. The exact nature of each symbol is not important for the current discussion, but in its simplest form it could simply be a digital word from an ADC. As mentioned in the introduction, in order to compress this source we need to make assumptions about the data using a probabilistic model. In fact, *all* lossless compression schemes depend either implicitly or explicitly upon some such model. Generally speaking, a more complex model will give more accurate probabilistic predictions and hence achieve greater compression. The key problem in lossless compression is therefore to obtain a good probabilistic model of the source. Compression will then be achieved by treating

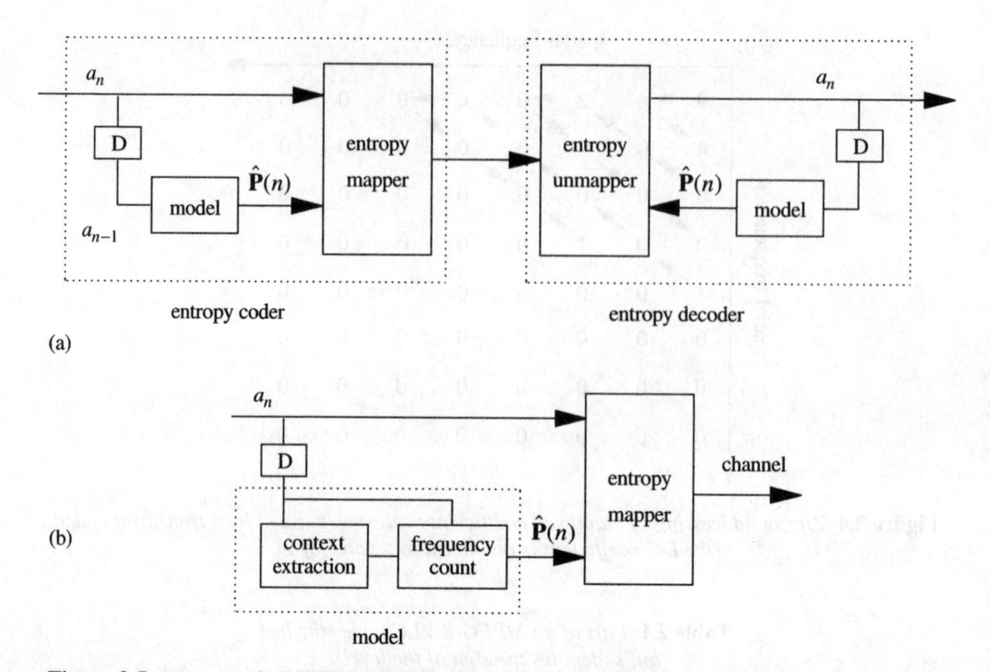

Figure 2.5 *(a) general model for adaptive symbol-by-symbol entropy coding and decoding; (b) detail of adaptive probability estimation.*

high-probability inputs differently from low-probability inputs. Unfortunately, for lossless schemes, given the random nature of information sources, there will also be some inputs that are a poor match to the model and result in *expansion* rather than compression.

From the above discussion we could view lossless compression as comprising two distinct stages, namely modelling and coding. A scheme using an explicit model is shown in Figure 2.5(a). Here, the source supplies a symbol a_n at time n, and using past data the model provides a corresponding estimate $\hat{\mathbf{P}}(n)$ of the current probability distribution:

$$\hat{\mathbf{P}}(n) = \left[\hat{P}_1(n), \hat{P}_2(n), \ldots, \hat{P}_j(n), \ldots, \hat{P}_M(n)\right]^{\mathrm{T}} \tag{2.1}$$

Here, $\hat{P}_j(n)$ is an estimate of the probability that symbol a_n will be value a_j. Note that the model provides a prediction vector $\hat{\mathbf{P}}(n)$ which can be based on any of the observed symbols a_0, ..., a_{n-1}, hence the unit delay, D. Also note that it must be possible for the decoder to generate exactly the same probability distribution.

Equation (2.1) corresponds to a general *adaptive* model in the sense that the probabilities may change for each new input symbol. To simplify matters we will initially assume a simple non-adaptive or *fixed* model. In this case the vector prediction reduces to

$$\hat{\mathbf{P}} = \left[\hat{P}(a_1), \hat{P}(a_2), \ldots, \hat{P}(a_j), \ldots, \hat{P}(a_M)\right]^{\mathrm{T}} \tag{2.2}$$

and the model input in Figure 2.5(a) would be removed. The probabilities in (2.2) could be predetermined by counting symbol frequencies in representative strings of source symbols. Clearly, $\hat{P}(a_j) \geq 0$, and the sum of all M probabilities must be unity. In effect,

(2.2) is assuming a *memoryless* model and the symbols are taken to be *statistically independent.* Put simply, $\hat{P}(a_j)$ is simply an estimate of the probability that symbol a_j is generated by the source at any instant in time.

Now, if a symbol a_j is unlikely to be generated ($\hat{P}(a_j)$ small), we are quite 'surprised' when it *is* generated, and we could interpret this as giving us more information than that conveyed by a commonly occurring symbol. Thus, information has an inverse relationship with $\hat{P}(a_j)$ and it is standard practice in information theory to define the *self-information* of symbol a_j as

$$I(a_j) = \log\left[\frac{1}{\hat{P}(a_j)}\right] \tag{2.3}$$

The units of information depend upon the base and it is usual to use base 2. This gives the information in *bits* (which must be distinguished from *logic* or *data* bits). We are particularly interested in the *average* information per symbol and this is easily calculated since we are assuming all M symbols to be statistically independent. In a long message of N symbols the jth symbol will then occur about $N\hat{P}(a_j)$ times, giving a total information of approximately

$$-\sum_{j=1}^{N} N\hat{P}(a_j)\log\hat{P}(a_j) \quad \text{bits}$$

assuming base 2. The statistical average information is referred to as the *source entropy* and is therefore given by

$$H(A) = -\sum_{j=1}^{M}\hat{P}(a_j)\log\hat{P}(a_j) \quad \text{bits/symbol} \tag{2.4}$$

As in statistical mechanics, $H(A)$ is a measure of the uncertainty or disorder of the source or system and it is low for an ordered source. For example, if $\hat{P}(a_j)=1$ and $\hat{P}(a_i)=0 \ \forall i \neq j$, then $H(A) = 0$. On the other hand, for a very disordered source all symbols would be equiprobable ($\hat{P}(a_j)=1/M \ \forall j$) and it can be shown that this leads to the maximum possible average information per symbol:

$$H(A)_{\max} = \sum_{j=1}^{M}\frac{1}{M}\log M = \log M \tag{2.5}$$

Thus, $0 \leq H(A) \leq \log M$. Finally, for a symbol rate k the average information rate (or entropy rate) is simply $R = kH(A)$ bits/s.

Example 2.3

Figure 2.6(a) shows a model of an ADC where it is assumed that the samples are quantised to just four levels or symbols. We also assume that the source statistics result in the indicated probabilities. The quantiser output is a discrete source ($M = 4$) and if the symbols are independent the source entropy is given by (2.4):

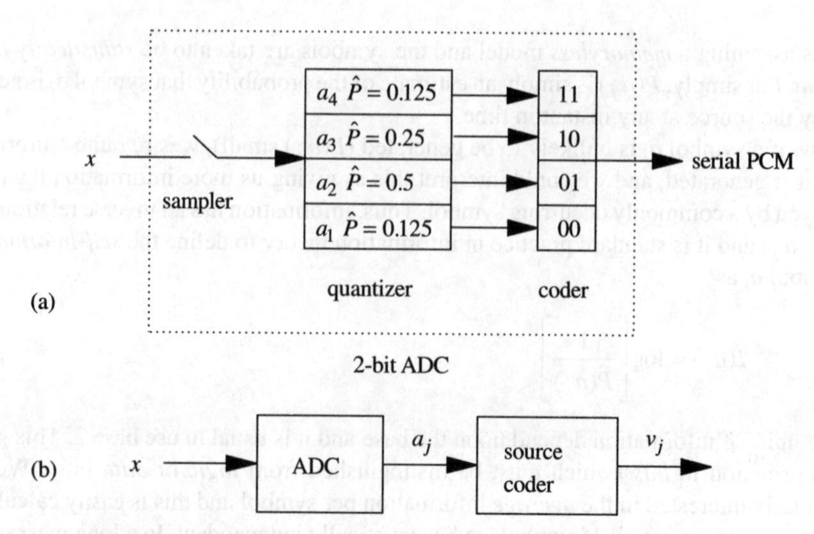

Figure 2.6 *(a) Model of a simple ADC; (b) addition of a source coder.*

Table 2.2 *Simple VLC table for the discrete source (Comma coding).*

a_j	$\hat{P}(a_j)$	v_j	l_j
a_2	0.5	0	1
a_3	0.25	10	2
a_1	0.125	110	3
a_4	0.125	1110	4

$$H(A) = \frac{1}{8}\log 8 + \frac{1}{2}\log 2 + \frac{1}{4}\log 4 + \frac{1}{8}\log 8 = 1.75 \quad \text{bits/symbol}$$

It is apparent that the average information rate of $R = 1.75$ bits/symbol differs from the average bit rate (or *data* rate) of $R_b = 2$ bits/symbol. The significant point here is that, in general, *the information rate will not be the same as the data rate*. Now suppose we follow the ADC with a *source encoder* with the objective of reducing R_b (Figure 2.6(b)). A simple approach would be to use the VLC shown in Table 2.2. This is, in fact, a *Comma code*, where the 0 corresponds to the comma at the end of a codeword and symbol a_j is now transmitted as codeword v_j. Clearly, the number of data bits/symbol, l_j, varies but the average length of a transmitted codeword is now reduced from $R_b = 2$ bits/symbol to

$$R_b = \bar{L} = \sum_j \hat{P}(a_j)l_j = 0.5 + 0.25\times 2 + 0.125\times 3 + 0.125\times 4 = 1.875 \quad \text{bits/symbol}$$

Thus, even *simple* source coding can reduce the data rate and achieve data compression. Moreover, *the data rate R_b is reduced towards the value given by H(A).*

Example 2.3 highlights several fundamental points in source coding. First, in order to apply entropy coding we need a source model and for a simple fixed model we have to assume some typical probability vector $\hat{\mathbf{P}}$. Such a fixed model is known *a priori* by both encoder and decoder. However, if the source is non-stationary (its statistics vary, as in a video signal for example) then a fixed model is likely to be inadequate. Secondly, in (2.4) we have assumed that the symbols are statistically independent in the sense that $\hat{P}(a_j|a_i) = \hat{P}(a_j)$, and for some practical sources such as two-tone facsimile this is reasonable (see Example 2.7). We say the source has *zero-memory* (a *memoryless* source) and (2.4) computes the *zero-order* entropy. Example 2.3 also highlights the fact that source coding can reduce the data rate to a value nearer to the source entropy as found from some source model. This important point is formally stated in Shannon's *noiseless coding theorem*:

Theorem 2.1 A discrete source of entropy $H(A)$ bits/sample or symbol can be binary coded with an average bit rate R_b that is arbitrarily close to $H(A)$, that is

$$R_b = H(A) + \varepsilon \tag{2.6}$$

where ε is an arbitrarily small positive quantity.

This theorem implies that $H(A)$, as found using a good source model, can be used as a target for data compression schemes. It also implies that the maximum achievable compression ratio is

$$CR = \frac{\text{average bit rate of source coder input}}{\text{average bit rate of source coder output}} \approx \frac{n}{H(A)} \tag{2.7}$$

where the PCM input is quantised to n bits/sample.

2.2.1 Source memory and context

So far we have assumed that the source symbols are statistically independent. However, in practice, we often have to account for the fact that the occurrence of a symbol depends upon some preceding symbols or some *context*. In this case, (2.4) needs to be modified in order to give a true measure of the source information. Also, we often need to compute the probability vector on the fly due to changing statistics or simply insufficient statistics. Combining both points, we conclude that in practice it will often be necessary to use an adaptive model which also accounts for context, as indicated in Figure 2.5(b). The model is then a method of adaptively computing $\hat{\mathbf{P}}(n)$ for a given context.

A classical approach is to model the source as an *mth-order Markov process* whereby the current source symbol a_n depends upon the m immediately preceding symbols. Let the m preceding symbols define a particular state or context, c_i; for an alphabet A of M symbols there will be $C = M^m$ such contexts. The probability of a_n being symbol a_j is then defined by the conditional probability $P(a_j|c_i)$ and we can define a *conditional* entropy for context c_i:

$$H(A|c_i) = -\sum_{j=1}^{M} P(a_j|c_i) \log P(a_j|c_i) \tag{2.8}$$

By definition the source entropy (the overall *conditional* entropy) is the statistical average of $H(A|c_i)$ over the C possible contexts:

$$H_c(A) = \sum_{i=1}^{C} P(c_i) H(A|c_i) \tag{2.9}$$

$$= -\sum_{i=1}^{C}\sum_{j=1}^{M} P(c_i) P(a_j|c_i) \log P(a_j|c_i) \tag{2.10}$$

As an example, for a first-order Markov source ($m = 1$) each context c_i is determined by only one previous symbol, say a_i, so $P(a_j|c_i) = P(a_j|a_i)$. In this case (2.10) reduces to

$$H_c(A) = -\sum_{i=1}^{M}\sum_{j=1}^{M} P(a_i, a_j) \log P(a_j|a_i) \tag{2.11}$$

For many sources this approach will give a lower target entropy and hence better compression. Put another way, given efficient source coding, (2.10) gives the average number of bits that will be used to code the current input symbol given a context of m preceding symbols.

Example 2.4

The importance of conditional entropy can be illustrated using English text. This has M = 27 characters including the space character and so

$$H(A)_{\max} = \log 27 = 4.8 \quad \text{bits/character}$$

Of course, not all characters are equiprobable, so $H(A)$ will be less than 4.8. A measure of character frequencies may for instance give $P(\text{space}) \approx 0.2$ and $P(z) \approx 0.001$, from which we conclude that thespacegiveslittleinformation! Taking estimated probabilities of all 27 characters (Shannon, 1951) and applying (2.4) gives $H(A) \approx 4.1$ bits/character. However, we also know that the letters are not statistically independent, so the true source entropy must be lower than 4.1. For instance, given letter 'q' the next letter is virtually certain to be 'u'. Taking account of the full statistical dependence between letters it can be shown that the conditional (true) source entropy is approximately 2 bits/character. Since direct PCM encoding of all 27 characters would require 5 bits/character, we see that lossless compression could yield a transmission rate close to 2 bits/character, and $CR \approx 2.5$.

Example 2.5

For English text Example 2.4 exposed a *statistical redundancy* of 0.7 bits/character and an actual or *correlation redundancy* of about 2.8 bits/character, and a similar example can be given for video signals. Linear quantisation of a high-contrast monochrome video signal will typically result in a discrete source with a nearly flat discrete pdf, $\hat{P}(a_j)$, as shown in

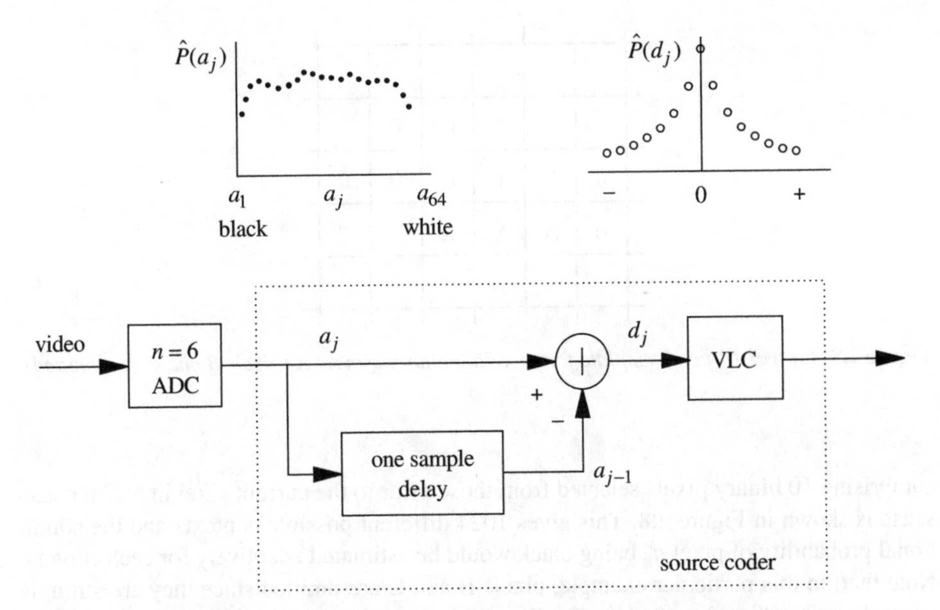

Figure 2.7 *Simple source coder for monochrome video signals.*

Figure 2.7. Since this is not perfectly flat, $H(A)$ as computed from (2.4) will be somewhat less than $H(A)_{\max}$, and typically $H(A) \approx 5$ bits/sample for 6-bit coding (Kretzmer, 1952). However, symbols a_j will not be independent, and we must account for this correlation redundancy in order to estimate the true source entropy. A simple but effective way of doing this for monochrome signals is to use *differential coding*, as indicated. The resulting difference samples, d_j, are approximately independent (and uncorrelated) and their probability distribution is highly non-uniform. Inserting probabilities $\hat{P}(d_j)$ into (2.4) will therefore give a reasonable estimate of the true entropy of the difference output, and for typical pictures this computation gives $H(A) \approx 3$ bits/sample (Kretzmer, 1952). It follows from (2.6) that efficient coding of the difference signal using VLC should result in an average bit rate close to 3 bits/sample, and $CR \approx 2$.

Example 2.5 highlights a fundamental point that can be summarised for a discrete source with a probability distribution given by (2.1) as follows:

Corollary 2.1 A significant entropy coding advantage is only achievable if **P**(n) is non-uniform with only a few of the symbols much more likely than the others.

Example 2.6

For images, source memory in the Markov sense is usually provided by a *template*. This provides a 'context' for the current pixel. Consider the coding (compression) of a two-tone black and white image. In this case the source model determines the probability of a pixel being black, conditioned on a template of pixels surrounding it. A typical template

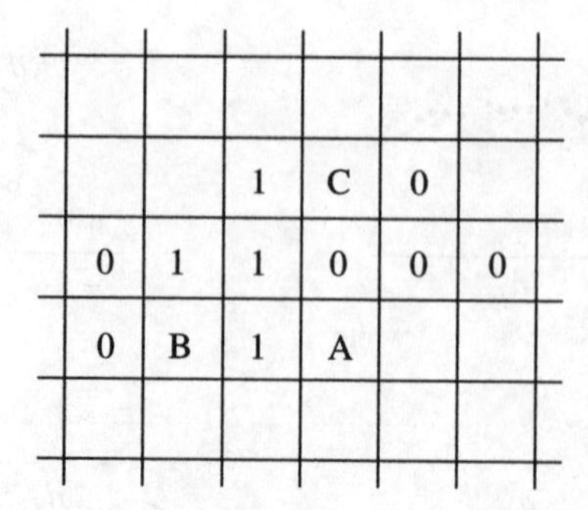

Figure 2.8 *Typical 10-pixel template for encoding binary symbol A (pixels B and C are omitted).*

comprising 10 binary pixels selected from those prior to the current pixel in a raster scan sense is shown in Figure 2.8. This gives 1024 different possible contexts and the conditional probability of pixel a_n being black would be estimated adaptively for each context. Note that, in this particular example, pixels B and C are omitted since they are strongly dependent upon the others and so contribute little to the conditional probability. In practice the exact template could be selected experimentally.

2.3 Variable-length coding

Examples 2.2 and 2.3 illustrated the concept of VLC. In this section we shall describe a procedure for deriving an efficient VLC (a Huffman code) given a source model, and discuss some of the applications and deficiencies of such coding.

The variable-length coding process is a mapping of a source symbol sequence to a code symbol sequence given a source model. Consider a fixed model and independent source symbols such that symbol a_j has estimated *a priori* probability $\hat{P}(a_j)$. Symbol a_j would then be coded to codeword v_j of length l_j code symbols, and the average length of such a VLC would be

$$\bar{L} = \sum_{j=1}^{M} \hat{P}(a_j) l_j \quad \text{code symbols/source symbol} \tag{2.12}$$

Having defined the average length of a code we can now restate Theorem 2.1 in more precise terms:

Theorem 2.2 For a code with c possible code symbols ($c = 2$ for a binary code) the average length of the codewords per source symbol can be made arbitrarily close to $H(A)/\log c$ by coding larger and larger blocks of symbols. Mathematically:

$$\frac{H(A)}{\log c} \leq \frac{\bar{L}}{n} \leq \frac{H(A)}{\log c} + \frac{1}{n} \tag{2.13}$$

where $\overline{L}$ is now the average length of the codeword associated with a block of n source symbols.

The assignment of a codeword to a *block* of *n* source symbols ($n > 1$) is called *block coding* and will be discussed later. It is more usual to code individual symbols to a binary code, in which case

$$H(A) \le \overline{L} \le H(A) + 1 \tag{2.14}$$

and the corresponding *coding efficiency* is

$$\eta = \frac{H(A)}{\overline{L}} \tag{2.15}$$

Note that, if

$$\hat{P}(a_j) = 2^{-l_j}, \quad j = 1, 2, \ldots, M \tag{2.16}$$

then

$$H(A) = -\sum_j \hat{P}(a_j) \log \hat{P}(a_j) = \sum_j \hat{P}(a_j) l_j = \overline{L} \tag{2.17}$$

and $\eta = 100\%$.

Example 2.7

The above concepts can be applied to facsimile (two-tone) image transmission. Lumping white and black run-lengths together for simplicity, we could obtain the run-length probabilities for a typical image:

Source symbol	*Run-length, r (pixels)*	*Run probability*
a_1	1	$\hat{P}_1$
a_2	2	$\hat{P}_2$
a_3	3	$\hat{P}_3$
$\vdots$	$\vdots$	$\vdots$
a_M	M	$\hat{P}_M$

Note that *M* now corresponds to the number of pixels in a scan line, so for standard facsimile this will be a *large* alphabet. Also, for this type of source it is usual to assume that successive symbols (run-lengths) are statistically independent (Meyr *et al.*, 1974) and so we can find the *run-length entropy* as

$$H(A) = -\sum_{i=1}^{M} \hat{P}_i \log \hat{P}_i \quad \text{bits/run}$$

The optimum variable-length binary code for this source will have an average wordlength $\bar{L}$ (in bits/run) bounded by (2.14), and the corresponding bit rate R_b (in bits/pixel) will be

$$\frac{H(A)}{\bar{r}} \le R_b \le \frac{H(A)+1}{\bar{r}}$$

Here, $\bar{r}$ is the average run-length

$$\bar{r} = \sum_{i=1}^{M} i\hat{P}_i \quad \text{pixels/run}$$

For typical facsimile weather maps $H(A)/\bar{r} \approx 0.2$ bits/pixel (Huang, 1977). This indicates that run-length/variable-length coding should yield an average bit rate close to 0.2 bits/pixel, in contrast to an uncoded rate of 1 bit/pixel ($CR \approx 5$). It is interesting to note that $H(A)/\bar{r}$ for weather maps can be well estimated using a binary first-order Markov model (see (2.11)); for example, given a white pixel it is highly probable that the next pixel is also white. This approach circumvents the need to measure run probabilities P_i.

2.3.1 Huffman coding

Here we describe a standard procedure for generating an efficient VCL, assuming the simple model of fixed symbol probabilities and M independent symbols (a zero-memory source). The method generates a *uniquely decodable*, *instantaneous code*, that is, each codeword can be unambiguously identified in the code sequence, and each codeword can be decoded without reference to any other codeword. The codewords can be concatenated and then decoded on a word-for-word basis. The simple Comma code in Example 2.3 is an example of such a code.

Theorem 2.3 A necessary and sufficient condition for a uniquely decodable code to be instantaneously decodable is that no codeword must be a prefix for another.

Starting from M source symbols, the coding procedure consists of a step-by-step reduction in the number of symbols until there are c source symbols, corresponding to c code symbols. Each reduction step reduces the number of symbols by $(c-1)$, so that for a total of α steps we find that M must satisfy

$$M = c + \alpha(c-1) \tag{2.18}$$

If there are insufficient symbols to satisfy (2.18) we simply add a dummy symbol of zero probability and discard it after coding. The Huffman coding algorithm for a zero-memory source is then as follows:

1. List the source symbols in decreasing probability.
2. Combine the c smallest probabilities and reorder. Repeat α times to achieve an ordered list of c probabilities.
3. Coding starts with the last reduction. Assign the first code symbol as the first digit in the codeword for all source symbols associated with the first probability of the reduced

source. Assign the second code symbol as the first digit in the codeword for all the second probabilities, etc.

4. Proceed to the next to the last reduction. For the c new probabilities the first code digit has already been assigned, so assign the first code symbol as the second digit for all source symbols associated with the first of these c new probabilities. Assign the second code symbol as the second digit for the second of these c new probabilities, and so on.
5. Proceed through to the start of the reductions in a similar manner.

Example 2.8 illustrates Huffman coding for a binary code (for a ternary Huffman code the final step would have three probabilities and the assigned code symbols could be 1, 0, −1). Note that no short codeword is a prefix of a longer codeword, which means that the code is instantaneously decodable. For example, using the code in Example 2.8, it is easily seen that the code sequence {11110101111110011111111110} decodes to the symbol sequence $\{a_5a_2a_8a_1a_7a_4\}$. Huffman decoders can be implemented using table-lookup or tree-follower methods.

Example 2.8
Example of Huffman coding.

a_j	$P(a_j)$	*Final code*	*Step 1*		*Step 2*		*Step 3*		*Step 4*		*Step 5*		*Step 6*	
a_1	0.35	0	0.35	0	0.35	0	0.35	0	0.35	0	0.35	0	0.65	1
a_2	0.3	10	0.3	10	0.3	10	0.3	10	0.3	10	0.35	11	0.35	0
a_3	0.15	110	0.15	110	0.15	110	0.15	110	0.2	111	0.3	10		
a_4	0.08	1110	0.08	1110	0.08	1110	0.12	1111	0.15	110				
a_5	0.05	11110	0.05	11110	0.07	11111	0.08	1110						
a_6	0.03	111110	0.04	111111	0.05	11110								
a_7	0.03	1111111	0.03	111110										
a_8	0.01	1111110												

$\bar{L} = 2.43 \quad H(A) = 2.34 \quad \eta = 96.25\%$

Example 2.9
Consider the compression of a monochrome video signal. A classical approach is to use DPCM followed by VLC and a buffer store (Figure 2.9). Here the quantiser applies non-uniform quantisation (nonlinear mapping) to the digital difference signal. The buffer smooths the irregular bit rate resulting from VLC, and feedback is required in order to minimise buffer overflow or underflow. It is found that 4-bit DPCM can be satisfactory, and for typical video material the quantiser output will have a peaked pdf, as indicated. Since the pdf is non-uniform it is reasonable to apply VLC to the quantiser output in order to achieve further compression (Nicol *et al.*, 1980). To illustrate the advantage of doing this, a typical discrete pdf has been assumed for just 11 quantiser output samples (5 negative, zero, 5 positive), giving $H(A) = 2.435$; see Example 2.10. If we simply transmitted the 4-bit output without VLC, the coding efficiency would be only 60.9%. However, Example 2.10 shows that VLC can give an efficiency of over 97%, and the corresponding bit rate

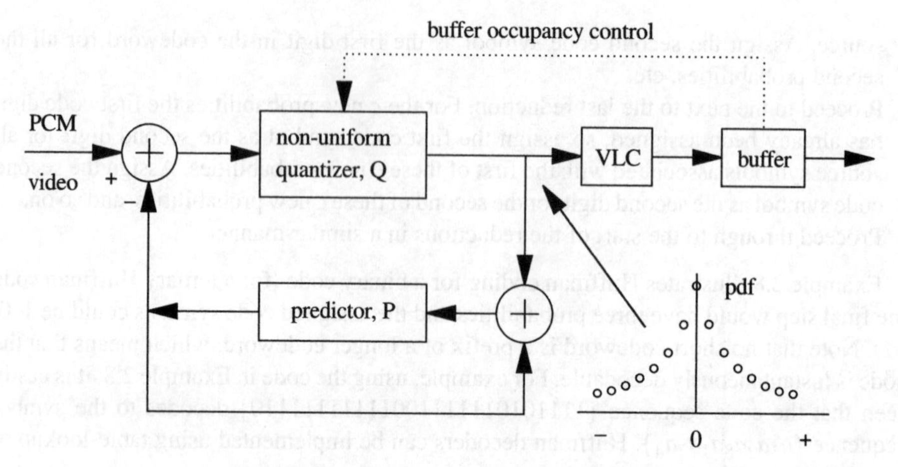

Figure 2.9 *DPCM video encoder with VLC.*

would be reduced by a factor 2.499/4 = 0.625 compared to straight 4-bit transmission. VLC is used for much the same reason at the output of an MPEG-2 encoder.

2.3.2 Modified Huffman coding for facsimile

The Huffman coding procedure assumes independent source symbols, so it is applicable to facsimile transmission (see Example 2.7). However, the CCITT Group 3 facsimile standard (Hunter and Robinson, 1980) has a normal resolution of 1728 samples/line for A4 documents, which means that the maximum run-length and number of codewords is 1728. Such a large number of codewords rules out the use of a pure Huffman code, so a modified and *fixed* Huffman code was designed. Every run-length greater than 63 is broken into two run-lengths; one is a *make-up* run-length having a value $N \times 64$ (N integer), and the other is a *terminating* run-length having a value between 0 and 63. In fact, since runs are approximately independent it is advantageous to code white and black runs independently, since this leads to a lower bit rate. The final code is shown in Table 2.3.

Redundancy removal via run-length/variable length coding makes the transmitted signal more susceptible to transmission errors (a problem common to many compression schemes). In general, but not always, a single transmission error will generate *error propagation* as the decoder loses synchronisation and generates a string of decoded runs different from those at the transmitter. Figure 2.10(a) shows an arbitrary transmitted sequence derived from Table 2.3, and Figure 2.10(b) shows the same sequence received with one error. In this particular case there is no loss of synchronisation, although there will be a slight displacement of the decoded scan line. Figure 2.10(c) illustrates typical error propagation and the generation of a string of false run-lengths. Note, however, that eventually the code *resynchronises* (this being the property of most instantaneous codes). The modified Huffman code tries to ensure that errors do not propagate beyond the current line by transmitting a unique synchronizing word at the end of each coded line. This is the end-of-line (EOL) codeword in Table 2.3.

Example 2.10

Huffman and comma coding for an assumed 4-bit DPCM source.

a_j	$P(a_j)$	*Comma code*	*Huffman code*																		
1	0.4	0	0	0.4	0	0.4	0	0.4	0	0.4	0	0.4	0	0.4	0	0.4	0	0.4	0	0.6	1
2	0.175	10	111	0.175	111	0.175	111	0.175	111	0.175	111	0.175	111	0.175	111	0.25	10	0.35	11	0.4	0
3	0.175	110	110	0.175	110	0.175	110	0.175	110	0.175	110	0.175	110	0.175	110	0.175	111	0.25	10		
4	0.09	1110	100	0.09	100	0.09	100	0.09	100	0.09	100	0.09	100	0.16	101	0.175	110				
5	0.09	11110	1011	0.09	1011	0.09	1011	0.09	1011	0.09	1011	0.09	1011	0.09	100						
6	0.028	111110	10100	0.028	10100	0.028	10100	0.028	10100	0.042	10101	0.07	1010								
7	0.028	1111110	101011	0.028	101011	0.028	101011	0.028	101011	0.028	10100										
8	0.005	11111110	1010100	0.005	1010100	0.009	1010101	0.014	101010												
9	0.005	111111110	10101011	0.005	10101011	0.005	1010100														
10	0.002	1111111110	101010101	0.004	10101010																
11	0.002	11111111110	101010100																		

$H(A) = 2.435$ $\quad \bar{L}_{\text{comma}} = 2.576$ $\quad \eta_{\text{comma}} = 94.54\%$ $\quad \bar{L}_{\text{Huffman}} = 2.499$ $\quad \eta_{\text{Huffman}} = 97.45\%$

Table 2.3 *Modified Huffman code table (Hunter and Robinson, 1980, IEEE).*

Terminating codewords

Run length	*White*	*Black*	*Run length*	*White*	*Black*
0	00110101	0000110111	32	00011011	000001101010
1	000111	010	33	00010010	000001101011
2	0111	11	34	00010011	000011010010
3	1000	10	35	00010100	000011010011
4	1011	011	36	00010101	000011010100
5	1100	0011	37	00010110	000011010101
6	1110	0010	38	00010111	000011010110
7	1111	00011	39	00101000	000011010111
8	10011	000101	40	00101001	000001101100
9	10100	000100	41	00101010	000001101101
10	00111	0000100	42	00101011	000011011010
11	01000	0000101	43	00101100	000011011011
12	001000	0000111	44	00101101	000001010100
13	000011	00000100	45	00000100	000001010101
14	110100	00000111	46	00000101	000001010110
15	110101	000011000	47	00001010	000001010111
16	101010	0000010111	48	00001011	000001100100
17	101011	0000011000	49	01010010	000001100101
18	0100111	0000001000	50	01010011	000001010010
19	0001100	00001100111	51	01010100	000001010011
20	0001000	00001101000	52	01010101	000000100100
21	0010111	00001101100	53	00100100	000000110111
22	0000011	00000110111	54	00100101	000000111000
23	0000100	00000101000	55	01011000	000000100111
24	0101000	00000010111	56	01011001	000000101000
25	0101011	00000011000	57	01011010	000001011000
26	0010011	000011001010	58	01011011	000001011001
27	0100100	000011001011	59	01001010	000000101011
28	0011000	000011001100	60	01001011	000000101100
29	00000010	000011001101	61	00110010	000001011010
30	00000011	000001101000	62	00110011	000001100110
31	00011010	000001101001	63	00110100	000001100111

Continues...

Table 2.3 *(continued)*

Make-up codewords					
Run length	*White*	*Black*	*Run length*	*White*	*Black*
64	11011	0000001111	960	011010100	0000001110011
128	10010	000011001000	1024	011010101	0000001110100
192	010111	000011001001	1088	011010110	0000001110101
256	0110111	000001011011	1152	011010111	0000001110110
320	00110110	000000110011	1216	011011000	0000001110111
354	00110111	000000110100	1280	011011001	0000001010010
448	01100100	000000110101	1344	011011010	0000001010011
512	01100101	0000001101100	1408	011011011	0000001010100
576	01101000	0000001101101	1472	010011000	0000001010101
640	01100111	0000001001010	1536	010011001	0000001011010
704	011001100	0000001001011	1600	010011010	0000001011011
768	011001101	0000001001100	1664	011000	0000001100100
832	011010010	0000001001101	1728	010011011	0000001100101
896	011010011	0000001110010	EOL	000000000001	000000000001

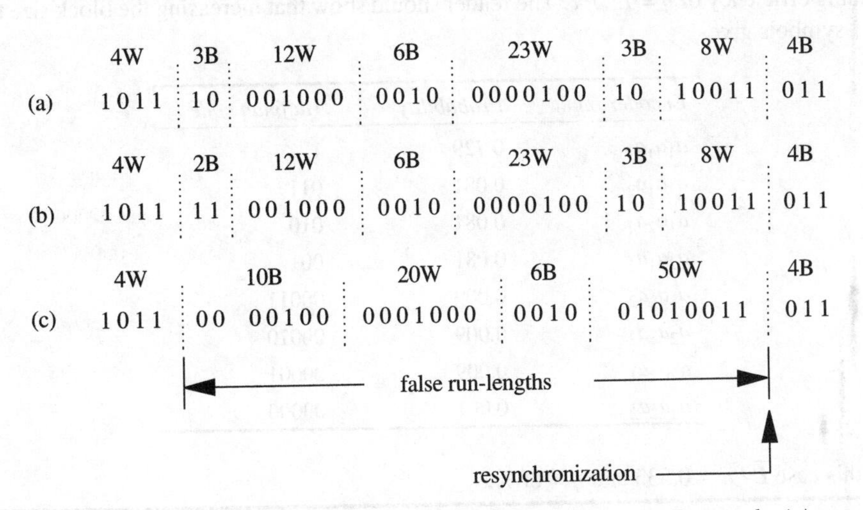

Figure 2.10 *Effect of a single transmission error upon the modified Huffman code: (a) transmitted sequence; (b) and (c) received sequences.*

Block coding

According to Theorem 2.2, the average codeword length/source symbol tends to $H(A)$ as the block length n increases, so the coding efficiency will tend to increase. This is true even for independent symbols and so is easily demonstrated.

Example 2.11
Consider a zero-memory source with a binary alphabet:

$$A = \{a_1, a_2\} \quad \hat{P}(a_1) = 0.9 \quad \hat{P}(a_2) = 0.1$$

Using (2.4) we have

$$H(A) = -(0.9 \log 0.9 \; + \; 0.1 \log 0.1) = 0.469 \;\; \text{bits/source symbol}$$

In this case, Huffman coding of individual symbols is trivial and corresponds to $\overline{L} = 1$ and a coding efficiency of $\eta = 46.9\%$. Now suppose the encoder input comprises independent blocks of $n = 2$ symbols rather than individual symbols. Using the approach in Example 2.8, the reader should show that binary Huffman coding gives:

Encoder input	*Probability*	*Huffman code*
a_1a_1	0.81	1
a_1a_2	0.09	00
a_2a_1	0.09	011
a_2a_2	0.01	010

The average length is now $\overline{L} = 1.29$, so $\overline{L} / n = 0.645$ bits/source symbol, corresponding to a coding efficiency of $\eta = 72.7\%$. The reader should show that increasing the block size to $n = 3$ symbols gives

Encoder input	*Probability*	*Huffman code*
$a_1a_1a_1$	0.729	1
$a_1a_1a_2$	0.081	011
$a_1a_2a_1$	0.081	010
$a_2a_1a_1$	0.081	001
$a_1a_2a_2$	0.009	00011
$a_2a_1a_2$	0.009	00010
$a_2a_2a_1$	0.009	00001
$a_2a_2a_2$	0.001	00000

In this case $\overline{L} / n = 0.533$ and $\eta = 88\%$.

2.3.3 A basic limitation of Huffman coding

We have seen that entropy coding is only really advantageous when the true probability distribution $\mathbf{P}(n)$ is highly non-uniform, with some symbols much more likely than others (Corollary 2.1). We have also seen that for such a source symbol-by-symbol Huffman coding can give very high efficiency (Example 2.10). In fact, for a large alphabet, M, Huffman coding (mapping) can come close to the 'ideal' and can be a viable choice for entropy coding. However, it does not follow that Huffman coding is *always* the best choice for the symbol-by-symbol coding of a source with non-uniform **P**(n). Consider the following Huffman coding example (for clarity, actual probabilities rather than estimates are assumed).

Example 2.12

a_j	$P(a_j)$	v_j	l_j
a_1	0.9	1	1
a_2	0.06	01	2
a_3	0.02	001	3
a_4	0.02	000	3

$$\bar{L} = \sum_j P(a_j) l_j = 1.14 \text{ bits/symbol}$$

$$H(A) = -\sum_j P(a_j) \log P(a_j) = 0.6061 \text{ bits/symbol}$$

For this source, symbol-by-symbol Huffman coding gives an efficiency of just 53%.

Example 2.12 simply highlights the fact that, when **P**(n) becomes highly non-uniform (that is, one symbol has a probability approaching 1) the source symbols convey negligible information on average ($H(A) \rightarrow 0$), *but Huffman coding still requires at least one bit/symbol to transmit the source!* To emphasise this point, consider a fixed, memoryless source of alphabet $A = \{a_1 a_2 ... a_j ... a_M\}$. When symbol a_j occurs it conveys $-\log P(a_j)$ bits of information and the *ideal* entropy coder will then generate

$$l_j = -\log P(a_j) \text{ code bits} \tag{2.19}$$

Given a specific model and a data string of N symbols in which symbol a_j occurs n_j times, the corresponding ideal code length for the string will be

$$l_s = \sum_j n_j l_j \text{ code bits} \tag{2.20}$$

The significant point is that Huffman coding requires l_j to be an integer, *so symbol-by-symbol Huffman coding will only be ideal or optimal when* $P(a_j) = 2^{-l_j}$, that is, when the source probabilities are binary weighted.

Corollary 2.2 Huffman coding is optimal at the block level, or when the encoder output translates to an integral number of code bits/symbol, but in general it is not optimal at the symbol level.

The classical solution to this dilemma is to use block coding to increase the efficiency (see (2.13)), but there is really no need to resort to such an artifice, as discussed in Section 2.4.

2.4 Arithmetic coding

In contrast to Huffman coding, arithmetic coding (AC) dispenses with the restriction that each symbol must translate into an integral number of code bits, and so it generally encodes more efficiently for symbol-by-symbol coding. In fact, AC can achieve the theoretical entropy bound for any source (not just one which obeys (2.16)), and it is particularly advantageous for sources with a *binary* alphabet, such as two-tone images. Put another way, given a perfect probabilistic model of the source, AC will provide perfect (optimal) compression without the need for blocking of the input symbols.

2.4.1 Encoding principle

Consider the fixed model in Table 2.4. Here, $\hat{P}(a_j)$ (which could be obtained from a relative frequency distribution) is in *binary*, and $\hat{P}_c(a_j)$ is the corresponding cumulative probability. Since the probabilities are binary weighted, (2.19) gives integer bits and Huffman coding happens to give an optimal code. Suppose that we wish to *arithmetically* code the sequence $\{a_2 a_3 a_1 \ldots\}$ using the model in Table 2.4. We commence with a unit interval, A_0, and a *codepoint* $C_0 = 0$ (Figure 2.11). Arithmetic coding is performed by successively subdividing this interval, and in order to achieve compression it is necessary to assign the more likely symbols to the larger subintervals. This follows since fewer bits are necessary to describe a point in a large interval than in a small one, and so, on average, the codelength is minimised. Interval A_0 is therefore subdivided according to $\hat{P}(a_j)$, as shown.

It is usual to assign a code point C and a subinterval A to a particular symbol. For example, the first symbol a_2 has $C_1 = 0.10$ and $A_1 = 0.01$. Formally,

$$C_1 = C_0 + A_0 \cdot \hat{P}_c(a_2) = 0 \ + 1 \cdot 0.1 = 0.10$$

$$A_1 = A_0 \cdot \hat{P}(a_2) = 1 \cdot 0.01 = 0.01$$

For decoding purposes it is important to strictly define A_1 as the interval $[0.10, 0.11)$, which denotes that 0.10 is included in the interval but 0.11 is not. To encode a_3 we subdivide A_1 into the same proportions as used for A_0. The new codepoint and interval are given by

$$C_2 = C_1 + A_1 \cdot \hat{P}_c(a_3) = 0.1 \ + \ 0.01 \cdot 0.11 = 0.1011$$

$$A_2 = A_1 \cdot \hat{P}(a_3) = 0.01 \cdot 0.001 = 0.00001$$

Similarly

$$C_3 = C_2 + A_2 \cdot \hat{P}_c(a_1) = 0.1011 + 0.00001 \cdot 0 = 0.1011$$

Table 2.4 *A fixed, binary-weighted model.*

a_j	$\hat{P}(a_j)$	$\hat{P}_c(a_j)$	*Huffman code*
a_1	0.1	0	0
a_2	0.01	0.10	10
a_3	0.001	0.110	110
a_4	0.001	0.111	111

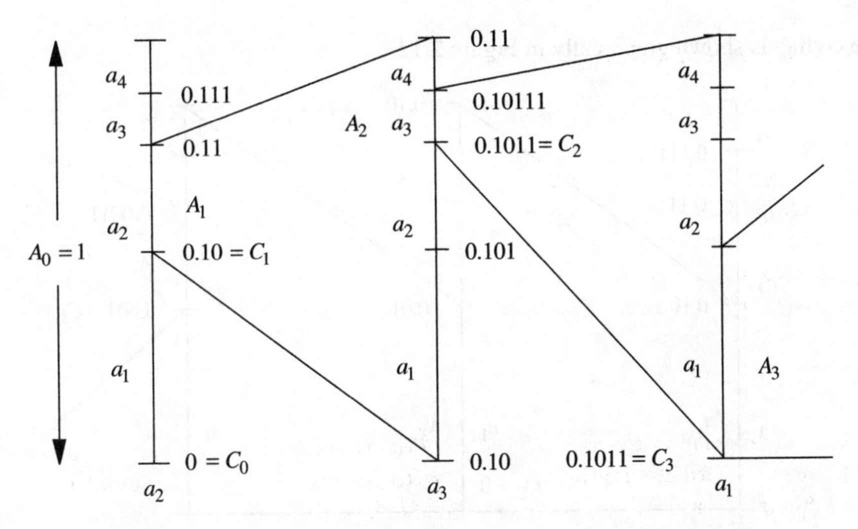

Figure 2.11 *Illustration of arithmetic coding for sequence* $\{a_2 a_3 a_1 ...\}$.

Generalising, if symbol a_j is to be coded at time interval n, we recursively compute the codepoint and code interval as

$$C_n = C_{n-1} + A_{n-1} \cdot \hat{P}_c(a_j) \tag{2.21}$$

$$A_n = A_{n-1} \cdot \hat{P}(a_j) \tag{2.22}$$

It is important to note that, whilst recursive equations (2.21) and (2.22) describe the principle of AC, in practice they suffer from a precision problem and the coding ideas must be refined.

Example 2.13
The encoding of sequence $\{a_1 a_1 a_2 ...\}$ using the model in Table 2.4 is shown below:

$$C_0 = 0 \qquad A_0 = 1$$

$$a_1: \quad C_1 = C_0 + A_0 \cdot \hat{P}_c(a_1) = 0 + 1 \cdot 0 = 0$$

$$A_1 = A_0 \cdot \hat{P}(a_1) = 1 \cdot 0.1 = 0.1$$

$$a_1: \quad C_2 = C_1 + A_1 \cdot \hat{P}_c(a_1) = 0 + 0.1 \cdot 0 = 0$$

$$A_2 = A_1 \cdot \hat{P}(a_1) = 0.1 \cdot 0.1 = 0.01$$

$$a_2: \quad C_3 = C_2 + A_2 \cdot \hat{P}_c(a_2) = 0 + 0.01 \cdot 0.1 = 0.001$$

$$A_3 = A_2 \cdot \hat{P}(a_2) = 0.01 \cdot 0.01 = 0.0001$$

The coding is shown graphically in Figure 2.12.

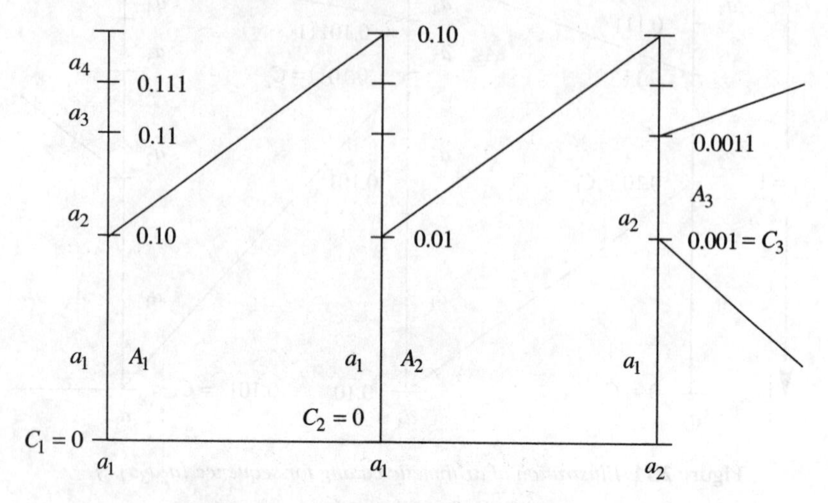

Figure 2.12 *Arithmetic coding for Example 2.13.*

Example 2.14

The model in Table 2.4 used an ordered set of probabilities, as required for Huffman coding. In AC we can rearrange the symbols, as for example in Table 2.5.

Table 2.5 *A fixed model with arbitrary symbol ordering.*

a_j	$\hat{P}(a_j)$	$\hat{P}_c(a_j)$
a_3	0.001	0
a_2	0.010	0.001
a_1	0.100	0.011
a_4	0.001	0.111

Consider the coding of sequence $\{a_1a_1a_2a_4\}$ using the model in Table 2.5:

$$C_0 = 0 \qquad A_0 = 1$$

$$a_1\colon \quad C_1 = C_0 + A_0 \cdot \hat{P}_c(a_1) = 0 + 1 \cdot 0.011 = 0.011$$

$$A_1 = A_0 \cdot \hat{P}(a_1) = 1 \cdot 0.1 = 0.1$$

$$a_1\colon \quad C_2 = C_1 + A_1 \cdot \hat{P}_c(a_1) = 0.011 + 0.1 \cdot 0.011 = 0.1001$$

$$A_2 = A_1 \cdot \hat{P}(a_1) = 0.1 \cdot 0.1 = 0.01$$

$$a_2\colon \quad C_3 = C_2 + A_2 \cdot \hat{P}_c(a_2) = 0.1001 + 0.01 \cdot 0.001 = 0.10011$$

$$A_3 = A_2 \cdot \hat{P}(a_2) = 0.01 \cdot 0.01 = 0.0001$$

$$a_4: \quad C_4 = C_3 + A_3 \cdot \hat{P}_c(a_4) = 0.10011 + 0.0001 \cdot 0.111 = 0.1010011$$
$$A_4 = A_3 \cdot \hat{P}(a_4) = 0.0001 \cdot 0.001 = 0.0000001$$

After the coding of a_4 the final interval is [0.1010011, 0.1010100), as shown in Figure 2.13. In practice it is not necessary to transmit all the interval information to the decoder and a single number within the interval suffices, such as the codepoint.

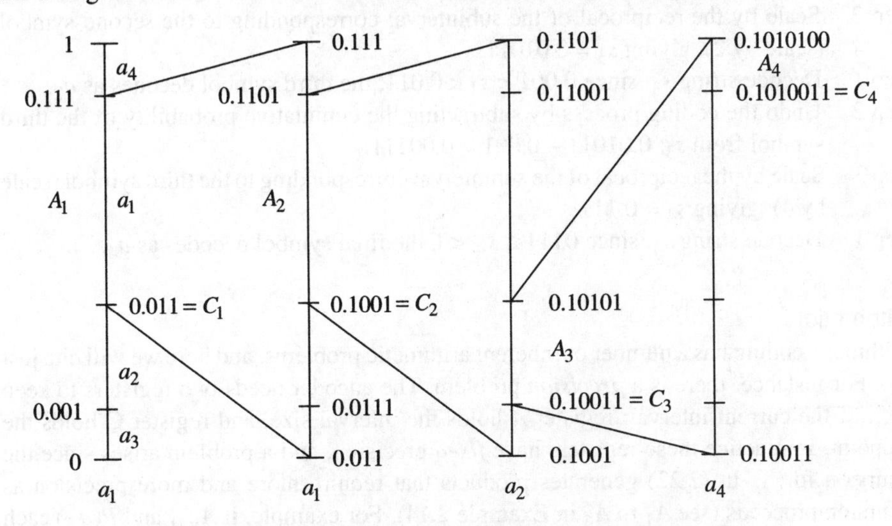

Figure 2.13 *Arithmetic coding for Example 2.14.*

2.4.2 Decoding principle

Suppose that a_4 in Example 2.14 is the last data symbol. As noted, the string s_1 = 0.1010011 would suffice for transmission, and we observe that this is the result of adding cumulative probabilities, that is

$$s_1 = C_4 = A_0\hat{P}_c(a_1) + A_1\hat{P}_c(a_1) + A_2\hat{P}_c(a_2) + A_3\hat{P}_c(a_4) \tag{2.23}$$

The transmitted code string therefore has a *base value* of 0.011, corresponding to $A_0\hat{P}_c(a_1)$, the cumulative probability of the first symbol, a_1. In other words, referring to Figure 2.13, s_1 lies in the code range [0.011, 0.111) of the first symbol. If a_2 had been the first symbol, then the code string would have been in the range [0.001, 0.011), and so on. The decoding algorithm for Figure 2.13 is then as follows:

Step 1 Examine the code interval in which string s_1 lies, and decode this interval using the initial cumulative probability scale. In this case $0.011 < s_1 < 0.111$, implying that the first symbol is a_1.

Step 2 To decode the second symbol we must 'undo' the encoder operations. We first subtract the cumulative probability of the first symbol from s_1: 0.1010011 – 0.011 = 0.0100011.

Step 3 Now scale by the reciprocal of the subinterval corresponding to the first symbol, giving the new string s_2. In this case we scale by 2, giving $s_2 = 0.100011$.

Step 1 Decode string s_2: since $0.011 < s_2 < 0.111$, we deduce that the second symbol is also a_1.

Step 2 Undo the coding process by subtracting the cumulative probability of the second symbol from s_2: $0.100011 - 0.011 = 0.001011$.

Step 3 Scale by the reciprocal of the subinterval corresponding to the second symbol (scale by 2), giving $s_3 = 0.01011$.

Step 1 Decode string s_3: since $0.001 < s_3 < 0.011$, the third symbol decodes as a_2.

Step 2 Undo the coding process by subtracting the cumulative probability of the third symbol from s_3: $0.01011 - 0.001 = 0.00111$.

Step 3 Scale by the reciprocal of the subinterval corresponding to the third symbol (scale by 4), giving $s_4 = 0.111$.

Step 1 Decode string s_4: since $0.111 \leq s_4 < 1$, the final symbol decodes as a_4.

Arithmetic

Arithmetic coding has a number of inherent arithmetic problems, and here we will cite just two. For instance, there is a *precision* problem. The encoder needs two registers to keep track of the current interval; register A holds the interval size, and register C holds the endpoint. In practice these registers have *fixed* precision, and a problem arises since the recursion for A_n in (2.22) generates products that require more and more precision as recursion proceeds (see A_1 to A_4 in Example 2.14). For example, if A_{n-1} and $\hat{P}(a_j)$ each have 8 bits of precision, then A_n requires 16 bits of precision.

A second arithmetic problem is as follows. Usually, as the sequence progresses, the code point (register C) has the same 'more significant' bits. For instance, C_2 and C_3 in Figure 2.13 have the same leading bits (0.100). This means that it is *usually* possible to transmit these bits long before the subdivision process is complete, and focus attention on the less significant bits. However, occasionally, the more significant bits that have left the C register change as the sequence progresses, for example, for C_4 in Figure 2.13 the bits change from (0.100) to (0.101). This is an example of the *carry-over* problem. Fortunately, the precision problem can be resolved by modifying the basic recursions in (2.21) and (2.22), and the carry-over problem can be resolved by using a combination of *spacer* bits and *stuff* bits in the C register (Jones, 1981; Arps *et al.*, 1988).

2.4.3 Binary Arithmetic Codes (BACs)

Arithmetic coding has been used to compress English text files, C programs, and bi-level (black and white) images, and for these applications it generally outperforms (adaptive) Huffman coding. In this section we are particularly interested in bi-level image coding and the class of arithmetic coders that handle only *binary* symbols. For example, the encoder input could be a string of black (B) and white (W) pixels in raster scan order from an image, and the encoder will represent this string as a real binary number in the interval [0.0, 1.0). The essential difference between a BAC and Huffman coding can be seen from (2.19). If a W pixel of probability P_W is to be encoded, then according to (2.19) the ideal codeword length is $l_W = -\log P_W$ bits. For a predominantly white facsimile image we may

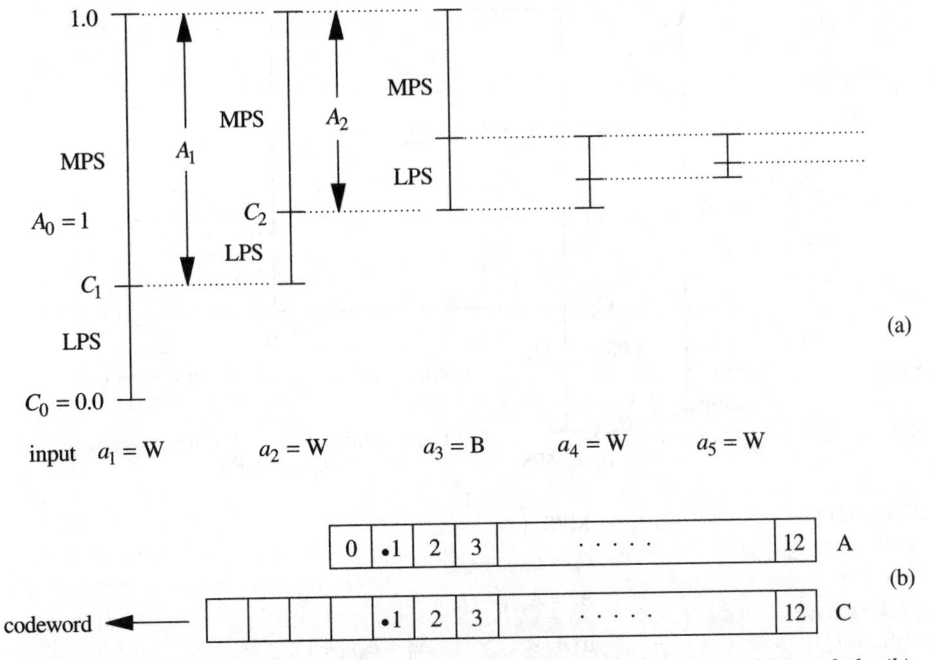

Figure 2.14 *Binary arithmetic coding: (a) interval subdivision for black and white symbols; (b) typical A and C registers.*

have $P_W = 7/8$, $P_B = 1/8$, corresponding to $l_W = 0.193$ bits, and $l_B = 3$ bits. Since one of the two ideal codewords is less than 1 bit, such ideal coding cannot be achieved through symbol-by-symbol Huffman coding (which generates 1-bit codewords), but fractional bits *can* be achieved using BACs.

The basic idea behind a BAC is illustrated in Figure 2.14(a). Suppose that we are compressing a predominantly white two-tone facsimile document, so that the 'more probable' symbol (MPS) would be W and the 'less probable' symbol (LPS) would be B. We then subdivide the current interval according to these probabilities, as in Section 2.4.1. For example, if for the first symbol the probability of a LPS is q, then the LPS is assigned the interval $A_0 \cdot q$. The first W pixel is then assigned interval $1 - A_0q$ and since the second pixel is also W this larger interval is further subdivided. Note that the smaller interval is selected for the B pixel, and that, in general, the subdivision ratio will vary from symbol to symbol, as determined by the adaptive model (Figure 2.5(a)). Figure 2.14(b) shows typical fixed precision registers, the codeword being shifted left out of the C register.

A basic BAC algorithm

Figure 2.15 shows a basic binary AC algorithm. For symbol a_n the current interval is A_{n-1} and this is subdivided using current statistics. For best compression, the subdivision corresponding to the LPS has value $Q_n = A_{n-1} \cdot P(a_n = \text{LPS})$, and the MPS has the interval $A_{n-1} - Q_n$ (note that the probability will depend on current context). In Figure 2.15 a_n is an MPS, so the current interval is formally written $[C_n, C_n + A_n)$. A practical encoder can be obtained by constraining Q_n to the form

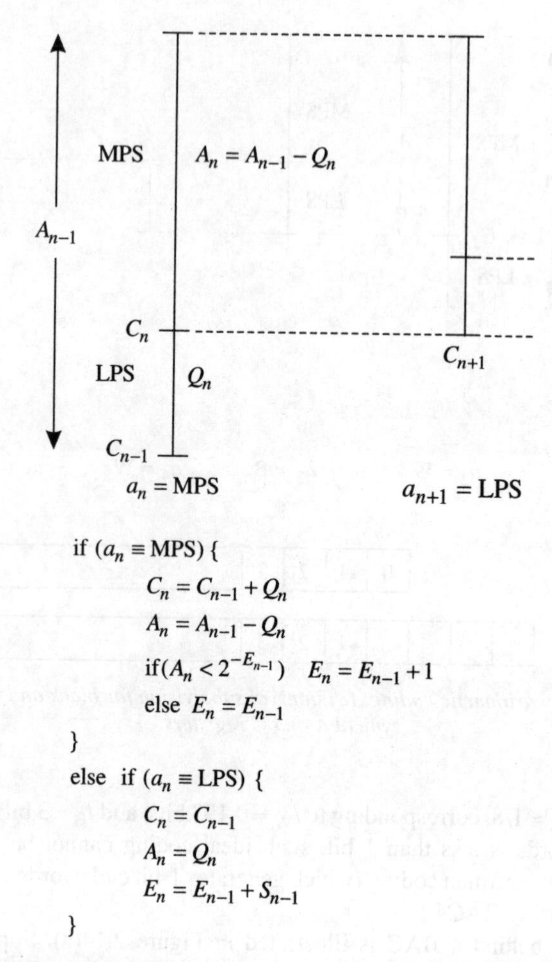

Figure 2.15 *A basic binary arithmetic coding algorithm.*

$$Q_n = 2^{-(E_{n-1}+S_n)} \tag{2.24}$$

where S_n is an integer ranging from typically 1 to 12 (see Figure 2.14(b)), and E_{n-1} represents the leading 0s of A_{n-1}; that is, E tracks the size of the current interval.

Example 2.15

Assume the BAC is initialised to $C_0 = 0.0$, $A_0 = 1.0$, $E_0 = 0$, and that the symbol sequence to be encoded is

$$s = \{ a_1 = W, a_2 = W, a_3 = B, a_4 = W, \ldots \}$$

with statistic $S_1 = 3$, $S_2 = 1$, $S_3 = 1$, $S_4 = 1$, ... We again assume a predominantly white background so that a W symbol is the MPS, and a B symbol is the LPS.

$$a_1:\quad Q_1 = 2^{-(E_0+S_1)} = 2^{-3} = 0.001$$
$$C_1 = C_0 + Q_1 = 0 + 0.001 = 0.001$$
$$A_1 = A_0 - Q_1 = 0.111$$
$$A_1 < 2^{-E_0} = 1;\quad \therefore E_1 = 0 + 1 = 1 \quad (A_1 \text{ has a leading } 0)$$

$$a_2:\quad Q_2 = 2^{-(E_1+S_2)} = 2^{-2} = 0.01$$
$$C_2 = C_1 + Q_2 = 0.001 + 0.01 = 0.011$$
$$A_2 = A_1 - Q_2 = 0.111 - 0.01 = 0.101$$
$$A_2 > 2^{-E_1} = 0.5;\quad \therefore E_2 = E_1 = 1$$

$$a_3:\quad Q_3 = 2^{-(E_2+S_3)} = 2^{-2} = 0.01$$
$$C_3 = C_2 = 0.011$$
$$A_3 = Q_3 = 0.01$$
$$E_3 = E_2 + S_2 = 1 + 1 = 2$$

$$a_4:\quad Q_4 = 2^{-(E_3+S_4)} = 2^{-(2+1)} = 0.001$$
$$C_4 = C_3 + Q_4 = 0.011 + 0.001 = 0.100$$
$$A_4 = A_3 - Q_4 = 0.01 - 0.001 = 0.001$$
$$A_4 < 2^{-E_3} = 0.01;\quad \therefore E_4 = E_3 + 1 = 3 \quad (A_4 \text{ has 3 leading zeros})$$

The coding is summarised in Figure 2.16.

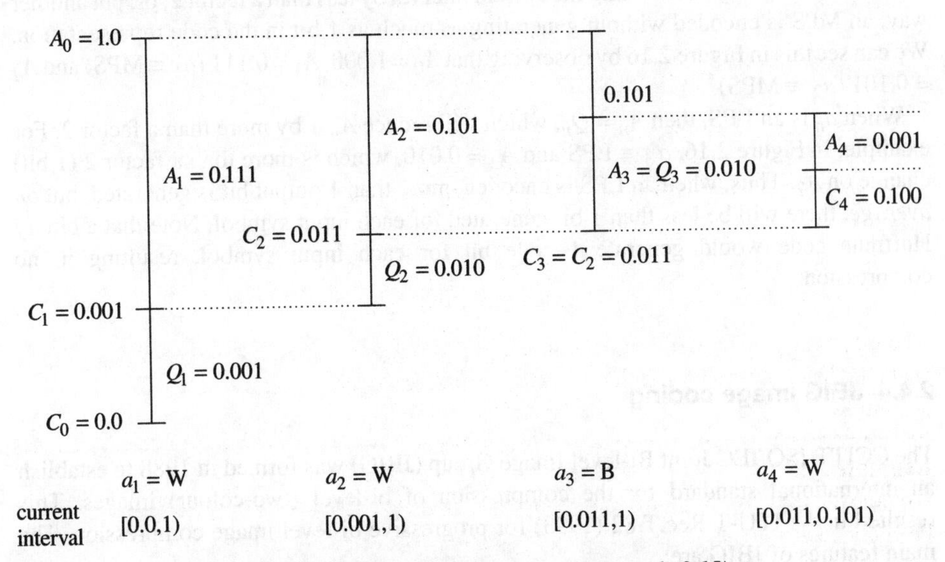

Figure 2.16 *Binary arithmetic coding (Example 2.15).*

The basic algorithm in Figure 2.15 suffers from the arithmetic precision problem, namely that as E increases, so does the length of registers A and C. As discussed in Section

2.4.2, in practice A and C have a fixed precision, and this can be achieved using a process of normalisation. The leading 0s of the A register are not required, so, after each coding event the A register is normalised or shifted left such that its leading bit (the bit left of the radix point) is 1. This normalisation process ensures that $a_3 \equiv$ LPS, that is, A is nearly constant. It also means that E_{n-1} in (2.24) is forced to 0 each time, so that for normalised operation

$$\text{LPS} \equiv Q_n = 2^{-S_n} \tag{2.25}$$

$$\text{MPS} \equiv A_{n-1} - 2^{-S_n} \tag{2.26}$$

Recalling that, for ideal performance, $Q_n = A_{n-1} \cdot P(a_n = \text{LPS})$, where $P(a_n = \text{LPS}) < 1/2$ by definition, then (2.25) provides a reasonable approximation when $A_{n-1} \approx 1$. Also, since $Q_n < 1$, no integer bits are required for the C register. This register, which defines the lower end-point of the interval, is shifted by the same number of places as the A register. During coding, bits shifted out of the left (MSB end) of the C register are buffered and eventually transmitted as the code string.

Compression mechanism of BAC

From the foregoing discussion we can say that $Q_n \approx P(a_n = \text{LPS})$ and $A_{n-1} \approx 1$. By definition, most of the time the current input symbol a_n will be the MPS, and it will be encoded to the interval $[C_n, C_n + A_n)$, where $A_n = A_{n-1} - Q_n \approx A_{n-1}$ (see Figure 2.15). An MPS is therefore encoded by subdividing the current interval by less than a factor 2, or, put another way, an MPS is encoded without generating as much as 1 bit in the code representation. We can see this in Figure 2.16 by observing that $A_0 = 1.000$, $A_1 = 0.111$ ($a_1 \equiv$ MPS) and $A_2 = 0.101$ ($a_2 \equiv$ MPS).

When a_n is an LPS, then $A_n = Q_n$, which will reduce A_{n-1} by more than a factor 2. For example, in Figure 2.16, $a_3 \equiv$ LPS and $A_3 = 0.010$, which is more than a factor 2 (1 bit) change on A_2. Thus, when an LPS is encoded, more than 1 output bit is generated, but *on average*, there will be less than 1 bit generated for each input symbol. Note that a binary Huffman code would generate 1 code bit for each input symbol, resulting in no compression.

2.4.4 JBIG image coding

The CCITT-ISO/IEC Joint Bi-level Image Group (JBIG) was formed in 1988 to establish an international standard for the compression of bi-level (two-colour) images. This resulted in the ITU-T Rec.T.82 (1993) for progressive bi-level image compression. The main features of JBIG are:

- Superior compression compared to the more established G3 and G4 algorithms.
- Adaptive, model-based arithmetic coding.
- Two coding modes: sequential and progressive. Progressive coding tends to give superior compression compared to sequential coding.

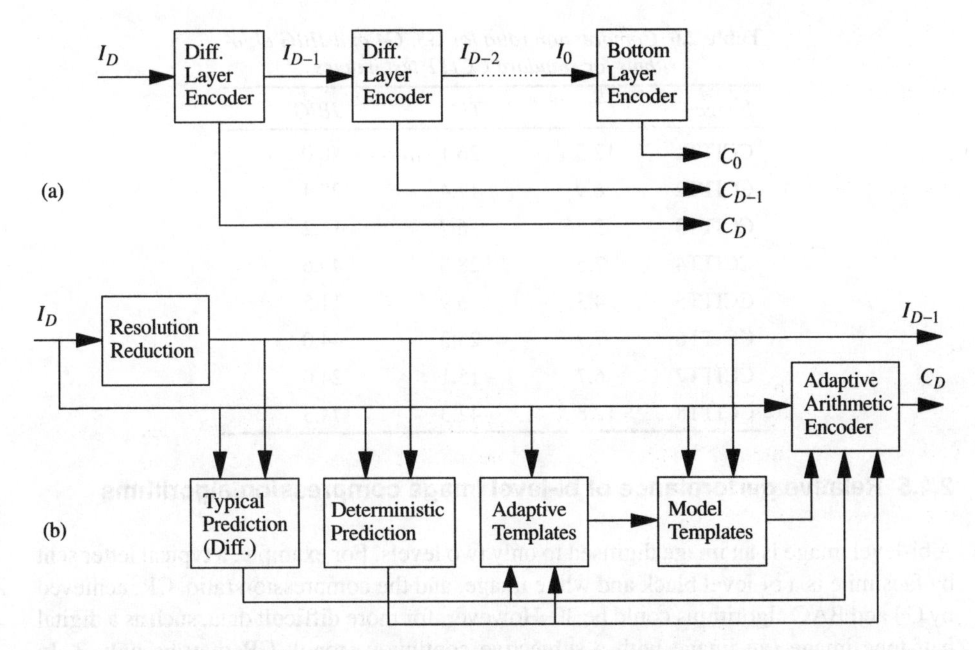

Figure 2.17 *JBIG algorithm, progressive coding mode: (a) differential layer structure; (b) detail of a differential layer encoder.*

- Although primarily a bi-level standard, JBIG offers a multilevel (greyscale) capability by simply coding each bit plane independently as though it were itself a bi-level image. JBIG has shown a compression advantage over JPEG for greyscale images.

The progressive coding mode captures images as a compression of a low-resolution or *bottom layer* image plus a sequence of *delta* files that each provide another layer of resolution enhancement. This is illustrated in Figure 2.17, where image I_D is the high-resolution image and C_D denotes its coding. Image I_{D-1} has approximately half as many rows and half as many columns as image I_D, or in other words, each delta file contains the information needed to double both the vertical and horizontal resolution. Progressive coding is attractive for database applications, where output devices (such as PC displays and laser printers) of varying resolution have to be served. Only the bottom layer and as many delta files as required are transmitted and processed. Progressive coding is also useful for image browsing over medium rate links (say, up to 64 kbit/s). In this case a low-resolution image is quickly available, and this is followed by as much resolution enhancement as desired. In sequential coding, the image is coded at full resolution from left to right, top to bottom, and there are no differential layers.

The JBIG *resolution reduction* algorithm in Figure 2.17(b) is more complex than simple sub-sampling of every column and line (which could generate aliasing, or simply delete thin lines). The algorithm is table-based, the colour (tone) of a given low resolution pixel being determined by the colours of nine particular high-resolution neighbours and three causally positioned low-resolution neighbours. The *model template* in Figure 2.17(b) provides the context for the coding of input image I_D (in a similar way to Figure 2.8). The context is then used to estimate the conditional probability of the current pixel.

Table 2.6 *Compression ratio for G3, G4 and JBIG algorithms for standard CCITT test images.*

Image	*G3*	*G4*	*JBIG*
CCITT1	12.5	26.1	40.0
CCITT2	6.9	17.4	28.4
CCITT3	3.7	6.7	11.2
CCITT4	7.5	28.8	49.6
CCITT5	4.3	6.9	11.5
CCITT6	7.7	25.3	44.0
CCITT7	6.7	15.1	24.0
CCITT8	12.8	42.3	74.3

2.4.5 Relative performance of bi-level image compression algorithms

A bi-level image is an image digitised to only two levels. For example, a typical letter sent by facsimile is a bi-level black and white image, and the compression ratio, CR, achieved by G4 and BAC algorithms could be 30. However, for more difficult data, such as a digital half-tone image (an image with a subjective continuous tone), CR may be only 3. In general therefore, the performance of a compression algorithm is highly data dependent, and the most difficult binary images to compress are those with approximately equal black and white pixels. This is not surprising since a main objective in source coding is to generate a 'random-like' source, and in the latter case the source may already have a significantly random characteristic.

Table 2.6 compares the compression performance of the G3 (Modified Huffman, MH, coding), G4 (MMR algorithm), and JBIG algorithms on eight standard CCITT test images comprised of black and white text and drawings. For these results the JBIG algorithm is in the progressive mode with both differential layer prediction and deterministic prediction activated. Note that CR is very image dependent and that the JBIG algorithm can be nearly 6 times better than the G3 algorithm.

2.4.6 A resolution reduction algorithm

Resolution reduction is required wherever a display device has lower resolution than the original image, as for example in mobile phone displays (strictly speaking this process is not lossless, but it is included here for completeness). For this type of application we restrict discussion to bi-level images, that is, images with just two distinct colour levels, as in black and white facsimile. As noted in Section 2.4.4, simply omitting lines and columns gives unsatisfactory results and a more complex sampling pattern is required. A simple but effective resolution reduction algorithm for a bi-level image is as follows.

Let R be the image reduction factor and divide the original image into $R \times R$ blocks. A threshold T is then determined as $\lfloor R^2/2 \rfloor$ and if the number N of black pixels in each block is greater than T, then the block is represented by a black pixel in the reduced image. If $N \leq T$ then the block is represented by a white pixel. This algorithm is illustrated in Figure 2.18 for $R = 2$, and Figure 2.19 is a practical example for $R = 3$.

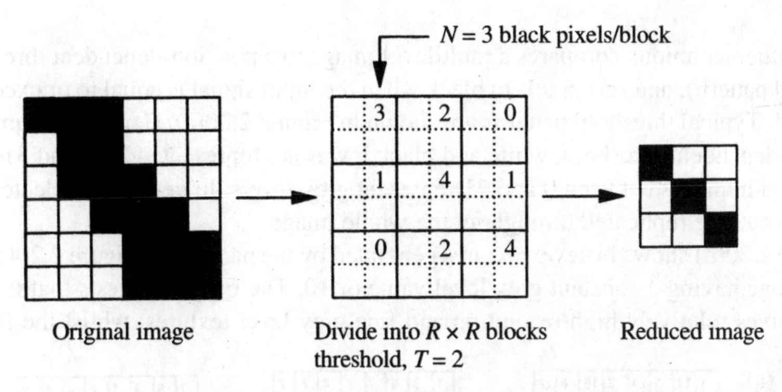

Figure 2.18 *Principle of resolution reduction (R = 2).*

Original

Special Eve

The Registration Receptic
tration, is being held on S
Center Hotel in Minneapo

1/3 size

Special Events/Social Program

The Registration Reception, following early conference r
tration, is being held on Sunday, May 19 at the Marriott
Center Hotel in Minneapolis from 6-8 p.m. A Monday eve
Reception, sponsored by 3M, takes place at the Science
seum of Minnesota in St. Paul from 6:30-9:00 p.m. This
ception includes a screening of the Academy Award nomir
IMAX film "Ring of Fire" at the William L. McKnight-3M C
Theater at 8:00 p.m. A chartered bus will provide transp
tion to the Museum. The Honors and Awards Banquet, on T
day at noon in the Marriott City Center, is a special e
honoring scientific achievement and valuable contributio
IS&T. The Wednesday evening Conference Reception is
Marriott City Center from 6-9:30 p.m. includes a light bu

Figure 2.19 *Illustration of resolution reduction for R = 3.*

2.4.7 Dither techniques

Dithering enables a single bit-plane (an array of binary pixels) to appear subjectively as a continuous-tone image, that is, as an image with multiple grey levels with no perceptible quantisation between them. The process is similar to *half-toning* used in printing in the sense that a half-tone image is composed of only two grey levels, black and white. Clearly, such techniques are applicable to mobile phone displays. In general there are two disadvantages of using dither:

1. The perceived increase in grey levels is usually accompanied by a perceived *decrease* in image resolution.
2. Dither patterns tend to generate short runs, and so reduce the effectiveness of run-length coding, such as modified Huffman (MH) coding. 'Line' type dither patterns have been proposed in an attempt to improve the compression ratio (Ochi and Tetsutani, 1987).

The dither technique compares a multilevel image to a position-dependent threshold (a threshold pattern), and sets pixels to black when the input signal is equal to or exceeds the threshold. Typical threshold patterns are shown in Figure 2.20(a). Here, each number is a position-dependent threshold, white and black levels are represented by 0 and 31 respectively, and numbers between 0 and 31 represent grey levels. In general, a selected dither pattern would be replicated throughout the whole image.

Figure 2.20(b) shows bi-level textures generated by the patterns in Figure 2.20(a) for an input image having a constant grey level value of 10. The figure suggests that the *Bayer* pattern gives relatively high resolution and fine grey level textures, whilst the line-type

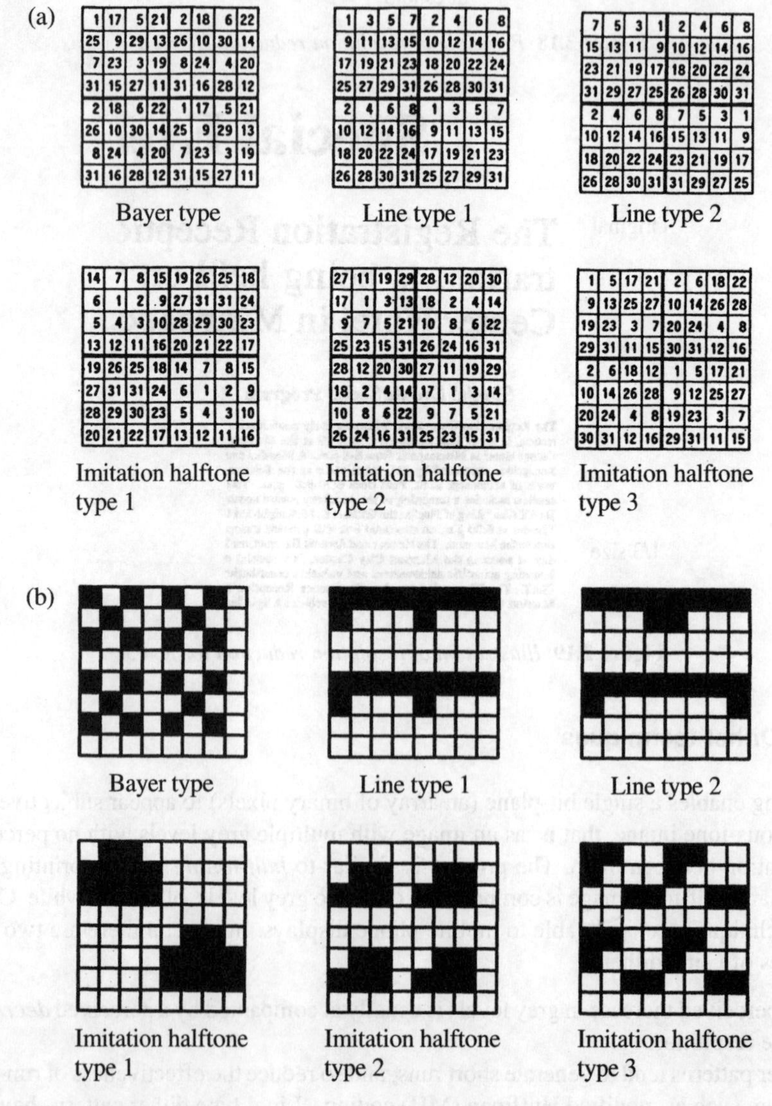

Figure 2.20 *(a) Dither threshold patterns; (b) corresponding bi-level textures generated by the patterns in (a) for an input grey level value of 10 (IEEE Trans. Comm.,* ***COM-35****(4), April 1987).*

patterns offer long runs of black and white pixels and so give a compression advantage when using RLC. Imitation half-tone patterns generate textures similar to the half-tone pattern of typical printed matter.

2.5 LZW compression

The lossless LZW compression algorithm (Welch, 1984) is a practical and relatively simple variant of the basic LZ algorithm (Ziv and Lempel, 1977). Originally designed for implementation by special hardware, it turned out to be highly suitable for efficient software realisation too. It is therefore the basis of compression techniques like PKZIP and the Unix *compress*, and is supported by the GIF and TIFF graphics formats.

An important practical point is that, like the LZ algorithm, LZW adapts to the source statistics, so no *a priori* knowledge of the source is required. In other words, the coding (decoding) process is interlaced with a learning process as the encoder (decoder) builds and dynamically changes its string table. This means that the LZW decoder does not require transmission of the decoding table, since it builds an identical table as it receives compressed data. Also, like Huffman coding, the LZ and LZW algorithms exploit the statistical redundancy of the source (frequently used symbols and symbol patterns) rather than any positional redundancy. On the other hand, unlike Huffman coding the basic form of these algorithms employs fixed-length (typically 12-bit) codewords.

2.5.1 Basic coding algorithm

The coding algorithm parses an input string of source symbols into successive substrings. If the input string is $S = s_1s_2s_3...$, where s_i is a source symbol, then the encoder parses it as $S_1S_2S_3...$, where S_i is the longest substring that can be recognised by the encoder at time i. Once a substring is found it is coded into a fixed-length uniquely decodable codeword, v_i. No real attempt is made to parse string S in an optimal way, so the LZW algorithm is less than optimal, but on the other hand it can be very fast.

The LZW string table has a prefix property in that for every string in the table its *prefix string* is also in the table. For example, if the table has string 'and' it also has a prefix string 'an'. Let us generalise this by defining the string format in the table as PE where P is a prefix string and E is an *extension symbol*. The basic LZW coding algorithm is then shown in Figure 2.21. Note that strings are assigned a unique codeword as they are added to the table.

Example 2.16

Let the source alphabet be $A = \{a,b,c\}$ so that the string table is initialised to

s_i	v_i	*Memory location*
a	1	0
b	2	1
c	3	2

```
Initialise string table to contain single symbol strings
Read first symbol s1
P = s1
for the next input symbol sk {
  E = sk
  if (PE is in the string table) P = PE
  else {
    transmit code for P
    add PE to string table
    P = E
  }
}
```

Figure 2.21 *Pseudocode for LZW coding.*

Here we assume a simple codeword representation for clarity, and load the codewords at memory locations 0–2. The first string would then be loaded into location 3, the second at location 4, and so on. LZW coding for the input string

$$S = ababcbababaaaaaaa$$

is then as shown in Figure 2.22. Coding starts by reading symbol a and equating P to a. Symbol b is then read, and since string ab is not in the table, the code for symbol a is transmitted, P is equated to b, and string ab is added to the table (at location 3) in a convenient storage format. The third symbol is now read and, since string ba is not in the table, the code for b is transmitted, ba is added to the table, and P is equated to a. For the fourth input symbol, ab is already in the table and so no codeword is transmitted and P is equated to ab.

If the input in Example 2.16 is considered to be a string of 8-bit ASCII characters, and if 12-bit codewords are used, then the CR for this particular input string is about 1.3. Clearly, good compression occurs when the encoder recognises a long input string (that is, when the table is well established), since it will then be replaced by a relatively short codeword.

Example 2.17

We now consider a more realistic coding example. Suppose that the encoder string table is initially loaded with single-character ASCII characters up to an including memory location 255. There are then 256 empty locations for 9-bit codewords, 768 for 10-bit codewords, and so on. Assuming this initial table loading, the coding for input string

$$S = \text{This is a ...}$$

is shown in Figure 2.23. The first character T is treated as the prefix P, and the second character h is the extension symbol E. Since Th is not in the table it is added at memory location 256 (the first available location). Coding continues like this until the second character s is encountered. At this point $PE = is$, and since this is already in the table no code is transmitted. The prefix now becomes $P = PE = is \equiv 258$. For the next character (a space) we have $PE = 258(sp)$, and since this is not in the table it is added at location 261.

input string	S	*a*	*b*	*a*	*b*	*c*	*b*	*a*	*b*	*a*	*b*	*a*	*a*	*a*	*a*	*a*	*a*	*a*
prefix string	P	*a*	*b*	*a*	*ab*	*c*	*b*	*ba*	*b*	*ba*	*bab*	*a*	*a*	*aa*	*a*	*aa*	*aaa*	*a*
output code			1	2		4	3		5			8	1		10			11
table addition			*ab*	*ba*		*abc*	*cb*		*bab*			*baba*	*aa*		*aaa*			*aaaa*
added codeword	v_i		4	5		6	7		8			9	10		11			12
stored format			1*b*	2*a*		4*c*	3*b*		5*b*			8*a*	1*a*		10*a*			11*a*

Figure 2.22 *Example of LZW coding.*

input string	S	T	h	i	s	(sp)	i	s	(sp)	a
prefix string	P	T	h	i	s	(sp)	i	is	(sp)	a
compressed string			T	h	i	s	(sp)	–	258	(sp)
table addition			Th	hi	is	s(sp)	(sp)i		258(sp)	(sp)a
memory location			256	257	258	259	260		261	262

Figure 2.23 *LZW coding assuming an alphabet of 256 ASCII characters.*

The significant point here is that the transmission of string *is* using character 258 *saves a codeword* compared with the transmission of the literal string.

Implementation

Clearly, the encoder needs to have an efficient table search procedure, as opposed to a simple, time-consuming linear search. One approach is to use a data structure called a *binary tree*. The basic idea is to construct a tree of nodes (one node for each distinct word) and at any node the left subtree contains words that are lexicographically less than the node word, while the right subtree contains words that are greater. To test for a word in the tree we start at the root and compare the new word with the node word. Assuming that the new word is greater than the node word, continue searching the right subtree, otherwise search the left subtree. If there is no subtree in the required direction the word is not in the tree and it should be added at the current empty location. This procedure forces the search down a particular set of subtrees and avoids an exhaustive search. An alternative and common approach to table search is to use *hashing*.

Another principal concern is that of string table size or codeword length. If the codeword length is too small then the table rapidly fills, while if it is too large the compression overhead for single character strings is excessive. An enhanced LZW algorithm might therefore use variable-length codewords, starting with a short-length codeword and extending the length as the table fills. For example, when the input string is a sequence of 8-bit ASCII characters, the wordlength could start at 9 bits and extend to 12 or 13 bits. Clearly, even then the table will eventually overflow and there are several strategies at this point. One approach is to monitor the CR after the table has filled, and if it deteriorates the table could be cleared (a special code being output to reinitialise the decoder table at the same point). This adaptive approach is highly effective for files composed of sections with differing statistics, such as executable files, although compression will be poor while the table is only sparsely filled. Alternatively, the table could be 'freshened' by clearing only some of the older strings.

It is also important to note that, like other compression schemes, the LZW algorithm is susceptible to error propagation in the presence of channel errors. In the context of data storage on disk or tape, the usual solution is to employ *cyclic redundancy checks* (Section 6.4.7).

2.5.2 Basic decoding algorithm

The decoding process is essentially the reverse of the coding process and uses the same string table as used for compression. The decoder table therefore commences with single-character strings of the basic alphabet, and the decoder then builds it from the compressed input data. Each received codeword is translated by the decoder's table into a prefix string and an extension symbol (which is pulled off and stored), and this is repeated in a recursive way until the prefix string is a single symbol. For example, given input code '9' from Figure 2.22, the decoder would perform the following steps:

$$9 \Rightarrow 8a \text{ (store } a) \Rightarrow 5b \text{ (store b)} \Rightarrow 2a \text{ (store a)} \Rightarrow b \tag{2.27}$$

giving the decoded string *baba*.

Figure 2.24 expresses the decoding algorithm in pseudocode form. It assumes that the source alphabet uses locations 0–255, so that the first available memory location is 256.

```
next_code = 256
code = inbuff[ ]                    /* read char from input buffer */
write(code)                         /* output first char */
lastcode = code                     /* start with this char */
while (input buffer not empty) {
        code = inbuff[ ]            /* read next input code */
        outstring(code)             /* output string for code */
        firstchar = (first char from output string)
        add_code(next_code,lastcode, firstchar)
                                    /* add new string to table */
        lastcode = code
        if (table full) {
                clear table
                next_code = 256
        }
        else next_code++
}
outstring(code)
{
        if (code is a char) {
                write(code)         /* write char to output */
                return
        }
        while (code is not a char) {
                push last char from code to stack
                code = code[base]
        }
        push last code char to stack
        pop stack
}
```

Figure 2.24 *Basic pseudocode for LZW decoding.*

input	coder table	compressed string	decoder table	decoder output string
T	—	—		
h	256 ← Th	T	—	T
i	257 ← hi	h	256 ← Th	h
s	258 ← is	i	257 ← hi	i
(sp)	259 ← s(sp)	s	258 ← is	s
i	260 ← (sp)i	(sp)	259 ← s(sp)	(sp)
s	—	—	—	—
(sp)	261 ← 258(sp)	258	260 ← (sp)i	is
a	262 ← (sp)a	(sp)	261 ← 258(sp)	(sp)
		a		a

Figure 2.25 *Example of LZW coding and decoding.*

This scheme could apply to ASCII text files and to 8-bit image coding, where, in the latter case, locations 0–255 would correspond to all 8-bit pixel values. Referring to Figure 2.24, each input code after the first code adds code/character strings to the table via the routine 'add_code'. This code comprises the last code plus the first character in the current string. Routine 'outstring' simply outputs the code if it is a single character. If the input code is a code/character string, the routine recursively pushes single characters from the code to stack and generates a new code/character string from its table (as in (2.27)). This continues until the routine reaches a single character, at which point the routine pops the stack to generate the output string.

The use of Figure 2.24 can be illustrated through the decoding of Example 2.17. To this end, the coding in Figure 2.23 is rewritten in Figure 2.25 to show both coding and decoding. According to Figure 2.24, the first character, *T*, is read in, output, and stored as the 'lastcode' for the first string. Character *h* is then read in, output, and the string *Th* is added to the table at location 256 (using routine 'add_code'). The same decoding process applies for compressed strings *i*, *s*, and (sp), corresponding to locations 257, 258, and 259. However, when code 258 is received, routine 'outstring' decodes this into the known code/character string, *is*, and the last character, *s*, is sent to the stack (in routine 'outstring', 'code' should be regarded as a *local* variable). The remaining or 'base' code is a single character in this case (character *i*), and so the routine pops the complete string *is* from the stack to the decoder output. Variable 'firstchar' is now loaded with the first character of the output string, that is, character *i*, and the code/character combination *(sp)i* is loaded into location 260 by routine 'add_code'. It is now readily apparent from Figure 2.25 that the decoder builds an identical table to that at the encoder using just the compressed string.

Performance of the LZW algorithm

The CR for lossless coding is bounded by the entropy of the source. For instance, in Example 2.4 we showed that, while nearly 5 bits/character are required for straight PCM coding of the basic English character set (27 characters), the actual average information is nearer to 2 bits/character. This corresponds to a redundancy of about 50%, so using lossless source coding we should expect a CR close to 2. In practice, a CR of 1.8 is typical for LZW compression of English text, while CR is typically in the range 1.5 to 3 for source code and executable files. The CR for image files can vary significantly, depending upon the image structure. An image can often be compressed by a factor of 2 and occasionally by a factor of 15. The technique tends to perform best on images that contain large areas of a single colour. On the other hand, LZW can give virtually no compression (or even expansion) on rather detailed binary images. In general, CR tends to improve when using a variable wordlength compared with the use of a single, fixed wordlength.

Bibliography

Arps, R., Truong, T., Lu, D., Pasco, R. and Friedman, T. (1988) A multipurpose VLSI chip for adaptive data compression of bilevel images, *IBM J. Res. Develop.* **32**(6), 775–794.

Huang, T. (1977) Coding of two-tone images, *IEEE Trans. Communications*, **COM-25**(11), 1406–1424.

Hunter, R. and Robinson, A. (1980) International digital facsimile coding standards, *Proc. IEEE*, **68**(7), 854–867.

ITU-T Rec. T.82 (1993) *Information technology – coded representation of picture and audio information – progressive bi-level image compression.*

Jones, C. (1981) An efficient coding system for long source sequences, *IEEE Trans. on Inform. Theory*, **IT-27**, May, 280–291.

Kretzmer, E. (1952) Statistics of television signals, *Bell Syst. Tech. J.*, **31**, 751–763.

Meyr, H. *et al.* (1974) Optimum run length codes, *IEEE Trans. Communications*, **COM-22**(6), 826–835.

Nicol, R. *et al.* (1980) Transmission techniques for picture viewdata, *1980 International Broadcasting Convention, IEE Conf. Publication*, **191**, 109–113.

Ochi, H. and Tetsutani, N. (1987) A new halftone reproduction and transmission method standard black and white facsimile code, *IEEE Trans. Communications*, **COM-35**(4), 466–470.

Pennebaker, W., Mitchell, J., Langdon, G. and Arps, R. (1988) An overview of the basic principles of the Q-coder adaptive binary arithmetic coder, *IBM J. Res. Develop.* **32**(6), 717–726.

Rubin, F. (1979) Arithmetic stream coding using fixed-precision registers, *IEEE Trans. on Inform. Theory*, **IT-25**, 672–675.

Shannon, C. E. (1951) Prediction and entropy of printed English, *Bell Syst. Tech. J.*, **30**, 50.

Wade, G. and Li, Z. (1991) Compression of Meteosat data for image dissemination, *Computer Communications*, **14**(8), 489–495.

Welch, T. A. (1984) A technique for high-performance data compression, *IEEE Computer*, **17**(6), 8–19.

Ziv, J. and Lempel, A. (1977) A universal algorithm for sequential data compression, *IEEE Trans. on Inform. Theory*, **IT-23**(3), 337–343.

CHAPTER 3

Lossy source coding

Important lossy compression techniques were highlighted in Figure 2.1. Generally speaking, lossy techniques give much higher compression than lossless techniques, the latter being limited by the entropy of the source (Theorem 2.1). The information loss can be slight (virtually undetectable), as for example in MPEG-2 broadcast video, but in other cases it can be significant. For example, some speech compression techniques generate 'communications' quality speech that is intelligible but yet has detectable distortion. We start by examining one of the more established lossy techniques: differential PCM.

3.1 Differential PCM (DPCM)

3.1.1 Principles of DPCM

In Example 2.5 and Figure 2.7 we saw that a simple way to reduce correlation redundancy (correlation between samples) is to generate and encode sample-to-sample *differences*. The difference source was a more optimum source to transmit, since its samples were less correlated. In terms of video signals, the 'sample delay' in Figure 2.7 could be the time span between pixel samples or it could be the time between video frames (pictures). In the former case the circuit would reduce the *spatial* correlation redundancy, while the latter case would reduce the *temporal* correlation redundancy (Section 3.2.7). In Example 2.5 the compression mechanism was difference transmission and VLC. An alternative approach is to generate a 'better' difference signal by some *prediction* circuit and then to apply non-uniform quantisation to the resulting difference signal. The resulting system is then called differential PCM (DPCM). Both Figure 2.7 and DPCM exploit the non-uniform pdf of the difference signal, and DPCM does this by coarsely quantising rare (large) error samples and finely quantising the more probable (small) samples. In other words, *basic DPCM schemes achieves compression by non-uniform quantisation of the difference signal*, and in doing this the quantiser loses information and the process becomes lossy or non-reversible.

The basic DPCM scheme is shown in Figure 3.1 and the diagram highlights a number of fundamental points.

1. The difference sample $e_n = x_n - \hat{x}_n$ depends upon a prediction $\hat{x}_n$ derived from k previous samples, where k is the order of the predictor, P.
2. In the absence of transmission errors the encoder develops an output signal which is identical to that at the receiver (the *locally decoded output*).

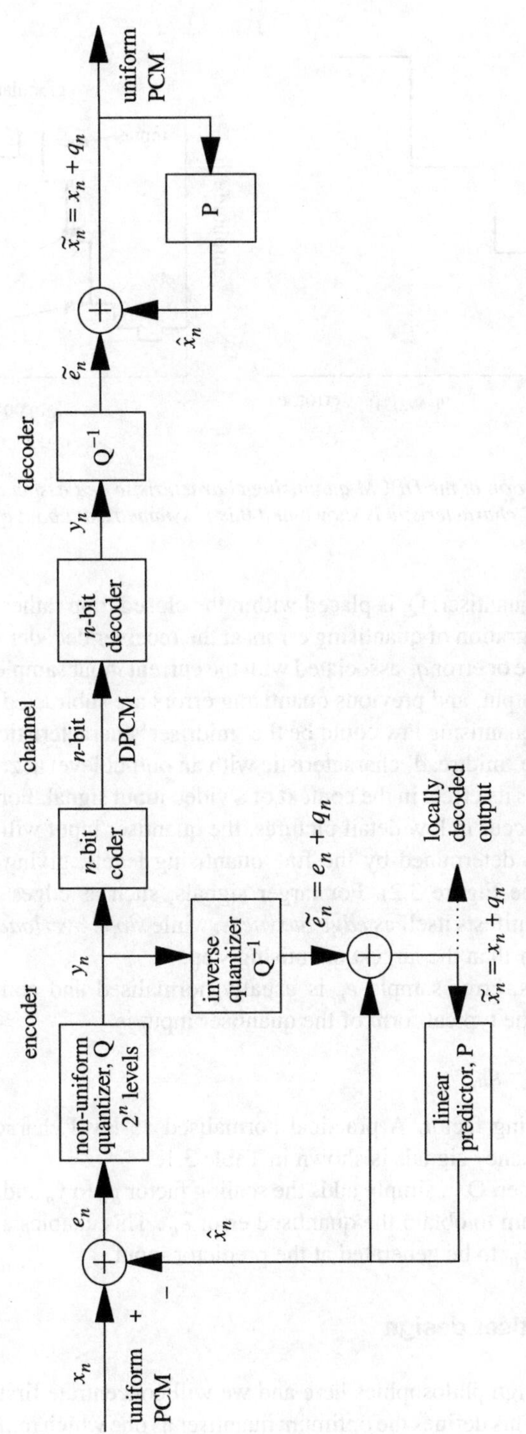

Figure 3.1 *Block diagram of a DPCM system (non-adaptive).*

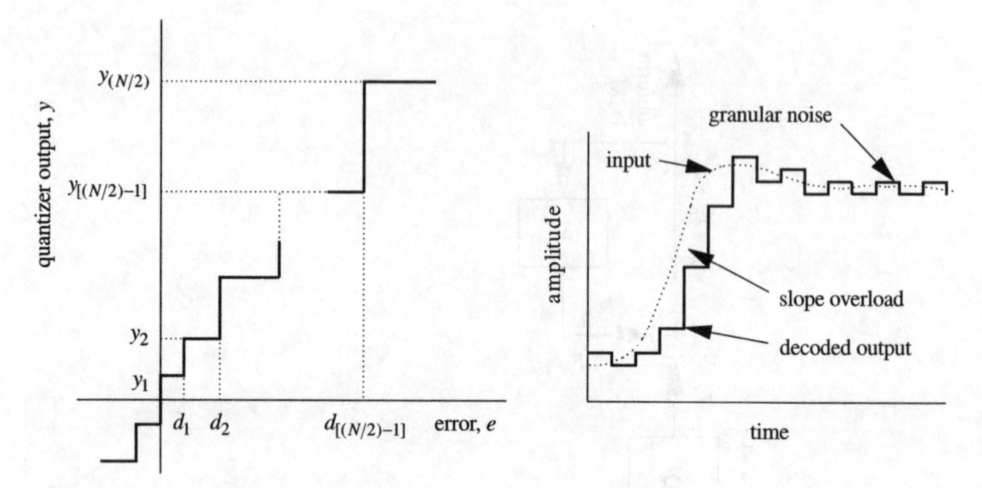

Figure 3.2 *General form of the DPCM quantising characteristic and associated distortions (a 'midriser' characteristic is shown and this is symmetrical about e = 0).*

3. The non-uniform quantiser, Q, is placed within the closed loop rather than outside it in order to avoid integration of quantising errors at the receiver decoder. In this way, only the quantising noise or error q_n associated with the current input sample x_n appears at the locally decoded output, and previous quantising errors are subtracted out.
4. The non-uniform quantising law could be the 'midriser' characteristic shown in Figure 3.2, or it could be a 'midtread' characteristic with an output level at zero. Either way, it is helpful to discuss its effect in the context of a video input signal. For slowly changing signals, as might occur in low detail pictures, the quantiser input will be small and the quantising error is determined by the fine quantising levels, giving rise to so-called *granular noise* (see Figure 3.2). For larger signals, such as edges in an image, the increased noise manifests itself as *edge busyness*, while *slope overload* occurs when the input error is larger than the largest quantising step.
5. For speech signals, error sample e_n is usually normalised and compressed prior to quantisation, and the typical form of the quantiser input is

   $$\log_2 |e_n| - g_n$$

   where g_n is a scaling factor. A practical normalised 15-level characteristic for 4-bit DPCM voice frequency signals is shown in Table 3.1.
6. The inverse quantiser, Q^{-1}, simply adds the scaling factor g_n to y_n and takes the inverse logarithm of the sum to obtain the quantised error $\tilde{e}_n$. This enables a reconstructed or quantised sample $\tilde{x}_n$ to be generated at the predictor input.

3.1.2 DPCM quantiser design

There are several design philosophies here and we will concentrate first upon a classical statistical approach. This defines the optimum quantiser as one which minimises the *mean-square quantising error*, so in terms of Figure 3.2 we would minimise $E[(e - y_k)^2]$, where E

Table 3.1 *Quantiser normalised input–output characteristic for 32 kbit/s 4-bit DPCM voice frequency signals; S is a sign bit (CCITT Rec. G.721, 1988).*

Quantiser input $\log_2\|e_n\| - g_n$	*Quantiser output* $y_n = \log_2\|\tilde{e}_n\| - g_n$	*Encoder output* v_n
3.12, ∞	3.32	S111
2.72, 3.12	2.91	S110
2.34, 2.72	2.52	S101
1.91, 2.34	2.13	S100
1.38, 1.91	1.66	S011
0.62, 1.38	1.05	S010
–0.98, 0.62	0.031	S001
–∞, –0.98	–∞	S000

denotes expectation and y_k represents one of a total of $N = 2^n$ output or *representative levels*. In this case the optimum values of y_k and input *decision levels* d_k are given by (Max, 1960)

$$\int_{d_{k-1}}^{d_k}(e - y_k)p(e)de = 0; \qquad k = 1,2,\ldots,\frac{N}{2} \tag{3.1}$$

$$d_k = \begin{cases} 0; & k = 0 \\ 0.5(y_k + y_{k+1}); & k = 1,2,\ldots,\frac{N}{2}-1 \\ \infty; & k = \frac{N}{2} \end{cases} \tag{3.2}$$

where $d_{k-1} < y_k < d_k$. Here, $p(e)$ is an even function representing the probability density of the quantiser input. *Note in particular that the optimum decision levels are half-way between the values of adjacent representative levels*. Equations (3.1) and (3.2) define the *Lloyd–Max* quantiser and solution of (3.1) is usually carried out numerically after assuming some appropriate form for $p(e)$. It is usual to assume that the effect of the quantiser on $p(e)$ is small (provided quantising is not too coarse), and for speech $p(e)$ is well approximated by the gamma density function. Computed results for N = 8, 16 and 32 are given in Table 3.2 (Paez and Glisson, 1972). Note that levels y_k and d_k should be scaled by the actual standard deviation σ_e of error e, which implies that a fixed quantiser will be optimised for only one specific variance!

For video systems we might assume $p(e)$ to be Laplacian, that is

$$p(e) = \frac{1}{\sigma_e\sqrt{2}}\exp\left(-\frac{\sqrt{2}}{\sigma_e}|e|\right) \tag{3.3}$$

This is a reasonable approximation for monochrome video (O'Neal, 1966), although a Cauchy pdf may be a better approximation for colour signals (Read, 1974). Assuming

Table 3.2 *Lloyd–Max quantisers assuming p(e) has a gamma density with* E[e] = 0, $\sigma_e^2 = 1$ *(Paez and Glisson, 1972, © IEEE).*

k	N					
	8		16		32	
	d_k	y_k	d_k	y_k	d_k	y_k
1	0.504	0.149	0.229	0.072	0.101	0.033
2	1.401	0.859	0.588	0.386	0.252	0.169
3	2.872	1.944	1.045	0.791	0.429	0.334
4	∞	3.799	1.623	1.300	0.630	0.523
5			2.372	1.945	0.857	0.737
6			3.407	2.798	1.111	0.976
7			5.050	4.015	1.397	1.245
8			∞	6.085	1.720	1.548
9					2.089	1.892
10					2.517	2.287
11					3.022	2.747
12					3.633	3.296
13					4.404	3.970
14					5.444	4.838
15					7.046	6.050
16					∞	8.043

(3.3), the mean-square quantising error will be minimised by selecting the y_k levels according to the function (Smith, 1957)

$$y(x) = -\frac{e_m}{r}\ln\left[1 - x(1 - e^{-r})\right]; \quad x \geq 0 \tag{3.4}$$

where

$$r = \frac{\sqrt{2}}{3}\frac{e_m}{\sigma_e} \tag{3.5}$$

Here, e_m is the maximum prediction error or, equivalently, the peak-to-peak value of the input signal, x_n. The input parameter x is a positive ratio, so (3.4) gives the positive part of the symmetrical characteristic in Figure 3.2, and x takes on uniformly spaced values between zero and unity. For a quantiser with a total of 2^n levels there must be 2^{n-1} equally spaced input points in this range, so that

$$x = \frac{1}{2^n}, \frac{3}{2^n}, \ldots, \frac{2^n - 3}{2^n}, \frac{2^n - 1}{2^n} \tag{3.6}$$

Once the representative levels y_k have been found, the decision levels d_k can be determined using (3.2).

Example 3.1

A 5-bit DPCM video system might typically have an 8-bit PCM input (256 quanta) and the parameters (Devereux, 1975)

$$e_m = 130 \text{ quanta}, \sigma_e = 12.3 \text{ quanta}$$

Using (3.4) this gives

$$y(x) = -26.0923 \ln(1 - 0.993142\, x)$$

and the centre of the 32-level quantising characteristic is therefore defined as follows:

Level, k	d_k	y_k
1	1.69 (1.81)	0.82 (0.89)
2	3.48 (3.71)	2.55 (2.73)
3	5.39 (5.74)	4.40 (4.70)
4	–	6.39 (6.78)
⋮	⋮	⋮

For comparison purposes, the corresponding values for a scaled Lloyd Max quantiser optimised to a Laplacian density are given in brackets.

A non-uniform quantising law is a natural consequence of applying the minimum-mean-square error (mmse) criterion to a non-uniform pdf, $p(e)$. Given $p(e)$ (or e_m and σ_e), mmse design is then fairly straightforward, as discussed. Clearly, this approach becomes more difficult if $p(e)$ varies, and in such cases the quantising characteristic could be found by using a random search to minimise the mean-square error (Cohn and Melsa, 1975).

Subjectively optimised quantisers

For video signals it is widely accepted that the mmse criterion is too crude and that quantisers should be designed on *psychovisual* grounds using the *human visual system* (HVS) (Netravali and Limb, 1980; Safranek and Johnson, 1989). This is true for basic DPCM video systems through to advanced hybrid differential systems based on the DCT (MPEG codecs) or on wavelets.

At the outset we might argue that, for static pictures at least, the viewer can tolerate larger quantising errors on rapid transitions (edges) than on quasi-uniform areas, so this in itself suggests a non-uniform quantising law. To determine this law we could determine the y_k values on a purely experimental basis (subjective testing to minimise noise visibility), and the d_k values could then be determined from (3.2). Using this technique it has been found that some optimal experimental values for y_k in Example 3.1 can differ from the mmse law by typically 40% (Devereux, 1975). More generally, subjective tests have shown that statistically optimised video quantisers tend to have too many levels for small prediction errors and too few levels for large prediction errors (Musmann *et al.*, 1985).

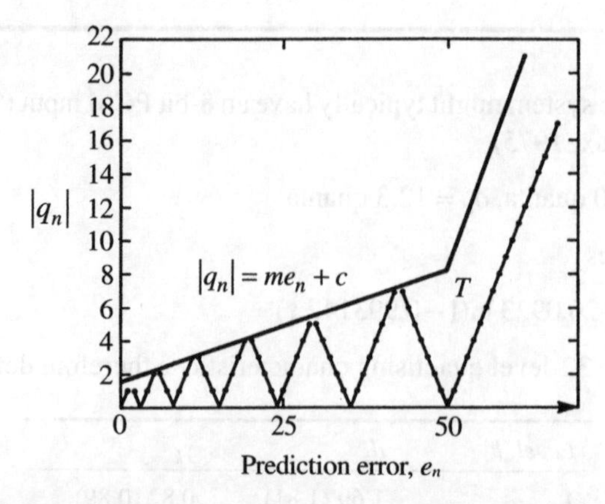

Figure 3.3 *Graphical design of a non-uniform quantiser.*

Also note that mmse design takes no account of any interaction between quantiser and predictor, whereas ideally these designs should be combined.

Graphical method for video quantisers

Non-uniform quantisers can also be designed using a simple graphical technique (Brainard *et al.*, 1982). This takes some account of subjective effects and has been found to give near optimum results. The basic idea is illustrated in Figure 3.3. The magnitude quantising error or noise $|q_n|$ is plotted against the prediction error e_n, and the points where $|q_n| = 0$ correspond to the quantiser output levels, y_k. A subjectively good quantiser can be constructed by arranging that the maximum quantisation error is always less than the line given by

$$q_n = m e_n + c; \quad e_n \leq T \tag{3.7}$$

The values of m, c and T are easily adjusted to obtain a quantising law with the required number of output levels. The value of T must be chosen such that slope overload effects are not visible. The values of c and m then affect the levels of granular noise and edge busyness respectively.

3.1.3 DPCM predictor design

For analysis purposes it is convenient to model the DPCM system as in Figure 3.4. We assume that a random signal x_n of zero mean and variance σ_x^2 is applied to the encoder input. It is usual to make prediction $\hat{x}_n$ a linear combination of previous quantised samples, as shown, and to assume that P can be designed independently of Q. We will therefore ignore the effects of quantisation when designing P, although the designs should ideally be combined.

In terms of Figure 3.4, a linear prediction of the next sample value x_0 given k previous sample values is defined as

$$\hat{x}_0 = \sum_{i=1}^{k} a_i \tilde{x}_i \tag{3.8}$$

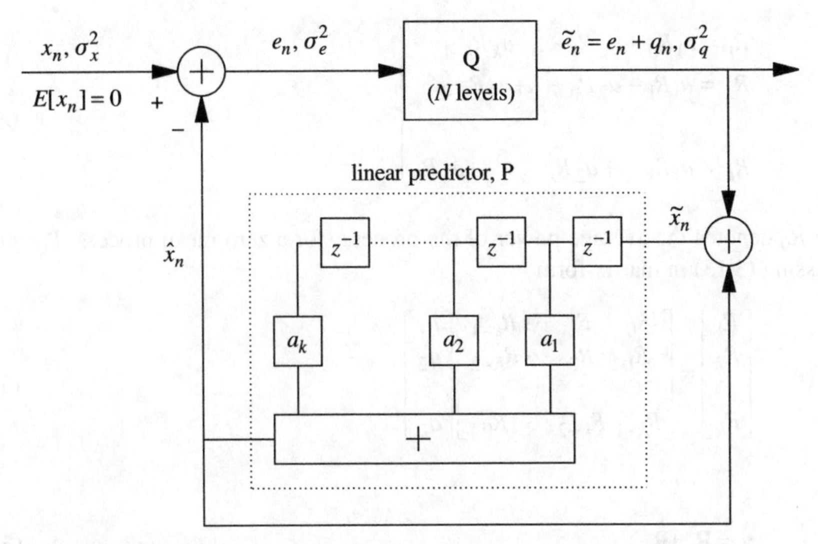

Figure 3.4 *Model for DPCM analysis assuming kth-order linear prediction.*

where $\tilde{x}_i$ is the quantised value of x_i and a_i is a predictor coefficient. Determination of the optimum coefficients is a 'least-squares' problem and we minimise the mean-square error or error variance σ_e^2:

$$\text{minimise } E[e_0^2]; \quad e_0 = x_0 - \hat{x}_0$$

Differentiating and ignoring the effects of quantisation

$$\frac{\partial E[e_0^2]}{\partial a_i} = E\left[\frac{\partial e_0^2}{\partial a_i}\right] = 2E\left[e_0 \frac{\partial e_0}{\partial a_i}\right] = -2E\left[e_0 x_i\right] \tag{3.9}$$

so the optimum coefficients are given by the equations

$$E\left[x_i(x_0 - \hat{x}_0)\right] = 0; \quad i = 1,\ldots,k \tag{3.10}$$

Expanding

$$\begin{aligned} E\left[x_0 x_i\right] &= E[\, x_i(a_1 x_1 + a_2 x_2 + \ldots + a_k x_k)\,] \\ &= a_1 E\left[x_1 x_i\right] + a_2 E\left[x_2 x_i\right] + \ldots + a_k E\left[x_k x_i\right]; \quad i = 1,\ldots,k \end{aligned} \tag{3.11}$$

In a strict statistical sense, x_n in (3.11) should be regarded as a *random variable* belonging to a *random process*. We will also assume that the random process is *wide-sense stationary*, which essentially means that the low-order statistics of the process (its mean and autocorrelation function) are independent of time. On this assumption, for a zero-mean process, terms $E[x_i x_j]$ in (3.11) can be equated to the autocorrelation function term R_{ij} and we can write

$$R_{0i} = a_1 R_{1i} + a_2 R_{2i} + \ldots + a_k R_{ki}; \quad i = 1,\ldots,k \tag{3.12}$$

Noting that R_{ij} can be written $R_{|i-j|}$, this gives a set of k simultaneous linear equations

$$\left.\begin{aligned} R_1 &= a_1R_0 + a_2R_1 + \ldots + a_kR_{k-1} \\ R_2 &= a_1R_1 + a_2R_0 + \ldots + a_kR_{k-2} \\ &\vdots \\ R_k &= a_1R_{k-1} + a_2R_{k-2} + \ldots + a_kR_0 \end{aligned}\right\} \tag{3.13}$$

where R_0 denotes the average power of the process (for a zero mean process, $R_0 = \sigma_x^2$). Expressing (3.13) in matrix form

$$\begin{bmatrix} R_1 \\ R_2 \\ \vdots \\ R_k \end{bmatrix} = \begin{bmatrix} R_0 & R_1 & \cdots & R_{k-1} \\ R_1 & R_0 & \cdots & R_{k-2} \\ & & \vdots & \\ R_{k-1} & R_{k-2} & \cdots & R_0 \end{bmatrix} \begin{bmatrix} a_1 \\ a_2 \\ \vdots \\ a_k \end{bmatrix} \tag{3.14}$$

or

$$\mathbf{a} = \mathbf{R}_{xx}^{-1}\mathbf{R} \tag{3.15}$$

Here $\mathbf{R}_{xx}$ is the $k \times k$ autocorrelation matrix for the input signal and it can always be inverted. Under the assumption of stationarity, (3.15) states that the optimum predictor coefficients can be determined from a knowledge of the signal autocorrelation function. Formally this is referred to as *Weiner prediction*, and (3.15) is a linear predictive form of the *Weiner–Hopf equations*. Efficient solution of (3.15) is discussed in Section 3.1.4, although the major computational load in (3.15) is usually the computation of the R_i terms. Given just N samples of the random signal, these can be well estimated using (4.39).

Types of predictor

In practice the exact form of the linear predictor varies significantly between applications. If, for example, the input sequence $\{x_n\}$ can be modelled as an mth-order Markov (or AR(m)) sequence, then only m samples of x_n are needed to form the best estimate of the input sample. In other words, the ideal predictor for an mth-order Markov source has order $k = m$, and it will fully decorrelate or 'whiten' the error sequence $\{e_n\}$. This is, after all, a main objective in DPCM. For example, the short-time spectral envelope for voiced speech can be well modelled by the spectrum of an AR(10) or larger process, so predictors of this order should give good decorrelation for such a source (Flanagan *et al.*, 1979).

Example 3.2

A second-order linear prediction of sample x_0 is

$$\hat{x}_0 = a_1\tilde{x}_1 + a_2\tilde{x}_2$$

and the coefficients are found from

$$R_1 = a_1R_0 + a_2R_1$$

$$R_2 = a_1R_1 + a_2R_0$$

Assuming unit variance ($R_0 = 1$) we have

$$a_1 = R_1 \frac{1-R_2}{1-R_1^2} \quad \text{and} \quad a_2 = \frac{R_2 - R_1^2}{1-R_1^2}$$

For a first-order Markov source (AR(1) model) it can be shown that $R_2 = R_1^2$, so $a_2 = 0$. Therefore, as expected, only a first-order predictor is required for a first-order Markov source. The general form of the required coefficient will then be

$$a_1 = \frac{R_1}{R_0} = \frac{E[x_0 x_1]}{E[x_0^2]} = \rho_1 \tag{3.16}$$

where ρ_1 is the adjacent sample correlation coefficient. It is interesting to note that the autocorrelation function of one line of a monochrome video signal can be close to a Laplacian function, that is $R_\tau = \mathrm{e}^{-\alpha|\tau|}$, τ being the spatial lag (in samples) along a scan line. Therefore, again, $R_2 = R_1^2$ and the optimum predictor uses only the most previous sample.

The simplified argument in Example 3.2 suggests that second or higher-order prediction in video systems is unnecessary if all samples are within the same scan line. Practical video systems do however use second-order prediction involving several scan lines (two-dimensional or 2D prediction), since this can greatly improve the subjective effect of transmission errors as well as improving prediction on large luminance transitions, especially when coding composite PAL and NTSC signals. Figure 3.5 illustrates two simple 2D predictors for NTSC signals sampled at three-times colour sub-carrier frequency. The idea here is to select samples closest to the current input sample x_0 which have the same sub-carrier phase.

Fixed prediction as designed by (3.15) is often unsuitable for video signals, since they cannot be assumed to be stationary. In other words, *adaptive* rather than fixed prediction tends to perform best for video signals, and adaptive techniques based upon the Least Mean Square (LMS) algorithm are particularly attractive. For example, if P in Figure 3.4 is considered to be an LMS predictor then the predictor coefficients could be updated using the recursion

$$a_{i,n+1} = a_{i,n} + \mu \tilde{e}_n \tilde{x}_{n-i}; \quad i = 1, 2, \ldots, k \tag{3.17}$$

Here, μ is a constant and $\tilde{e}_n$ is the quantised prediction error in Figure 3.4. Using simple 2D second-order prediction on video signals (that is, $\tilde{x}_{n-1}$ and $\tilde{x}_{n-2}$ are not adjacent in

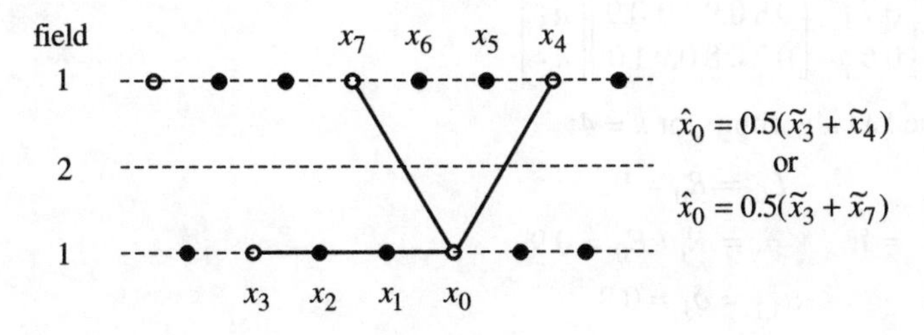

Figure 3.5 *Simple 2D predictors for NTSC signals sampled at three-times colour sub-carrier frequency.*

time), and with $\mu \approx 10^{-5}$, this adaptive approach can give significant advantage compared to fixed prediction (Alexander and Rajala, 1985). More sophisticated LMS prediction for DPCM is described by Knee and Wells (1989).

Finally, spatial predictors (*intraframe predictors*) of the form in Figure 3.5 account only for the spatial *correlation redundancy* in a video sequence. Most video encoders also use *interframe prediction* to account for the *temporal correlation redundancy*. Interframe coding (Section 3.2.7) transmits the difference between frames and can drastically reduce the bit rate.

3.1.4 Levinson–Durbin (LD) recursion

The autocorrelation matrix $\mathbf{R}_{xx}$ in (3.15) is symmetric. Moreover, it is a *Toeplitz* matrix, since the elements along any diagonal are identical. The Toeplitz property means that the matrix can be inverted very efficiently using the LD recursion algorithm (LD solution of (3.15) requires only order k^2 arithmetic operations compared with order k^3 operations for Gaussian elimination). The algorithm for a kth-order predictor is as follows:

Step 1: $E_0 = R_0$

Step 2: $\delta_i = \left[R_i - \sum_{j=1}^{i-1} \alpha_{j,i-1} R_{i-j} \right] / E_{i-1}; \quad 1 \le i \le k$

Step 3: $\alpha_{i,i} = \delta_i$

Step 4: $\alpha_{j,i} = \alpha_{j,i-1} - \delta_i \alpha_{i-j,i-1}; \quad 1 \le j \le i-1$

Step 5: $E_i = (1 - \delta_i^2) E_{i-1}$

Step 1 is initialisation. Steps 2 to 5 are solved recursively, giving the solution $a_j = \alpha_{j,k}$, $1 \le j \le k$. The summation in step 2 is omitted for $i = 1$, and step 4 is omitted for $i = 1$.

Example 3.3

Suppose (3.14) has the form

$$\begin{bmatrix} 0.9 \\ 0.8 \\ 0.7 \\ 0.6 \end{bmatrix} = \begin{bmatrix} 1.0 & 0.9 & 0.8 & 0.7 \\ 0.9 & 1.0 & 0.9 & 0.8 \\ 0.8 & 0.9 & 1.0 & 0.9 \\ 0.7 & 0.8 & 0.9 & 1.0 \end{bmatrix} \begin{bmatrix} a_1 \\ a_2 \\ a_3 \\ a_4 \end{bmatrix}$$

Applying the LD algorithm for $k = 4$:

$$E_0 = R_0 = 1$$

$$i = 1: \quad \delta_1 = R_1 / E_0 = 0.9$$

$$\alpha_{1,1} = \delta_1 = 0.9$$

$$E_1 = (1 - \delta_1^2) E_0 = 0.19$$

$$
\begin{aligned}
i=2: \quad & \delta_2 = (R_2 - \alpha_{1,1}R_1)/E_1 = -0.05263 \\
& \alpha_{2,2} = \delta_2 = -0.05263 \\
& \alpha_{1,2} = \alpha_{1,1} - \delta_2\alpha_{1,1} = 0.94737 \\
& E_2 = (1-\delta_2^2)E_1 = 0.18947
\end{aligned}
$$

$$
\begin{aligned}
i=3: \quad & \delta_3 = (R_3 - (\alpha_{1,2}R_2 + \alpha_{2,2}R_1)/E_2 = -0.05557 \\
& \alpha_{3,3} = \delta_3 = -0.05557 \\
& \alpha_{1,3} = \alpha_{1,2} - \delta_3\alpha_{2,2} = 0.94444 \\
& \alpha_{2,3} = \alpha_{2,2} - \delta_3\alpha_{1,2} = 0.000015351 \\
& E_3 = (1-\delta_3^2)E_2 = 0.18889
\end{aligned}
$$

Continuing in this way gives

$$
\begin{aligned}
a_1 &= \alpha_{1,4} = 0.94117 \\
a_2 &= \alpha_{2,4} = 0.000016254 \\
a_3 &= \alpha_{3,4} = -0.000036928 \\
a_4 &= \alpha_{4,4} = -0.05880
\end{aligned}
$$

3.1.5 Adaptive DPCM (ADPCM)

As discussed, adaptive prediction tends to perform best, especially for speech signals. A fully adaptive DPCM system will adapt both the predictor, P, and the quantiser, Q, as shown in Figure 3.6. In general, adaptive information can be either separately transmitted to the decoder (forward adaption), or it can be derived from recent quantised samples (backward adaption), and the latter is assumed in Figure 3.6. In other words, the use of previous samples for adaption enables the decoder to operate without being sent any adaption information.

Predictor design for ADPCM can be considered in terms of the *decoder transfer function*, $D(z)$ (Figure 3.7). The predictor in Figure 3.7(a) corresponds to (3.8), and using basic digital filter theory it is easily shown that

$$D(z) = \left(1 - \sum_{i=1}^{k} a_i z^{-i}\right)^{-1} \tag{3.18}$$

This is an 'all-pole' filter and its poles must lie in the unit circle for a stable decoder. However, in ADPCM the coefficients will adapt to minimise the mean-square error, and transmission errors may force the poles outside the unit circle (creating instability). In fact, stability can be a crucial issue for high-order all-pole predictors.

A practical solution is to express $D(z)$ as a cascade of two filters as shown in Figure 3.7(b), giving

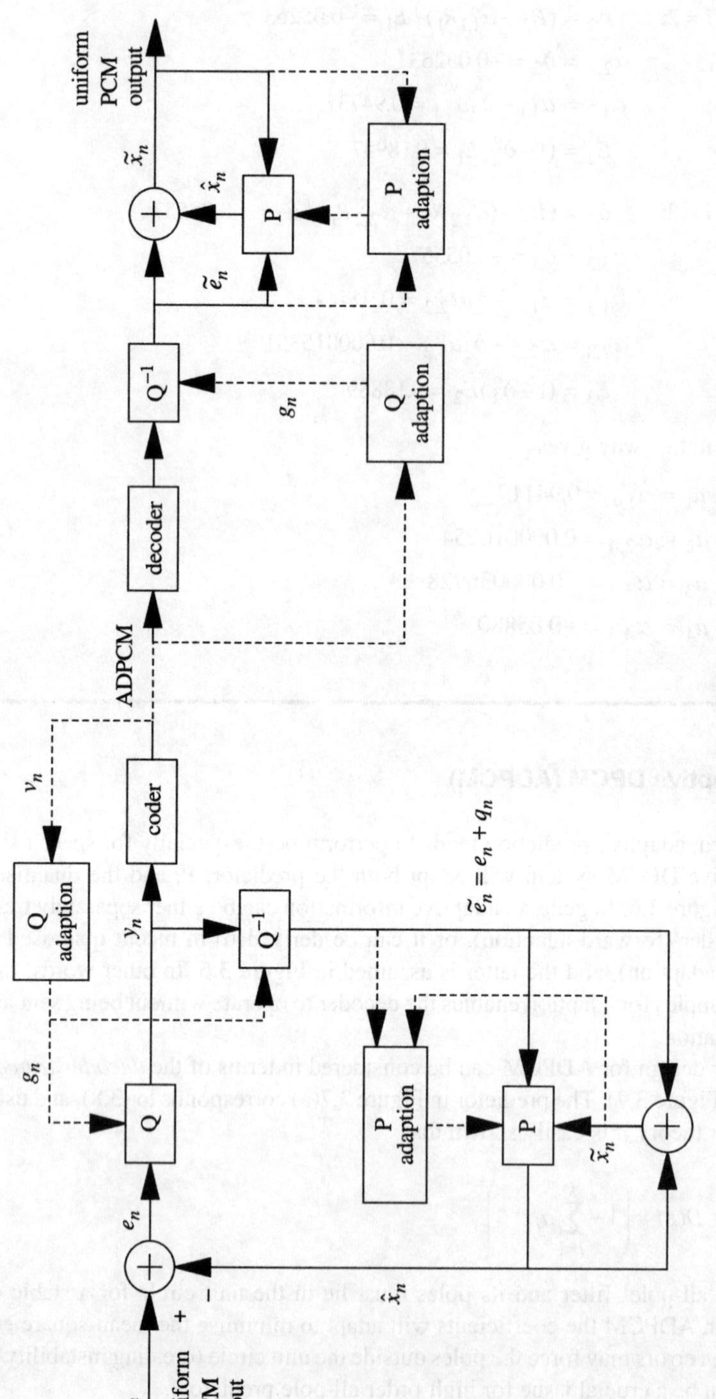

Figure 3.6 *General block diagram of an ADPCM system (backward adaption).*

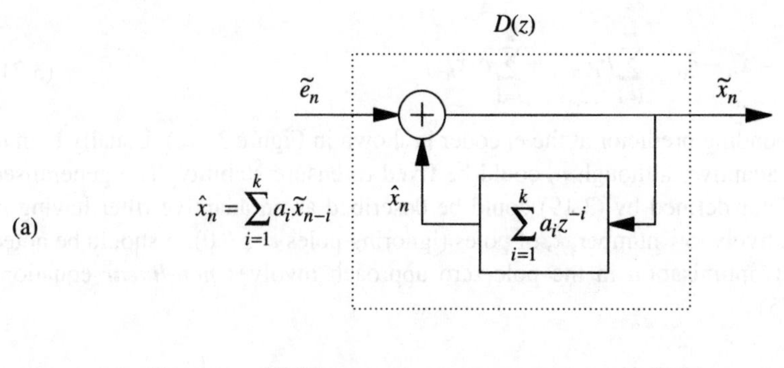

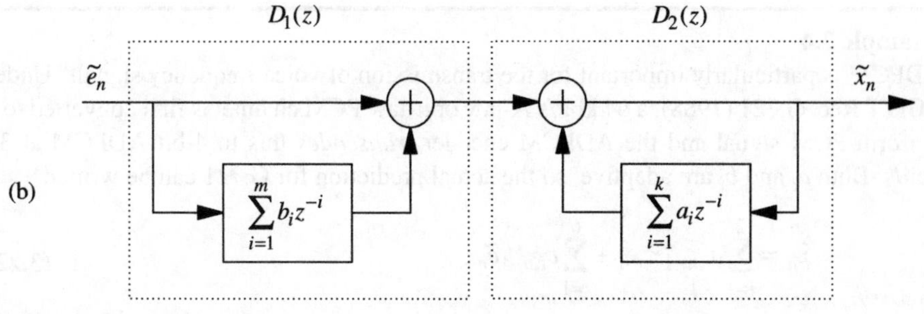

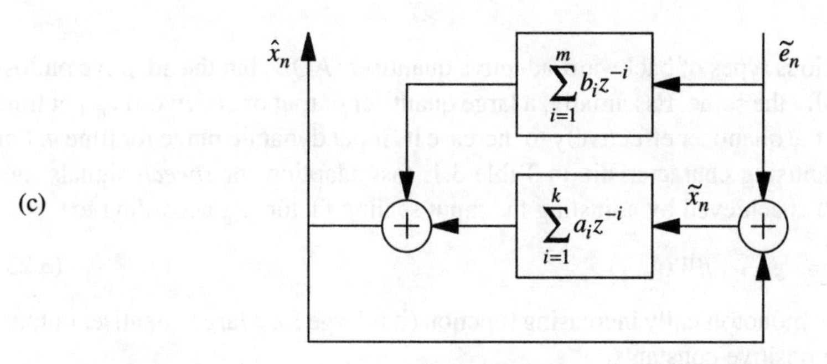

Figure 3.7 *Prediction filters: (a) all-pole (decoder); (b) pole-zero (decoder); (c) pole-zero (encoder).*

$$D(z) = D_1(z)D_2(z) = \left(1 + \sum_{i=1}^{m} b_i z^{-i}\right)\left(1 - \sum_{i=1}^{k} a_i z^{-i}\right)^{-1} \tag{3.19}$$

Filter $D_1(z)$ provided zeros in the transfer function and helps to stabilise the filter in the presence of transmission errors. It is easily seen from Figure 3.7(b) that

$$\tilde{x}_n = \tilde{e}_n + \sum_{i=1}^{m} b_i \tilde{e}_{n-i} + \sum_{i=1}^{k} a_i \tilde{x}_{n-i} \tag{3.20}$$

Therefore, since $\tilde{x}_n = \hat{x}_n + \tilde{e}_n$ (Figure 3.7(a))

$$\hat{x}_n = \tilde{x}_n - \tilde{e}_n = \sum_{i=1}^{m} b_i \tilde{e}_{n-i} + \sum_{i=1}^{k} a_i \tilde{x}_{n-i} \tag{3.21}$$

and the corresponding predictor at the encoder is shown in Figure 3.7(c). Usually both a_i and b_i will be adaptive, although a_i could be fixed to ensure stability. The generalised ADPCM predictor defined by (3.19) could be described as an adaptive filter having m zeros and a relatively low number, k, of poles (ignoring poles at $z = 0$). It should be noted that coefficient optimisation in the pole-zero approach involves *non-linear* equations (Makhoul, 1975).

Example 3.4

ADPCM is particularly important for the transmission of voice frequency signals. Under CCITT Rec. G.721 (1988), a 64 kbit/s A-law or μ-law PCM channel is first converted to a uniform PCM signal and the ADPCM encoder *transcodes* this to 4-bit ADPCM at 32 kbit/s. Both a_i and b_i are adaptive, so the actual prediction for G.721 can be written

$$\hat{x}_n = \sum_{i=1}^{2} a_{i,n-1} \tilde{x}_{n-i} + \sum_{i=1}^{6} b_{i,n-1} \tilde{e}_{n-i} \tag{3.22}$$

The coefficients are updated using a gradient algorithm.

There are various types of backward-adaptive quantiser (AQB) but the adaptive philosophy is essentially the same. Essentially, a large quantiser output or codeword v_{n-1} at time $n-1$ will cause the quantiser effectively to increase its input dynamic range for time n. For the specific quantising characteristic in Table 3.1, fast adaption for *speech* signals (not voiceband *data*) is achieved by adjusting the input scaling factor, g_n, according to

$$g_n = \alpha g_{n-1} + \beta W(v_{n-1}) \tag{3.23}$$

where W is some monotonically increasing function (it is large for a large quantiser output) and α and β are positive constants.

Effect of transmission errors

Since the all-pole decoder filter in Figure 3.7(a) is a linear system we can consider its input to be a superposition of the signal and an error sequence, and treat the error sequence separately. Clearly, a single error sample at the filter input will generate an infinite sequence of decaying error samples at its output and so generate *error smearing* or *error propagation*. On the other hand, the all-zeros filter $D_1(z)$ in Figure 3.7(b) (an FIR filter) will generate only $(1 + m)$ error samples for a single input error, so its inclusion helps to stabilise the decoder, as previously noted.

The *subjective* performance of DPCM relative to PCM in the presence of channel errors depends upon the type of input signal. In the case of speech or music, DPCM is subjectively *more* robust than PCM, typically by several orders of magnitude (Yan and Donaldson, 1972). This is because in speech or music a single PCM error is subjectively more annoying (as in 'spiky clicks') than a relatively small error smeared over a long period. More precisely, in DPCM a short noise burst will cause an all-pole filter to resonate at its current resonant frequency, thereby giving an 'acceptable' sound. Typically, DPCM speech and music systems can tolerate channel bit error rates (BER) of 10^{-3} and 10^{-4} respectively, and FEC is often unnecessary.

In contrast, for image/video coding DPCM is subjectively *less* robust than PCM because error propagation is subjectively more annoying. The problem is worst for 1D prediction, since then an error generates a streak across the picture, and for this case the decoder loops could be reset during line blanking.

Sub-band coding

ADPCM is sometime used in conjunction with sub-band coding (SBC) in order to optimise compression. In a sub-band encoder, the linearly quantised PCM signal is processed through a bank of filters and each sub-band is separately coded (quantised) according to the input signal energy. It is particularly applicable to audio signals and here the objective is to exploit the spectral redundancies within the audio spectrum. For example, the ear is particularly sensitive to low frequencies, so the lowest frequency band can be assigned the largest number of bits/sample. Conversely, quantisation can be less accurate for bands to which the ear is more tolerant.

SBC of speech might typically be based on 11 bandpass filters (each 250 Hz wide) spanning the range 300 Hz to 3 kHz. Heuval *et al.* (1991) have shown that, out of these, the four most energetic bands generally contain some 95% of the energy 50% of the time, or 50% of the energy 88% of the time. They also showed by subjective testing that, essentially, it was only necessary to select the four most energetic bands for subsequent adaptive quantisation and transmission in order to obtain acceptable speech quality. SBC of compact disc (CD) quality music typically uses four sub-bands, as shown in Figure 3.8(a). This figure shows that excellent quality stereo music can be transmitted at 384 kbit/s using a combination of ADPCM and SBC, the corresponding compressed rate being just 4 bits/sample.

Example 3.5

SBC offers an opportunity to employ bandpass sampling (section 1.4.1). Suppose a bandwidth B is split into M equal sub-bands, as shown in Figure 3.8(b). The bandwidth of each sub-band is therefore B/M and the maximum frequency of the nth band is nB/M. According to the bandpass sampling theorem we could therefore sample each band at a rate

$$f_s = 2\frac{B}{M}\frac{n}{\lfloor n \rfloor} = \frac{2B}{M}$$

rather than at rate $2B$, prior to independent coding operations.

3.1.6 DPCM coding gain

It is important to establish formally the compression advantage of DPCM w.r.t. linear PCM, and to do this we will refer to Figure 3.1 and Figure 3.4. First note that the decoded

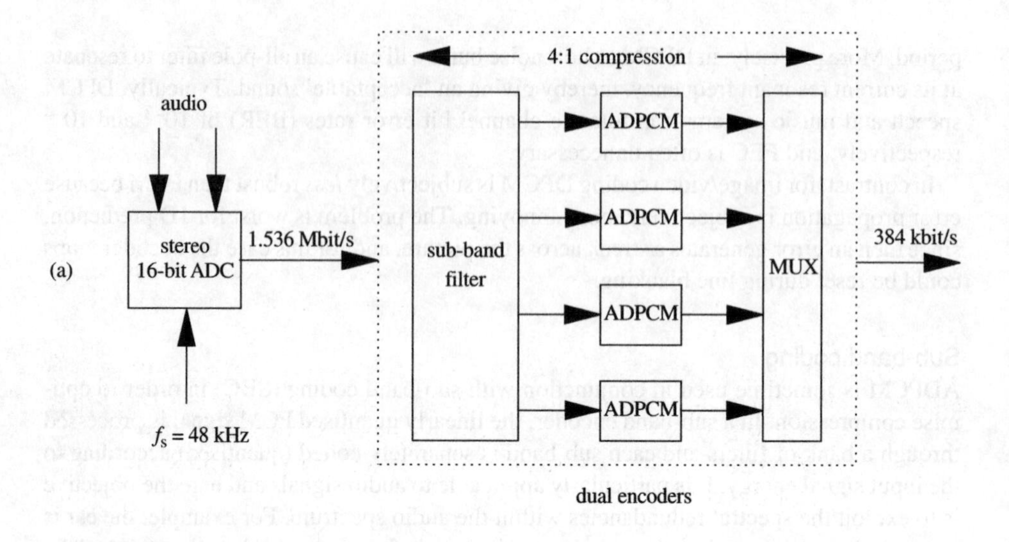

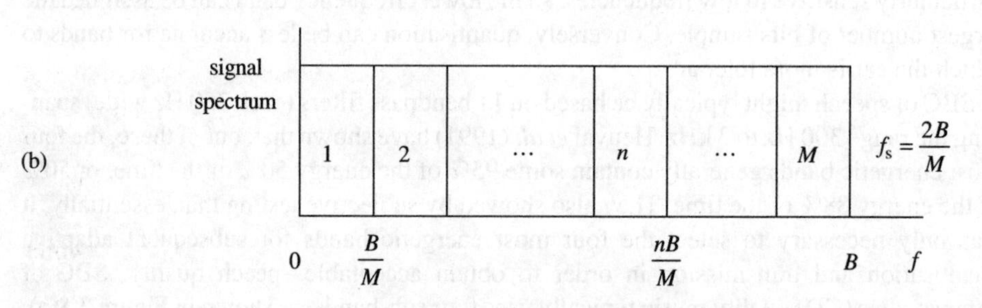

Figure 3.8 *SBC: (a) use of ADPCM and sub-band coding for CD-quality stereo music; (b) bandpass sampling for equal width sub-bands.*

signal $\tilde{x}_n$ contains a normalised signal power σ_x^2 and a normalised quantising noise power σ_q^2, so we can define an output SNR. By making plausible assumptions, such as a Laplacian pdf for $p(e)$, reasonably fine quantising ($N \geq 8$) and small probability of slope overload, it can be shown that the maximum SNR at the decoder output is (O'Neal, 1966)

$$\begin{aligned} \mathrm{SNR}_{\mathrm{DPCM}} &= 10 \log \frac{\sigma_x^2}{\sigma_q^2\big|_{\min}} \\ &= 10 \log \frac{2N^2 \sigma_x^2}{9\sigma_e^2} \qquad (3.24) \\ &= 6n - 6.5 + 10 \log \frac{\sigma_x^2}{\sigma_e^2} \ \mathrm{dB} \end{aligned}$$

Now consider a limiting case in which all the input samples x_n are uncorrelated, that is, all the autocorrelation terms R_i, $i \neq 0$, are zero. In this case the optimum prediction $\hat{x}_n$ is the mean of the input sequence, or zero, and the feedback path can be removed. Effectively the DPCM system has become a PCM system ($\sigma_x^2 = \sigma_e^2$) and so we can rewrite (3.24) as

$$\mathrm{SNR}_{\mathrm{DPCM}}(\mathrm{dB}) = \mathrm{SNR}_{\mathrm{PCM}}(\mathrm{dB}) + 10\log\frac{\sigma_x^2}{\sigma_e^2} \tag{3.25}$$

The term

$$G_p = \frac{\sigma_x^2}{\sigma_e^2} \tag{3.26}$$

is called the *prediction gain* and it can be evaluated for a given predictor. Looked at in a simple way, quantising error variance σ_q^2 tends to be proportional to quantiser input variance (for a given n), and so if we reduce the quantiser input variance by a factor G_p the output SNR must increase by G_p. The simple expression for G_p in (3.26) is approximately true for both voice frequency signals and video signals. Also, it is usual to take G_p (dB) as the approximate SNR improvement in going from PCM to DPCM assuming that both the PCM and DPCM systems code to n bits/sample. Alternatively, for a given SNR, the bit rate can be reduced for DPCM.

Example 3.6

Consider simple first-order prediction. According to (3.8), and ignoring quantisation, we can define first-order prediction as $\hat{x}_0 = a_1 x_1$. Therefore, since $E[e_n] = 0$ in Figure 3.4 we can write

$$\sigma_e^2 = E[(x_0 - a_1 x_1)^2]$$
$$= E[x_0^2] - 2a_1 E[x_0 x_1] + a_1^2 E[x_1^2]$$

Using (3.16) this reduces to

$$\sigma_e^2 = (1 - 2a_1\rho_1 + a_1^2)\sigma_x^2$$

giving

$$G_p = (1 - \rho_1^2)^{-1} \tag{3.27}$$

Measurements of the long-time autocorrelation function for speech gives $\rho_1 \approx 0.85$ (Flanagan *et al.*, 1979), in which case $G_p \approx 5.6$ dB. Higher gain can be achieved by using higher-order prediction. Similarly, for monochrome video signals ρ_1 may be typically 0.96 for a 'head and shoulders' portrait, and 0.75 for a 'crowd' scene, giving prediction gains of 11.1 and 3.6 dB respectively. A lower gain for high detail scenes is to be expected since there is less redundancy to remove.

So far we have assumed the SNR to be a valid criterion for comparing PCM and DPCM systems, and for video systems at least, we might question this given that the subjective effect of noise depends upon the noise spectrum. Fortunately, it turns out that the quantising noise in both PCM and DPCM video systems has a substantially flat spectrum, so PCM and DPCM systems with the same SNR will have similar subjective impairment. Figure 3.9 is therefore a reasonable comparison of colour video systems and indicates that substantial prediction gain is possible. For video systems it is generally accepted that

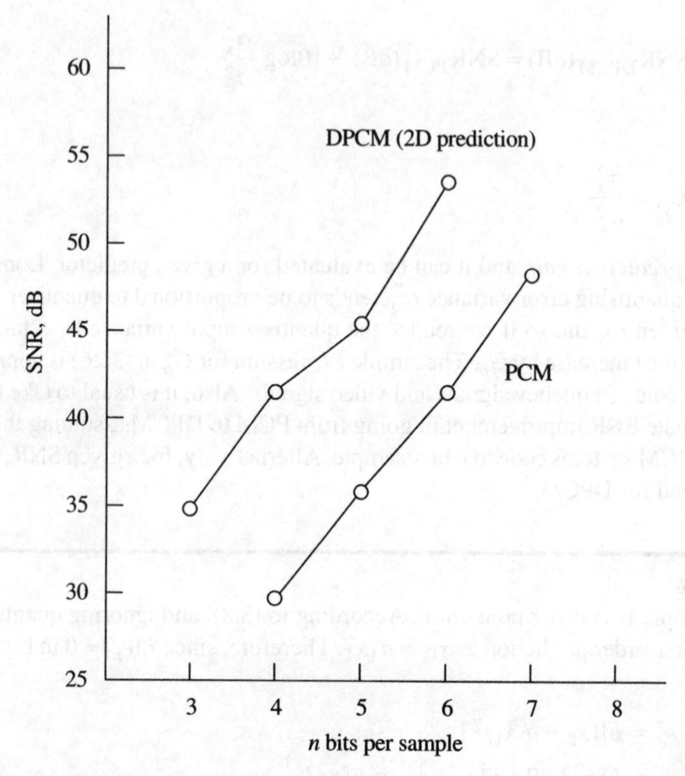

Figure 3.9 *Measured SNR for PCM and DPCM PAL colour signals (Devereux, 1975).*

DPCM on its own gives approximately 12 dB SNR improvement for a given value of n, or, for a given SNR, DPCM requires about 2 bits/sample less than PCM (see (3.24)). Similar prediction gains are possible for speech using ADPCM and high-order prediction (typically $k = 10$).

Despite its significant advantage, we have to admit that, for video systems, DPCM *on its own* offers only 2:1 compression or less. For video, the real advantage of DPCM occurs when it is combined with *transform coding* and advanced interframe prediction. This approach leads to MPEG-2 coding for example, with a typical compression ratio of 30, even for broadcast quality signals.

3.2 Principles of transform coding

Transform coding (TC) is an alternative compression scheme to DPCM and is particularly applicable to video signals. For video signals, TC generally has greater potential for compression than DPCM, (although the signal processing tends to be more complex). When combined with DPCM in a *hybrid* coding scheme, it finds application in video telephony (H.261 and H.263), and in MPEG-2 and MPEG-4 video encoders. Also, TC offers flexibility, as for example when progressive picture build-up (fast reception and display of a low-resolution picture with increasingly finer detail being added subsequently) is employed in low bit-rate still picture systems. As in DPCM, an underlying objective in TC is to transmit independent (and therefore uncorrelated) samples and, generally speaking, for video signals both systems can achieve compression by quantising not only on statistical grounds, but also in such a way as to exploit the viewer's perception. Clearly, the process of quantising makes TC a lossy or non-reversible process.

3.2.1 *N*th-order linear transformation

A transform coder performs a sequence of two operations. The first is a linear transformation that, ideally, transforms a set of *N* samples into a set of *N* *independent* coefficients. This ideal is intractable for several reasons and the closest we can get is a linear transform that generates *uncorrelated* coefficients (which are not necessarily independent). In practice we use a linear transform that transforms a set of *N* correlated samples into a set of *N* 'more uncorrelated' samples or *coefficients*. Assuming independence, the coefficients can then be quantised independently in order to achieve compression. A general TC system is shown in Figure 3.10. Here, *N* PCM input samples are represented as a vector

$$\mathbf{X}^{\mathrm{T}} = \left[x_0\, x_1 \ldots x_{N-1}\right] \tag{3.28}$$

where T denotes the transpose. This vector is transformed to *N* coefficients in the 'frequency' domain using the general *forward transformation*

$$\mathbf{G} = \mathbf{A}\,\mathbf{X} \tag{3.29}$$

where $\mathbf{G}^{\mathrm{T}} = [G_0 G_1 \ldots G_{N-1}]$. The term frequency is used loosely here, since in general the transform domain is not a strict frequency domain as measured in cycles/second. In (3.29), **A** is an $N \times N$ *transform matrix* whose elements are specified by the selected transform (in general they will be complex).

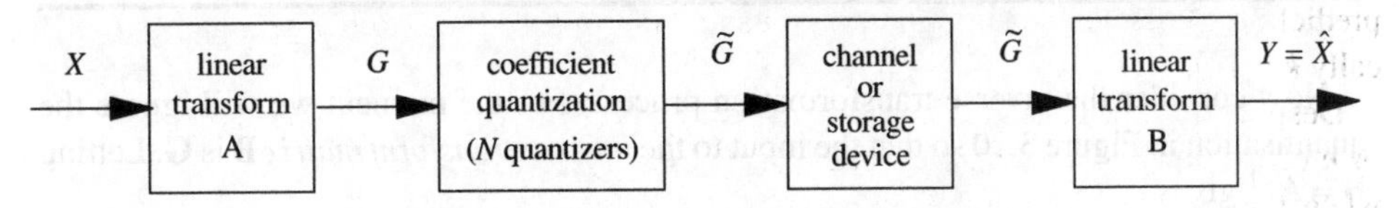

Figure 3.10 *Model of an Nth-order TC system.*

Example 3.7
The *N*th-order *Walsh–Hadamard matrix*, $\mathbf{H}_N$ is a specific form for **A** that leads to the *discrete Walsh–Hadamard transform* (DWHT). For $N = 8$ the transform coefficients are given by

$$\mathbf{a} = \frac{1}{8}\mathbf{H}_8\mathbf{X}$$

which can be expanded as

$$\begin{bmatrix} a_0 \\ a_7 \\ a_3 \\ a_4 \\ a_1 \\ a_6 \\ a_2 \\ a_5 \end{bmatrix} = \frac{1}{8} \begin{bmatrix} +&+&+&+&+&+&+&+ \\ +&-&+&-&+&-&+&- \\ +&+&-&-&+&+&-&- \\ +&-&-&+&+&-&-&+ \\ +&+&+&+&-&-&-&- \\ +&-&+&-&-&+&-&+ \\ +&+&-&-&-&-&+&+ \\ +&-&-&+&-&+&+&- \end{bmatrix} \begin{bmatrix} x_0 \\ x_1 \\ x_2 \\ x_3 \\ x_4 \\ x_5 \\ x_6 \\ x_7 \end{bmatrix}$$

where + and – denote ±1. For the eight PCM input samples

$$\mathbf{X}^{\mathrm{T}} = [x_0\ x_1\ x_2\ x_3\ x_4\ x_5\ x_6\ x_7] = [202\ 231\ 190\ 174\ 137\ 110\ 115\ 130]$$

the transformer output (or sequency spectrum) is easily shown to be

$$\mathbf{a}^{\mathrm{T}} = [161.125\ 38.125\ 8.375\ 8.875\ -0.375\ -10.875\ -3.125\ -0.125]$$

It is apparent that the elements of the transformer output are less correlated than the elements of the transformer input, which is one objective in TC. Also note that the transformation has concentrated most of the signal energy into a relatively few coefficients. This 'energy compaction' is another objective of TC. Perfect decorrelation could be contrived by careful selection of the PCM input samples. For example, suppose a sine wave is sampled at exactly four times its frequency and in a particular phase to give the input vector

$$\mathbf{X}^{\mathrm{T}} = [E+e\ \ E+e\ \ E-e\ \ E-e\ \ E+e\ \ E+e\ \ E-e\ \ E-e]$$

where E and e are arbitrary. The reader should show that the DWHT transforms this to $\mathbf{a}^{\mathrm{T}} = [E00e0000]$. Clearly, the transformer output samples are now completely uncorrelated since E can be varied independently of e.

Now consider the inverse transformation process. For the moment we will ignore the quantisation in Figure 3.10 so that the input to the *inverse transform matrix* **B** is **G**. Letting $\mathbf{B} = \mathbf{A}^{-1}$ gives

$$\mathbf{Y} = \mathbf{B}\mathbf{G} = \mathbf{A}^{-1}\mathbf{G} = \mathbf{A}^{-1}\mathbf{A}\mathbf{X} = \mathbf{X} \tag{3.30}$$

which means that in the absence of quantisation the TC process is completely reversible. In practice the inverse $\mathbf{A}^{-1}$ is not a problem since practical TC systems invariably use *unitary* transforms, that is

$$\mathbf{A}^{-1} = \mathbf{A}^{*\mathrm{T}} \tag{3.31}$$

where * denotes the complex conjugate. Practical sinusoidal transforms, such as the cosine, sine and Fourier transforms, and non-sinusoidal ('sequency' domain) transforms such as the Walsh–Hadamard, Haar and Slant transforms, are all unitary. Therefore, ignoring quantisation and assuming a unitary transform we can write

$$\mathbf{X} = \mathbf{A}^{*\mathrm{T}}\mathbf{G} \tag{3.32}$$

which means that the inverse transformation is as easy to implement as the forward transform.

Example 3.8
One of the most useful transforms is the *discrete cosine transform*, or DCT. The 1D *N*th-order forward DCT could be written

$$\mathbf{C} = \mathbf{W}\mathbf{X} = \begin{bmatrix} w_{00} & w_{01} & \cdots & w_{0N-1} \\ w_{10} & w_{11} & \cdots & w_{1N-1} \\ & & \vdots & \\ w_{N-10} & w_{N-11} & \cdots & w_{N-1N-1} \end{bmatrix} \mathbf{X} \tag{3.33}$$

where w_{kn} is a real cosine function (see (3.51)). Since $\mathbf{W}$ has real coefficients, the inverse DCT (IDCT) is given by $\mathbf{X} = \mathbf{W}^{\mathrm{T}}\mathbf{C}$, or

$$\begin{bmatrix} x_0 \\ x_1 \\ \vdots \\ x_{N-1} \end{bmatrix} = \begin{bmatrix} w_{00} & w_{10} & \cdots & w_{N-10} \\ w_{01} & w_{11} & \cdots & w_{N-11} \\ & & \vdots & \\ w_{0\,N-1} & w_{1\,N-1} & \cdots & w_{N-1N-1} \end{bmatrix} \begin{bmatrix} C_0 \\ C_1 \\ \vdots \\ C_{N-1} \end{bmatrix} \tag{3.34}$$

Clearly, the IDCT can also be written

$$x_n = \sum_{k=0}^{N-1} w_{kn} C_k; \quad n = 0, 1, \ldots, N-1 \tag{3.35}$$

3.2.2 Types of discrete transform

Before discussing data compression using TC, it is helpful to expand a little on the mathematics of a discrete transform. Let the general element of the forward transformation matrix in (3.29) be $g(k,n)$. It then follows from (3.29) that the forward transform can also be written

$$G_k = \sum_{n=0}^{N-1} g(k,n) x_n; \quad k = 0, 1, \ldots, N-1 \tag{3.36}$$

Equation (3.36) denotes a general *N*th-order linear transform and, as already stated, a particular discrete transform is generated by assigning a function to $g(k,n)$. More precisely, if we make the set $\{g(k,n)\}$ a set of *mutually orthogonal functions*, that is, the inner product of any two functions is zero, then these functions can form an orthogonal *basis* for a Fourier-type expansion. Under these conditions, (3.36) can be considered to be an expression for the coefficients of some generalised Fourier series. In fact, the corresponding inverse transform will be the generalised discrete Fourier series, this being the expansion of a function as a linear combination of mutually orthogonal functions. This 'Fourier series' concept applies to many types of discrete transform, including the familiar *discrete Fourier transform* (DFT) and the *discrete wavelet transform* (DWT) (Section 3.2.8).

The discrete Fourier transform (DFT)

The most common discrete transform uses an exponential set of orthogonal functions defined by

$$g(k,n) = \mathrm{e}^{-\mathrm{j}2\pi kn/N} \tag{3.37}$$

Substituting into (3.36) gives the DFT

$$F_k = \sum_{n=0}^{\infty} x_n \mathrm{e}^{-\mathrm{j}2\pi kn/N}; \quad k = 0, 1, \ldots, N-1 \tag{3.38}$$

and the corresponding inverse DFT is the Fourier series

$$x_n = \frac{1}{N} \sum_{k=0}^{N-1} F_k \mathrm{e}^{\mathrm{j}2\pi kn/N}; \quad n = 0, 1, \ldots, N-1 \tag{3.39}$$

The discrete cosine transform (DCT)

The DCT is particularly useful for data compression and can be defined by letting

$$g(k,n) = \frac{2}{N} a_k \cos\left[\frac{(2n+1)k\pi}{2N}\right] \tag{3.40}$$

where

$$a_k = \begin{cases} 1/\sqrt{2} & k = 0 \\ 1 & k = 1,2,\ldots,N-1 \end{cases} \tag{3.41}$$

The 1D DCT can then be written

$$C_k = \frac{2}{N} a_k \sum_{n=0}^{N-1} x_n \cos\left[\frac{(2n+1)k\pi}{2N}\right]; k = 0, 1, \ldots, N-1 \tag{3.42}$$

This means that the sequence x_n is represented in the spectral domain by a set of cosine functions with special phase and amplitude characteristics. The IDCT is simply

$$x_n = \sum_{k=0}^{N-1} a_k C_k \cos\left[\frac{(2n+1)k\pi}{2N}\right]; \quad n = 0, 1, \ldots, N-1 \tag{3.43}$$

It is worth noting that (3.42) is one of four DCT expressions that have appeared in the literature (Wang, 1984). Also, the normalisation factor 2/*N* could be split as $\sqrt{2/N}$ between the DCT and the IDCT.

3.2.3 DCT algorithms

Many fast algorithms for a 1D DCT have been developed using, for example, recursive methods (Hou, 1987), or indirect, FFT-based methods (Narasimha and Peterson, 1978), or direct, matrix-based methods (Suehiro and Hatori, 1986; Brebner and Ritchings, 1988).

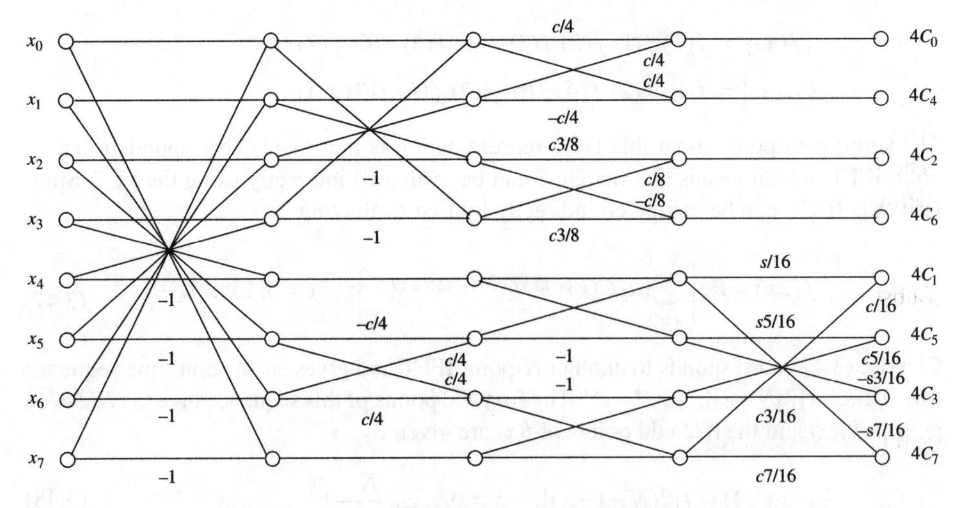

Figure 3.11 *A fast 8-point DCT algorithm (Chen et al., 1977).*

Eight-point DCT algorithms are of particular interest since image and video compression algorithms usually perform 8 × 8 DCTs. A popular 8-point fast DCT (FDCT) algorithm is shown in Figure 3.11 (Chen *et al.*, 1977). In this diagram a circle denotes addition, *c*/4 denotes $\cos(\pi / 4)$, and *s*5/16 denotes $\sin(5\pi / 16)$. The flowgraph output gives $4C_k$ where C_k is given by (3.42), and the IDCT is obtained by simply working from right to left. This particular algorithm requires 26 additions and 16 multiplications, compared with 56 additions and 64 multiplications for direct 8-point evaluation of (3.42). A somewhat more efficient 8-point FDCT algorithm requiring 29 additions and 13 multiplications is described by Wang (1984).

Another efficient DCT algorithm is described by Narasimha and Peterson (1978), and we will use this to describe a 2D FDCT, as required for image or video processing. With images in mind, we will change the variables in (3.42) and for clarity ignore the normalisation factor $2/N$. For a signal $f(x)$ the 1D DCT can then be written

$$C_k = a_k \sum_{x=0}^{N-1} f(x) \cos\left[\frac{(2x+1)k\pi}{2N}\right] \equiv a_k \beta_k \tag{3.44}$$

where

$$\beta_k = \mathrm{Re}\left[\mathrm{e}^{\mathrm{j}\pi k/2N} \sum_{x=0}^{N-1} f_{\mathrm{r}}(x) \mathrm{e}^{\mathrm{j}2\pi kx/N} \right] \tag{3.45}$$

The term $f_{\mathrm{r}}(x)$ is simply a reordered form of $f(x)$ according to the rule

$$\left.\begin{aligned} f_{\mathrm{r}}(x) &= f(2x) \\ f_{\mathrm{r}}(N-1-x) &= f(2x+1) \end{aligned}\right\} \quad x = 0, 1, \ldots, \frac{N}{2} - 1 \tag{3.46}$$

Thus, for $N = 8$:

$$\{f(x)\} = f(0)\, f(1)\, f(2)\, f(3)\, f(4)\, f(5)\, f(6)\, f(7)$$
$$\{f_r(x)\} = f(0)\, f(2)\, f(4)\, f(6)\, f(7)\, f(5)\, f(3)\, f(1)$$

The significant point about this DCT representation is that (3.45) corresponds to an N-point IFFT, which means that the DCT can be evaluated indirectly using the FFT. Similarly, the IDCT can be computed indirectly by first evaluating

$$f(2x) = \mathrm{Re}\left[\sum_{k=0}^{N-1}[a_k C(k)\mathrm{e}^{\mathrm{j}\pi k/2N}]\mathrm{e}^{\mathrm{j}2\pi kx/N}\right]; \quad x = 0, 1, \ldots, N-1 \tag{3.47}$$

Clearly, (3.47) corresponds to another N-point IFFT and gives an N-point time sequence $f(0), f(2), \ldots, f[2(N-2)], f[2(N-1)]$. The first $N/2$ points of this sequence give the $N/2$ even points of $f(x)$ and the $N/2$ odd points of $f(x)$ are given by

$$f(2x+1) = f[2(N-1-x)]; \quad x = 0, 1, \ldots, \frac{N}{2}-1 \tag{3.48}$$

A 2D FDCT

The 2D DCT of an $N \times N$ image $f(x,y)$ can be expressed as

$$C(u,v) = c(u)c(v)\sum_{x=0}^{N-1}\sum_{y=0}^{N-1} f(x,y)\cos\left[\frac{(2x+1)u\pi}{2N}\right]\cos\left[\frac{(2y+1)v\pi}{2N}\right] \tag{3.49}$$

Here, $c(0) = 1/\sqrt{2},\ c(j) = 1,\ j \neq 0$, and for clarity a normalisation factor $4/N^2$ has been associated with the IDCT. The significant point is that the kernel in (3.49) is separable, *which means that the 2D DCT can be performed in two steps using two 1D DCTs*. A 1D DCT could be applied to the rows of an image, and this would be followed by a 1D operating on columns. Figure 3.12 illustrates the *row–column* approach for a fast DCT based on the IFFT method in (3.44) and (3.45). Here, a row of data in image $f(x,y)$ is transformed to a row of data in $C(x,v)$, and N such row transforms are required to complete $C(x,v)$. The value of v is then fixed in $C(x,v)$ and the IFFT is evaluated for all values of u (a column transform). N such column transforms are required to complete $C(u,v)$. A C routine for an 8 × 8 fast DCT based upon the row–column approach is given in Appendix E.

Pipelined 2D DCT

An alternative approach to 2D DCT computation is simply to evaluate the problem as two matrix multiplications. This is equivalent to a row–column algorithm but is well suited to hardware realisation. Let (3.42) be expressed in the form

$$C_k = \sum_{n=0}^{N-1} w_{kn} x_n; \quad k = 0, 1, \ldots, N-1 \tag{3.50}$$

$$w_{kn} = \frac{2}{N} a_k \cos\left[\frac{(2n+1)k\pi}{2N}\right] \tag{3.51}$$

Expressing (3.50) in matrix form

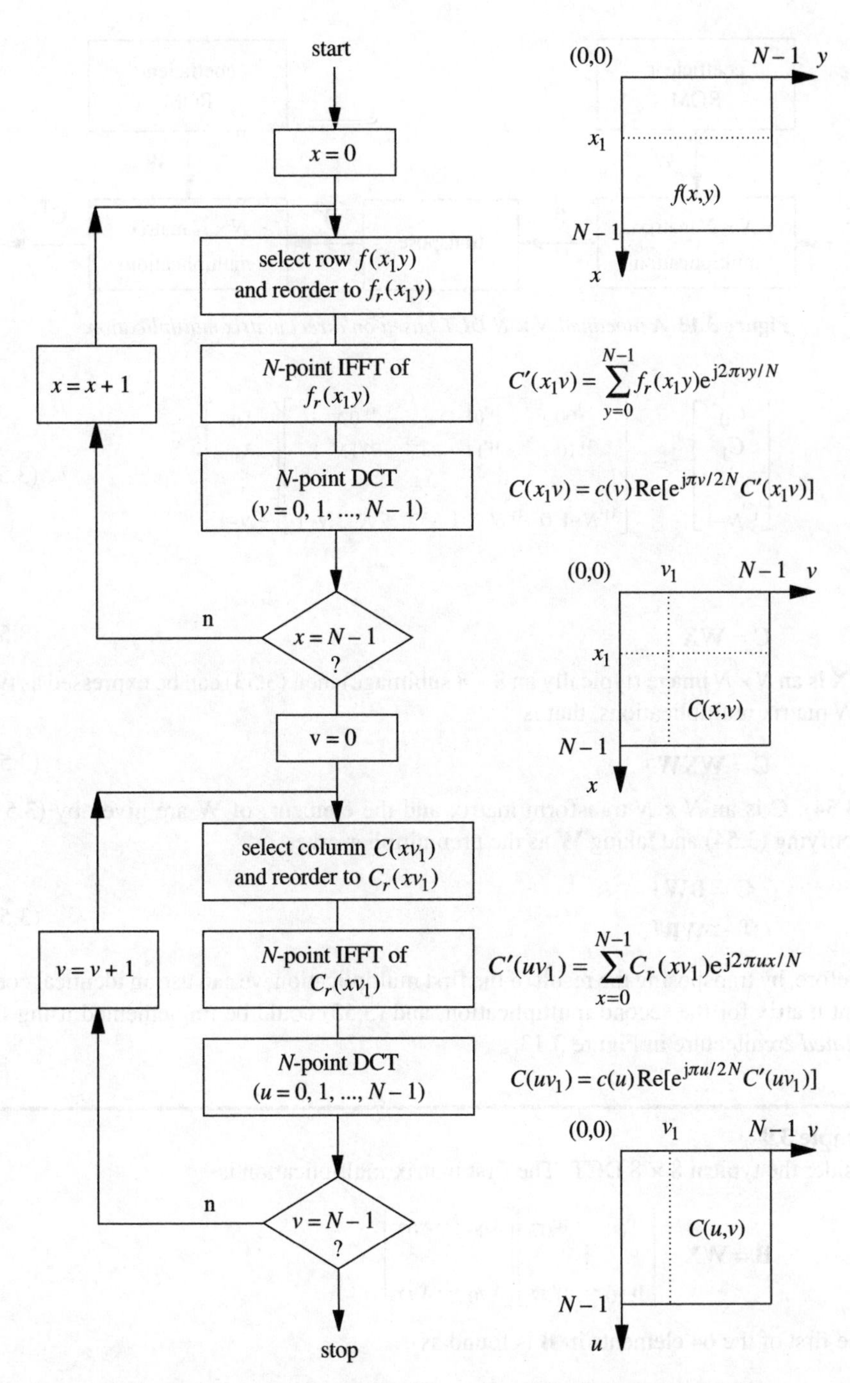

Figure 3.12 *A fast 2D row–column DCT based on the N-point IFFT.*

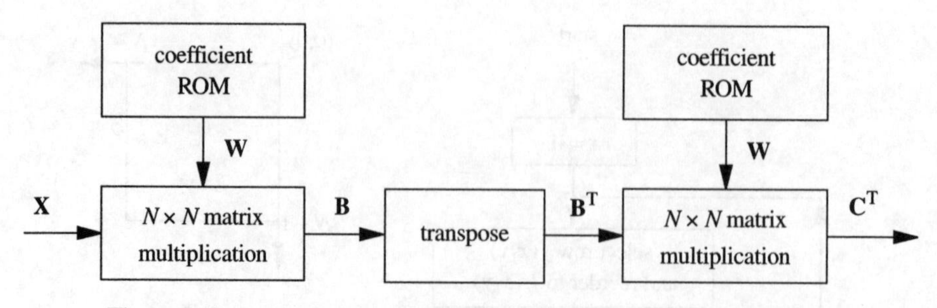

Figure 3.13 *A pipelined N × N DCT based on direct matrix multiplication.*

$$\begin{bmatrix} C_0 \\ C_1 \\ \vdots \\ C_{N-1} \end{bmatrix} = \begin{bmatrix} w_{00} & w_{01} & \cdots & w_{0N-1} \\ w_{10} & w_{11} & \cdots & w_{1N-1} \\ & & \vdots & \\ w_{N-1\,0} & w_{N-1\,1} & \cdots & w_{N-1N-1} \end{bmatrix} \begin{bmatrix} x_0 \\ x_1 \\ \vdots \\ x_{N-1} \end{bmatrix} \tag{3.52}$$

or

$$\mathbf{C} = \mathbf{WX} \tag{3.53}$$

If $\mathbf{X}$ is an $N \times N$ image (typically an 8×8 subimage) then (3.53) can be expressed as two $N \times N$ matrix multiplications, that is

$$\mathbf{C} = \mathbf{WXW}^{\mathrm{T}} \tag{3.54}$$

In (3.54), $\mathbf{C}$ is an $N \times N$ transform matrix and the elements of $\mathbf{W}$ are given by (3.51). Simplifying (3.54) and taking $\mathbf{W}$ as the premultiplier gives

$$\begin{aligned} \mathbf{C} &= \mathbf{BW}^{\mathrm{T}} \\ \mathbf{C}^{\mathrm{T}} &= \mathbf{WB}^{\mathrm{T}} \end{aligned} \tag{3.55}$$

Therefore, by transposing the result of the first multiplication we can use an identical coefficient matrix for the second multiplication, and (3.55) could be implemented using the *pipelined* architecture in Figure 3.13.

Example 3.9

Consider the typical 8×8 DCT. The first matrix multiplication is

$$\mathbf{B} = \mathbf{WX} = \begin{bmatrix} w_{00} & \cdots & w_{07} \\ & \vdots & \\ w_{70} & \cdots & w_{77} \end{bmatrix} \begin{bmatrix} x_{00} & \cdots & x_{07} \\ & \vdots & \\ x_{70} & \cdots & x_{77} \end{bmatrix}$$

so the first of the 64 elements in $\mathbf{B}$ is found as

$$w_{00}x_{00} + w_{01}x_{10} + \cdots + w_{07}x_{70}$$

This could be computed using a single multiplier-accumulator (MAC) and eight machine (MAC) cycles. If we used eight MACs to evaluate the eight row elements of $\mathbf{B}$ in parallel, then $\mathbf{B}$ could be computed in just 64 cycles. The pipelined architecture in Figure 3.13 could therefore deliver an 8 × 8 DCT every 64 cycles or 3.2 μs, assuming a typical MAC cycle time of 50 ns.

3.2.4 Energy compaction

Apart from computational efficiency, the FDCT algorithm in Figure 3.11 highlights another practical point. In general, each addition requires an extra bit to represent the sum and so the wordlength will increase through the transformer. To take a practical example, given an 8 × 8 block of 8-bit pixels, the DCT in an MPEG-2 encoder will generate an 8 × 8 block of 11-bit coefficients (12-bits including the sign bit). In order to achieve compression, it is therefore necessary to follow the transformer with a coefficient quantisation (or 'bit allocation') process, as indicated in Figure 3.10. Quantisation of the 12-bit coefficients in MPEG-2 is discussed in Section 3.3.3. If transformation does not in itself generate compression we might ask 'what does it achieve'? *It turns out that transformation provides great potential for compression in the sense that most of the signal energy is usually compacted into relatively few coefficients.* This is particularly true for highly correlated inputs, as illustrated in Table 3.3. Taking the extreme case, if all the input samples have the same value (perfect correlation), then the transform will compact all the signal energy into the zero frequency coefficient. It is interesting to note that such energy compaction would *not* occur if the *discrete sine transform* (DST) were to be used. Since

Table 3.3 *DCT coefficients for two input vectors (N = 16).*

$\mathbf{X}^T$ = [9139350126912048]		$\mathbf{X}^T$ = [5467677878978989]	
k	C_k	k	C_k
0	63.0	0	115.0
1	5.3	1	–18.4
2	11.5	2	–5.1
3	4.5	3	–4.0
4	3.3	4	–0.8
5	–25.4	5	–3.0
6	26.5	6	–0.5
7	9.6	7	0.3
8	7.0	8	–1.0
9	13.3	9	4.5
10	17.6	10	5.5
11	3.8	11	–2.4
12	10.5	12	5.8
13	4.0	13	1.1
14	–18.0	14	–1.9
15	–2.4	15	–1.4

the DCT tends to concentrate the energy onto a relatively few low-frequency coefficients, bit rate reduction can be achieved by quantising and coding these large coefficients, and by quantising the small coefficients to 0 (and thereby not transmitting them).

To summarise, a good transform concentrates most of the energy from commonly occurring signals into as few coefficients as possible. The theoretically optimal transform is the *Hotelling* or *Karhunen–Loeve transform* (KLT) in the sense that it yields uncorrelated coefficients and packs the maximum average energy into the first M coefficients ($M \leq N$). In practice, the complexity and data dependent properties of the KLT mean that it is invariably replaced by a good sub-optimal transform such as the DCT. In fact, the performance of the DCT becomes equivalent to the KLT as the input vector size $N \Rightarrow \infty$.

3.2.5 TC of images and coefficient quantisation

So far we have established that a good transform both decorrelates the signal and compacts the signal energy into a relatively few coefficients. It remains to quantise the coefficients in order to achieve compression and this is usually done in the context of image (or frame) compression, Figure 3.14.

In practice an $L \times L$ image is usually partitioned into small $N \times N$ blocks and each block is transformed independently. Block coding is necessary in order to make the TC of images computationally efficient in terms of storage and speed, that is, the DSP cost tends to decrease rapidly with decreasing block size, as indicated in Figure 3.15. On the other hand, independently transform coding each block neglects the redundancies (pixel correlations) that exist between blocks, and on this basis the block size should be large. In fact, decreasing the block size leads to increased mean-square error (MSE) between the original image and the reconstructed image (Figure 3.15), so, in practice, the block size is a compromise (usually 8 × 8).

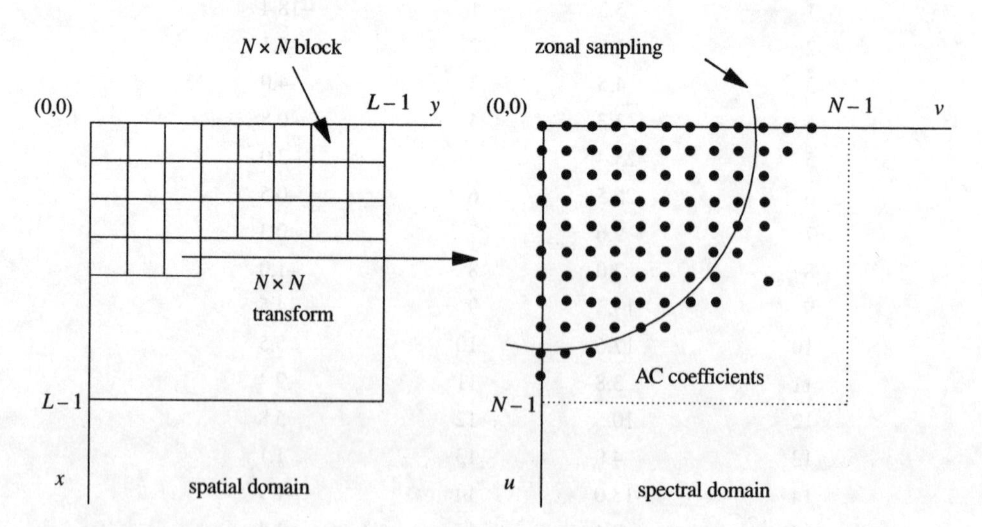

Figure 3.14 *TC: block coding for image compression (u and v are spatial frequencies).*

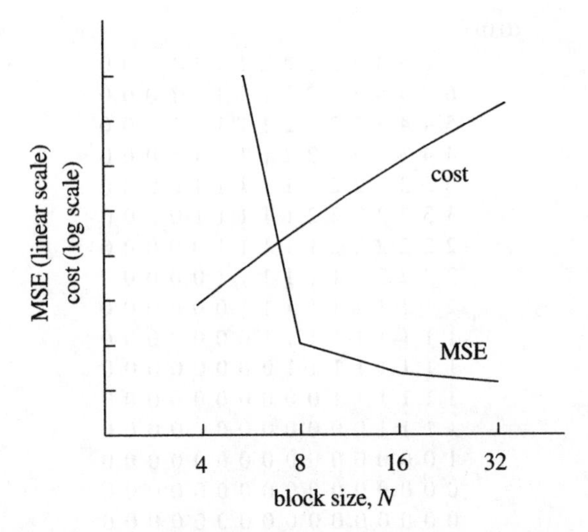

Figure 3.15 *TC: relative mean-square error and DSP cost vs. block size.*

Generally speaking, compression can be achieved by a combination of coefficient selection, quantisation and efficient coding (run-length coding/variable-length coding). Coefficient selection can be deliberate, as in *zonal sampling* and *threshold sampling*. Zonal sampling is a simple technique based on the assumption that only coefficients within a certain fixed geometric zone (namely those concentrated around the DC coefficient) need be quantised and transmitted (Figure 3.14). The zone could be selected by evaluating the coefficient variances on a set of 'average' pictures, and then discarding all coefficients whose variances fall below a threshold. At the decoder the discarded coefficients could be set to zero. Clearly, large errors can occur if the image statistics vary significantly. Better results are obtained when the retained coefficients are not fixed *a priori*, as in zonal sampling, but are adaptively selected to match the signal statistics. Threshold sampling is one such approach, whereby a coefficient has to exceed an adaptively determined threshold before being transmitted. The threshold could be based on some measure of block activity. In this way the number and location of the coefficients are adapted to the local block structure, but at the expense of having to transmit the indices of the selected coefficients to the decoder. Having selected the coefficients to transmit, one of a number of quantisation/bit allocation strategies could then be applied. Figure 3.16 illustrates typical bit allocation for a 16×16 coefficient block, and it is apparent that many coefficients will not be transmitted since they are allocated zero bits.

The usual compression approach is not to deliberately select coefficients prior to quantisation. Rather, all coefficients are quantised according to *psychovisual thresholds* (the net result still being that many of the higher frequency coefficients are quantised to zero). This is the approach adopted in JPEG and MPEG coding. Essentially, quantisation is achieved by dividing the ith DCT coefficient, a_i, by a *quantiser step size* $\hat{Q}_i$ derived from the ith element Q_i in a *quantising matrix* $\mathbf{Q}$. To obtain psychovisual thresholds, *the degree of quantisation applied to* a_i *is weighted according to the visibility of the resulting quantisation noise to a human observer.* In matrix terms, this means that the higher frequency elements in $\mathbf{Q}$ have larger integer values than the low-frequency terms, which in

```
(0,0)                                      v
        7 6 5 4 3 3 2 2 2 1 1 1 1 1 0 0
        6 5 4 4 3 3 2 2 1 1 1 1 1 0 0 0
        5 4 4 3 3 2 2 2 1 1 1 1 1 0 0 0
        4 4 3 3 3 2 2 2 1 1 1 1 1 0 0 0
        3 3 3 3 2 2 2 1 1 1 1 1 0 0 0 0
        3 3 2 2 2 2 2 1 1 1 1 1 0 0 0 0
        2 2 2 2 2 2 1 1 1 1 1 0 0 0 0 0
        2 2 2 2 1 1 1 1 1 1 1 0 0 0 0 0
        2 1 1 1 1 1 1 1 1 1 0 0 0 0 0 0
        1 1 1 1 1 1 1 1 1 0 0 0 0 0 0 0
        1 1 1 1 1 1 1 1 0 0 0 0 0 0 0 0
        1 1 1 1 1 1 0 0 0 0 0 0 0 0 0 0
        1 1 1 1 0 0 0 0 0 0 0 0 0 0 0 0
        1 0 0 0 0 0 0 0 0 0 0 0 0 0 0 0
        0 0 0 0 0 0 0 0 0 0 0 0 0 0 0 0
        0 0 0 0 0 0 0 0 0 0 0 0 0 0 0 0
  u
```

Figure 3.16 *Typical bit allocation after applying the DCT to a 16 × 16 image block. The average compression rate corresponds to approximately 1 bit/pixel (Jain, 1981, © IEEE).*

turn means that there is a high probability of high-frequency coefficients being quantised to zero. The selection of **Q** depends upon the picture source, display characteristics and the viewing distance. However, if these are reasonably well defined then a standard quantising matrix can be given for a particular coding technique (see Example 3.13).

Example 3.10

Figure 3.17 shows a simple DCT compression scheme for static colour images in component format. The approach is simple in the sense that it does not have the refinements of JPEG coding and it does not use VLC (Huffman coding). The omission of VLC removes the need for detailed image statistics and coding tables. However, a modest degree of extra compression is still obtained by zigzag scanning the DCT coefficient block and encoding the resulting long sequences of zeros using RLC (Section 2.1). The quantising matrix is shown in Table 3.4, and with appropriate scaling the matrix is similar to that used for MPEG-2 coding. For this example the AC coefficient a_i is quantised by dividing it by the

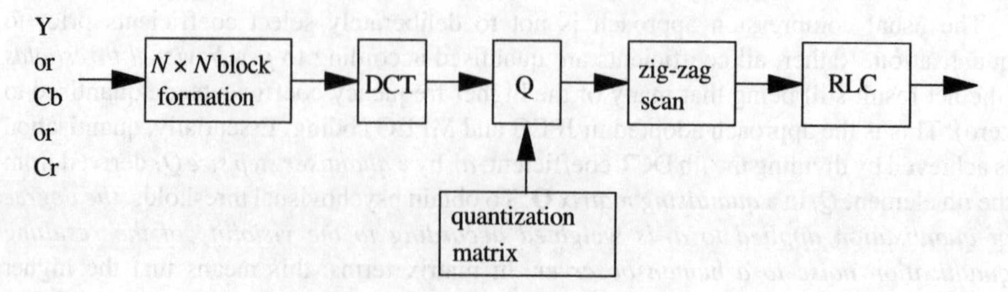

Figure 3.17 *Simple DCT-based image compression scheme.*

Table 3.4 *Quantising matrix Q used for Figure 3.17.*

1	8	9	11	12	13	14	15
8	9	11	12	13	14	15	16
9	11	12	13	14	15	16	17
11	12	13	14	15	16	17	20
12	13	14	15	16	17	20	24
13	14	15	16	17	20	24	28
14	15	16	17	20	24	28	35
15	16	17	20	24	28	35	42

product of Q_i and a user-defined scaling factor. The DC coefficient is quantised by dividing by a user-defined DC scaling factor. Figure 3.18 illustrates the typical performance of the system in Figure 3.17 for 16 × 16 blocks: coding artefacts are visible at 29:1 compression, and the block structure is apparent at 62:1.

Figure 3.18 *Performance of the system in Figure 3.17: (a) original; (b) 29:1 compression; (c) 62:1 compression.*

3.2.6 JPEG image coding

The JPEG international standard ISO 10918 defines image compression techniques for continuous-tone (multilevel) still images, both greyscale and colour. It is a general-purpose tool with multiple modes of operation, and it is aimed at applications as diverse as photovideotex, desktop publishing, graphic arts, colour facsimile, newspaper wire-photo transmission and medical systems. JPEG is designed for 'natural real-world scenes', but it is not particularly suitable for simple 'cartoon' type images (GIF is better here). Images are typically 720 × 576 pixels, corresponding to the CCIR 601 digital studio TV standard. A JPEG objective was to obtain excellent picture quality at around 1 bit/pixel: at this compression rate a 720 × 576 pixel image would take 6.5 seconds to transmit on a 64 kbit/s channel. Good image quality is obtained at 0.5–0.75 bit/pixel. Another good reason for using JPEG is that it permits images to be compressed so that, for example, 24 bits/pixel colour data (16 million colours) can be stored instead of 8 bits/pixel data (256 colours). JPEG is therefore particularly attractive for use on the Internet, since it allows for widely varying display hardware. On this basis, compression down to 1 bit/pixel corresponds to a compression ratio of 24. Note that the JPEG algorithm would be applied equally to each component of a colour image.

The main features of JPEG are:

- The main compression mechanism is lossy, since it is based upon the quantisation of DCT coefficients.
- It has a core mode known as the *Baseline System*. This system provides a default mode and is intended to limit codec costs. The Baseline System limits operation to images with 8-bits/pixel/component, and places limits on the VLC used. It is a restricted version of the *sequential build-up* mode.
- In the sequential build-up mode the image is encoded in a single left-right, top-bottom scan; that is, the entire image is transmitted in a single scan.
- It has a *progressive build-up* mode whereby a first crude image is quickly available and is then refined by additional image scans. Each additional scan combines with data from previous scans to give a progressively better image. This mode was deemed necessary, since a 6.5 second delay for image transmission at 64 kbit/s was considered excessive for frequent use. Progressive JPEG is particularly useful for example when browsing the Internet over a slow modem link.
- It has a *hierarchical coding* mode whereby the image is encoded at multiple spatial resolutions for adaption to network speed and to terminal resolution ability.
- It has a *lossless* compression mode giving a compression of typically 2:1. This mode is independent of any DCT processing and uses standalone DPCM-based lossless compression. The prediction difference is encoded losslessly using either Huffman or Arithmetic coding. A later lossless compression standard is called *JPEG-LS*.

Baseline JPEG system

The Baseline System is shown in Figure 3.19, where $f(x,y)$ corresponds to a particular pixel block from a particular colour component. First note that, since we are encoding *images* rather than video, there is no 'temporal' redundancy to remove. In other words, the spatial redundancy within an image can be largely removed by transform coding and there is no

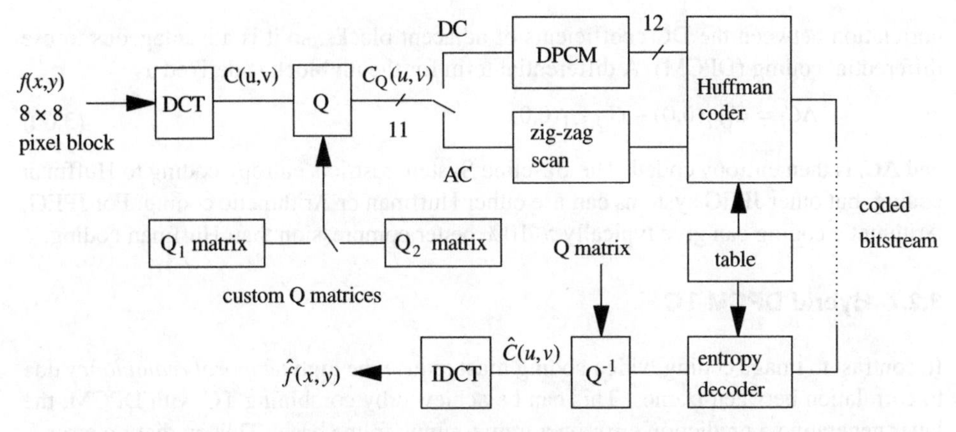

Figure 3.19 *The Baseline JPEG System.*

need for a 'DPCM loop'. The reader should compare Figure 3.19 with the corresponding diagram for an MPEG codec (Figure 3.29). The JPEG system is therefore simply based on the DCT of 8 × 8 pixel blocks, and JPEG has chosen a 'symmetrical' definition for the DCT and the IDCT:

$$C(u,v) = \frac{1}{4}c(u)c(v)\sum_{x=0}^{7}\sum_{y=0}^{7} f(x,y)\cos\left[\frac{(2x+1)u\pi}{16}\right]\cos\left[\frac{(2y+1)v\pi}{16}\right] \tag{3.56}$$

$$f(x,y) = \frac{1}{4}\sum_{u=0}^{7}\sum_{v=0}^{7} c(u)c(v)C(u,v)\cos\left[\frac{(2x+1)u\pi}{16}\right]\cos\left[\frac{(2y+1)v\pi}{16}\right] \tag{3.57}$$

Here, $c(0) = 1/\sqrt{2}$, $c(j) = 1$, $j \neq 0$. These equations define the idealised, infinite-precision DCT, and each DCT coefficient is then quantised using an 8 × 8 quantisation matrix, **Q**. Ideally each matrix element (here denoted $Q(u,v)$) should be selected as the perceptual threshold for the visual contribution of the corresponding cosine basis function. As discussed, this will depend upon the picture source, display characteristics, and the viewing distance. For JPEG, the quantisation process can then be expressed as

$$C_Q(u,v) = \left\lfloor \frac{C(u,v)}{Q(u,v)} + 0.5 \right\rfloor \tag{3.58}$$

where $\lfloor\ \rfloor$ performs integer rounding and $Q(u,v)$ is here regarded as the step size. The inverse quantisation process at the decoder gives

$$\hat{C}(u,v) = C_Q(u,v)Q(u,v) \tag{3.59}$$

The degree of compression can be controlled by multiplying elements $Q(u,v)$ by a *quality factor*, g. Setting $g = 0.5$ gives low compression and low distortion, while increasing g increases visible distortion. Since the perceptual contribution of each colour component varies, the encoder could use multiple quantisation matrices for different colour components, as indicated in Figure 3.19. Note that the quantised DC coefficient $C_Q(0,0)$ is treated differently from the quantised AC coefficients. This is because there is usually a strong correlation between the DC coefficients of adjacent blocks, so it is advantageous to use differential coding (DPCM). A differential term for the *i*th block is derived as

$$\Delta C_i = C_{Q,i}(0,0) - C_{Q,i-1}(0,0) \tag{3.60}$$

and ΔC_i is then entropy coded. The Baseline System restricts entropy coding to Huffman coding, but other JPEG systems can use either Huffman or Arithmetic coding. For JPEG, Arithmetic coding can give typically 5–10% better compression than Huffman coding.

3.2.7 Hybrid DPCM-TC

In contrast to image coding, video coding must reduce the high *temporal redundancy* due to correlation between frames. This can be achieved by combining TC with DPCM, the latter generating a prediction error on a frame-minus-frame basis. This prediction error is then applied to the transform coder. Figure 3.20 illustrates such a hybrid DPCM-TC scheme as used for very low rate video. CCITT Rec. H.261 (1990) supports video conferencing and video telephony in integer multiples of 64 kbit/s up to 1920 kbit/s. It is therefore particularly suitable for the ISDN rates of 64 kbit/s and 128 kbit/s.

Referring to Figure 3.20, the video input format can be arbitrary and so must first be converted to component signals *Y, Cr, Cb* in CIF or QCIF format. This enables codecs to interwork between regions of the world with 625-line and 525-line standards. In the *intraframe* mode, input pixels in 8 × 8 blocks are passed to an 8 × 8 DCT, while in the *interframe* mode the DPCM loop is invoked and the prediction error is passed to the DCT,

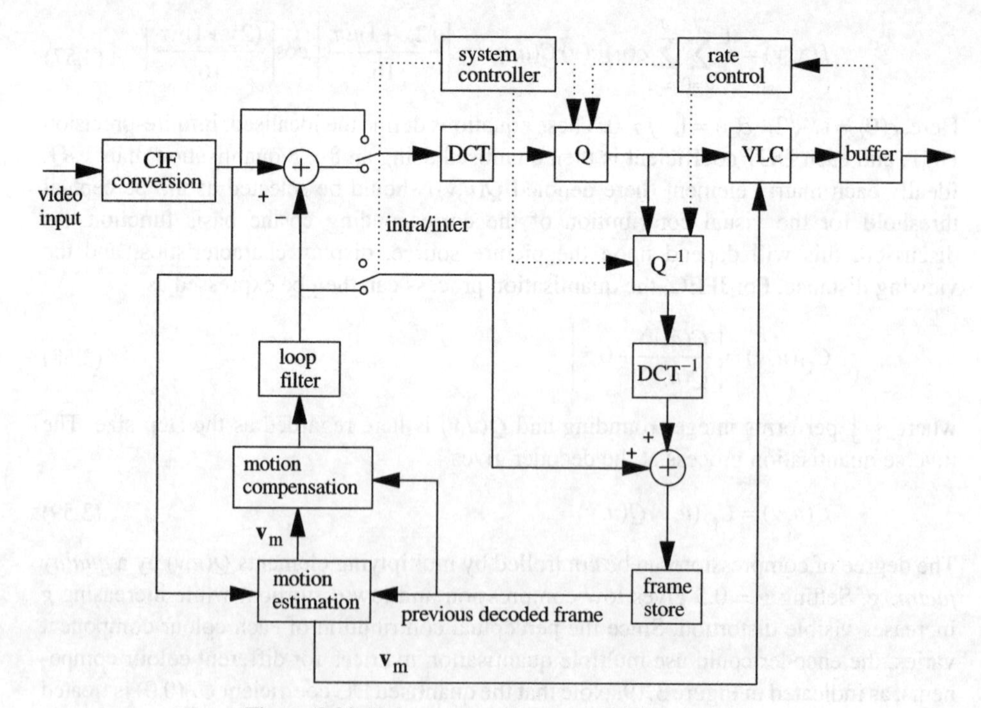

Figure 3.20 *Video processing in the H.261 encoder.*

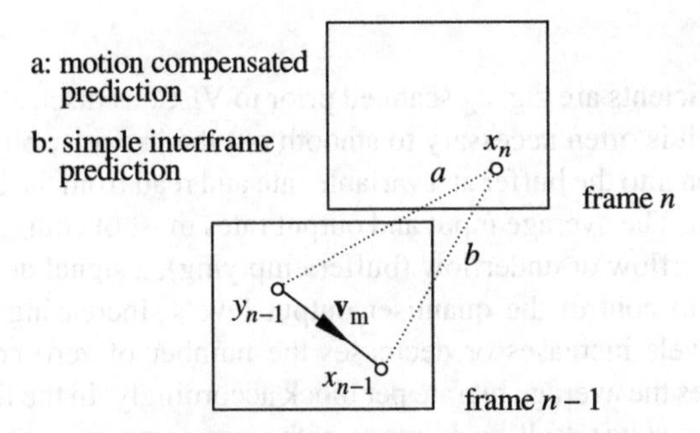

Figure 3.21 *Types of interframe prediction.*

also in 8 × 8 blocks. This brings us to the concept of *motion-compensated* interframe prediction and the use of *motion vectors* (Figure 3.21). For still pictures the simple interframe prediction $\hat{x}_n = \tilde{x}_{n-1}$ performs well, where $\tilde{x}_{n-1}$ is a quantised version of x_{n-1}. However, it is a very poor prediction on moving pictures, and in this case it is necessary to introduce motion compensation. Suppose that a point y_{n-1} on a moving object in frame $n-1$ appears in position x_n on frame n. In this case a better prediction is $\hat{x}_n = \tilde{y}_{n-1}$, and it can be achieved by estimating the motion vector $\mathbf{v}_m$ and effectively using it to control a variable prediction delay of approximately one frame.

This type of encoder has to make two major decisions: namely, whether to use intraframe or interframe compression, and when using the latter, whether to use motion compensation; that is, whether to suppress or not suppress the motion vector $\mathbf{v}_m$. Both decisions use computations made by the motion estimator in Figure 3.20. In Rec.H.261, motion vector estimation is made by using 16 × 16 luminance blocks and by searching in the previous frame. Consider a luminance block k in the current frame n having pixel intensity $b(k,n)$. We could compute the difference between this and the 'zero-displaced' block k in the previous frame as

$$d(k,n) = b(k,n) - b(k,n-1) \tag{3.61}$$

where the actual computation finds some mean absolute error (MAE). Assuming lateral translation, we could also search for the 'best match' block k' in the previous frame by invoking a vector shift – typically the 16 × 16 pixel luminance block is shifted ±7 pixels in both the horizontal and vertical directions. The new MAE value would then correspond to the 'displaced' block difference

$$d'(k,n) = b(k,n) - b(k',n-1) \tag{3.62}$$

and both MAE values can then be used to make the motion compensation/no motion compensation decision. If motion compensation is selected, the corresponding vector is used for prediction, as indicated in Figure 3.21. One of the MAE values together with a measure of the variance of the current block can also be used to make the interframe/intraframe decision. The lowpass filter in the DPCM loop performs a 3 × 3 convolution on an 8 × 8 pixel block in order to reduce artefacts introduced by motion compensation.

Output buffer

The quantised coefficients are zigzag scanned prior to VLC, as discussed in Example 2.2. When VLC is used it is often necessary to smooth out the irregular bit rate occurring per block. Data is written into the buffer at a variable rate and read from the buffer at a constant rate for transmission. The average input and output rates must of course be equal. In order to prevent buffer overflow or underflow (buffer emptying), a signal describing the buffer occupancy is used to control the quantiser output levels. Increasing or decreasing the spacing between levels increases or decreases the number of zero coefficients respectively, which changes the average bit rate per block accordingly. In the limit, for very difficult pictures, there must be a 'fall-back' mode with a coarse quantiser which is guaranteed not to cause buffer overflow. Similarly, there must be a strategy for avoiding buffer underflow for plain picture material. A similar buffer is required at the decoder; in this case the data is written into the store at a constant rate from the transmission channel and read out from the store at a variable rate by the VLC decoder.

The H.263 video standard.

The H.263 standard is an improved version of H.261 for very low bit rate video coding applications such as videophone and surveillance. It is optimised for relatively low motion, and produces substantially better quality than H.261, especially at low bit rates. The main changes are:

- Half-pixel precision in motion estimation (giving better matching).
- Advanced motion estimation (in 8 × 8 blocks) and unrestricted motion vectors (pointing out of the picture).
- Support for SQCIF, 4CIF and 16CIF input formats.
- Syntax-based Arithmetic coding.

H.263 can be 50% more computation intensive then H.261, much of the extra computation arising from the advanced motion estimation.

3.2.8 Wavelets and the Discrete Wavelet Transform

This section outlines the wavelet concept prior to a discussion of the use of wavelets in image compression (Section 3.2.9). Wavelets can be used to give a frequency domain representation of a signal in a similar way to conventional Fourier analysis (although the term 'frequency' is replaced by 'scale' in wavelet terminology). The essential difference is that wavelets give a frequency domain representation *as a function of time*, whereas in the usual Fourier analysis all time dependence is apparently lost. In order to analyse a signal in terms of both frequency *and* time, the wavelet functions must have two indices, one for frequency (strictly, scale) and one for time or position. Moreover, the functions must be *localised in time* in contrast to the infinitely long orthogonal functions used in Fourier analysis. A set of functions satisfying this criterion could be expressed

$$\psi_{jk}(t) = 2^{-j/2}\psi(2^{-j}t - k) \tag{3.63}$$

where j and k are integers. It is apparent that the function $\psi_{jk}(t)$ is derived from a 'mother' wavelet $\psi(t) \equiv \psi_{00}(t)$ by a process of scaling (parameter j) and translation (parameter k).

Also, the mother wavelet must be some rapidly decaying pulse-type waveform in order to be localised in time, and have an oscillatory nature to be localised in frequency. Besides the term 'scale', wavelet theory also uses the term 'resolution', where resolution can be defined as $r = 2^{-j}$. This leads to the concept of *multiresolution analysis* (Section 3.2.9).

Example 3.11

The function $\psi_{10}(t) = 2^{-1/2}\,\psi(t/2)$ spans twice the time-scale of $\psi(t)$, that is, it is *stretched* in time (dilation). In this case, changing the scale of $\psi(t)$ results in a wavelet having lower frequencies than $\psi(t)$. Similarly, function $\psi_{01}(t)$ is simply the mother wavelet shifted or delayed by one unit of time.

Many wavelets possess the property of *orthogonality*; that is, their inner product is zero:

$$<\psi_{jk}, \psi_{lm}> = \int_{-\infty}^{\infty} \psi_{jk}(t)\,\psi_{lm}(t)\,dt = 0; \quad j \neq l, k \neq m \tag{3.64}$$

This is interesting since we know that a general Fourier series representation of a signal has a straightforward expression for its coefficients if the corresponding basis is orthogonal. In other words, given a mathematically 'complete' and orthogonal set of functions $\{\psi_{jk}(t)\}$, we can employ a *wavelet series* to represent a signal $x(t)$:

$$x(t) = \sum_{j,k=-\infty}^{\infty} w_k^j \psi_{jk}(t) \tag{3.65}$$

Here, w_k^j is a *wavelet coefficient* and $\psi_{jk}(t)$ is a *synthesis wavelet*. Due to orthogonality, w_k^j is simply given by the inner product of $x(t)$ with $\psi_{jk}(t)$:

$$w_k^j = <x(t), \psi_{jk}(t)> = \int_{-\infty}^{\infty} x(t)\,\psi_{jk}(t)\,dt \tag{3.66}$$

Equation (3.66) is a type of *wavelet transform*, and coefficient w_k^j measures the similarity of $\psi_{jk}(t)$ to a section of $x(t)$. For well-correlated waveforms, w_k^j will be relatively large, and vice versa. We refer to (3.66) as a form of *continuous-time* transform, since it is applied to a continuous-time signal $x(t)$. Generally speaking, wavelet analysis involves *two* functions: the mother wavelet $\psi(t)$ and a *scaling function* $\phi(t)$, which is mathematically linked to $\psi(t)$. The processes of scaling and translation can be defined for $\phi(t)$ in a similar way to (3.63).

The Discrete Wavelet Transform (DWT)

In practice, wavelet signal analysis and reconstruction is performed using *digital filter banks*, and a 2-band filter bank (corresponding to $j = 1$) is shown in Figure 3.22. Here, g' and h' are lowpass and highpass *analysis* filters respectively, and g and h are lowpass and highpass *synthesis* filters, respectively. Thus, g'_n denotes the impulse response of the lowpass analysis filter. Symbols $\downarrow 2$ and $\uparrow 2$ denote downsampling and upsampling by a

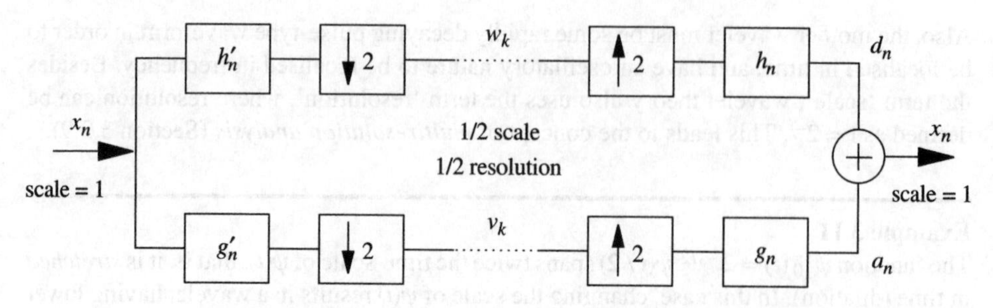

Figure 3.22 *Perfect reconstruction in a 2-band filter bank.*

factor 2 respectively. The filtering and sample rate changes are usually combined in a single filtering operation that can be expressed by the convolutions

$$w_k = \langle x_n, h'_{2k-n} \rangle = \sum_n x_n \, h'_{2k-n} \tag{3.67}$$

$$v_k = \langle x_n, g'_{2k-n} \rangle = \sum_n x_n \, g'_{2k-n} \tag{3.68}$$

Equations (3.67) and (3.68) represent the DWT for a single level of decomposition; the full DWT involves J levels of decomposition, as discussed in Section 3.2.9. There are a number of important points here:

- The filters are usually simple FIR filters with finite roll-off so that downsampling introduces inevitable aliasing. *The remarkable point about wavelets is that it is possible to select the filter coefficients to give perfect reconstruction of the signal,* as indicated in Figure 3.22. All aliasing effects can be cancelled!
- Reconstruction is via the inverse DWT (IDWT), and for a single level of reconstruction this can be expressed as

$$x_n = \sum_k w_k h_{n-2k} + \sum_k v_k g_{n-2k} \tag{3.69}$$

For the discussion on compression in the following section it can be useful to rewrite this in terms of operators:

$$x = H{\uparrow}{\downarrow}H'x + G{\uparrow}{\downarrow}G'x \tag{3.70}$$

- A plot of filter coefficients or impulse response is sometimes referred to as a discrete wavelet; we could refer to h' as the (discrete) analysis wavelet and to h as the synthesis wavelet. Similarly, g' and g are the analysis and synthesis scaling functions respectively.
- Perfect reconstruction requires careful selection of g', h', g and h, and generates relationships between them. For example, in an orthogonal filter bank, impulse response g_n is a 'flipped' version of g'_n, and h'_n is an 'alternating flip' of g'_n. This is illustrated in Figure 3.23 for the *Daubechies* D_4 wavelet.
- Equation (3.67) effectively assumes $j = 1$, so Figure 3.22 is really only a first level of analysis or decomposition. This first level of analysis has halved the scale and resolu-

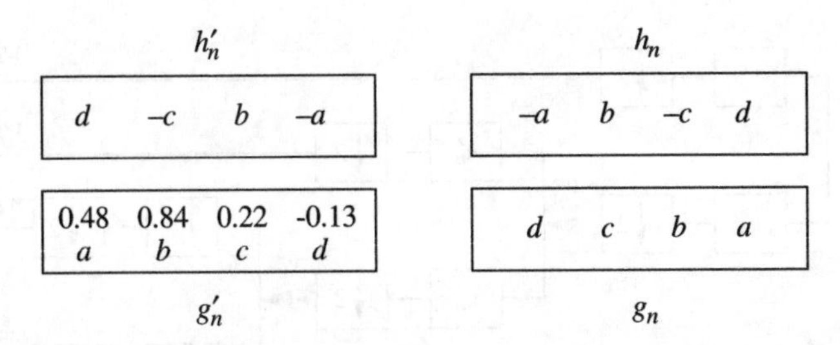

Figure 3.23 *The Daubechies 4-tap wavelet (FIR filter impulse responses).*

tion of the signal; the scale change is due to subsampling, while the resolution change is due to lowpass filtering.

- Lowpass filter g' loses high-frequency detail, and this remains lost even at the output of filter g. The upsampling prior to g (insertion of a zero between every sample) has simply restored the scale, but it has not restored the resolution. Signal a_n is therefore referred to as an *approximation* (or lowpass) signal and, conversely, signal d_n is referred to as the *detail* signal.

3.2.9 Image compression using the DWT

Compared to the DCT, the DWT is a later development in the field of transform coding. As for the DCT, the compression approach first concentrates a large percentage of the image energy in the low-frequency terms of the transform, then quantises them, and then applies entropy coding (see Figure 3.24). The classical approach to entropy coding is to use the Huffman technique, but arithmetic coding may be preferred since it adapts to the data and so does not require *a priori* information. As for the DCT, the DWT has a decorrelation property (a primary objective in compression). Again as for the DCT, simple compression could be achieved by forcing to zero some or all coefficients below a certain threshold. An important distinction between the DCT and the DWT lies in the nature of the wavelet transform. For wavelets, errors arising from information loss tend to be subjectively less annoying (more noise-like) than those for the DCT. A second distinction is that it is not necessary to perform the DWT on blocks of data as in the DCT (wavelet analysis is done on the entire image). In other words, DCT-based encoders suffer from blocking effects when operating at low bit rates, while the wavelet approach eliminates this type of distortion. For example, at 0.2 bits/pixel (CR = 40) the DWT can give an acceptable image, whereas the JPEG coded image can show objectionable blocking. The DWT has therefore

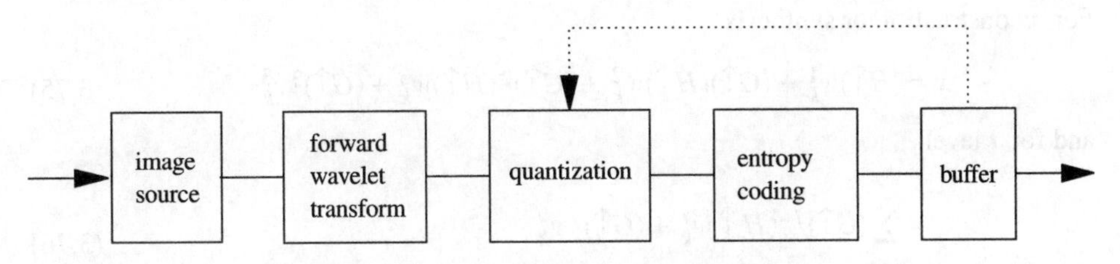

Figure 3.24 *Image compression using the DWT.*

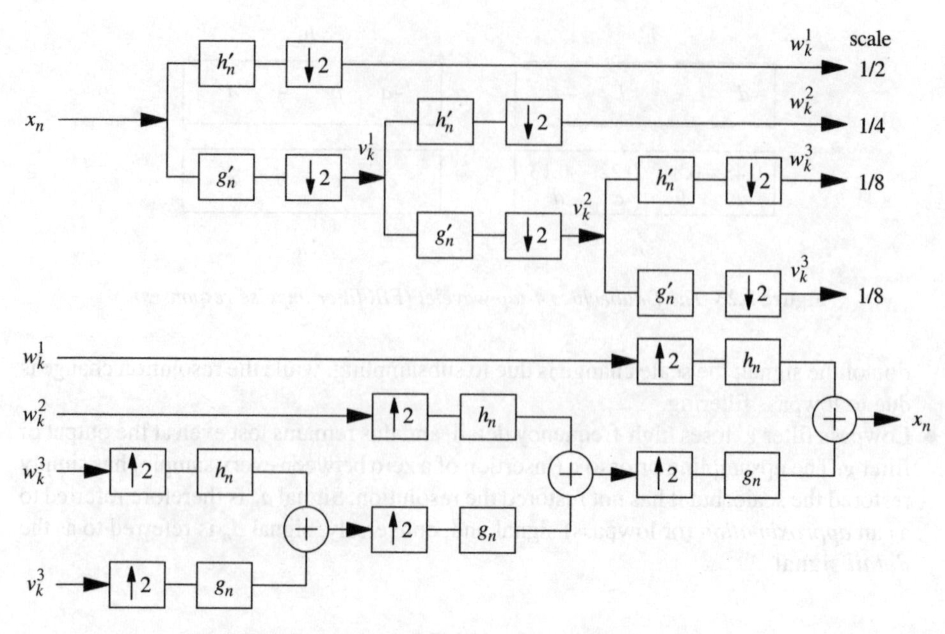

Figure 3.25 *1D decomposition and reconstruction for J = 3.*

been applied successfully to both still image coding (Antonini *et al.*, 1992) and to very low bit rate (10 kbit/s) video coding (Cinkler, 1998).

So far, wavelets have been discussed in terms of a single filter bank and single resolution (Figure 3.22). In practice, signal analysis and data compression systems use multiple filter banks by recursively using the basic filter bank J times, and this is shown in Figure 3.25 for $J = 3$. Each level reduces the resolution by a factor 1/2 leading to a *multiresolution* analysis. Consider the resolution level corresponding to $j = 2$. In this case, g' and h' operate upon input sequence v_k^1 which is at half the sample rate of x_n. In terms of operators

$$w_k^2 = \downarrow H'(\downarrow G')^1 x \tag{3.71}$$

$$v_k^2 = (\downarrow G')^2 x \tag{3.72}$$

Similarly

$$w_k^3 = \downarrow H'(\downarrow G')^2 x \tag{3.73}$$

$$v_k^3 = (\downarrow G')^3 x \tag{3.74}$$

For reconstruction or synthesis

$$x = (H\uparrow)w_k^1 + (G\uparrow)(H\uparrow)w_k^2 + (G\uparrow)^2(H\uparrow)w_k^3 + (G\uparrow)^3 v_k^3 \tag{3.75}$$

and for J levels

$$x = \sum_{j=1}^{J} (G\uparrow)^{j-1} H\uparrow w_k^j + (G\uparrow)^J v_k^J \tag{3.76}$$

The equivalent continuous-time reconstruction could be expressed in terms of scaled and shifted wavelet functions and scaling functions (as in (3.63)). Also note that at level j there are j detail signals and one approximation signal, and some of these can be useful in identifying features in signal analysis.

Image transformation

Image transformation using the DWT is carried out using the usual row–column technique. Consider image x in Figure 3.26. First a 1D convolution is performed on each of the rows to give two 1/2 scale, 1/2 resolution subimages (or subbands) g and h. The 1D convolution is then applied to each column of each subimage thereby generating four subimages

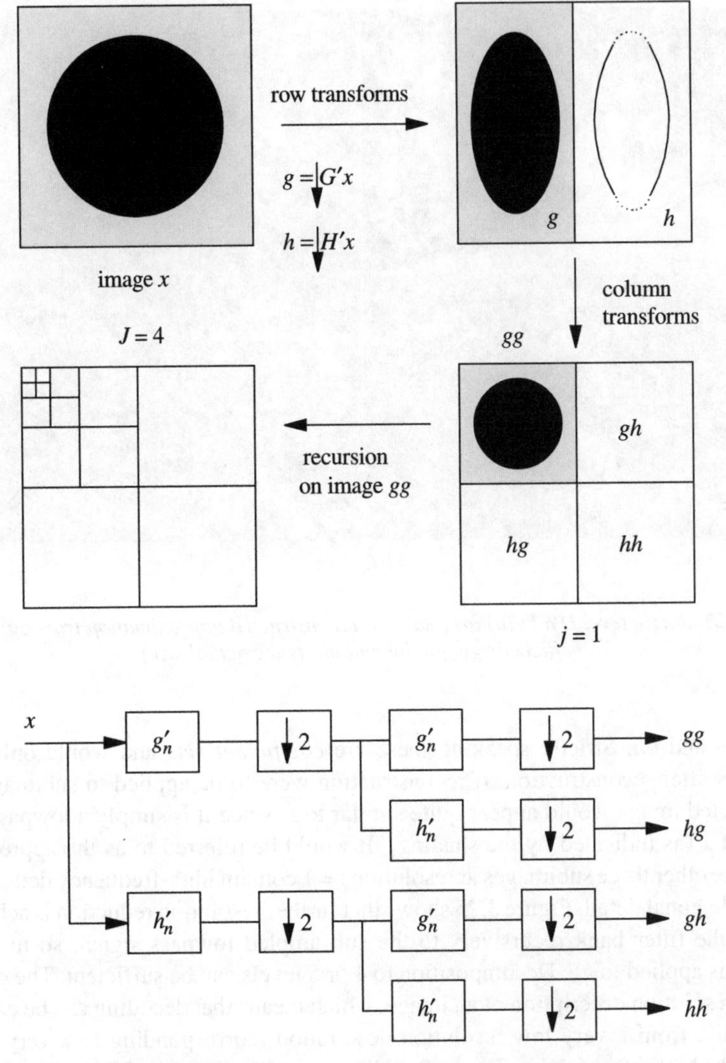

Figure 3.26 *Image transformation using the DWT.*

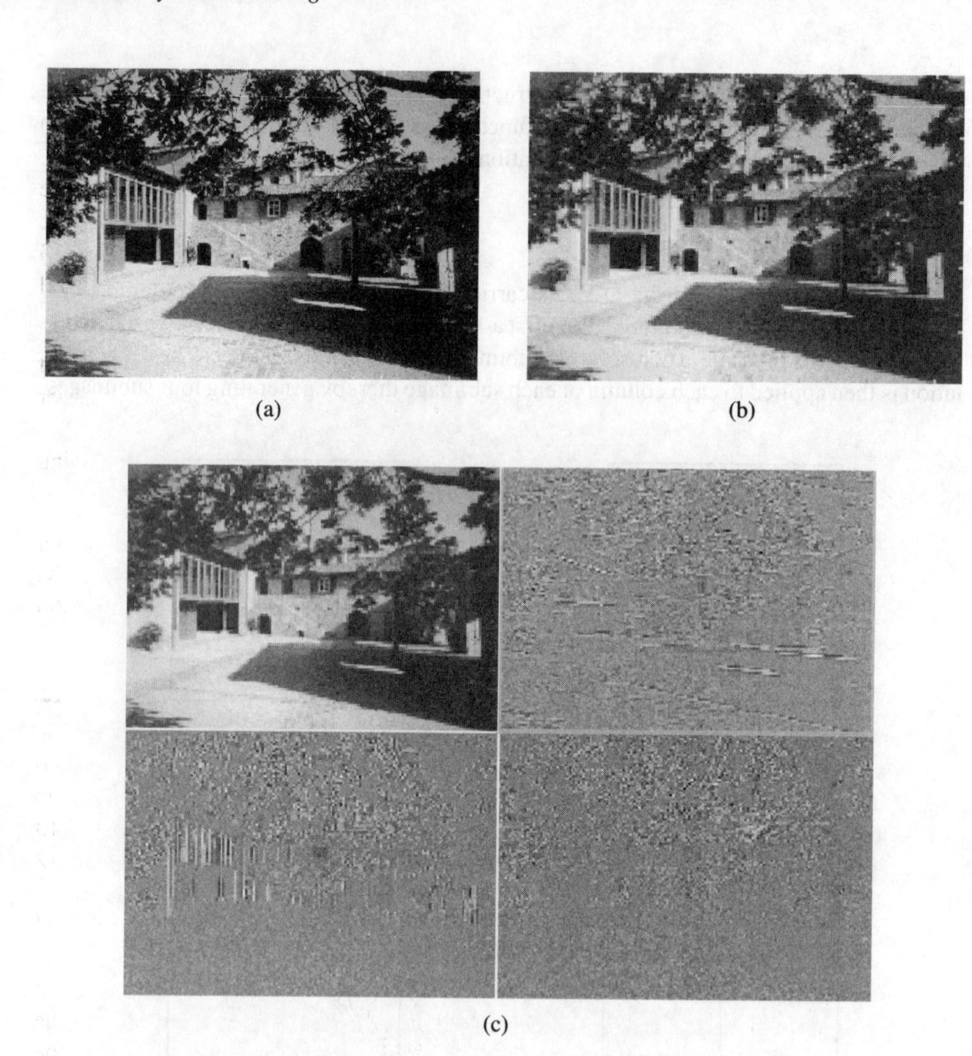

Figure 3.27 *Single level DWT: (a) original; (b) reconstructed approximation from gg; (c) DWT coefficients gg, gh, hg and hh (twice actual size).*

gg, *gh*, *hg* and *hh*. Strictly speaking these are *coefficient sets* and would only be true subimages after reconstruction. If reconstruction were to be applied to subimage *gg* the reconstructed image would appear quite similar to x, since it is simply a lowpass filtered version of x (as indicated by the shading). It would be referred to as the approximation image. The other three subimages at resolution $j = 1$ contain high-frequency detail, with *hh* showing diagonal detail. Figure 3.25 shows that further resolution reduction is achieved by applying the filter bank recursively to the subsampled lowpass signal, so in this case recursion is applied to *gg*. Decomposition to 4 or 5 levels can be sufficient. The end result is a multiresolution description of an image, which means that decoding can be carried out sequentially, from a very low resolution description (corresponding to a very compact code), to the highest resolution. Figure 3.27 illustrates the use of the 2D DWT for $J = 1$ and the Daubechies D_4 wavelet. It is interesting to observe that the original image can be seen in the transform *coefficients*! Also observe that the approximation signal (Figure 3.27(b)) reconstructed from coefficients *gg* is a lowpass filtered version of the original.

As for the DCT, compression is achieved by quantisation of the coefficients. Suppose that image *x* in Figure 3.26 comprises 256 × 256 8-bit pixels (524 288 bits) and that we require $CR = 32$. This means that 524 288/32 = 16384 bits must be distributed over the 13 subimages shown in Figure 3.26 in some optimal way. At the outset we could make several basic observations:

1. For the 16 × 16 subimage in Figure 3.26, assigning one extra bit per coefficient does not increase the total number of bits as much as assigning the extra bit to the larger subimages.
2. At level $j = 4$ most of the energy will be in the lowpass subimage and the remaining subimages (those with relatively low energy) should have fewer bits.

A formal quantisation strategy could aim to minimise a squared error criterion, although it is well known that such techniques are not well matched to the human visual system. As with DCT coefficient quantisation, it is therefore important to weight the quantisation noise according to visual perception.

Example 3.12

Figure 3.28 shows the elements of a simple wavelet-based encoder for low bit rate video applications (ISDN rate). Spatial redundancy is first reduced by using the DWT on a video frame (as in image coding). Typically a 4-level ($J = 4$) DWT could be performed to give 13 subimages or subbands. The decomposition would then comprise a lowpass image and 12 highpass images, as in Figure 3.26. The use of the Daubechies D_4 wavelet leads to particularly simple hardware (Lewis and Knowles, 1991). The coefficients in the lowpass band could be transmitted in full since they are only 1/256 of the image size. The highpass coefficients can be quantised by constructing spatially local trees whose roots cover the lowest frequency band and whose branches reach upward in frequency. A node at any level of the tree is then a 2 × 2 coefficient block, and for optimal results these coefficients should be subjectively quantised according to the HVS. For example, the energy of each block could be compared with a HVS weighted threshold and a binary token sent to indicate whether

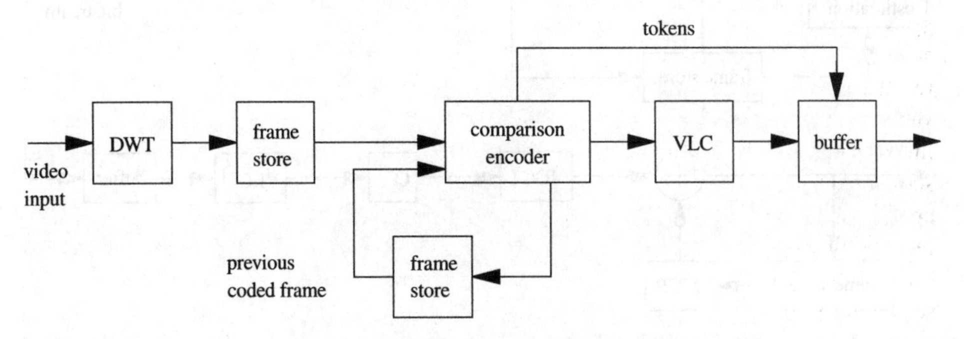

Figure 3.28 *Low bit rate video encoder.*

the block is important or not (in the latter case the encoder would assume the remainder of the tree to be zero).

In order to achieve a low bit rate it is necessary to reduce the temporal redundancy, and this is done via frame differencing. To simplify the encoder (no motion compensation) we could assume that both the transmitter and receiver have identical copies of the previous frame in coded form. Generation of the difference between the new transformed frame and the previous transformed and coded frame will then enable both transmitter and receiver to construct the new transformed coded frame. The interframe encoder operates in a similar way to intraframe coding in that it recursively scans all trees which make up the decomposition. It is worth noting that a more conventional DWT-based video encoder would use motion compensation and essentially replaces the DCT with the DWT in an H.261-type encoder (Cinkler, 1998).

3.3 MPEG video coding

The various bit rates, coding levels and coding profiles available in the generic MPEG-2 video coding standard have been discussed in Section 1.7.8. In this section we examine the signal processing concepts employed by an MPEG-2 encoder.

The general compression mechanism of transform coding (TC) has been discussed in Section 3.2, and applies equally well to MPEG-2 coding. MPEG-2 uses a hybrid DCT/DPCM encoder, as used in the H.261 video codec. With reference to Figure 3.29, we might summarise the signal processing requirements as follows:

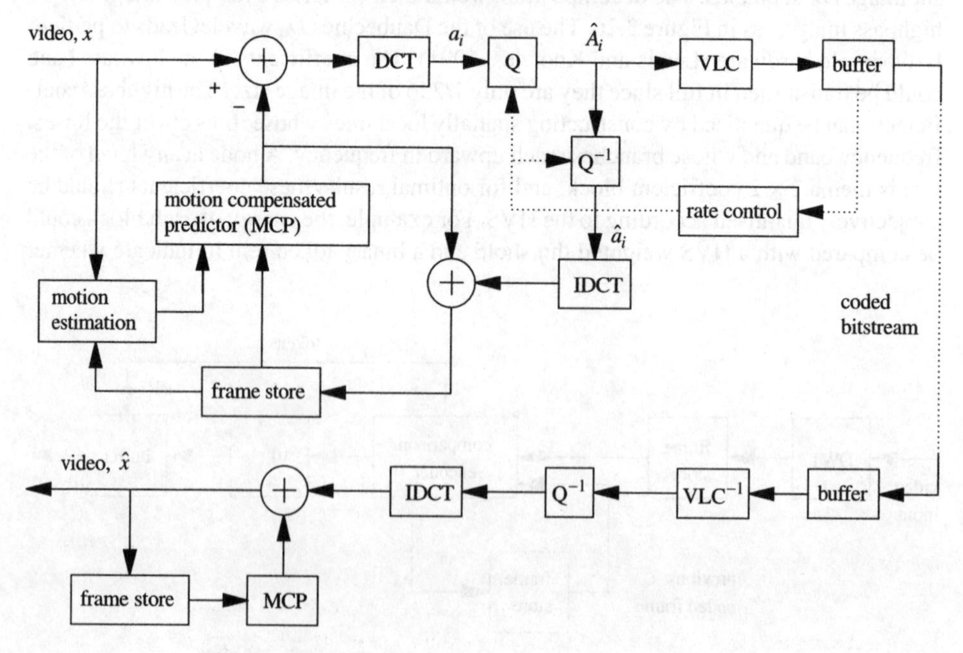

Figure 3.29 *MPEG-2 video codec (scalability not shown).*

1. The encoder input pixels are not independent but are correlated with other pixels both within the same frame and across frames. We refer to this as *spatial* and *temporal redundancy* and the basic objective is to remove the redundancy (decorrelate) in order to achieve compression. We therefore need some form of spatial and temporal compression.
2. A good sub-optimal transform that almost achieves perfect decorrelation is the DCT. We could regard the quantisation of the DCT coefficients as performing the required spatial compression within a frame.
3. The quantisation process in turn exploits the *psychovisual redundancy* of the human visual system (HVS). More explicitly, the degree of quantisation applied to each coefficient could be weighted according to the visibility of the resulting quantisation noise to the human observer (in MPEG-2 this approach is applied to 'intra' blocks). Broadly speaking, this means that the highest frequency components tend to be more coarsely quantised.
4. The required temporal compression is achieved by encoding only the difference or prediction error between successive frames. In other words, all that is encoded is the part of the frame that contains movement or illumination changes. This involves *motion-compensated prediction* (MCP) and the generation of motion vectors, as discussed in Section 3.2.7.
5. Transform coding and quantisation results in weakly correlated coefficients clustered around the origin, with many high-frequency coefficients being zero. By carefully ordering the coefficients (zigzag coding), this can be exploited to provide a further compression mechanism through the use of run-length/variable-length coding (VLC); see Example 2.2. Note that buffering is required for fixed bit rate applications, although more efficient variable bit rate systems are possible.

3.3.1 I, P and B frames

Temporal compression means that, in general, a frame is constructed from the prediction from an adjacent frame, but there is also a need to have frames that are independent of other frames for random access purposes. MPEG-2 therefore defines three types of frame:

- I (intra) frames: these are coded without reference to other frames, and so provide random access points in the bitstream where decoding can begin. In other words, only intraframe correlations are exploited. Since I frames use only spatial compression (this being achieved by the DCT and quantising) they achieve less compression than P or B frames.
- P (predictive) frames: these are coded using motion-compensated prediction on the previous reference frame (which could be an I or P frame). For good prediction the distance between predictive frames must be short, so typically a P frame occurs every two or three frames.
- B (bidirectional) frames: these are coded using motion compensated prediction on the immediately previous I or P reference frame and on the immediately future I or P reference frame. As for P frames, the motion compensation used for B frames removes temporal redundancy, and since B frames are well predicted they give the highest compression. Typically, the bit allocation in P and B frames will be only 50% and 20% respectively of that of an I frame (the actual percentages depend upon picture complexity).

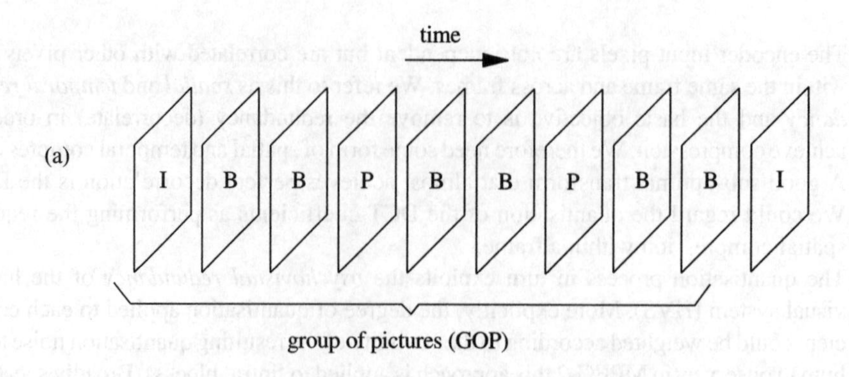

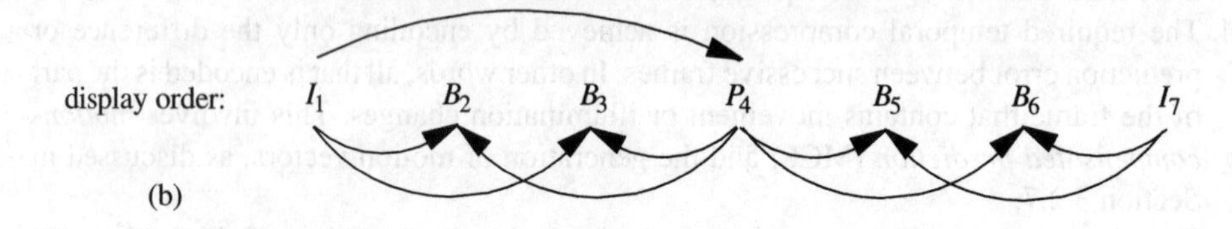

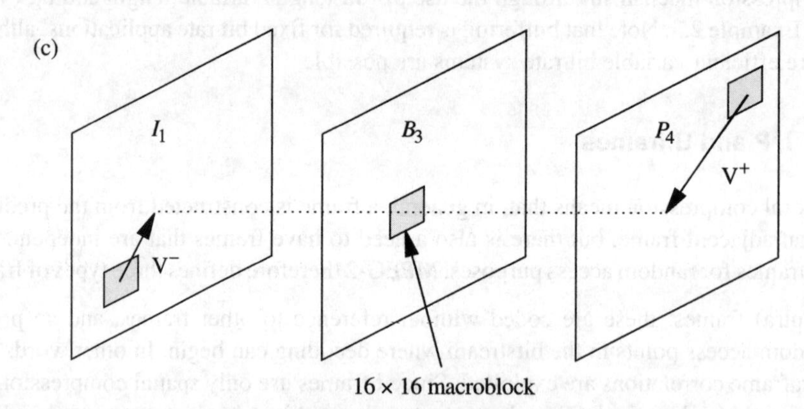

Figure 3.30 *MPEG-2 frames: (a) a GOP; (b) frame dependency; (c) motion compensated bidirectional prediction.*

I, P and B frames are formed into a group of pictures, or *GOP* (Figure 3.30(a)). A GOP starts with an I frame to allow random access to the group, and its length is typically 12 frames for 50 Hz systems and 15 frames for 60 Hz systems. Generally speaking, the GOP length is made to suit the type of video being coded; for example, it is better to use a short GOP for a video sequence containing fast-moving action. The order in which frames are displayed is not the order of the frames in the bitstream (Figure 3.30(b)). The encoder reorders the frames so that a B frame is transmitted *after* the past and future frames it uses for reference; for instance, frame B_2 is transmitted after I_1 and P_4, since it uses both for reference.

Table 3.5 *Typical MPEG-2 bitstream illustrating 'block dropping' in P and B frames (frame size: 352 × 240 pixels).*

Frame	*Frame type*	*Luminance blocks*
1	I	1320
2	B	526
3	B	551
4	P	1134
5	B	558
6	B	502
7	P	1159
8	B	508
9	B	600
10	P	1157
11	B	652
12	B	615
13	I	1320
14	B	365
15	B	437
16	P	1122
17	B	526

Table 3.5 illustrates the GOP structure and the number of transmitted 8 × 8 luminance blocks in a typical MPEG-2 bitstream (the frame size is 352 × 240 pixels). For I frames all 1320 possible luminance blocks are transmitted, while the number of transmitted blocks/frame is significantly lower for P and B frames. A block is not transmitted when all the coefficients of the transformed interframe difference signal are quantised to zero.

3.3.2 Motion compensation

The principle of motion estimation in MPEG-2 is similar to that discussed in Section 3.2.7, so here we will summarise the process and add a few details specific to MPEG-2. The basic idea is to estimate and compensate the motion of objects *before* calculating pixel-by-pixel differences. This is done by using motion vectors to create a *prediction picture* and it is this picture that is compared to the current picture on a pixel-by-pixel basis. Note from Figure 3.29 that prediction is a separate process from motion estimation. It uses the motion vectors and the reference frame(s) to construct the prediction picture. The prediction picture is then subtracted from the current picture to form the prediction error. Coding of the prediction error requires far fewer bits compared with the coding of the original video block. Also note that motion compensation is not used on I frames, although it is possible for an I frame to have a motion vector for error concealment purposes.

In MPEG-2, motion vector estimation is achieved by dividing a frame into blocks of 16 × 16 pixels, called *macroblocks*. Suppose for example that the current frame is a P frame.

In this case the encoder searches an area in the previous reference frame (I or P) for the closest matching macroblock. Since MPEG-2 defines only the *decoding* process and not the *encoding* process, the exact motion estimation algorithm is left to the designer. For instance, in the MPEG-2 reference encoder TM5 (Eckart and Fogg, 1995) an exhaustive search to one-pixel accuracy is first performed, followed by a search to half-pixel accuracy, and when a match is found the motion vector is calculated. In effect, the motion vector is used to shift a block from the reference frame in order to use it as the prediction. Note that if the current frame is a B frame then the prediction macroblock will be an average of forward and backward predictions, as shown in Figure 3.30(c). Also note that I frames are encoded by making a zero-valued prediction such that the source block rather than the prediction error block is DCT encoded.

Once calculated, a vector fully describes the prediction macroblock and is therefore sent to represent it. The prediction error macroblock (the difference between the current macroblock and its prediction) goes through the DCT, quantisation and VLC process and is then sent to refine the prediction. The decoder receives the motion vector and looks for the prediction macroblock in the preceding picture. Once the error macroblock has been decoded, it is added to the translated prediction to reconstruct the current macroblock. Calculation of the motion vectors is the most time-consuming part of the coding process, and of course they incur a bit rate overhead. For instance, when using MPEG-2 to compress standard definition video to 6 Mbit/s, the motion vector overhead could account for about 2 Mbit/s during a picture making heavy use of motion compensation.

Finally, in contrast to MPEG-1, MPEG-2 uses four different types of motion compensation:

- Frame motion compensation: one motion vector for both fields.
- Field motion compensation: one motion vector plus a field select flag for each field.
- 16 × 8 motion compensation: one motion vector plus a field select each for the upper half field and the lower half field.
- Dual prime motion compensation: one motion vector and one differential motion vector.

As indicated in Figure 3.30, each of these motion compensation types can be based upon past frames (forward prediction), future frames (backward prediction), or both past and future frames (bidirectional prediction). Dual prime motion compensation is, however, limited to forward prediction.

3.3.3 Coefficient quantisation

Given an input of 8-bit pixels, the DCT will produce an 8 × 8 block of 12-bit coefficients (including the sign bit). Recall that word growth across a discrete transform is to be expected due to the need to represent the sum of additions by increased precision (Section 3.2.4). It is therefore *essential* to quantise the DCT coefficients in order to achieve compression. The degree of quantisation applied to each coefficient is weighted according to the visibility of the resulting quantising noise, and this results in the high-frequency coefficients being more coarsely quantised than the low-frequency coefficients. In fact, quantisation eliminates many high-frequency coefficients by setting them to zero.

First note from Figure 3.29 that both quantisation, Q, and inverse quantisation, Q^{-1}, are required. The former gives quantised (integer) values $\hat{A}_i$ in the range [–2048, 2047], and the latter reconstructs the real-valued DCT coefficients a_i, but with quantisation error. For all coefficients except intra DC coefficients, the inverse quantisation process for intra blocks (I frames) is formally defined in the MPEG-2 specification as

$$\hat{a}_i = [\,2 \times \hat{A}_i \times Q_i \times quant\_scale\,]\,/\,32 \tag{3.77}$$

Here, Q_i is the ith element of a *quantisation matrix* **Q** (an 8 × 8 matrix of positive integers) and it is either read from the bitstream or is the appropriate matrix as defined in the MPEG-2 specification. The term *quant_scale* varies between 1 and 112 and is used to adjust the required precision during quantising, and hence adjusts the bit rate (the bitstream can be either fixed bit rate or variable bit rate). For implementation purposes it is helpful to rewrite (3.77) as

$$\hat{a}_i = \hat{Q}_i \cdot \hat{A}_i \tag{3.78}$$

$$\hat{Q}_i = \frac{quant\_scale}{16} \cdot Q_i \tag{3.79}$$

Clearly, the inverse quantisation process is essentially the multiplication of the quantised DCT coefficients $\hat{A}_i$ by the *quantiser step size* $\hat{Q}_i$.

In contrast to the inverse quantisation process in the decoder, an MPEG-2 encoder is free to perform quantisation in any suitable way. It could simply be a mirror of the inverse quantiser, in which case for intra blocks we can write the ith quantised coefficient as

$$\hat{A}_i = \left\lfloor \frac{16\,a_i}{Q_i \times quant\_scale} + \frac{\operatorname{sign}(a_i)}{2} \right\rfloor \tag{3.80}$$

$$= \left\lfloor \frac{a_i}{\hat{Q}_i} + \frac{\operatorname{sign}(a_i)}{2} \right\rfloor \tag{3.81}$$

In other words, quantising simply amounts to division by $\hat{Q}_i$ followed by rounding.

Example 3.13

The MPEG-2 default quantisation matrix for intra blocks (both luminance and chrominance) is

$$\mathbf{Q} = \begin{bmatrix} 8 & 16 & 19 & 22 & 26 & 27 & 29 & 34 \\ 16 & 16 & 22 & 24 & 27 & 29 & 34 & 37 \\ 19 & 22 & 26 & 27 & 29 & 34 & 34 & 38 \\ 22 & 22 & 26 & 27 & 29 & 34 & 37 & 40 \\ 22 & 26 & 27 & 29 & 32 & 35 & 40 & 48 \\ 26 & 27 & 29 & 32 & 35 & 40 & 48 & 58 \\ 26 & 27 & 29 & 34 & 38 & 46 & 56 & 69 \\ 27 & 29 & 35 & 38 & 46 & 56 & 69 & 83 \end{bmatrix}$$

This matrix was used to quantise a particular 8 × 8 coefficient block in an I frame, giving

Table 3.6 *Matrices* $\hat{\mathbf{a}}$ *and* $\hat{\mathbf{Q}}$ *for Example 3.13.*

	+976.0000	−38.0000	−90.2500	−52.2500	−123.5000	+0.0000	+0.0000	+0.0000
	−304.0000	−76.0000	+104.5000	−57.0000	+64.1250	+0.0000	+0.0000	+0.0000
	+45.1250	−104.5000	+61.7500	+128.2500	+0.0000	+0.0000	+0.0000	+0.0000
$\hat{a}$	−104.5000	+52.2500	+0.0000	+0.0000	+0.0000	−80.7500	+0.0000	+0.0000
	+0.0000	+123.5000	−64.1250	−68.8750	+0.0000	+0.0000	+0.0000	+0.0000
	+61.7500	+0.0000	+0.0000	+0.0000	+0.0000	+0.0000	+0.0000	+0.0000
	+0.0000	+64.1250	+0.0000	+0.0000	+0.0000	+0.0000	+0.0000	+0.0000
	+0.0000	+0.0000	+0.0000	+0.0000	+0.0000	+0.0000	+0.0000	+0.0000
	+8.0000	+38.0000	+45.1250	+52.2500	+61.7500	+64.1250	+68.8750	+80.7500
	+38.0000	+38.0000	+52.2500	+57.0000	+64.1250	+68.8750	+80.7500	+87.8750
	+45.1250	+52.2500	+61.7500	+64.1250	+68.8750	+80.7500	+80.7500	+90.2500
$\hat{Q}$	+52.2500	+52.2500	+61.7500	+64.1250	+68.8750	+80.7500	+87.8750	+95.0000
	+52.2500	+61.7500	+64.1250	+68.8750	+76.0000	+83.1250	+95.0000	+114.0000
	+61.7500	+64.1250	+68.8750	+76.0000	+83.1250	+95.0000	+114.0000	+137.7500
	+61.7500	+64.1250	+68.8750	+80.7500	+90.2500	+109.2500	+133.0000	+163.8750
	+64.1250	+68.8750	+83.1250	+90.2500	+109.2500	+133.0000	+163.8750	+197.1250

$$\hat{\mathbf{A}} = \begin{bmatrix} 122 & -1 & -2 & -1 & -2 & 0 & 0 & 0 \\ -8 & -2 & 2 & -1 & 1 & 0 & 0 & 0 \\ 1 & -2 & 1 & 2 & 0 & 0 & 0 & 0 \\ -2 & 1 & 0 & 0 & 0 & -1 & 0 & 0 \\ 0 & 2 & -1 & -1 & 0 & 0 & 0 & 0 \\ 1 & 0 & 0 & 0 & 0 & 0 & 0 & 0 \\ 0 & 1 & 0 & 0 & 0 & 0 & 0 & 0 \\ 0 & 0 & 0 & 0 & 0 & 0 & 0 & 0 \end{bmatrix}$$

Clearly, quantisation has eliminated many high-frequency coefficients. The corresponding reconstructed DCT coefficients, $\hat{a}_i$, and quantiser step size, $\hat{Q}_i$, are shown in Table 3.6. The fixed scaling parameter for this processing is normally found from the bitstream, but it can also be deduced from (3.79); for example, taking the last entry in **Q** and $\hat{\mathbf{Q}}$ gives

$$quant\_scale = \frac{197.125 \times 16}{83} = 38$$

The reconstructed intra DC coefficient is defined separately from (3.77) and is given by

$$\hat{a}_{00} = intra\_dc\_mult \times \hat{A}_{00}$$

where, typically, $intra\_dc\_mult = 8$.

3.3.4 Scalability

The broad objective of scalable coding is to provide interoperability between different services and to support receivers with different display capabilities. The basic idea is shown in Figure 3.31. Here two *layers* or bitstreams are provided, each layer supporting

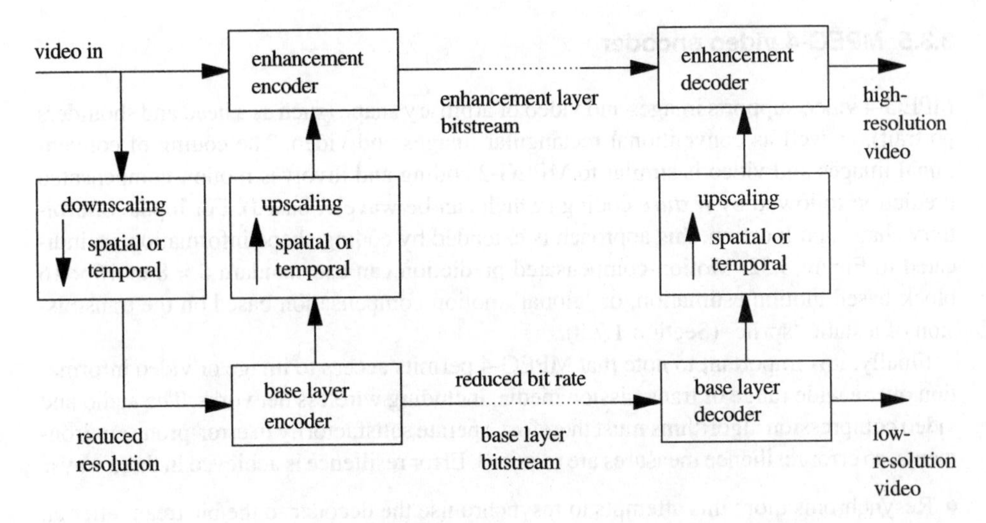

Figure 3.31 *Scalable coding of video (spatial and temporal scalability).*

video at a different scale. A low-resolution video signal is obtained by downsampling spatially or temporally and this is transmitted as a *base layer* signal at a reduced bit rate. The upscaled reconstructed base layer video is used as a prediction for the coding of the original input signal, and the prediction error is then encoded into an *enhancement* layer bitstream. This scalable system can be applied in several ways:

- The different bit rates allocated to each layer can meet the specific bandwidth requirements of different transmission channels.
- A decoder could decode subsets of a layered bitstream in order to display video at a lower spatial or temporal resolution. Put another way, a base layer decoder would decode the base layer bitstream to give low resolution video, while another decoder could use an enhancement layer to give high resolution video (Figure 3.31).

There are actually three modes of scalability in MPEG: spatial, temporal and signal to noise ratio (SNR) scalability, plus *data partitioning* (see Table 1.5). Spatial scalability supports displays with different spatial resolution, the lower spatial resolution being constructed from the base layer. Temporal scalability has similar aims as spatial scalability, and can have different frame rates between layers. SNR scalability can provide video at the same picture resolution but at different *quality* levels. For example, the base layer at 3–4 Mbit/s could provide a picture quality equivalent to PAL or NTSC video, but by using both layer bitstreams an enhanced decoder can deliver studio quality video at a total bit rate of 7–12 Mbit/s. SNR scalability is achieved through the use of an extra quantisation stage in the codec, while spatial scalability is achieved by using multiple hybrid DPCM/DCT loops. Finally, data partitioning partitions the macroblock data into multiple layers; essentially low-frequency DCT coefficients (partition 0) and high-frequency DCT coefficients (partition 1). Good tolerance to channel errors can be achieved if the errors are distributed so that partition 1 receives most errors.

3.3.5 MPEG-4 video encoder

MPEG-4 video supports images and video of arbitrary shape (such as a head and shoulders portrait), as well as conventional rectangular images and video. The coding of conventional images and video is similar to MPEG-2 coding and involves motion-compensated prediction followed by *texture* coding (which can be wavelet based). For inputs of arbitrary shape and location, this approach is extended by coding shape information, as indicated in Figure 3.32. Motion-compensated prediction can use standard 8 × 8 or 16 × 16 block-based motion estimation, or 'global' motion compensation based on the transmission of a static 'sprite' (Section 1.7.8).

Finally, it is important to note that MPEG-4 permits access to image or video information over a wide range of transmission media, including wireless networks. The audio and video compression algorithms must therefore operate satisfactorily in error-prone environments, so error resilience measures are required. Error resilience is achieved in three ways:

- Resynchronisation: this attempts to resynchronise the decoder to the bitstream after an error has occurred.
- Data recovery: after synchronisation has been re-established, data recovery tools attempt to recover data that in general would be lost. These tools are not simply error-correcting codes, but instead techniques that encode the data in an error-resilient manner (a possibility here is to use *Reversible Variable Length Codes* (RVLC)).

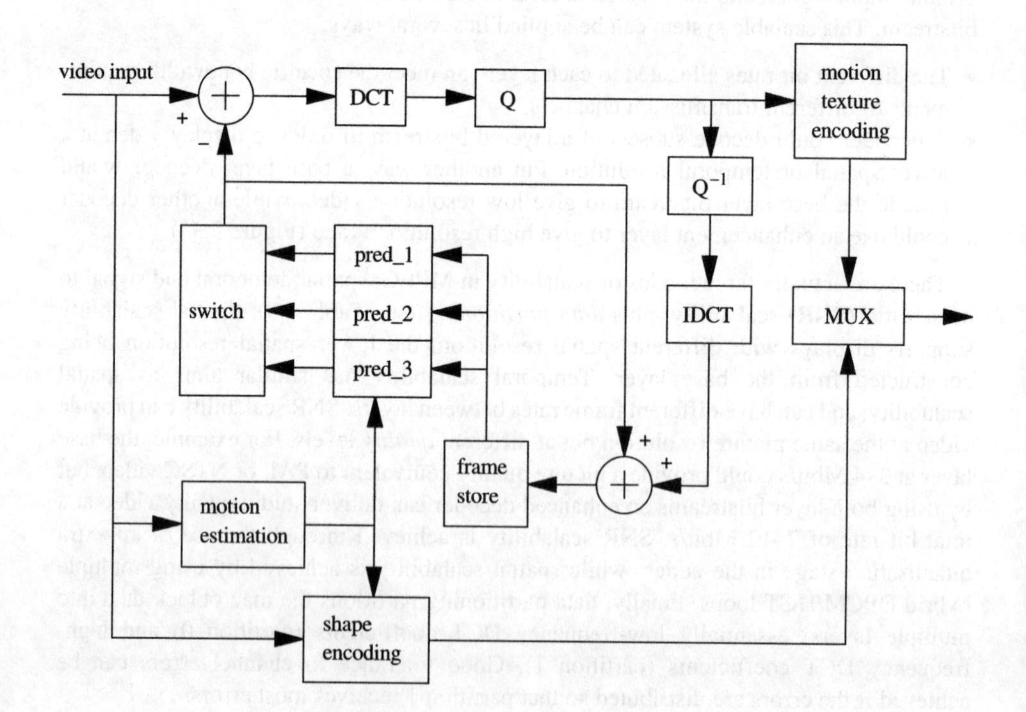

Figure 3.32 *MPEG-4 video encoder.*

- Error concealment: if resynchronisation can localise the error, then error concealment can be quite simple, such as copying blocks from the previous frame.

3.4 Speech coding

Common speech coding techniques spanning a range of bit rates were listed in Section 1.7.3. The bit rate is in fact only one of a number of speech codec attributes that must be considered by the applications engineer. The attributes of *delay*, *complexity* and *quality* must also be considered, and the less important attributes can be relaxed so that the more important attributes can be met.

The delay of a speech encoder can have a great impact on its suitability for a particular application. For instance, if the one-way delay is greater than about 300 ms a real-time conversation will become more like a 'push to talk' experience rather than a normal conversation. Relative to other methods, the PCM and ADPCM speech techniques in Table 1.4 have the lowest delay but also the highest bit rate. To achieve greater compression the speech must be divided into *frames* and then encoded a frame at a time (ADPCM does not have frames). For example, the low-delay CELP standard (G.728) has 5-sample (0.625 ms) frames and the first generation cellular encoders have 20 ms frames. A rule of thumb for the total delay through a speech encoder/decoder (codec) is 3 × frame size + lookahead, where 'lookahead' uses some samples from the following frame to improve codng. Thus, first generation cellular systems (VSELP and GSM) have a one-way codec delay of more than 60 ms. In contrast, the 8 kbit/s G.729 standard has a 10 ms frame size and a total one-way codec delay of about 30 ms.

A speech encoder can be implemented with dedicated DSP hardware or via a host CPU, but the heart of its complexity is always the raw number of computational instructions required by the coding algorithm. In practice this is usually taken as the number of MIPS required by the encoder. The 16 kbit/s G.728 standard has very low delay but it has relatively high complexity (about 30 MIPS). In contrast, the 16 kbit/s G.726 standard requires only 2 MIPS. The high complexity of G.728 was one of the incentives for developing G.729, which requires about 20 MIPS.

Speech quality measurement should not only measure the performance of the system for 'clean' speech. It is also important to measure the 'robustness' of the system, that is, the performance in the presence of background noise and the performance in the presence of transmission errors. This is particularly important for low bit rate encoders, since these use a *speech production model* which is usually not capable of modelling the combination of speech plus a background signal (or music). The result is that when the input speech contains background noise, the encoder performance degrades significantly. It is interesting to note that the first generation digital cellular systems were not tested for robustness to noisy input speech, while later generation encoders handle background noise much better. Transmission errors are a problem on noisy radio channels and a standard intended for such channels (such as G.729) should be able to handle both random bit errors and burst errors/frame erasures. The G.729 standard can handle a 0.1% random error rate, although some mobile codecs (such as the AMBE system – Table 1.4) use FEC and can handle random error rates as high as 5%.

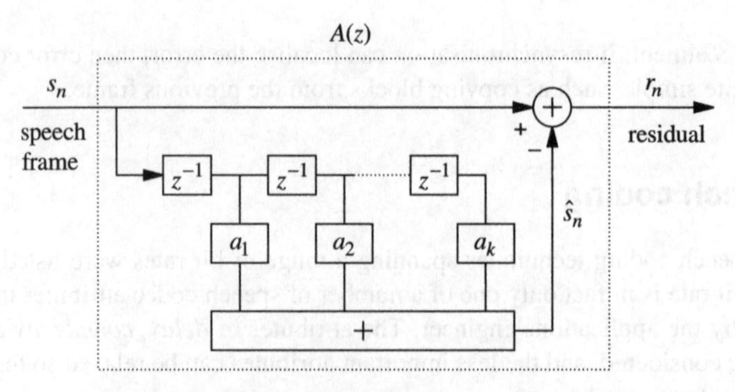

Figure 3.33 *Linear prediction (LP) for a speech frame.*

3.4.1 Linear predictive coding (LPC)

We now examine the signal processing concepts underlying one of the earliest and simplest speech compression techniques. The technique is still important since it plays a major role in later, more complex speech standards, such as CELP and RPE-LTP (GSM). It is also the basis of the 2.4 kbit/s US military standard LPC10E.

Over a short period of time (or speech frame) a speech sample s_n can be well estimated as a linear combination of k previous samples, as shown in Figure 3.33. Here, filter $A(z)$ is called a *short-term analysis filter* or *inverse filter*. Due to the non-stationary nature of speech the coefficients a_i, $i = 1, ..., k$ should be computed on a frame-by-frame basis. Given good prediction there will be a small *residual signal* given by

$$r_n = s_n - \sum_{i=1}^{k} a_i s_{n-i} \tag{3.82}$$

Taking the z-transform we have

$$R(z) = S(z) - \sum_{i=1}^{k} a_i z^{-i} S(z) \tag{3.83}$$

giving

$$A(z) = \frac{R(z)}{S(z)} = 1 - \sum_{i=1}^{k} a_i z^{-i} \tag{3.84}$$

In practice the coefficients can be found for a short speech frame by using a least-square-error criterion, that is,

$$\text{minimize} \sum_n r_n^2$$

where the summation is over a speech frame. As in DPCM predictor design, this minimisation problem reduces to a set of k linear equations involving just the a_i terms and the signal autocorrelation function (equation (3.13)). This is referred to as the *autocorrelation method* for determining the coefficients, and it requires the speech samples to be windowed to avoid transient effects arising from an abrupt start and finish. Conventional LPC analysis is therefore as follows:

1. Segment the speech samples s_n into quasi-stationary frames by applying a window function w_n. It is usual to use a Hamming window and to overlap adjacent windows.
2. Use the windowed samples to compute the *short-time* autocorrelation function, R_i, as defined in (4.39).
3. Compute coefficients a_i by matrix inversion (equation 3.15). This could be done using Levinson–Durbin recursion (Section 3.1.4) or Schur recursion (ETSI, 1991).
4. Extract the source characteristics for the frame, that is, characterise the frame as either *voiced* or *unvoiced*, and if voiced, determine the *pitch*.

The practical significance of $A(z)$ lies in the fact that it is widely used in a powerful 'source-filter' approach to speech synthesis. In fact, by separating the fine spectral structure of the sound source from the overall spectral envelope description (as provided by the filter), synthetic quality speech is possible at rates around 1 kbit/s. The basic idea is shown in Figure 3.34. The *excitation sequence* v_n essentially corresponds to the physical sound source (lungs and vocal chords), and $H(z) = A(z)^{-1}$ essentially models the short-term resonances of the pharynx–mouth channel. Clearly, if $v_n \approx r_n$ then simple inverse filtering gives

$$\hat{S}(z) = H(z)V(z) \approx H(z)A(z)S(z) = S(z) \tag{3.85}$$

and synthesis is very accurate. Unfortunately, the simple synthesis model in Figure 3.34 does have its limitations. The 'all-pole' form of $H(z)$ can model physical resonances or *formants*, but fails to model 'anti-resonance' (nasal) sounds. Also, some sounds cannot be clearly classified as either 'voiced' or 'unvoiced' (noise-like) sounds. In addition, the periodic pulse generator in Figure 3.34 gives a 'flat' amplitude spectrum, but in practice the spectrum of a voiced source decreases by some 12 dB/octave (Hart *et al.*, 1982). This implies that $H(z)$ should model not only the pharynx–mouth channel, but also the spectral

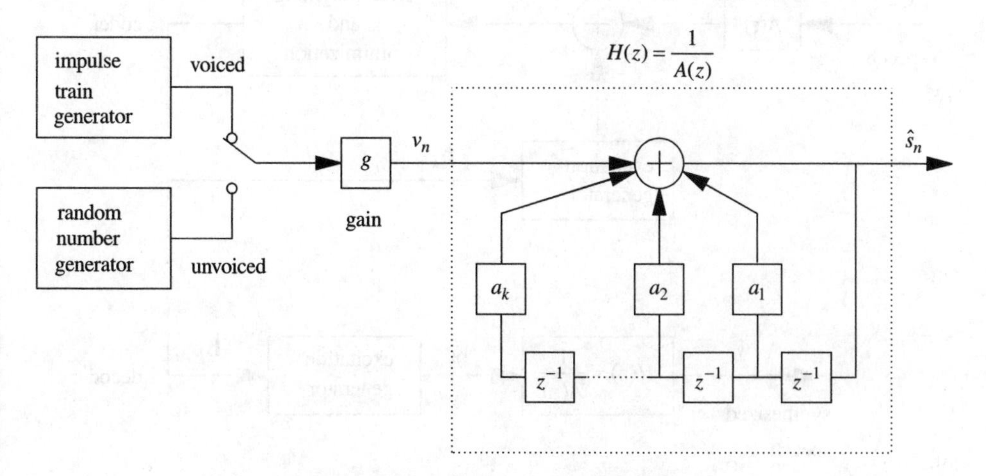

Figure 3.34 *Use of LPC for speech synthesis.*

amplitude changes of the voiced source. Also, as already pointed out, simple speech models like this are not capable of modelling music or the combination of speech plus some background signal. Nevertheless, the synthesis technique is still powerful and is used to synthesise speech frames in advanced speech systems like the CELP codec.

Example 3.14
For an LPC speech compression system the main concern is to transmit the filter and excitation information. The LPC10E (1984) standard samples speech at 8 kHz and segments it into 180 samples/frame. The frame size is therefore 22.5 ms. It employs the basic source and all-pole filter synthesis technique in Figure 3.34. Ten coefficients are computed, although when an unvoiced decision is made only four coefficients are used. The transmitted bitstream comprises 41 bits for the coefficients, 7 for the pitch and voiced/unvoiced decision, 5 for gain and one for synchronisation. This gives a total of 54 bits/frame and a bit rate of 54/0.0225 = 2400 bits/sec. As mentioned, the simple speech model in LPC10E cannot handle background noise and doesn't model nasal sounds. In addition, the binary voiced/unvoiced decision is sometimes poor.

3.4.2 RPE-LTP (GSM)

The specification for the Regular Pulse Excited linear predictive encoder with Long-Term (pitch) Prediction (RPE-LTP) was finalised in 1987 and was the first standardised cellular radio system to use digital transmission (Vary *et al.*, 1987; ETSI, 1991). A conceptual model of the RPE-LTP codec is shown in Figure 3.35. The speech frame is 20 ms and so comprises 160 samples at the 8 kHz sample rate. As for LPC, the decoder generates synthesised speech by applying a sequence v_n to a time-varying filter $H(z) = A(z)^{-1}$ which

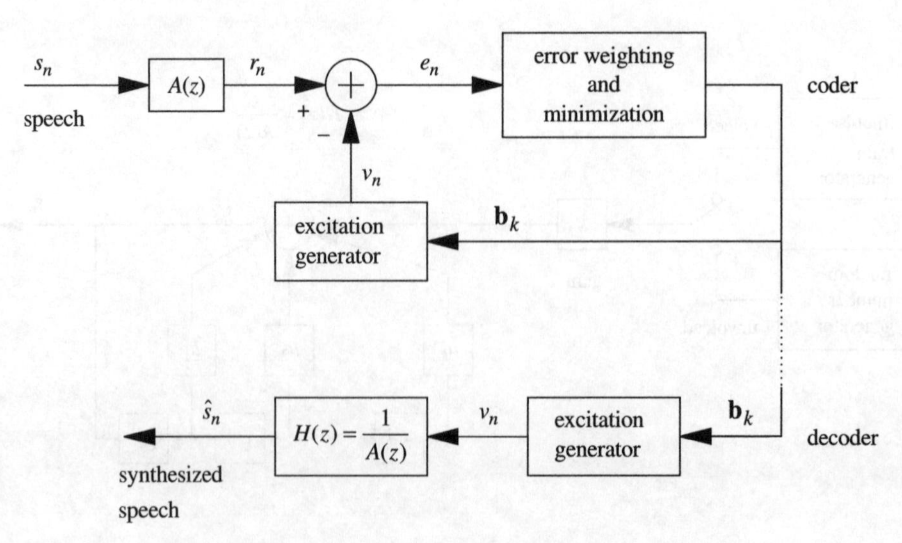

Figure 3.35 *A conceptual model for RPE-LTP speech coding.*

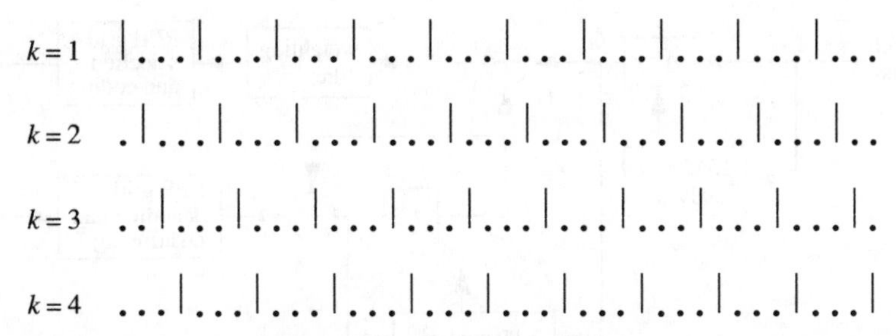

Figure 3.36 *Candidate excitation vectors for $L = 40$, $N = 4$, $Q = 10$ (non-zero valued samples are represented by vertical lines).*

models the short-time spectral envelope (formants or resonances) of the speech signal. Filter $A(z)$ is designed as before, that is, some form of LPC analysis is performed on a speech frame and involves computation of the short-time autocorrelation function and some recursion rule to find the filter coefficients. For GSM the filter order is 8.

Having obtained $A(z)$, the remaining problem is to convey the short-term residual signal r_n that contains information describing the fine structure of the underlying spectrum. Essentially we need to transmit r_n in compressed form, and for GSM this source coding process is performed on 5 ms sub-frames of length $L = 40$ samples. In Figure 3.35 r_n is modelled over a sub-frame by a sequence v_n also of length L samples. The key point is that compression is achieved by constraining v_n to the form of Q ($Q < L$) *equispaced* non-zero samples (of different amplitude) spaced by zero-valued samples. In other words, v_n is an upsampled version of a certain optimal vector $\mathbf{b}_k$, and a pulse train or *regular-pulse excitation* (RPE) occurs at the input to $H(z)$ at the decoder. Looked at another way, a block of L residual samples e_n are represented by a vector $\mathbf{b}_k$ comprised of Q pulses.

The presence of zero-valued samples leads naturally to a set of N *candidate excitation vectors* $\mathbf{v}_k$, $k = 1, 2, ..., N$, where k denotes the position of the first non-zero sample in a particular sub-frame. Figure 3.36 is a simple illustration for $L = 40$, $N = 4$ and $Q = 10$; that is, there are four possible phases or positions of an excitation vector. The constraint on v_n enables it to be expressed as the excitation row vector

$$\mathbf{v}_k = \mathbf{b}_k \mathbf{M}_k, \quad k = 1, 2, \ldots, N \tag{3.86}$$

where $\mathbf{v}_k$ is of order L, $\mathbf{b}_k = [b_k(0) b_k(1) ... b_k(Q-1)]$ and $\mathbf{M}_k$ is a $Q \times L$ position matrix having entries of 1 or 0. In effect this means that $\mathbf{v}_k$ is entirely characterised by its position k and by the smaller vector $\mathbf{b}_k$. The coding objective is to select for each value of k the amplitudes $b_k(\cdot)$ that minimise some function of the (long-term) residual error e_n. The vector $\mathbf{b}_k$ that yields minimum error is then selected and transmitted, as indicted in Figure 3.35. Given the optimum value of $\mathbf{b}_k$, the decoding procedure is then straightforward.

Figure 3.35 simply illustrates the RPE concept, and a more detailed diagram for (full-rate) GSM is shown in Figure 3.37. Note at the outset that the encoder generates and transmits three different groups of data. As before, the 20 ms speech frame is divided into four 40 sample sub-frames, but now $Q = 13$ (corresponding to 52 RPE pulses over the 20 ms frame). Each pulse is quantised to 3 bits, and with additional side information, such as a 2-bit code identifying on of four grid positions, the net RPE bit rate is 9.4 kbit/s, as shown in Figure 3.37. Also note that the decoder uses a long-term or pitch synthesis filter

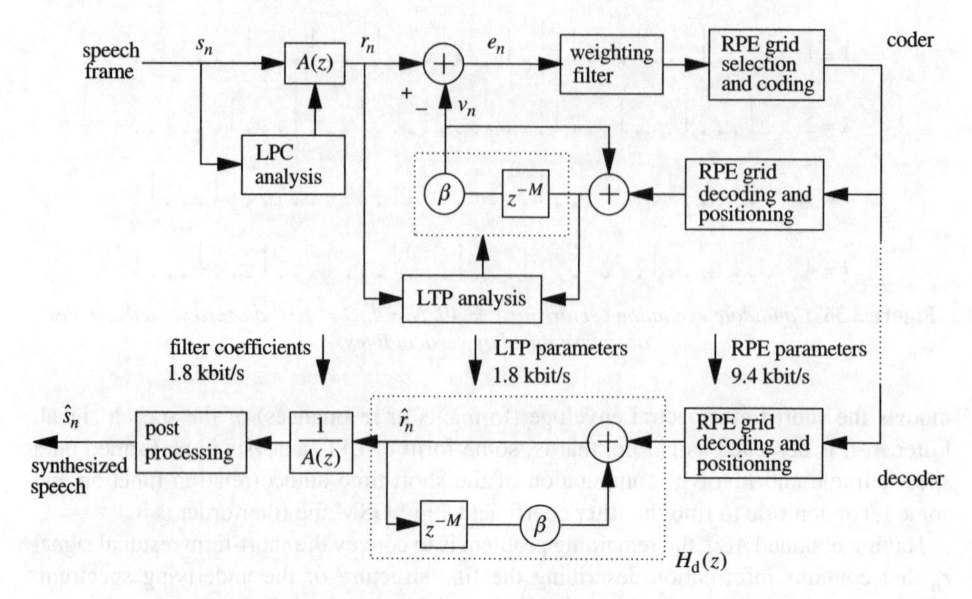

Figure 3.37 *Signal paths for a full-rate RPE-LTP (GSM) speech codec.*

$$H_{\rm d}(z) = (1 - \beta z^{-M})^{-1} \tag{3.87}$$

in order to model the major pitch pulses. In other words, this generates the periodic structure of the excitation for voiced sounds. The use of this, together with corresponding long-term prediction (LTP) at the encoder has the beneficial effect of reducing the speech error $s_n - \hat{s}_n$ (Kroon *et al.*, 1986). Transmission of the LTP parameters together with the filter coefficients requires a further 3.6 kbit/s, giving a net average bit rate for full rate GSM of 13 kbit/s. In fact, the GSM encoder can be defined as a transcoder between 13-bit uniform 8 kHz PCM (104 kbit/s) and 13 kbit/s. In a back-to-back configuration the transcoder delay is defined as the interval between the instant a speech frame of 160 samples has been received at the encoder input and the instant the corresponding 160 reconstructed samples have been output by the speech decoder at an 8 kHz sample rate. The theoretical minimum delay is 20 ms. The system gives communications quality speech (an MOS of about 3.5) and a performance that degrades 'gracefully' as channel bit-error rate increases.

3.4.3 Principles of CELP coding

Code Excited Linear Predictive (CELP) coding (Schroeder and Atal, 1985) is one of the most efficient approaches to high-quality speech coding at low bit rates, and it bridges the gap between waveform encoders and conventional *analysis/synthesis* techniques (vocoders). Table 1.4 shows that there are a number of versions of CELP with bit rates varying from 4.8 kbit/s to 16 kbit/s, and toll quality is achievable at 8 kbit/s for CS-ACELP (Conjugate Structure-Algebraic Code Excited Linear Prediction).

The fundamental concepts can be described with the aid of Figure 3.38 (this was in fact the configuration used in early CELP experiments). Over a short speech frame prediction

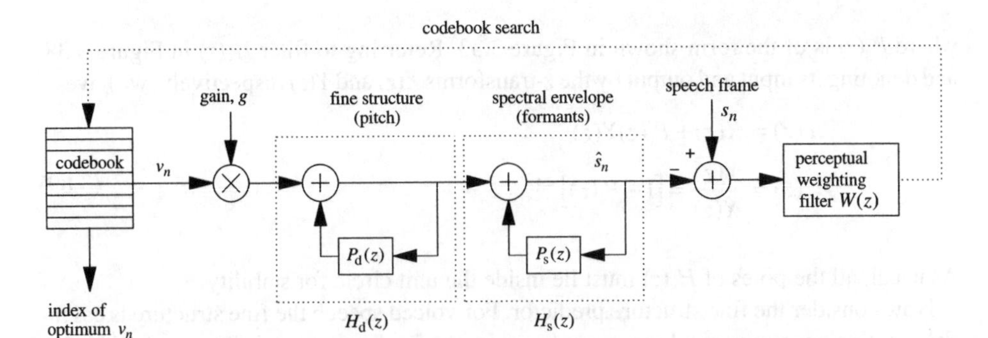

Figure 3.38 *Illustrating the prediction, codebook search and error-shaping concepts of CELP coding.*

$\hat{s}_n$ is derived from filtering a sequence v_n derived from a *codebook*. As in RPE-LTP coding we call this an excitation or *innovation* sequence, and it can be a succession of independent Gaussian random numbers having zero mean and unit variance (a *stochastic* codebook). It is common to refer to the different possible sequences from the codebook as code vectors. In the original CELP proposal the optimal excitation or codevector is determined by an exhaustive search, as indicated in Figure 3.38. Codebook samples v_n are then scaled by a gain factor g to produce an optimum match between the true signal s_n and the synthetic signal $\hat{s}_n$. To illustrate the basic concept, suppose speech sampled at 8 kHz is analysed over a 5 ms frame. The 40 samples s_n in this frame are compared to samples $\hat{s}_n$ derived from filtering a sequence v_n of 40 samples from the codebook. Sequence v_n is typically one of 1024 possible code vectors, so the optimal vector is identified by a 10-bit index. Transmission of this index therefore corresponds to a basic rate of 0.25 bit/sample or 2 kbit/s. This low bit rate is necessary in order to bring the total bit rate down to 4.8 kbit/s, a rate low enough to permit digital speech transmission over an analogue voice channel.

Determination of the optimum codevector is done using an analysis by synthesis approach. In other words, we assume that sequence v_n is filtered in such a way as to restore the essential spectral components of the speech. The reason for a cascade of two filters in Figure 3.38 is embedded in the classical assumptions behind speech synthesis. As discussed in Section 3.4.1, a speech segment can be either voiced or unvoiced, and its spectrum can be regarded as being shaped by the acoustic resonant system of the vocal tract. As in Figure 3.34, the speech source defines the *fine structure* of the spectrum (which could be line harmonics of a fundamental frequency or a continuous 'noise-like' spectrum) and this is then shaped to give the actual spectral envelope. Therefore, in Figure 3.38, predictor $P_d(z)$ essentially restores the spectral fine structure, while $P_s(z)$ restores the spectral envelope. As for LPC, prediction based on the spectral envelope involves relatively short delays and is typically 10th-order, that is

$$P_s(z) = \sum_{i=1}^{10} a_i z^{-i} \tag{3.88}$$

where $P_s(z)$ is of the form shown in Figure 3.33. Referring to filter $H_s(z)$ in Figure 3.38, and denoting its input and output by the *z*-transforms $X(z)$ and $Y(z)$ respectively, we have

$$Y(z) = X(z) + P_s(z)Y(z)$$

$$H_s(z) = \frac{Y(z)}{X(z)} = [1 - P_s(z)]^{-1} \tag{3.89}$$

As usual, all the poles of $H_s(z)$ must lie inside the unit circle for stability.

Now consider the fine structure predictor. For voiced speech the fine structure is essentially a line spectrum resulting from harmonics of a fundamental frequency, and this suggests that the associated filter $H_d(z)$ should have regular peaks in its frequency response corresponding to the line harmonics. We therefore need a transfer function of the form

$$H_d(z) = [1 - P_d(z)]^{-1} \tag{3.90}$$

where $P_d(z) = \beta z^{-M}$ and β and M are positive constants. In this case it is readily shown that $H_d(z)$ has poles at

$$z = \beta^{1/M} e^{j2\pi m/M}, \quad m = 0, 1, \ldots, M-1 \tag{3.91}$$

corresponding to resonant peaks at frequencies m/MT, where T is the sampling period. In other words, the fundamental or pitch period (typically 2–20 ms) will correspond to M sample periods. For unvoiced speech, which has a noise-like spectrum, β will be small and M would be random.

The objective of the *perceptual weighting filter*, $W(z)$, is to minimise the subjective distortion in the synthetic signal by shaping the speech error spectrum. This is of crucial importance for realising low bit rates. Error components falling in frequency regions where speech energy is concentrated (formant regions) can have higher energy relative to other components, since they will be masked by the speech itself. Filter $W(z)$ therefore attenuates those frequencies where the error is perceptually less important (that is, the formant or high-energy frequencies) and amplifies those lower energy frequencies where the error is perceptually more important.

Summary of CELP signal processing

We can now summarise the essential signal processing required in a typical CELP encoder. For example, the US DoD FS-1016 CELP standard (Table 1.4) samples at 8 kHz, has a 30 ms frame size (240 samples/frame), and reserves 10 bits for a 1024 vector codebook (Campbell *et al.*, 1989). The CELP analysis consists of three basic functions:

1. Short delay spectral envelope prediction once per frame for $P_s(z)$. This uses a 30 ms Hamming window and 10th-order LPC analysis.
2. Pitch search four times per frame (every 7.5 ms).
3. 1024 vector codebook search (a 'closed-loop' search four times per frame to select the optimum codebook index by minimising the weighted mean-square error).

CELP decoding (Figure 3.39) consists of the same three functions in reverse order.

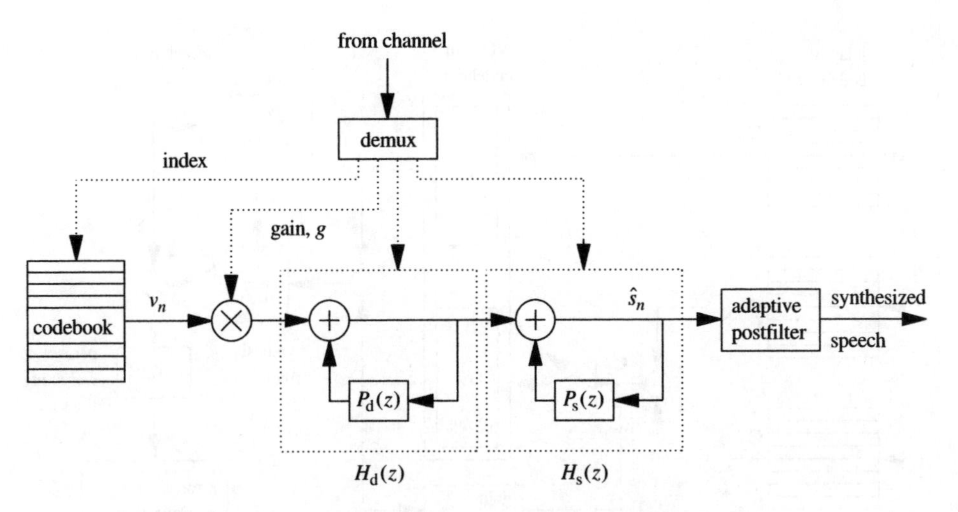

Figure 3.39 *Signal processing concepts used for CELP decoding.*

3.4.4 Practical CELP systems

The encoder in Figure 3.38 and the decoder in Figure 3.39 are only conceptual, and practical CELP codecs have a refined structure – see Figure 3.40. The main changes occur in the codebook and filtering. First note that the pitch filter $H_d(z)$ (the source of periodic excitation) has been replaced by an *adaptive codebook.* This provides the necessary periodic excitation vector, $\mathbf{C}_p$, corresponding to the vibration of the vocal chords. The codebook method is found to be superior to the pitch filter approach, especially for high-pitched speakers. A *random codebook* provides the necessary *random excitation.* Also, it is usual to refer to the spectral envelope filter $H_s(z)$ in Figure 3.38 as a *synthesis* filter $1/A(z)$, as in LPC. As discussed, the coefficients for this filter are provided by conventional LPC analysis. The input to this filter is therefore a sum of the pitch excitation and the random excitation, as in conventional LPC synthesis. Looked at another way, we could view polynomial $A(z)$ as providing short-term prediction and the adaptive codebook as providing the long-term prediction (LTP).

In practice the random codebook is usually split into two parts and the corresponding random excitation vector, $\mathbf{C}_r$, is the sum of two sub-excitation vectors:

$$\mathbf{C}_r = \theta_1 \mathbf{C}_1 + \theta_2 \mathbf{C}_2 \tag{3.92}$$

where θ_1 and θ_2 are signs. This is termed a *conjugate structure* and gives rise to the term conjugate structure CELP (CS-CELP). The two excitation gains g_p and g_r are held in a *vector quantised* (VQ) codebook (which also has a conjugate structure). A conjugate codebook structure gives increased robustness since the sum of two code vectors is less sensitive to channel errors than a single code vector. It also reduces memory requirements and helps to reduce the complexity of the random codebook search. The best combination of $\mathbf{C}_p$, $\mathbf{C}_r$, g_p and g_r is searched for by minimising the weighted mean squared error of the synthesised speech to the input speech.

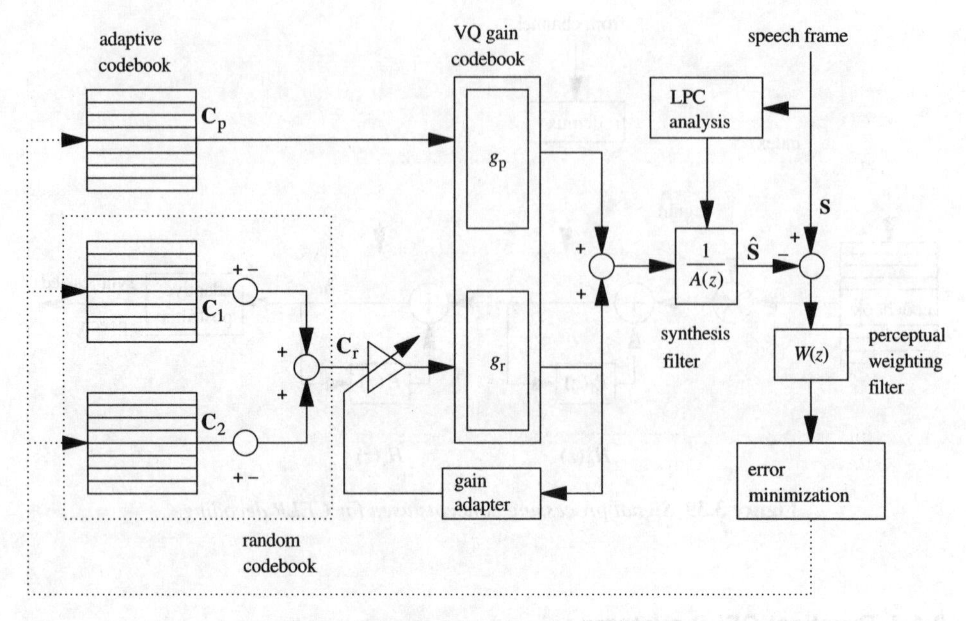

Figure 3.40 *The CS-CELP speech encoder.*

Bibliography

Alexander, S. and Rajala, S. (1985) Image compression results using the LMS adaptive algorithm, *IEEE Trans. Acoustics, Speech, and Signal Proc.*, **ASSP-33**(3), 712–714.

Antonini, M., Barlaud, M., Mathieu, P. and Daubechies, I. (1992) Image coding using wavelet transform, *IEEE Trans. on Image Processing*, **1**(2), 205–220.

Atal, B. and Schroeder, M. (1984) Stochastic coding of speech signals at very low bit rates, *Science, Systems & Services for Communications*, IEEE/Elsevier Science Publishers, North Holland, pp. 1610–1613.

Brainard, R. Netravali, A. and Pearson, D. (1982) Predictive coding of composite NTSC color television signals, *SMPTE Journal*, March, 245–252.

Brebner, G. and Ritchings, R. (1988) Image transform coding: a case study involving real time signal processing, *IEE Proc.*, **135** Pt. E (1), 41–48.

Campbell, J., Welch, V. and Tremain, T. (1989) An expandable error-protected 4800 bps CELP coder (US Federal Standard 4800 bps voice coder), *Proc. ICASSP, IEEE*, 735–737.

Chen, W., Smith, C. and Fralick, S. (1977) A fast computational algorithm for the discrete cosine transform, *IEEE Trans. Communications*, **COM-25**, 1004–1009.

Cinkler, K. (1998) Very low bit-rate wavelet video coding, *IEEE Journal on Selected Areas in Communications*, **16**(1), 4–11.

Cohn, D. and Melsa, J. (1975) The residual coder – an improved ADPCM system for speech digitisation, *IEEE Trans. Communications*, **COM-23**(9), 935–941.

Corset, I. (1994) MPEG2: A standard for Digital Moving Picture Coding, *Audio and Video Digital Radio Broadcasting Systems and Techniques*, Elsevier Science BV, Amsterdam.

Devereux, V. (1975) Digital video: differential coding of PAL colour signals using same-line and two-dimensional prediction', *Report* **1975/20**, British Broadcasting Corporation Research Department.

Eckart, S. and Fogg, C. (1995) ISO/IEC MPEG-2 software video codec, *SPIE*, **2419**, 100–109.

ETSI (1991) *GSM full rate speech transcoding*, European Telecommunications Standards Institute, Valbonne Cedex, France, **GSM 06.10** (Version 3.2.0).

Flanagan, J. *et al.* (1979) Speech coding, *IEEE Trans. Communications*, **COM-27**(4), 710–736.
Hart, J. *et al.* (1982) Manipulation of speech sounds, *Philips Tech. Review*, **40**(5), 134–145.
Heuval, A. *et al.* (1991) A spectrum efficient combined speech and channel coding method providing high voice quality for land mobile radio systems, *IEE Colloquium on Future Mobile Radio Trunking and Data Systems*, Savoy Place, February, Digest **1991/051.**
Jain, A. (1981) Image data compression: a review, *Proc. IEEE*, **69**(3), 349–381.
Kataoka, A., Moriya, T. and Hayashi, S. (1996) An 8-kb/s conjugate structure CELP (CS-CELP) speech coder, *IEEE Trans. on Speech and Audio Processing*, **4**(6), 401–411.
Knee, M. and Wells, N. (1989) Comparison of DPCM prediction strategies for high-quality digital television bit-rate reduction, *Report* **1989/8**, British Broadcasting Corporation Research Department.
Kroon, P. *et al.* (1986) Regular-pulse excitation – a novel approach to effective and efficient multipulse coding of speech, *IEEE Trans. Acoustics, Speech and Signal Proc.*, **ASSP-34**(5), 1054–1063.
Lewis, A. and Knowles, G. (1991) A VLSI architecture for the 2D Daubechies wavelet transform without multipliers, *Electronics Letts.*, **27**(2), 171–173.
Makhoul, J. (1975) Linear prediction: a tutorial review, *Proc. IEEE*, **63**(4), 561–580.
Max, J. (1960) Quantising for minimum distortion, *IRE Trans. Inform. Theory*, **IT-6**, 7–12.
MPEG Software Simulation Group (MSSG) Test Model 5 (TM5).
Musmann, H. G. *et al.* (1985) Advances in picture coding, *Proc. IEEE*, **73**(4), 523–547
Narasimha, M. and Peterson, A. (1978) On the computation of the discrete cosine transform, *IEEE Trans. Communications*, **COM-26**(6), 934–936.
Netravali, A. and Limb, J. (1980) Picture coding: a review, *Proc. IEEE*, March, 366–406.
O'Neal Jr, J. (1966) Predictive quantising systems (differential pulse code modulation) for the transmission of television signals, *Bell Syst. Tech. J.*, May–June, 689–721.
Paez, M. and Glisson, T. (1972) Minimum mean-squared-error quantisation in speech PCM and DPCM systems, *IEEE Trans. Communications*, April, 225–230.
Paris, J. *et al.* (1997) Low bit rate software-only wavelet video coding, *IEEE First Workshop on Multimedia Signal Processing*, pp. 169–174.
Read, D. (1974), Digital video: some bit-rate reduction methods which preserve information in broadcast-quality digital video signals, *Report* **1974/37**, British Broadcasting Corporation Research Department.
Rioul, O. (1993) A discrete-time multiresolution theory, *IEEE Trans. on Signal Processing*, **41**(8), 2591–2606.
Safranek, R. and Johnson, J. (1989) A perceptually tuned sub-band image coder with image dependent quantisation and post-quantisation data, *Proc. IEEE ASSP89*, **3**, 1945–1948.
Schroeder, M. and Atal, B. (1985) Code-excited linear prediction (CELP): high-quality speech at very low bit rates, *IEEE Proc. ICASSP'85 – International Conf. On Acoustics, Speech and Signal Processing*, pp. 937–940.
Sikora, T. (1997) MPEG Digital Video-Coding Standards, *IEEE Signal Processing Magazine*, September, 82–100.
Smith, B. (1957) Instantaneous companding of quantised signals, *Bell Syst. Tech. J.*, **36**, 653–709.
Suehiro, N. and Hatori, M. (1986) Fast algorithms for the DFT and other sinusoidal transforms, *IEEE Trans. Acoustics, Speech, and Signal Proc.*, **ASSP-34**, June, 642–644.
Vary, P., Sluyter, R., Galand, C. and Rosso, M. (1987) RPE-LTP codec – the candidate for the GSM radio communication system, *Internat. Conf. On Digital Land Mobile Radio Communications*, Venice, 30 June–3 July, pp. 507–516.
Vary, P. *et al.* (1988) A regular-pulse excited linear predictive codec, *Speech Communication, North Holland*, **7**, 209–215.
Wang, Z. (1984) Fast algorithms for the discrete *W* transform and for the Discrete Fourier Transform, *IEEE Trans. on Acoustics, Speech, and Signal Processing*, **ASSP-32**(4), 803–816.
Yan, J. and Donaldson, R. (1972) Subjective effects of channel transmission errors on PCM and DPCM voice communication systems, *IEEE Trans. Communications*, June, 281–290.

CHAPTER 4

Transmission codes and pseudorandom codes

In this chapter we consider aspects of channel coding other than error control coding. The broad objective of channel coding is to 'match' the source coder output to the channel, and error control coding is just one of several techniques used to achieve this objective. In practice we also need to shape the signal spectrum, and this can be achieved by using pulse shaping filters, or *transmission codes* (such as *line* codes or *modulation* codes), or by simple coding techniques such as *partial response* (PR) signalling.

In addition to the use of error control coding and transmission coding, a communications system will frequently use *pseudorandom* or *pseudonoise* (PN) sequences. The objective here may be to provide *timing recovery*, *conditional access*, *code division multiple access* (CDMA) or *spread spectrum* communications.

4.1 Pulse shaping

Figure 4.1(a) shows a typical QPSK modulation/demodulation scheme for use on a satellite channel. The role of the transmit filters $T(f)$ is to shape the power spectral density of the transmitted signal, and filters $V(f)$ could simply be detection filters. The analogue satellite channel is generally *nonlinear* (due to TWTA nonlinearity), but for analysis purposes we will *assume* that the channel is linear, so that the overall system can be considered to be a bandlimited channel comprised of a cascade of linear filters. The satellite channel is bandpass, but as a further simplification we can consider its equivalent *lowpass* transfer characteristic, $C(f)$, and dispense with the digital modulation. This brings us to the general *baseband* model of a communications system in Figure 4.1(b). The baseband input signal $x(t)$ could be written

$$x(t) = \sum_n a_n g(t - nT) \tag{4.1}$$

where, for example, $a_n = \pm 1$ for BPSK, and $g(t)$ is a rectangular gate pulse of unit amplitude and duration equal to the symbol or signalling period T. The basic pulse shape $g(t)$ is now changed by the overall filtering function of Figure 4.1(b). Suppose for example that the input to the transmit filter in Figure 4.1(b) is a single pulse $g(t) \Leftrightarrow G(f) = T\sin(\pi fT)/\pi fT$, where $G(f)$ is the Fourier transform (or spectrum) of $g(t)$. The pulse spectrum at the detector input is then

$$R(f) = G(f)T(f)C(f)V(f) \tag{4.2}$$

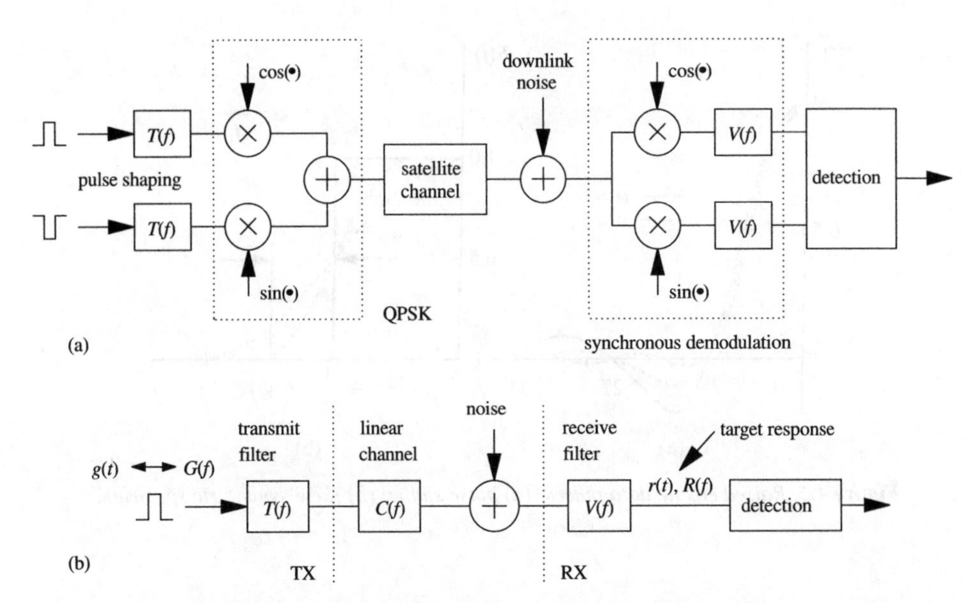

Figure 4.1 *Pulse shaping: (a) in a QPSK system; (b) in a general baseband communications system.*

where the corresponding pulse shape is $r(t) \Leftrightarrow R(f)$. Clearly, if we assume that the equivalent lowpass response $C(f)$ is 'flat' or ideal (often a reasonable assumption in broadcasting channels and satellite channels), then (4.2) implies it is possible to adjust $T(f)$ and $V(f)$ in order to obtain a specific pulse shape/spectrum at the detector input. Usually the pulse shape $r(t)$ is designed according to Nyquist's *first criterion*, that is

$$r(t) = 0 \qquad t = \pm kT \qquad k = 1, 2, 3, \ldots \tag{4.3}$$

In this way, subsequent pulses centred on $t = T, 2T, 3T, \ldots$ will not suffer *intersymbol interference* (ISI) from the pulse at $t = 0$ since the latter will have zero amplitude at the optimum sampling instants. The constraint in (4.3) translates into constraints in the frequency domain. In particular, Nyquist showed that if the pulse spectrum $R(f)$ is *real* then it must have odd (or skew) symmetry about a frequency

$$f_1 = 1/2T \tag{4.4}$$

For a binary system, f_1 corresponds to half the bit rate.

4.1.1 Raised cosine filters

There are many spectra that can satisfy the odd-symmetry criterion, but the class of functions that is often (but not always) used in practice is shown in Figure 4.2. This class of functions is defined by

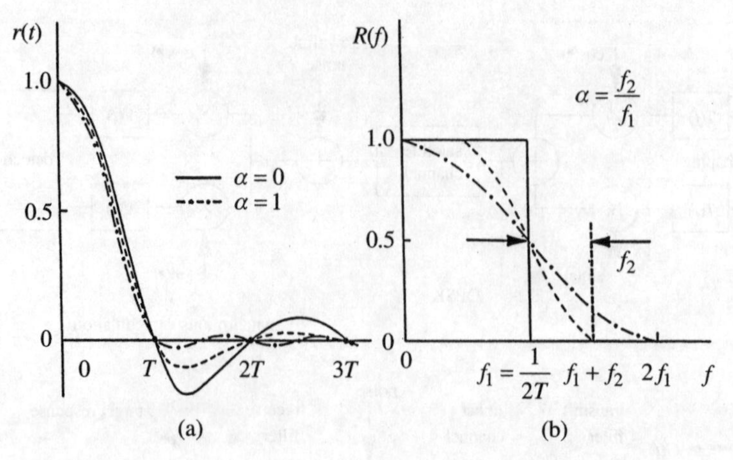

Figure 4.2 *Raised cosine data pulses: (a) pulse shape; (b) skew-symmetric spectrum.*

$$R(f) = \begin{cases} \dfrac{1}{2f_1} & f < f_1 - f_2 \\ \dfrac{1}{4f_1}\left\{1 - \sin\left[\dfrac{\pi}{2f_2}(f - f_1)\right]\right\} & f_1 - f_2 \le f \le f_1 + f_2 \\ 0 & f > f_1 + f_2 \end{cases} \tag{4.5}$$

$$r(t) = \frac{\sin(2\pi f_1 t)}{2\pi f_1 t} \cdot \frac{\cos(2\pi f_2 t)}{1 - (4 f_2 t)^2} \tag{4.6}$$

These pulses are referred to as *raised cosine* pulses due to the general shape of $R(f)$ (note that $R(f)$ is *real*). The significant point is that these pulses are *bandlimited* and so could be transmitted over an ideal but finite bandwidth channel without loss of pulse shape. This being the case, (4.3) means that zero ISI and 100% eye height could be achieved at the detector input.

Strictly speaking, (4.6) spans all time and so is unrealisable, although it can be well approximated in practice. Note that, when $f_2 = 0$, the transmitted pulses are of $(\sin x)/x$ shape and the bandwidth requirement is minimal; only f_1 Hz is required to transmit at a rate of $2f_1$ symbols/s (see Nyquist's bandwidth–signalling speed relationship, Theorem 1.2). Unfortunately, the rectangular spectrum is also unrealisable and, in the absence of transmission coding techniques, practical systems have to settle for a finite roll-off factor $\alpha = f_2/f_1$, that is, finite *excess bandwidth* (typically 50%). For 100% roll-off pulses ($\alpha = 1$), equation (4.5) reduces to a true raised cosine shape, that is

$$R(f) = 1 + \cos(\pi f T) \tag{4.7}$$

Practical shaping filters usually take the form of quite simple transversal FIR structures.

Example 4.1

Raised cosine pulses are often used in data broadcasting. For example, in teletext systems it is common for all deliberate pulse shaping to be concentrated at the transmitter, as shown in Figure 4.3. The idea is to shape and bandlimit the rectangular binary data pulses so that minimal pulse distortion occurs during transmission. Using raised cosine pulses, the spectrum $P(f)$ of a transmitted pulse could be limited to the channel bandwidth f_c, corresponding to a roll-off factor of $(f_c - f_1)/f_1$. If the equivalent lowpass channel is ideal, the data at the detector input will then have zero ISI, that is, $r(t) = p(t)$. As an example, the data rate in the UK teletext system is 6.9375 Mbit/s, and this corresponds to a maximum roll-off of 44.1% (taking $f_c = 5$ MHz to be compatible with other video systems). Alternatively, we could use a nominal 70% roll-off and truncate the spectrum before transmission; this may be beneficial despite a slight increase in ISI (Kallaway and Mahadeva, 1977).

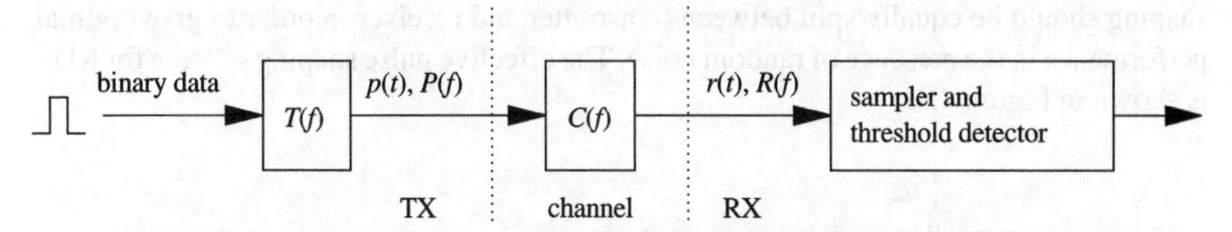

Figure 4.3 *A typical pulse-shaping scheme for teletext (P(f) is often a raised cosine function).*

4.1.2 Root-raised cosine filters

So far we have assumed that the role of the transmit and receive filters is to minimise ISI at the receiver. A more fundamental objective is to select these filters to minimise the combined effect of ISI and channel noise. If we insist on a particular response $R(f)$ at the detector in order to minimise ISI, the remaining question is how to divide the overall filtering between $T(f)$ and $V(f)$ in order to minimise the effect of noise. It is well known that the optimum detector for an isolated pulse is the *matched filter*, since it maximises the SNR at the detector sampling instant, and hence minimises the probability of error. In fact, if we assume an ideal channel this concept can be extended to a sequence of pulses because there is no risk of ISI, that is, the optimum receive filter will still be a matched filter. More specifically, under the assumptions of an impulse input to transmit filter $T(f)$, an ideal channel ($C(f) = 1$), a flat channel noise spectrum, and a *real* target response $R(f)$, maximum SNR at the detector sampling instant can be achieved by dividing the overall magnitude filtering *equally* between transmitter and receiver (Lucky *et al.*, 1968). In other words, given an impulse input in Figure 4.1(b), the error probability is minimised if half the pulse shaping is done by the transmit filter and half by the receive matched filter, that is

$$T(f) = V(f) = \sqrt{R(f)} \tag{4.8}$$

We stress that (4.8) is applicable to narrow, 'impulse-like' data pulses. Since the gate pulse $g(t)$ in Figure 4.1(b) has a $(\sin x)/x$ type transform, then a compensating filter of response $x/\sin(x)$ could precede $T(f)$ in order to apply (4.8). As discussed, it is usual to make $R(f)$ a raised cosine response, so the transmit and receive filters are *root* raised cosine ($\sqrt{RC}$) filters. For roll-off α the impulse response of an $\sqrt{RC}$ filter is given by

$$h(t) = \frac{\alpha}{T}\cos\left(\frac{\pi t}{T}+\frac{\pi}{4}\right)\operatorname{sinc}\left(\frac{\pi\alpha t}{T}+\frac{\pi}{4}\right)+\frac{1-\alpha}{T}\operatorname{sinc}\left(\frac{\pi(1-\alpha)t}{T}\right) + \frac{\alpha}{T}\cos\left(\frac{\pi t}{T}-\frac{\pi}{4}\right)\operatorname{sinc}\left(\frac{\pi\alpha t}{T}-\frac{\pi}{4}\right) \tag{4.9}$$

where sinc(x) = sin(x)/x.

Example 4.2

The European radio data system or RDS (CENELEC, 1990) specifies that spectrum shaping should be equally split between transmitter and receiver in order to give optimal performance in the presence of random noise. The effective pulse shaping scheme for RDS is shown in Figure 4.4.

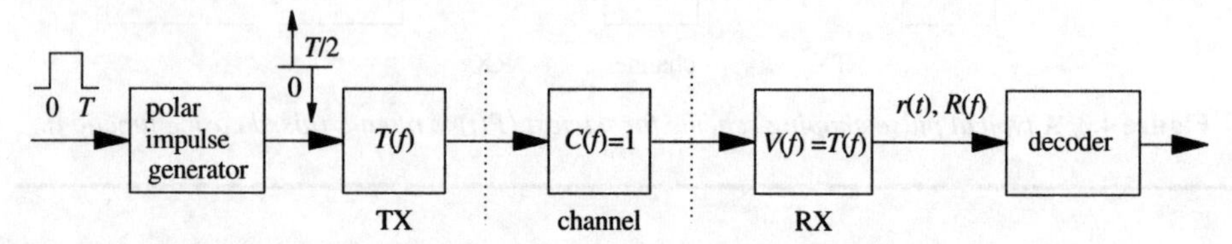

Figure 4.4 *Effective pulse shaping scheme for the European RDS (r(t) is a biphase symbol).*

If the source has a signalling period T, each filter has a transfer function of the form

$$T(f) = V(f) = \begin{cases} \cos\left(\dfrac{\pi f T}{4}\right) & f \le 2/T \\ 0 & f > 2/T \end{cases}$$

so the overall spectrum shaping ($T(f)V(f)$) is of the form $1+\cos(\pi f T/2)$. For RDS the basic data rate is 1187.5 bits/s, so the combined response of the shaping filters will fall to zero at 2375 Hz. Note that the effective signalling period is $T/2$ rather than T, since a binary '1' pulse at the source generates an odd impulse pair separated by $T/2$ seconds, as indicated. This, in turn, generates a *biphase* or *Manchester* coded symbol $r(t)$ of nominal duration T (see Section 4.2.2). If we regard the signalling period as being $T/2$ then, according to (4.7), the overall data channel spectrum shaping corresponds to 100% cosine roll-off.

Example 4.3

Orthogonal frequency division multiplexing (OFDM) splits N complex data symbols $c_k = a_k + jb_k$, $k = 0, 1, ..., N$ of period T amongst N equally spaced carriers, as shown in Figure 4.5 (for QPSK-OFDM, a_k, $b_k \in \{\pm 1\}$). This scheme extends the effective symbol period to $T_s = NT$, resulting in a system that is very effective at combating ISI arising from multipath reception (and hence is very attractive for mobile environments). In order to reduce out of band spectral components one approach is to use $\sqrt{RC}$ filtering for each of the N channels, as shown in Figure 4.5. For impulse inputs this gives an overall raised cosine response at the receiver (delays $T_s/2$ being necessary to maintain orthogonality). In practice, all the $\sqrt{RC}$ filters and modulation/demodulation in Figure 4.5 could be efficiently implemented

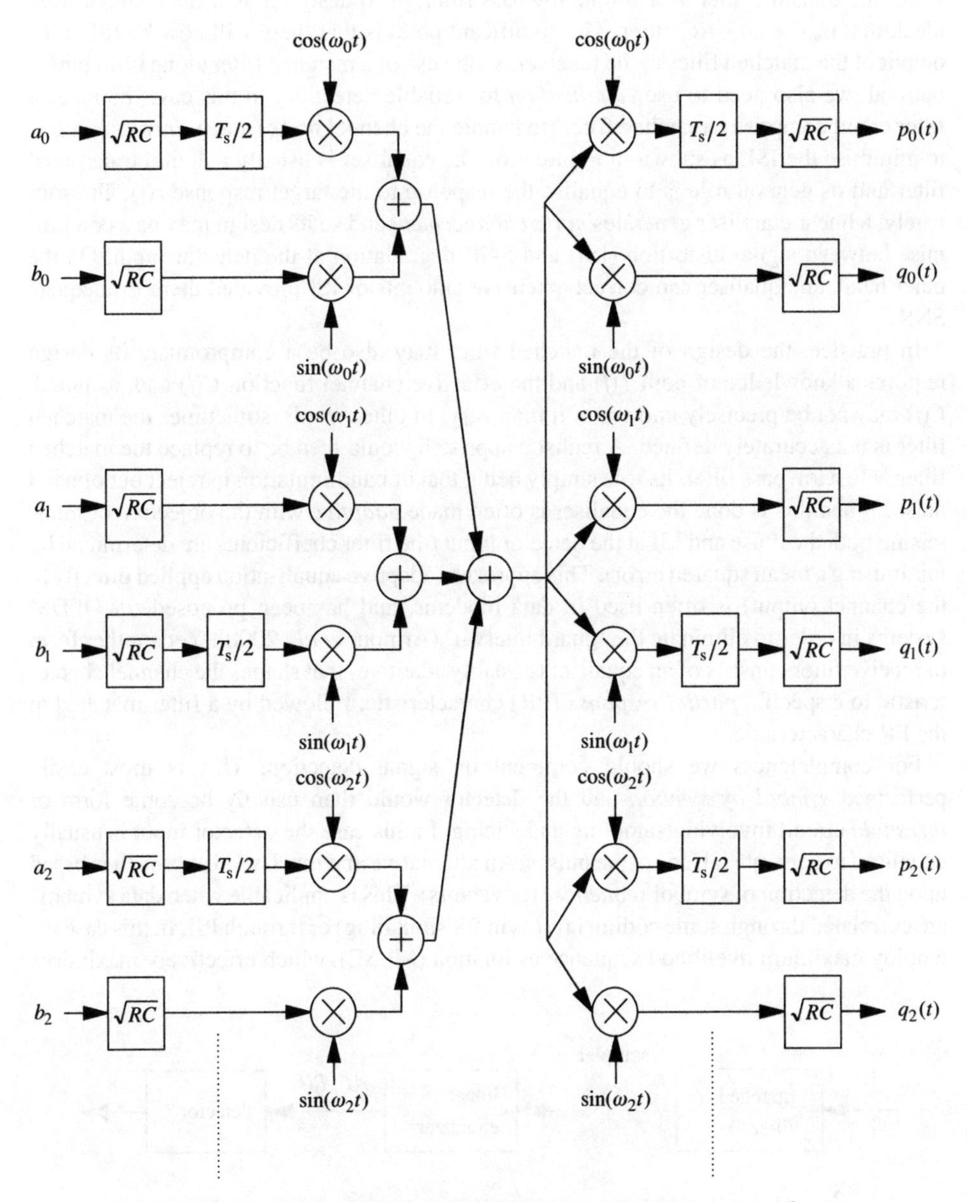

Figure 4.5 *Analogue model of an OFDM system with pulse/spectrum shaping* ($\omega_k - \omega_{k-1} = 2\pi / T_s$).

using DSP techniques based upon the FFT and polyphase filtering (Bhatoolaul and Wade, 1995).

The need for adaptive equalisation

So far we have assumed an ideal channel. When the channel is non-ideal it is common to select the transmit filter as a simple lowpass filter, or to design it as if the channel *were* ideal, that is, use an $\sqrt{RC}$ filter. The significant point is that there will now be ISI at the output of the matched filter in the receiver, so the use of a matched filter alone is no longer optimal; we also need to use *equalisation* for reliable detection. In this case the receive filter could comprise a matched filter (to handle the channel noise) and a *linear equaliser* to minimise the ISI, as shown in Figure 4.6. The equaliser is usually a digital transversal filter and its general role is to equalise the response to the target response $r(t)$. Unfortunately, a linear equaliser generates *noise enhancement* and so its design may be a compromise between signal distortion (ISI) and SNR degradation at the detector input. On the other hand, an equaliser can correct extensive amounts of ISI provided there is adequate SNR.

In practice, the design of the matched filter may also be a compromise. Its design requires a knowledge of both $T(f)$ and the effective channel function $C(f)$ and, as noted, $C(f)$ may not be precisely known, or it may vary. In other words, sometimes the matched filter is not accurately defined. A realistic approach would then be to replace the matched filter with a lowpass filter, its role simply being that of bandlimitation to reject out of band noise. When this is done the equaliser is often made *adaptive* with the objective of minimising both the noise and ISI at the detector input (the filter coefficients are determined by minimising a mean squared error). This approach (adaptive equalisation applied directly to the channel output) is often used in data modems, and has been proposed for OFDM systems in order to eliminate the 'guard interval' (Armour *et al.*, 2000). Yet another form of receive filter consists of an equaliser (possibly adaptive) that shapes the channel characteristic to a specific *partial response* (PR) characteristic, followed by a filter matched to the PR characteristic.

For completeness we should comment on signal detection. This is most easily performed *symbol by symbol*, and the detector would then usually be some form of *threshold* circuit involving sampling and slicing. In this case the detector input is usually equalised to generate raised cosine pulses. An alternative and much used approach is based upon the detection of symbol *sequences* (or vectors). This is applicable when data symbols are correlated through some coding law (as in PR signalling) or through ISI. In this case we employ maximum likelihood sequence estimation (MLSE), which effectively maximises

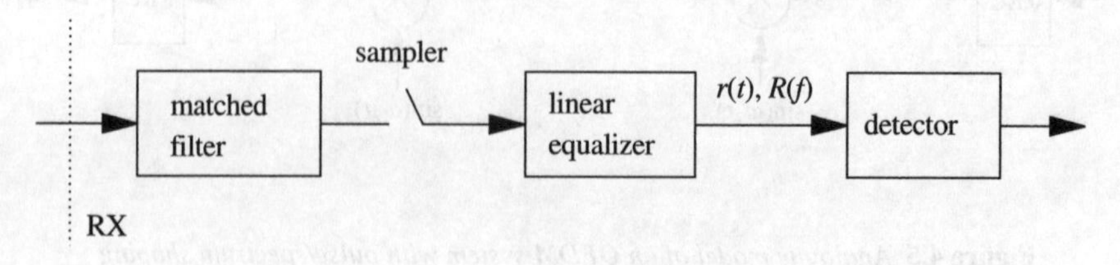

Figure 4.6 *Use of both matched filtering and equalisation.*

the probability density $p(\mathbf{r}|\mathbf{u})$, where $\mathbf{r}$ is the received sequence (or vector) and $\mathbf{u}$ is an assumed transmitted sequence. For equally probable sequences, this type of detection yields optimum performance (minimum sequence error probability) in the presence of ISI, albeit at the expense of increased complexity. The detector would be implemented via the Viterbi algorithm discussed in Section 7.3; essentially, it decides on an entire sequence of inputs, or path through a 'trellis', rather than on a symbol by symbol basis.

4.2 Line codes

Figure 4.7 shows another technique for shaping the transmitted spectrum. Broadly speaking it involves code translation and this is usually a *binary-to-binary* translation, or a *binary-to-ternary* translation. Among other advantages, this opens up the possibility of practical systems operating near the Nyquist limit of $2f_1$ symbols/s in a bandwidth of only f_1 Hz. When we are dealing with an actual communications system the final translated code is usually called an *interface code* or a *line code*, depending upon the application. On the other hand, the 'communications system' could be a magnetic recording channel for example, in which case the code used for transmission (writing) is usually called a *modulation* or *recording* code.

At this point it is helpful to examine the power spectral density (PSD) or power spectrum for a random data stream prior to transmission coding. Usually this is in non-return to zero (NRZ) unipolar format, and Figure 4.8 shows that the PSD is essentially limited to $1/T$ Hz. This means, for example, that a 100 Mbit/s NRZ signal could be transmitted without serious degradation down a channel of 100 MHz bandwidth. However, this format is poorly matched to most transmission channels for several reasons.

First, the PSD is zero at the bit frequency, so it is impossible to phase lock a receiver clock *directly* to the signal. However, a timing component and therefore locking *can* be achieved by first performing some non-linear operation on the NRZ signal (a common approach is to generate a short pulse at each transition or zero-crossing). Unfortunately there will still be no clock or timing information during long periods of 1 or 0. Secondly, for random data, a *polar* NRZ code (for example, $1 \equiv +1$ V, $0 \equiv -1$ V) will remove the DC component and the need for a DC coupled channel, but the PSD will still be significant near 0 Hz. This is inappropriate since most PCM and recording channels have a *bandpass* characteristic, with small low-frequency response and a null at zero frequency.

Generally speaking, the polar NRZ code is also unsuitable for recording on error propagation grounds. Consider the basic longitudinal magnetic recording mechanism in Figure 4.9. A polarity reversal in the NRZ write current occurs only if the current bit is not equal

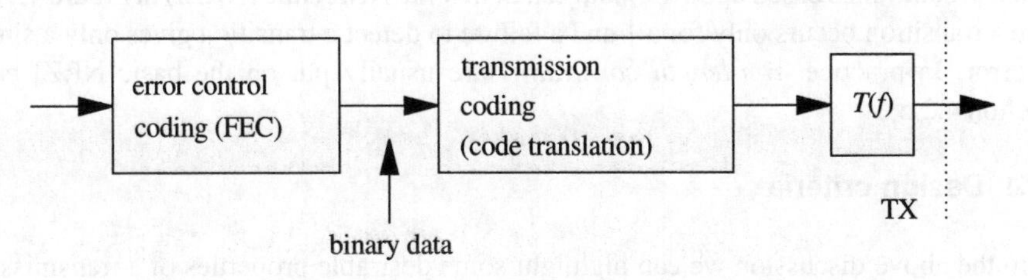

Figure 4.7 *The use of transmission coding for spectral shaping.*

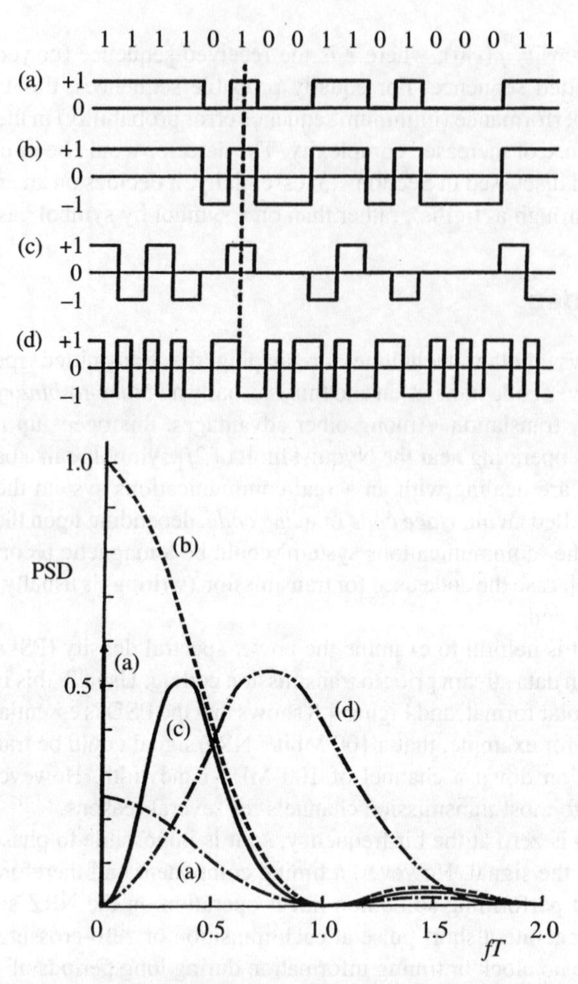

Figure 4.8 *PSD of some binary data formats: (a) NRZ unipolar; (b) NRZ polar; (c) AMI (bipolar); (d) Manchester (biphase).*

to the previous bit. A reversal generates a change in the direction of saturation and, upon readback, this transition generates a voltage pulse in the read head. Since a transition represents a switch from a string of 1s to a string of 0s (or vice versa) then failure to detect a readback pulse will give indefinite error propagation. For magnetic and optical recording, the usual solution is based upon the *modified* or *inverse* NRZ code (NRZI) in Figure 4.9(d). Here a transition occurs only for a 1 and a failure to detect a transition gives only a single bit error. In practice, *run-length constraints* are usually put on the basic NRZI code (Section 4.2.5).

4.2.1 Design criteria

From the above discussion we can highlight some desirable properties of a transmission code:

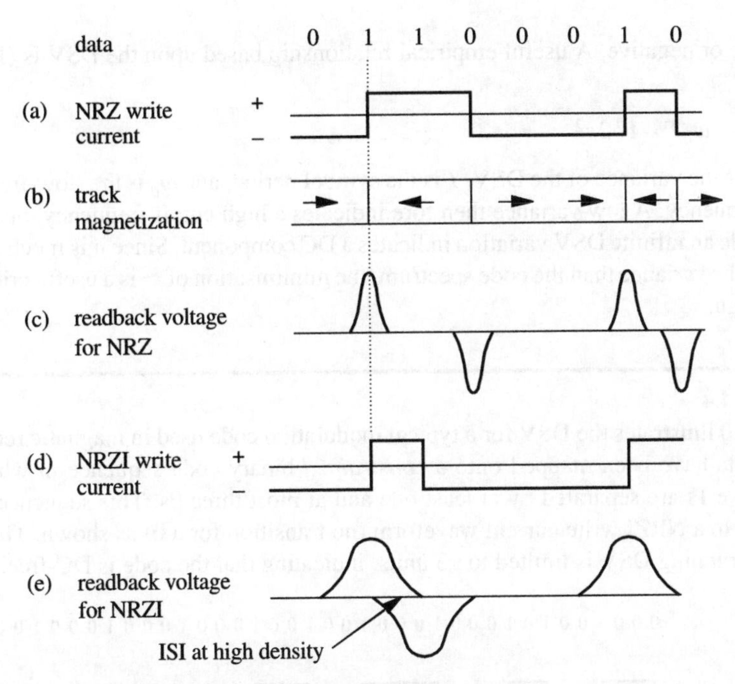

Figure 4.9 *Basic mechanism of saturated magnetic recording.*

1. A favourable PSD: it is often conjectured that the PSD of the coded data stream should 'match' the transfer function of the channel. For baseband (that is, non-carrier, AC-coupled) channels this usually means that the PSD should be small at low frequencies with preferably a null at DC. Otherwise, a code with significant low-frequency content, or a DC component, can give 'baseline (or DC) wander', which can generate errors if the detection system is sensitive to amplitude variation. However, baseline wander is less of a problem in *optical fibre* systems, since they can convey a DC component (corresponding to light constantly on or constantly off). Also, due to the inherently wide bandwidth of optical fibre, coding to optimise the signalling rate (and hence bandwidth) is not generally applicable.
2. Adequate timing content: the transmission code should enable the data clock to be extracted from the signal itself, irrespective of the binary data pattern from the source. This is a factor to be considered in both metallic and optical fibre systems.
3. Error detection capability: a good transmission code will have some error detection property.
4. Low error propagation: the average number of errors in the decoded binary sequence per transmission error should be low.

DC-free codes

A useful criterion for the design of DC-free and low-frequency-suppressed codes is based on the concept of running *digital sum value* (DSV), or *disparity count*. Basically, the DSV is increased or decreased by one unit depending whether the transmission code waveform is positive or negative. A useful empirical relationship based upon the DSV is (Justesen, 1982)

$$\omega_c T \approx 1/2s^2 \tag{4.10}$$

where s^2 is the variance of the DSV, T is the symbol period, and ω_c is the (low-frequency) cutoff frequency. A low variance therefore indicates a high cutoff frequency, or DC-free code, while an infinite DSV variation indicates a DC component. Since it is much easier to compute the variance than the code spectrum, the minimisation of s^2 is a useful criterion in code design.

Example 4.4

Figure 4.10 illustrates the DSV for a typical modulation code used in magnetic recording. Binary data have been mapped onto a *constrained* binary code sequence in which two consecutive 1s are separated by at least one and at most three 0s. This sequence is then converted to a NRZI write current waveform (no transition for a 0) as shown. The corresponding running DSV is limited to ±3 units, indicating that the code is DC-free.

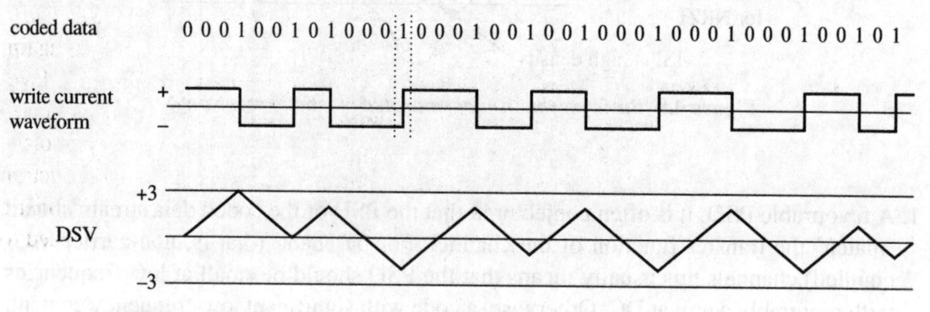

Figure 4.10 *Illustrating the running DSV for a DC-free modulation code.*

Example 4.5

The Miller[2] code is another example of a DC-free code, and is used in high bit rate instrumentation recording and in digital video recording systems. The *basic* Miller code (also known as the MFM code) encodes data by inserting a transition in the middle of a 1, and between two zeros, as shown in Figure 4.11. Unfortunately the Miller code is not DC-free,

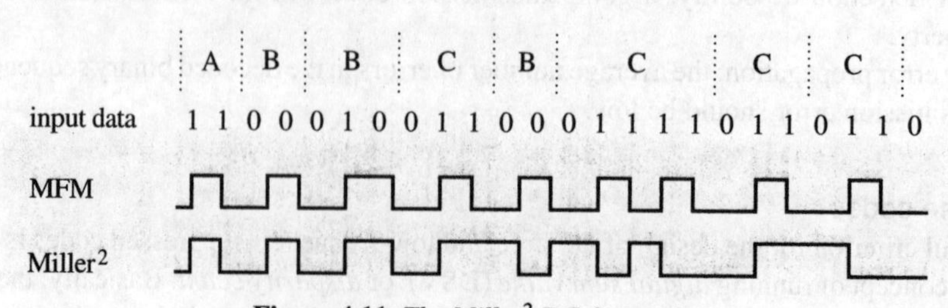

Figure 4.11 *The Miller[2] DC-free code.*

but it can be modified in order to remove the DC component, giving the Miller^2 code. Essentially, this is done by omitting the transition of the final 1 whenever a 0 is followed by an even number of 1s. This rule must be applied as part of a sequential coding procedure which breaks the input bit pattern into three sequence types:

Type A: any number of consecutive 1s.
Type B: two 0s separated by either no 1s, or an odd number of 1s.
Type C: one 0 followed by an even number of 1s.

Sequence types A and B are coded as in the basic Miller code, but the final 1 transition is omitted for type C sequences. Note from Figure 4.11 that a basic Miller transition occurs even when two 0s span adjacent sequence types.

Multilevel signalling

For some systems transmission coding is restricted to a binary-to-binary translation of the form nBmB, that is, n binary data bits are translated to m binary code bits ($m > n$) using a higher clock frequency. For instance, most optical fibre systems modulate a light carrier using simple on-off keying (OOK) and so require a binary transmission code. Similarly, digital magnetic recording magnetises the media into just two stable states and so, again, requires a binary transmission code. More generally, we can employ a binary to L level translation ($L \geq 3$) and hence change the symbol (baud) rate. In simplistic terms, a binary data stream of R bits/s could be translated via an n-bit DAC to a stream of R/n symbols/s, that is, an analogue signalling waveform of 2^n levels. This is attractive since the reduction in symbol rate by a factor n also reduces the required bandwidth by a factor n (Nyquist's bandwidth–signalling speed relationship). The number of levels is limited by the need to maintain acceptable error rate performance.

4.2.2 Manchester (biphase) code

This code is found in the European RDS, local area networks, low-bit-rate weather satellite systems and some digital recording systems. In order to understand the code it is helpful to model the transmitted baseband waveform as the output of a filter whose unit impulse response is the required pulse shape $p(t) \Leftrightarrow P(\omega)$ (Figure 4.12). The filter input would be a random impulse train (corresponding to the data sequence) having PSD $S_\delta(\omega)$. The impulse train would comprise positive and negative impulses and the strength of each

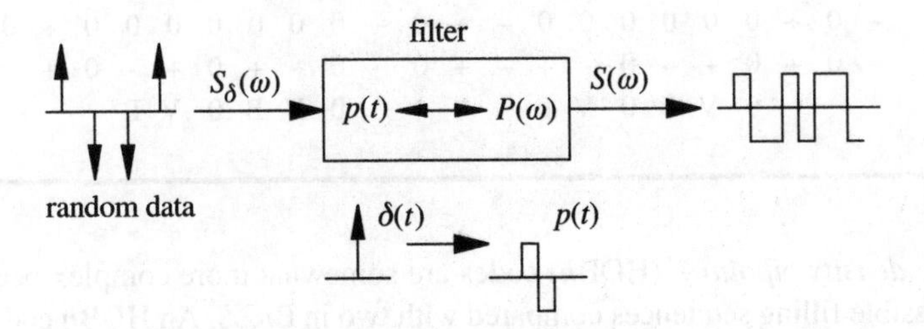

Figure 4.12 *Model for Manchester code generation.*

impulse scales $p(t)$ to give the required transmission waveform. The PSD of the transmitted waveform is then

$$S(\omega) = S_\delta(\omega)\,|P(\omega)|^2 \tag{4.11}$$

It is apparent from (4.11) that $S(\omega)$ can be controlled by changing either the transmission code or the pulse shape, and Manchester coding uses the latter approach. In particular, since $P(\omega)$ is the Fourier transform of $p(t)$, then a null at DC can be achieved *irrespective of the incoming data sequence* by setting

$$P(0) = \int_{-\infty}^{\infty} p(t)\,\mathrm{d}t = 0 \tag{4.12}$$

The Manchester code achieves this by using a 'biphase' pulse shape $p(t)$ that has a positive to negative transition in the middle of a 1; see Figure 4.8. The shape $-p(t)$ is assigned to each binary 0. Figure 4.8 shows that the high timing content is paid for by a high bandwidth requirement.

4.2.3 AMI and HDB*n* codes

The alternate mark inversion (AMI) code in Figure 4.8 (a *pseudo* ternary code) removes the DC component by inverting alternate 1s or marks. In addition, it is easy to generate and decode, it is bandwidth-efficient relative to the Manchester code, and it has a single error detection capability (a single error causes a violation of the AMI rule). Unfortunately, while timing information can generally be recovered via a non-linear operation (as for NRZ), this mechanism will fail for long sequences of 0s.

In order to obtain higher timing content the AMI code can be modified by replacing strings of zeros with a *filling sequence*. The B*n*ZS code for example follows the AMI rule unless *n* zeros occur together, in which case a filling sequence is used. Taking $n = 6$, this sequence is 0VB0VB, where V denotes a violation of the AMI rule and B denotes a mark obeying the AMI rule. The B*n*ZS codes are specified by the CCITT for use in systems based on 1.544 Mbit/s.

Example 4.6

(+ and – represent positive and negative marks).

Binary	1	0	1	0	0	0	0	0	0	1	1	0	1	0	0	0	0	0	0	0	1	0	1	1	1
AMI	–	0	+	0	0	0	0	0	0	–	+	0	–	0	0	0	0	0	0	0	+	0	–	+	–
B6ZS	–	0	+	0	+	–	0	–	+	–	+	0	–	0	–	+	0	+	–	0	+	0	–	+	–
				0	V	B	0	V	B					0	V	B	0	V	B						

The *high-density bipolar n* (HDB*n*) codes are somewhat more complex because there are four possible filling sequences compared with two in B*n*ZS. An HDB*n* code permits a string of *n* zeros but fills a string of $n + 1$ zeros with one of two sequences, each $n + 1$

symbols long. The sequence will be B0...0V or 0...0V and the choice of sequence is determined by the rule that successive violations should be of alternate polarity. This rule ensures a null at DC in the PSD. In fact, the PSD has nulls at both DC and at $\omega = 2\pi / T$, where T is the bit period. Clearly, the timing content increases as n is reduced and usually $n = 3$.

Example 4.7

Binary	1	0	1	0	0	0	0	0	1	0	1	1	0	0	0	0	1	1	1	0	0	0	0	0	0	0	0	0	1	0	1
AMI	+	0	−	0	0	0	0	0	+	0	−	+	0	0	0	0	−	+	−	0	0	0	0	0	0	0	0	0	+	0	−
HDB3	+	0	−	0	0	0	−	0	+	0	−	+	0	**0**	0	+	−	+	−	0	0	0	−	+	0	0	+	0	−	0	+
				0	0	0	V						0	0	0	V				0	0	0	V	B	0	0	V				
						a									b							c				d					

The HDBn code retains the error detection property of the AMI code. For example, a negative mark in place of the bold symbol (**0**) in Example 4.7 effectively removes filling sequence 'b', and sequence 'c' will now have the same violation polarity as the immediately preceding sequence 'a'. As with other AMI-based codes, the HDB3 code is specified as an *interface code*, that is, for connections within buildings.

4.2.4 Block (alphabetic) codes

Here we introduce another class of transmission code. Rather than being based on the AMI code, this class is block-based and is sometimes referred to as *alphabetic coding*. In contrast to AMI-based codes, block codes are usually used as line codes for long distance transmission. The technique divides the binary input sequence into blocks of specified length and typically assigns a block of binary or ternary symbols to each binary block using a look-up table.

A binary-to-ternary translation takes advantage of the relatively large number of possible ternary combinations in order to reduce the symbol rate, and therefore the bandwidth. For example, the 6B4T code translates a block of six binary digits to four ternary symbols, while the MS43 and 4B3T codes translate blocks of four binary digits to three ternary symbols. Considering the 4B3T code, three ternary symbols (27 combinations) are more than adequate to represent four binary digits (16 combinations) and ideally the transmission bandwidth will be scaled by ¾. The ability to reduce bandwidth made codes like 4B3T and 6B4T attractive for use as a line codes on long distance (repeatered) coaxial cable.

Table 4.1 shows a typical 4B3T coding alphabet and a PSD null at DC is achieved by selecting the mode in such a way as to minimise the running DSV. For example, if the signal has recently contained an excess of positive marks, then ternary words with a negative bias are chosen to restore the balance (Example 4.8). Also note that the ternary character 000 is absent from Table 4.1; this improves the timing content of the code, and can be utilised to check for loss of block synchronisation. Should synchronisation be lost, then the illegal character 000 will occasionally occur and synchronisation can be corrected.

Table 4.1 *A 4B3T coding alphabet.*

Binary input	*Ternary output* *Mode 1*	*Mode 2*	*DSV*
0000	0 − +	0 − +	0
0001	− + 0	− + 0	0
0010	− 0 +	− 0 +	0
1000	0 + −	0 + −	0
1001	+ − 0	+ − 0	0
1010	+ 0 −	+ 0 −	0
0011	+ − +	− + −	1
1011	+ 0 0	− 0 0	1
0101	0 + 0	0 − 0	1
0110	0 0 +	0 0 −	1
0111	− + +	+ − −	1
1110	+ + −	− − +	1
1100	+ 0 +	− 0 −	2
1101	+ + 0	− − 0	2
0100	0 + +	0 − −	2
1111	+ + +	− − −	3

Example 4.8

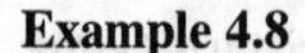

1 0 1 1 : 1 1 0 1 : 1 1 0 0 : 1 1 1 1 : 0 1 0 1 : 1 1 0 1 : 0 0 0 1

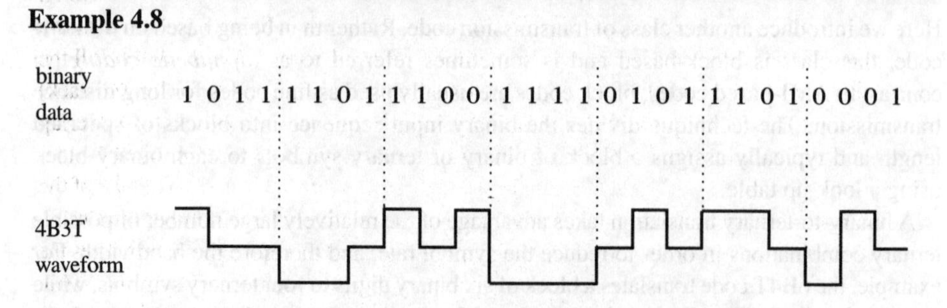

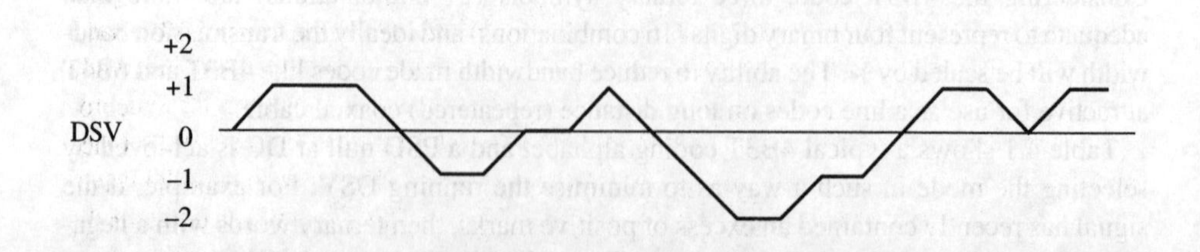

Table 4.2 *A 3B4B coding alphabet.*

Binary input	*Binary output*	*DSV*
001	− − + +	0
010	− + − +	0
100	+ − − +	0
011	− + + −	0
101	+ − + −	0
110	+ + − −	0
000	− − + − or + + − +	±2
111	− + − − or + − + +	±2

The 3B4B code

The 3B4B code is a relatively simple binary block code that can be used as a line code on optical fibre links. It translates a block of 3 binary digits to a block of 4 binary symbols and a typical coding alphabet is shown in Table 4.2. The alphabet is constructed so as to give frequent data transitions and so give adequate timing content. For (000) and (111) inputs the coder selects the word that minimises the running DSV (as for 4B3T). For high bit rate fibre optic links this code may be replaced by the more efficient 24B1P block code, comprised of 24 binary bits and an odd parity bit for error detection.

4.2.5 Run-length limited (RLL) codes

This family of codes is used almost universally in optical and magnetic digital recording. Consider the basic NRZI recording code in Figure 4.9(d). Here each logical 1 generates a flux transition on the medium, while a logical 0 corresponds to no transition, that is, all the information content of the code is in the *position* of the transitions. Clearly, as the (on-track) recording density increases, the readback pulses from the adjacent 1s will start to overlap (Figure 4.9(e)) and, in general, this ISI can result in a significant change in both pulse position and pulse amplitude. In severe cases, this on-track *peak-shift* will cause the peaks of the readback pulses to occur outside the clock window and the raw error rate will increase.

The usual solution for high-density recording is to use a modulation code that preserves the minimum distance between transitions, thereby minimising ISI. Assuming two-state recording, this will be a binary-to-binary translation where n data bits are mapped to m code bits, corresponding to a code rate $R = n/m$ ($R \leq 1$). In an $R(d,k)$ RLL code sequence, two logical 1s are separated by a run of at least d consecutive zeros and any run of consecutive zeros has a length of at most k. Constraint d controls the highest transition density and resulting ISI, while constraint k ensures a *self-clocking* property or adequate frequency of transitions for synchronisation of the read clock (this is a major failing of the *basic* NRZI code in Figure 4.9(d)). Once a constrained or coded sequence has been generated it is recorded in the form of a NRZI waveform, as described earlier. In practice this requires a simple precoding step which converts the constrained coded sequence to a RLL *channel sequence* (a string of ±1s) representing the positive and negative magnetisation of the recording medium, or pits and lands in optical recording. The RLL channel sequence will then be constrained by minimum and maximum run-lengths of $d + 1$ and $k + 1$ channel bits respectively.

An important parameter in digital recording is the ratio of data bits to recorded transitions, since this represents the efficiency of the code in utilising recording space. This is called the *density ratio* (*DR*) and is given by

$$DR = R(d+1) \tag{4.13}$$

The ratio increases as d increases but, on the other hand, the required timing accuracy for the detection of transitions increases exponentially with $d + 1$, while the clock content of the code decreases. In practice, d is typically 0, 1 or 2. Generally speaking, the code rate R should be as high as possible in order to maximise DR and other factors. The maximum value of R for specified values of d and k is called the *capacity* $C(d,k)$ of the code, and is measured in information bits per code bit. Rate R is therefore usually selected as a rational number n/m close to $C(d,k)$.

Example 4.9

The ½(2,7) RLL code (Franaszek, 1972) is a popular high-performance variable-length block code. Coding essentially amounts to partitioning the data sequence into blocks of 2, 3 or 4 bits, and applying the following translation:

Data	*Codeword*
10	0100
11	1000
000	000100
010	100100
011	001000
0010	00100100
0011	00001000

Figure 4.13 shows the ½(2,7) sequence for arbitrary data, together with the precoding step to turn this sequence into a RLL NRZI channel sequence. Given an input sequence $\{a_k\}$, the precoder will generate the channel sequence as $b_k = a_k + b_{k-1}$ modulo-2, with a level shift $0 \Rightarrow -1$, $1 \Rightarrow +1$. Precoding ensures that a 1 in the RLL code sequence indicates the position of a $1 \rightarrow -1$ or $-1 \rightarrow 1$ transition in the RLL channel sequence.

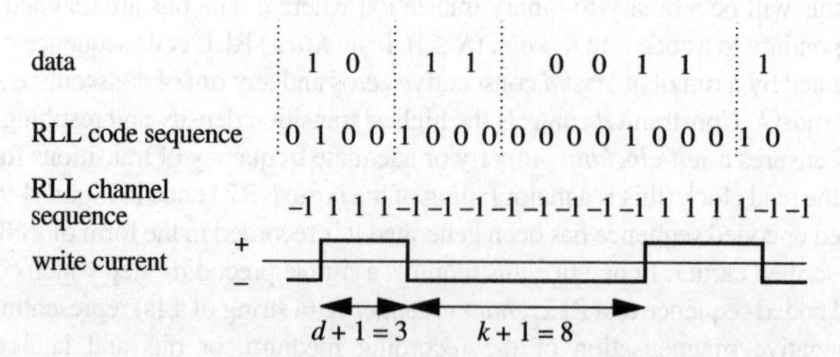

Figure 4.13 *Typical recording waveform for the ½(2,7) RLL code.*

It is worth noting that Example 4.4 is actually describing a RLL(1,3) modulation code. Similarly, both the Miller and Miller^2 codes in Example 4.5 are actually RLL codes of the form $R(d,k)$. An examination of long code sequences reveals that the Miller code is a ½(1,3) code, and the Miller^2 code is a ½(1,5) code. The code rate is 1/2 since one data bit is mapped to two code bits. The density ratio for the Miller^2 code is 1 compared to 1.5 for the ½(2,7) code.

4.3 Partial response signalling

In practice a data communications system is often designed for a specific or *target* response at the detector, given a rectangular binary symbol (pulse) at the input (Figure 4.14(a)). We could regard $r(t)$ as the overall response per binary symbol. In a data broadcasting system (such as teletext, for example), the target response $R(f)$ is often a raised cosine spectrum, as discussed in Example 4.1. In a magnetic recording system (Figure 4.14(b)), the input is a rectangular write current pulse and an equaliser of response $E(f)$ delivers the target (pulse) response $R(f)$. Also, the magnetic channel itself is often assigned a *pulse response* $h_c \Leftrightarrow H_c(f)$, that is, $h_c(t)$ is the head/medium response to a rectangular current pulse. This means we can write

$$E(f) = R(f) / H_c(f) \tag{4.14}$$

and the equaliser can be designed if $H_c(f)$ can be determined, for example, experimentally.

In order to develop the theory of PR signalling, it is helpful to model the binary input as an impulse train, as in Figure 4.14(c). Here we will assume the efficient polar signalling format and so take $x(n)$ as a sequence of independent binary symbols having values ±1. The signalling rate is $1/T$ symbols/s. Spectrum shaping is modelled by filter $H(f)$ and the target

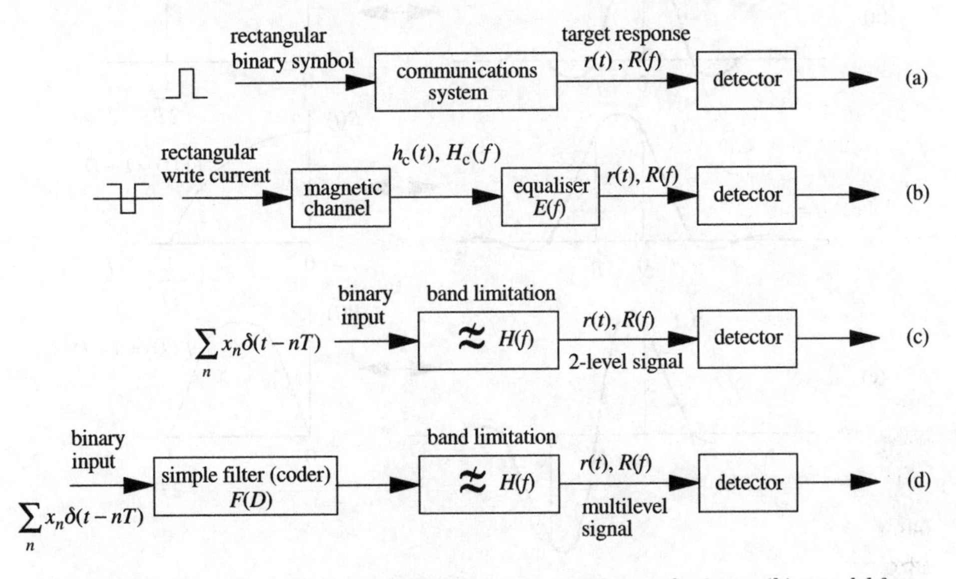

Figure 4.14 *Channel models: (a) defining target response for a pulse input; (b) a model for magnetic recording; (c) a model for straight binary signalling; (d) a model for PR signalling.*

response $r(t) \Leftrightarrow R(f)$ could be defined by (4.5) and (4.6). The limiting case in Figure 4.14(c) occurs when $R(f)$ has a rectangular spectrum cutting off at $f = 1/2T$, but, as previously discussed, this is impractical due to the discontinuity at $f = 1/2T$.

Now suppose we precede the bandlimiting filter in Figure 4.14(c) with a simple linear filtering (coding) operation (Figure 4.14(d)). In effect, Figure 4.14(d) models the transmit filter, channel and receive filter of Figure 4.1(b), and is applicable to various modulation schemes as well as to baseband systems. As before we will assume $x_n \in \{\pm 1\}$ and that $H(f)$ is a raised cosine filter, as in (4.5). The objective here is to generate a target spectrum $R(f)$ which is essentially bandlimited to $f = 1/2T$, *and yet is realizable*. It has been shown (Lender, 1966) that this is possible provided we can accept a pulse shape $r(t)$ which is non-zero at *several* sampling instants – a shape radically different from the Nyquist pulse defined in (4.3). In fact, $r(t)$ in Figure 4.14(d) will be a linear superposition of a number of closely spaced elementary impulse responses. Figures 4.15(b) and (c) illustrate two useful examples of this form of signalling and clearly show the deliberate ISI. *The significant point is that it is now practical to signal at the limiting (or Nyquist) rate of* $1/T$ *symbols/s in a bandwidth of just* $1/2T$ *Hz!* In other words, PR signalling offers 0% excess bandwidth.

At this point it is useful to highlight some important features of PR signalling:

- Since $r(t)$ is a linear superposition of impulse responses, a binary input results in a *multi-level* output. For basic threshold/peak detection systems this generally requires an increase in SNR compared to straight binary signalling for the same error rate, and PR systems exhibiting the smallest SNR degradation tend to be preferred (Kabal and Pasupathy, 1975). Typically three levels are used in practice.

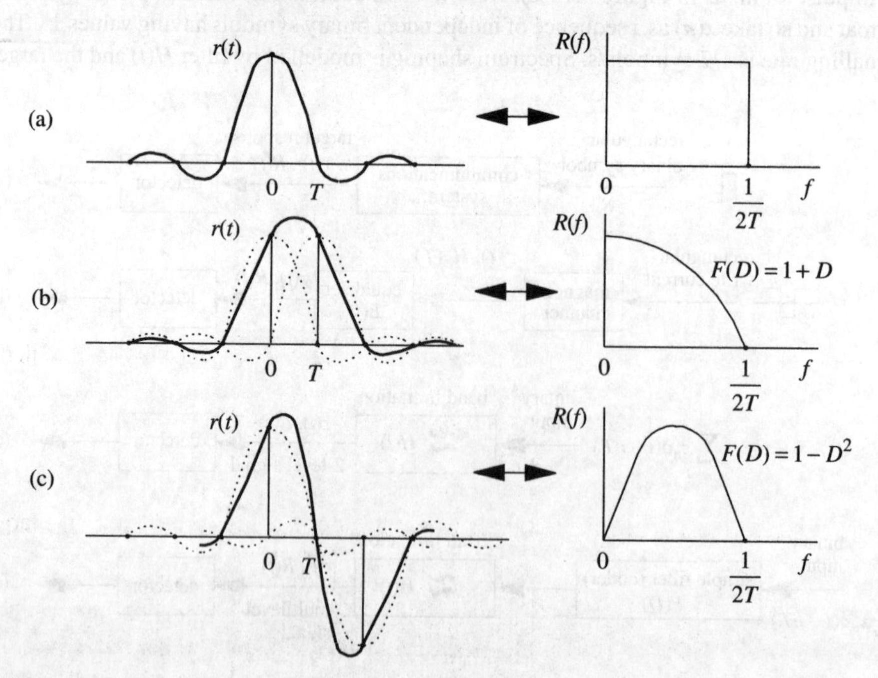

Figure 4.15 *Comparison of straight binary and PR signalling: (a) straight binary (ideal); (b) PR1; (c) PR4.*

- PR signalling deliberately generates ISI and a degree of *correlation* between transmitted symbols (it is sometimes referred to as *correlative coding*). This correlation or constraint on level transitions can be exploited by detectors based on MLSD (Viterbi detection) to give improved noise performance compared to symbol-by-symbol decoding of PR signals. For instance, PR signalling with Viterbi decoding to perform MLSD is particularly attractive for high-density magnetic recording (Dolivo *et al.*, 1989).
- PR signalling is not just applicable to baseband systems, and it has been used in carrier schemes such as FM and SSB.

It is usual to define the main spectrum shaping filter or coder in Figure 4.14(d) by a polynomial $F(D)$, where D is the delay operator (equivalent to z^{-1}). Most practical systems use a polynomial of the form

$$F(D) = (1-D)^m(1+D)^n \tag{4.15}$$

where either m or n can be zero, but not both. Also, since $F(D)$ essentially defines the target response of the PR system, it is usually referred to as a *system polynomial*. In other words, using the impulse input model of Figure 4.14(d), the target partial response, or end-to-end system function, can be written

$$R(f) = H(f)\,F(D)\big|_{D=\exp(-j2\pi fT)} \tag{4.16}$$

where $H(f)$ is a raised cosine bandlimiting filter; see (4.5). For minimum bandwidth PR systems, $H(f)$ will cut off close to $f = 1/2T$. In practice, $R(f)$ will also be a function of channel and receive filter (equaliser) characteristics.

Example 4.10
Consider the PR system

$$F(D) = 1 + D$$

This is shown in Figure 4.16 and its transfer function is obtained as

$$H_F(f) = F(D)\big|_{D=\exp(-j2\pi fT)} = 1 + e^{-j2\pi fT}$$

$$|H_F(f)| = 2|\cos(\pi fT)|$$

This is a comb filter response with a first zero at $f = 1/2T$. When this is cascaded with the bandlimiting filter $H(f)$, the ideal minimum bandwidth target response becomes

$$R(f) = \begin{cases} 2\cos(\pi fT) & |f| \le 1/2T \\ 0 & \text{otherwise} \end{cases}$$

In this ideal case the impulse response of $H(f)$ will be of $\mathrm{Sa}(x) = \sin(x)/x$ form, so the target impulse response will be

$$r(t) = \mathrm{Sa}\left(\frac{\pi}{T}t\right) + \mathrm{Sa}\left[\frac{\pi}{T}(t-T)\right]$$

Note that $r(t)$ has a baseline width equal to three clock periods (Figure 4.16) and so generates considerable ISI. The reader should show that a random sequence of positive and

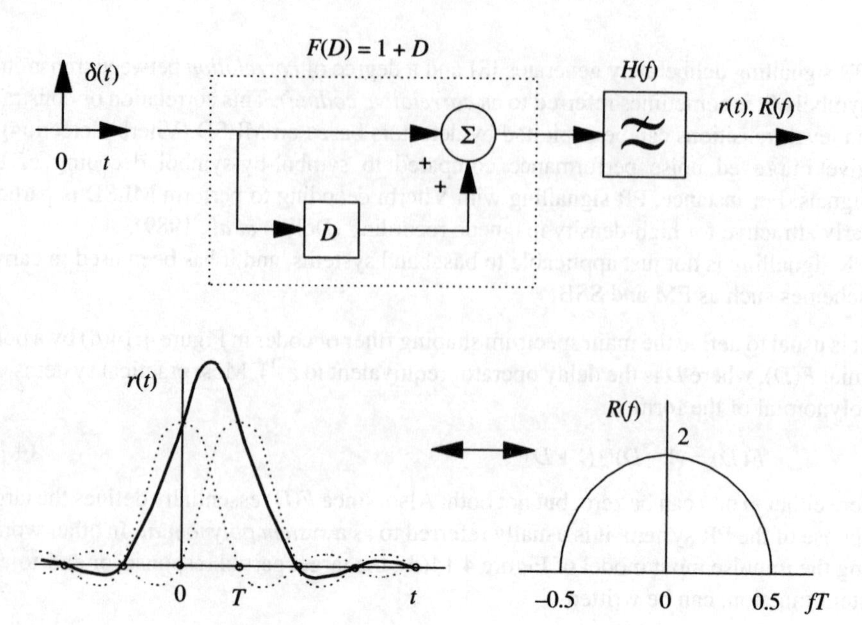

Figure 4.16 *Principle of duobinary coding (PR1).*

negative impulses at the input to $F(D)$ in Figure 4.16 will result in a 3-level output signal, with transitions occurring only between adjacent levels. For ± 1 input symbols, the output levels will be 0, ± 2.

The significant point about this form of signalling is that, even for a practical, finite roll-off filter $H(f)$, the PSD for a random binary input will be negligible above $f = 1/2T$. This means that, compared with 100% roll-off binary signalling, PR1 offers half the bandwidth or *double* the bit rate (which gave rise to the term 'duobinary').

4.3.1 Precoding

Figure 4.17(a) represents the basic duobinary coding process in terms of signal samples. The coder output is the arithmetic sum

$$c_k = b_k + b_{k-1} \tag{4.17}$$

and this suggests that the binary data can be decoded as

$$\hat{b}_k = c_k - \hat{b}_{k-1} \tag{4.18}$$

where $\hat{b}_k$ is an estimate of b_k obtained by binary slicing the term $c_k - \hat{b}_{k-1}$. Unfortunately, if $\hat{b}_{k-1}$ is decoded incorrectly, then the error will tend to propagate – a property we wish to avoid in transmission codes. An alternative decoder could use the MLSD technique (Viterbi decoding), but practical PR systems may employ a much simpler solution to the error propagation problem. Clearly, error propagation will be impossible if the decoder is made 'memoryless' and this can be achieved by employing a *precoder*, as in Figure 4.17(b).

(a) +1 −1 b_k 2-level $F(D) = 1 + D$ (coder) c_k 3-level

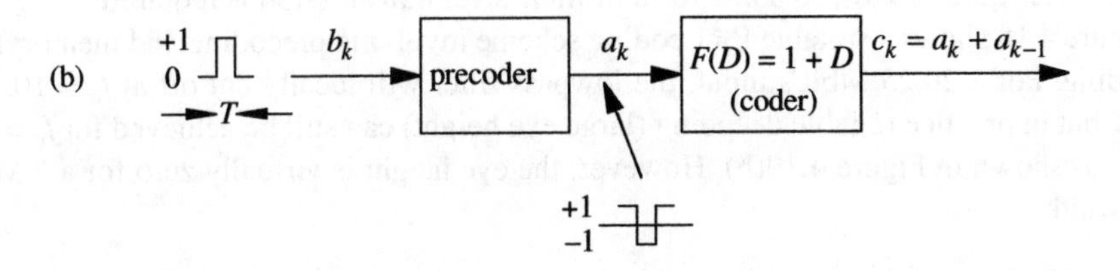

(c)

b_k	$\bar{b}_k$	a_k	a_{k-1}	c_k
1	0	+1	+1	2
0	1	+1	−1	0
0	1	−1	+1	0
1	0	−1	−1	−2

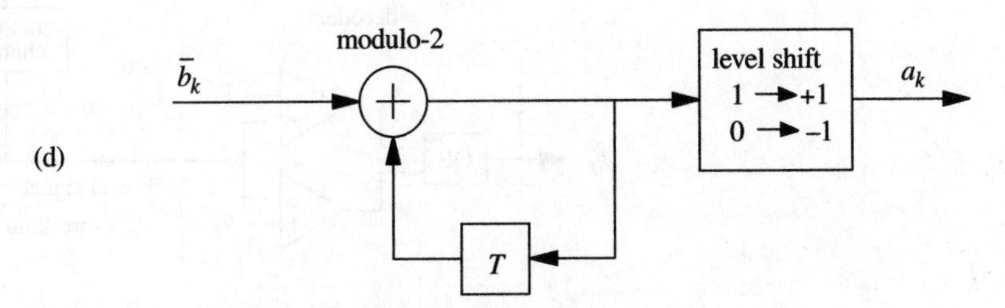

Figure 4.17 *Design of a precoder for duobinary signalling: (a) basic PR1 system; (b) use of a precoder; (c) precoder design; (d) precoder circuit.*

Consider the design of a precoder for PR1. We know that, given symbols ±1 at the coder input (a polar NRZ input) this coder generates symbols 0, ±2 at its output. However, each output symbol c_k only conveys 1 bit of information and so we could use the memoryless decoding rule

c_k	b_k
±2	1
0	0

This rule was used to construct the table in Figure 4.17(c), where a_k is now the polar input to the coder. Ignoring the polar format (regarding −1 as 0) we can see from this table that $a_k = \bar{b}_k + a_{k-1}$ modulo-2, and this gives the precoder circuit in Figure 4.17(d).

Example 4.11

The D-MAC television standard (now abandoned) specified that data and sound packets were to be transmitted at 20.25 Mbit/s in a baseband bandwidth of 8.5 MHz. The use of straight binary signalling via raised cosine pulses would require typically 15 MHz bandwidth (see Figure 4.2(b)), so some form of multilevel transmission is required.

Figure 4.18 shows a suitable PR1 coding scheme involving precoding and memoryless decoding. For a 20.25 Mbit/s input, the lowpass filter will ideally cut off at f_c = 10.125 MHz, but in practice reliable decoding (large eye height) can still be achieved for f_c = 8.5 MHz, as shown in Figure 4.19(b). However, the eye height is virtually zero for a 6 MHz bandwidth.

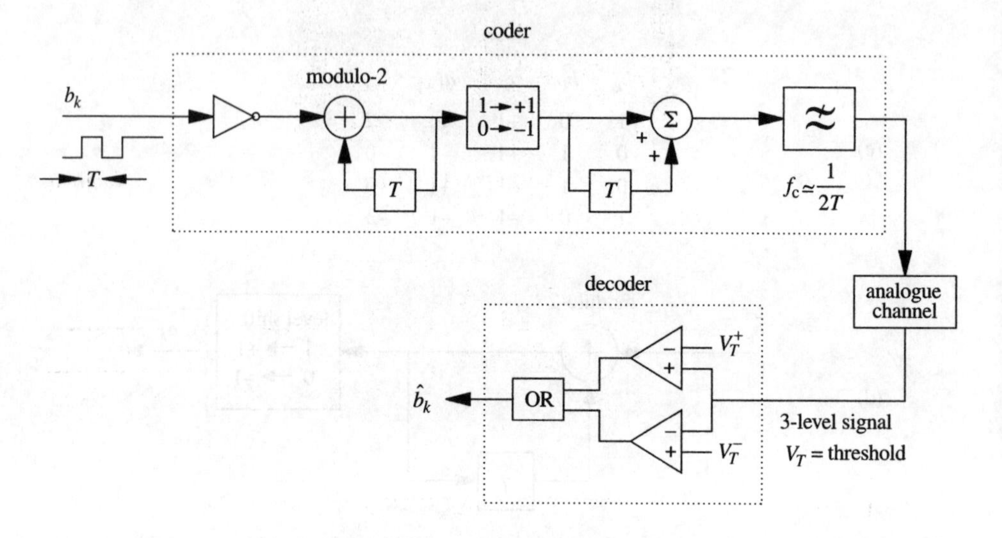

Figure 4.18 *Practical PR1 system.*

4.3.2 PR systems

The selection of polynomial $F(D)$ obviously depends upon the required spectrum shaping, but it can also depend upon the permissible degradation in SNR and data eye relative to binary transmission. Table 4.3 gives some typical polynomials cited in the literature. Polynomials PR1, PR4 and EPR4 are particularly useful, the latter two having a spectral null at DC as well as at $f = 1/2T$.

Example 4.12

The PR1 spectral response in Figure 4.16 has a large low-frequency content. While this is of no significance in carrier systems, for baseband signalling it would require a DC coupled channel or some form of DC restoration. If a null at DC is required we could use PR4, where $F(D) = 1 - D^2$. The corresponding transfer function is

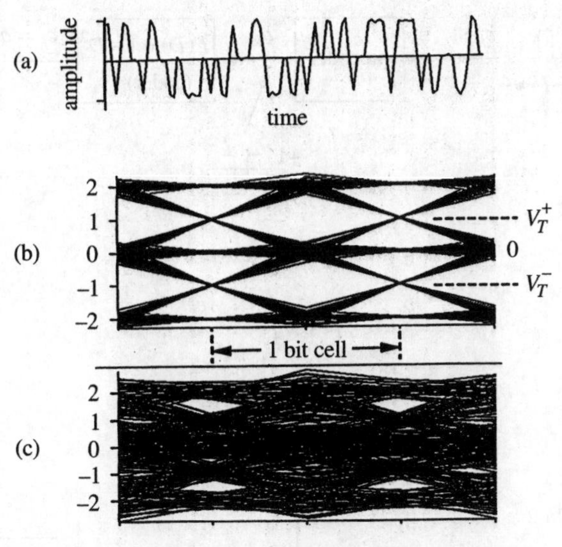

Figure 4.19 *Simulated results for Figure 4.18: (a) PR1 waveform, f_c = 8.5 MHz; (b) eye diagram for f_c = 8.5 MHz; (c) eye diagram for f_c = 6 MHz.*

Table 4.3 *Some PR system polynomials.*

F(D)	*PR classification*	*Output levels*
$1 + D$	PR1 (duobinary)	3
$(1 + D)^2 = 1 + 2D + D^2$	PR2	5
$(1 + D)(2 - D) = 2 + D - D^2$	PR3	5
$(1 + D)(1 - D) = 1 - D^2$	PR4 (modified duobinary)	3
$(1 + D)^2(1 - D) = 1 + D - D^2 - D^3$	EPR4 (extended PR4)	5
$(1 + D)^2(1 - D)^2 = 1 - 2D^2 + D^4$	PR5	5

$$H_F(f) = 1 - e^{-j4\pi fT}$$
$$= j2\sin(2\pi fT)e^{-j2\pi fT}$$
$$|H_F(f)| = 2|\sin(2\pi fT)|$$

and the bandlimited target response *R(f)* will appear as in Figure 4.15(c). Clearly, *F(D)* is realised by subtracting a two-sample delayed version of the binary signal from the current sample, so the minimal bandwidth target impulse response for PR4 is

$$r(t) = \mathrm{Sa}\left(\frac{\pi}{T}t\right) - \mathrm{Sa}\left(\frac{\pi}{T}(t - 2T)\right)$$

The overall shape of *r(t)* for both PR4 and PR1 is shown in Figure 4.15. Since the number of output levels is still 3, the precoder for PR4 can be designed using the decoding rule used for PR1; see Figure 4.20(b). Using the same argument as before (ignoring level

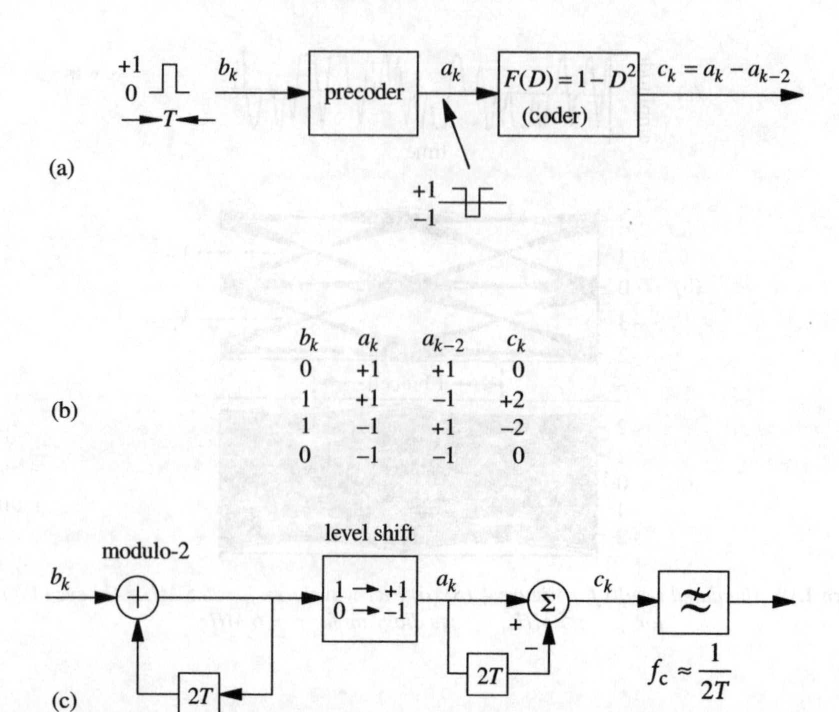

Figure 4.20 *PR4 signalling: (a) basic system; (b) precoder design; (c) practical coder.*

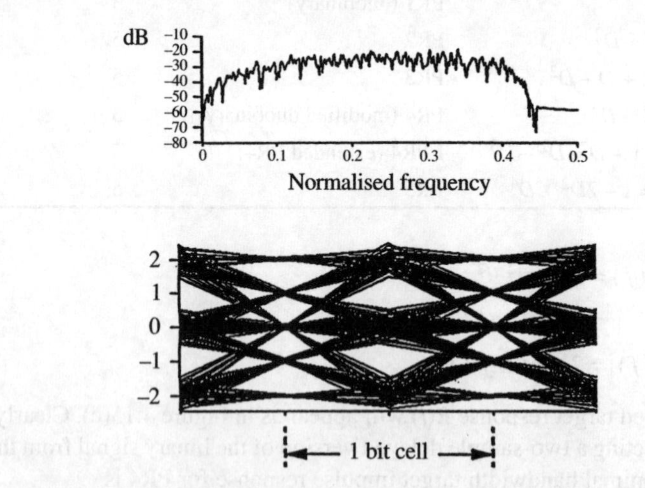

Figure 4.21 *Simulated PSD and eye diagram for PR4 data (f_c = 0.42/T).*

shifting) we can write $a_k = b_k + a_{k-2}$ modulo-2, so the precoder is as shown in Figure 4.20(c). In fact, to design a precoder for a general multilevel PR system, all we need to do is to divide the output levels into two sets, one corresponding to $b_k = 0$, and the other to $b_k =$

1. Figure 4.21 shows typical power spectrum and eye diagram measurements for the output of the coder in Figure 4.20(c), given random binary data at the input.

Example 4.13
Figure 4.22 illustrates the use of PR4 for magnetic recording. Here, the precoder is placed before the analogue channel and generates an NRZ write current waveform. The equaliser modifies the response of the analogue channel to give the target PR4 response at the detector. As already pointed out, this is ideally a Viterbi detector in order to exploit the correlations in the PR signal. For a pulse input and low recording density, the analogue channel response is sometimes approximated by $1 - D$, in which case a $1 + D$ equaliser response would be required for PR4.

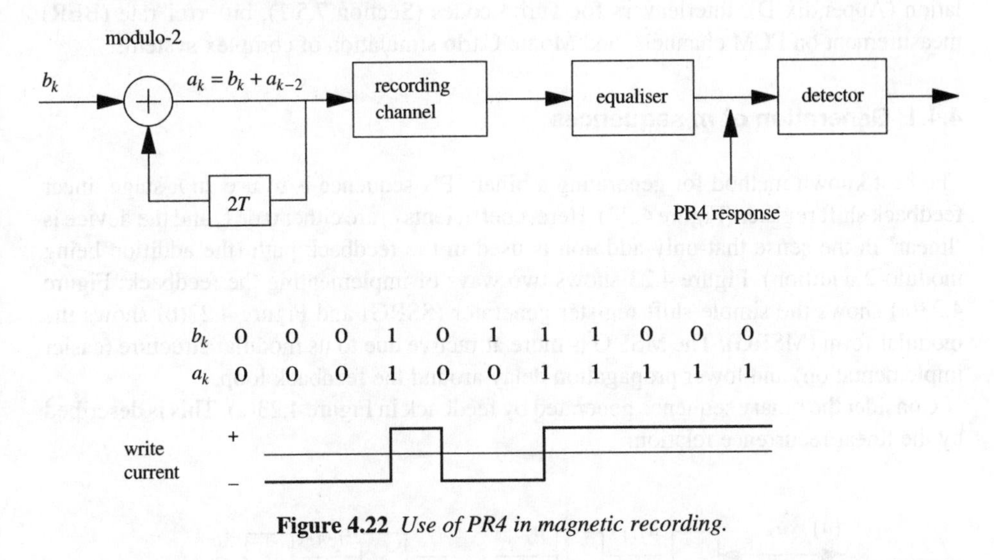

Figure 4.22 *Use of PR4 in magnetic recording.*

4.4 Pseudorandom sequences

Binary sequences that have a random-like nature but are nevertheless deterministic are used extensively in communications systems. Since they are deterministic they are called *pseudorandom*, and because they have noise-like properties they are often called pseudonoise (PN) sequences. One reason for studying them here is that they can be used to aid signal transmission by providing a form of channel coding. For example, PN sequences find extensive use in data modems in the form of *scramblers*. As the name implies, a scrambler uses a PN sequence to break up data patterns and so provide enough data transitions to ensure receiver clock stability irrespective of the modem input (which could be all-zeros for example). Scrambling also breaks up bit patterns which might otherwise cause accumulative clock jitter in a PCM repeater chain (Byrne *et al.*, 1963). Note that all this is achieved without increasing the bit rate (or bandwidth).

Another major application of PN sequences is in the area of spread spectrum. Here a PN sequence is used to spread the signal bandwidth, and a synchronised sequence is used by the receiver to de-spread the signal and recover the data. The larger bandwidth provides an inherent robustness to interference, (including intentional interference such as jamming), so spread spectrum could be regarded as another form of channel coding. A carefully selected PN sequence is also frequently used as a *unique word* (*UW*) or *preamble* prior to data transmission in order to provide frame synchronisation (it could also be embedded in data as a 'midamble'). Detection of the UW at the receiver (via *autocorrelation* or matched filtering) then marks the start of the data frame.

Communications systems also use PN sequences for *conditional access*. Here, an encryption algorithm controls scrambling in order to provide a secure data transmission system. Other applications of PN sequences include multiple access (CDMA), noise simulation (Appendix D), interleavers for Turbo codes (Section 7.5.1), bit error rate (BER) measurement on PCM channels, and Monte Carlo simulation of complex systems.

4.4.1 Generation of *m*-sequences

The best known method for generating a binary PN sequence is to use an n-stage linear feedback shift register (Figure 4.23). Here, coefficients c_i are either 0 or 1, and the device is 'linear' in the sense that only addition is used in the feedback path (the addition being modulo-2 addition). Figure 4.23 shows two ways of implementing the feedback; Figure 4.23(a) shows the simple shift register generator (SSRG) and Figure 4.23(b) shows the modular form (MSRG). The MSRG is more attractive due to its modular structure (easier implementation) and lower propagation delay around the feedback loop.

Consider the binary sequence generated by feedback in Figure 4.23(a). This is described by the linear recurrence relation:

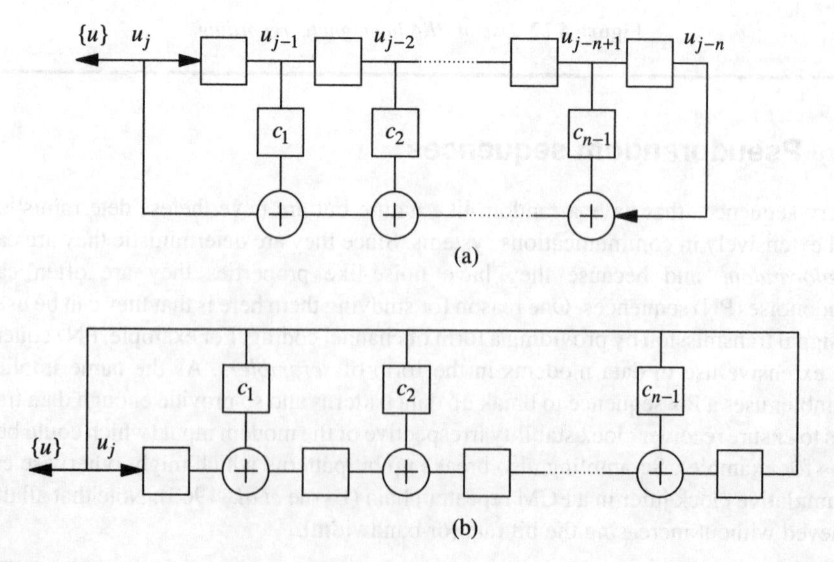

Figure 4.23 *Generalised linear feedback shift register generators: (a) SSRG; (b) MSRG.*

$$u_j = c_1 u_{j-1} + c_2 u_{j-2} + \ldots + c_{n-1} u_{j-n+1} + u_{j-n} \quad \text{modulo-2} \tag{4.19}$$

Clearly, if the initial register state is zero the sequence $\{u\}$ will be all-zeros and is of no interest. If the initial state is one of the $2^n - 1$ non-zero states, then $\{u\}$ will start to repeat when the vector in the register returns to the initial state. It follows that the *maximum possible* sequence length will be $N = 2^n - 1$, and thereafter the sequence will repeat. Binary sequences with this maximum length are called *m*-sequences, and they pass several statistical tests for randomness. For example, inspection of an *m*-sequence shows that there are always 2^{n-1} ones and $2^{n-1} - 1$ zeros, implying equal probabilities for large n, that is, $P(1) \approx P(0) \approx 0.5$. Also, for large n, the autocorrelation function approximates to that of white noise (Section 4.5.1).

The linear feedback shift registers in Figure 4.23 are described by a *characteristic polynomial*, $\phi(x)$. This simply describes the shift register taps, and for the notation in Figure 4.23(a) it is defined as

$$\phi(x) = x^n + c_{n-1}x^{n-1} + \ldots + c_1 x + 1 \tag{4.20}$$

Sometimes $\phi(x)$ is defined as

$$\phi(x) = x^n + c_1 x^{n-1} + \ldots + c_{n-1} x + 1 \tag{4.21}$$

This is compatible with Figure 4.23(a) providing we assign $j \rightarrow j + n$. Either way, the significant points here are that (a) $\phi(x)$ must be *primitive* over GF(2) in order to generate an *m*-sequence, and (b) the primitive requirement is a *sufficient* condition for generation of an *m*-sequence.

Definition 4.1 A polynomial $\phi(x)$ of degree n is primitive over GF(2) if it is irreducible (that is, it has no factors except 1 and itself) and if it divides $x^k + 1$ for $k = 2^n - 1$ and does not divide $x^k + 1$ for $k = 2^n < 1$.

Corollary 4.1 If $\phi(x)$ divides $x^k + 1$ for $k < 2^n - 1$ then $\phi(x)$ is not primitive and the smallest integer k for which $\phi(x)$ divides $x^k + 1$ determines the shift register period.

From this definition it is apparent that, even if a polynomial is irreducible, it is not necessarily primitive.

Example 4.14

Consider the polynomial $\phi(x) = x^4 + x^3 + 1$. Assuming the general form in (4.20), the corresponding SSRG circuit is shown in Figure 4.24(a). Suppose that the initial state is

$$(u_{-1}\, u_{-2}\, u_{-3}\, u_{-4}) = (0001)$$

The reader could use a clock-by-clock analysis to show that the feedback sequence $\{u\}$ will be $\{100110101111000\}$, where the first digit is generated by the initial state, and thereafter the sequence repeats. Since the sequence is of maximum possible length we conclude it must be an *m*-sequence. This being the case, $\phi(x)$ must be irreducible and must divide $x^{15} + 1$ (see Definition 4.1). Indeed, we find that

$$(x^4 + x^3 + 1)(x^{11} + x^{10} + x^9 + x^8 + x^6 + x^4 + x^3 + 1) = x^{15} + 1 \quad \text{over GF(2)}$$

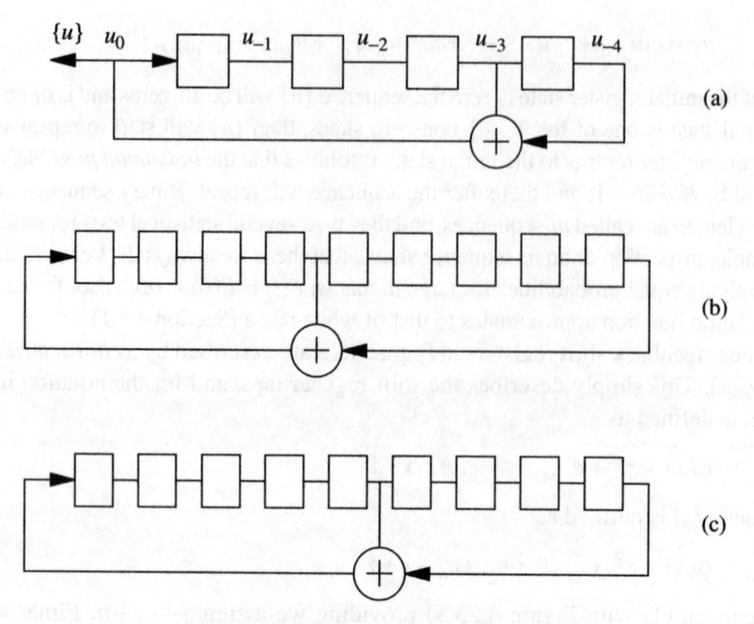

Figure 4.24 *SSRG examples (see text).*

In addition, it can be shown that $\phi(x)$ does not divide $x^k + 1$ for $k < 15$, so $\phi(x)$ must be primitive.

Example 4.15

Consider the polynomial $\phi(x) = x^6 + x^4 + x^2 + x + 1$. This is irreducible, but we also find that

$$(x^6 + x^4 + x^2 + x + 1)(x^{15} + x^{13} + x^{10} + x^7 + x^6 + x^3 + x + 1) = x^{21} + 1 \quad \text{over GF(2)}$$

According to Corollary 4.1 this means that the feedback shift register would generate three short cycles of length 21 in the ideal cycle length of 63.

The sequence in Example 4.14 can be deduced algebraically as follows. Using the terminology in Figure 4.24(a), let the initial shift register state be expressed by the polynomial

$$\begin{aligned} h(x) &= \sum_{k=1}^{n} c_k \sum_{m=0}^{k-1} u_{m-k} x^m \\ &= c_1 u_{-1} + c_2\left[u_{-2} + u_{-1}x\right] + c_3\left[u_{-3} + u_{-2}x + u_{-1}x^2\right] + \ldots \\ &\quad + c_n\left[u_{-n} + u_{-n+1}x + \cdots + u_{-1}x^{n-1}\right] \end{aligned} \tag{4.22}$$

where $c_n = 1$. Given initial condition $h(x)$, the binary sequence $\{u\} = (u_0 u_1 u_2 \ldots u_{N-1})$, that is, one period of the register, can be determined as

$$\{u\} \Leftrightarrow u(x) = \frac{h(x)(1+x^N)}{\phi(x)} \tag{4.23}$$

where N is the period of the *m*-sequence and $\phi(x)$ is defined by (4.20). Note that if the initial state is (000...01), that is, $u_{-n} = 1$, then $h(x) = 1$.

Example 4.16
In Example 4.14 the initial state is (0001), so

$$u(x) = (x^{15}+1)/(x^4+x^3+1) = 1+x^3+x^4+x^6+x^8+x^9+x^{10}+x^{11}$$

This corresponds to $\{u\}$ = (100110101111000), as deduced earlier (the reader unfamiliar with polynomial long division is referred to Section 6.4.2).

When designing an *m*-sequence generator it is helpful to consult a table of primitive polynomials, as in Table 4.4. These polynomials directly provide the taps for designing an *m*-sequence generator of the form in Figure 4.23(a) or Figure 4.23(b), and are interpreted as in the following examples (taking the first entry in each case):

$$n=5: \quad 100101 \equiv 45 \equiv x^5+x^2+1$$

$$n=8: \quad 100011101 \equiv 435 \equiv x^8+x^4+x^3+x^2+1$$

$$n=9: \quad 1000010001 \equiv 1021 \equiv x^9+x^4+1$$

Note that octal notation is often used for convenience. As a further illustration, the polynomial 1021 gives the SSRG in Figure 4.24(b) when using (4.20). Clearly, in general there are many alternative polynomials for a particular value of n, although some applications lead to criteria for selecting a particular polynomial (see later). It is interesting to note that a 60-stage register using the primitive polynomial $x^{60}+x+1$ would have a period of over 3600 years when clocked at 10 MHz!

Reciprocal polynomials and the MSRG

Suppose that we had interpreted polynomial 1021 for $n=9$ using (4.21). This would have given the SSRG in Figure 4.24(c), which is the same circuit generated by polynomial

$$1000100001 \equiv 1041 \equiv x^9+x^5+1$$

using (4.20). Polynomials $\phi_u(x) \equiv 1021$ and $\phi_v(x) \equiv 1041$ are an example of a *reciprocal pair*. Formally, the *reciprocal polynomial* $\phi_v(x)$ of $\phi_u(x)$ is defined as

$$\phi_v(x) = x^n \phi_u(x^{-1}) \tag{4.24}$$

Thus,

$$\phi_v(x) = x^9(x^{-9}+x^{-4}+1) = x^9+x^5+1 \tag{4.25}$$

Careful examination of Table 4.4 reveals other reciprocal pairs for specific values of n. It is easily demonstrated that the reciprocal of a polynomial $\phi(x)$ will generate a sequence which is the *time-reverse* of the sequence generated by $\phi(x)$. Thus, in Table 4.4 it could be

Table 4.4 *Primitive polynomials of degree n.*

n		n		n		n	
1	11	7	10011101	8	110000111		
2	111		10100111		110001101	11	$1+x^2+x^{11}$
3	1011		10101011		110101001	12	$1+x+x^4+x^6+x^{12}$
	1101		10111001		111000011		
4	10011		10111111		111001111	13	$1+x+x^3+x^4+x^{13}$
	11001		11000001		111100111		
5	100101		11001011		111110101	14	$1+x+x^6+x^{10}+x^{14}$
	101001		11010011	9	1000010001		
	101111		11010101		1000011011	15	$1+x+x^{15}$
	110111		11100101		1000100001		
	111011		11101111		1000101101	16	$1+x+x^3+x^{12}+x^{16}$
	111101		11110001		1000110011	17	$1+x^3+x^{17}$
6	1000011		11110111		1001011001	18	$1+x^7+x^{18}$
	1011011		11111101		1001011111		
	1100001	8	100011101		1001101001	19	$1+x+x^2+x^5+x^{19}$
	1100111		100101011		...		
	1101101		100101101			20	$1+x^3+x^{20}$
	1110011		101001101	10	10000001001	21	$1+x^2+x^{21}$
7	10000011		101011111		10000011011		
	10001001		101100011		10000100111	22	$1+x+x^{22}$
	10001111		101100101		10000101101		
	10010001		101101001		10001100101	23	$1+x^5+x^{23}$
			101110001		10001101111	24	$1+x+x^2+x^7+x^{24}$
					10010000001		
					10010001011		
					...		

argued that there are actually only three *unique* m-sequences for $n = 5$, even though there are six polynomials. Moreover, consider the general form of (4.24):

$$\phi(x) = x^n \phi_M(x^{-1}) \tag{4.26}$$

and define

$$\phi_M(x) = x^n + c_{n-1}x^{n-1} + \ldots + c_1 x + 1 \tag{4.27}$$

as the primitive polynomial for the MSRG in Figure 4.23(b). It then follows that the SSRG that generates the *same* sequence as the MSRG is defined by (4.26) and (4.20). Put simply, the SSRG corresponding to a MSRG is obtained by reversing the order of the coefficients.

Example 4.17
Consider the primitive polynomial $\phi_M = 1 + x + x^3$ and its MSRG in Figure 4.25(a). If the initial state is (001) the reader could use clock-by-clock analysis to show that $\{u\}$ = (1011100). The SSRG that generates the same sequence is given by

$$\phi(x) = x^3 \phi_M(x^{-1}) = x^3(1 + x^{-1} + x^{-3}) = 1 + x^2 + x^3$$

Interpreting this using (4.20) gives the SSRG circuit in Figure 4.25(b). If the initial state in Figure 4.25(b) is (001), then from (4.23) we have

$$u(x) = \frac{1+x^7}{1+x^2+x^3} = 1 + x^2 + x^3 + x^4$$

that is, $\{u\}$ = (1011100), and thereafter the sequence repeats.

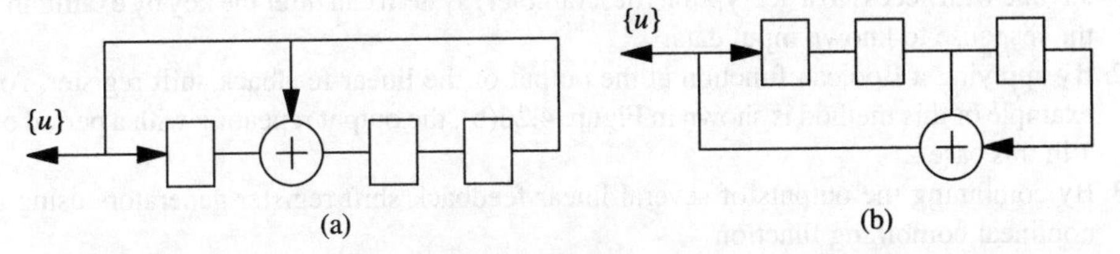

Figure 4.25 *(a) MSRG; (b) SSRG that generates the same sequence as (a).*

4.4.2 PN sequences with improved security

Although *m*-sequences are widely used in communications systems, it is important to realise that they do not in themselves make a system *secure*. For secure communications, as in spread spectrum systems, we require PN sequences which are difficult to reconstruct from a short segment of the sequence. Unfortunately, linear feedback shift register sequences do not posses this property. Consider again the recursion for Figure 4.23(a):

$$u_j = \sum_{k=1}^{n} c_k u_{j-k} \quad \text{modulo - 2} \tag{4.28}$$

where $c_n = 1$. For specific time indices we have the following sets of symbols:

$$j = n\text{:} \qquad u_n = \sum_{k=1}^{n} c_k u_{n-k} \Rightarrow u_n u_{n-1} u_{n-2} \cdots u_0$$

$$j = n+1\text{:} \qquad u_{n+1} = \sum_{k=1}^{n} c_k u_{n-k+1} \Rightarrow u_{n+1} u_n u_{n-1} \cdots u_1$$

$$\vdots$$

$$j = 2n-2\text{:} \quad u_{2n-2} = \sum_{k=1}^{n} c_k u_{2n-k-2} \Rightarrow u_{2n-2} u_{2n-3} u_{2n-4} \cdots u_{n-2}$$

Therefore, from time $j = n$ to $j = 2n - 2$ inclusive, we have $n - 1$ linear equations in $n - 1$ unknowns (the coefficients c_1 to c_{n-1}), and the equations span a total of $2n - 1$ binary symbols. In other words, a jammer can determine the shift register taps and hence the whole PN sequence by observing just $2n - 1$ consecutive bits! Clearly, the vulnerability of such sequences arises from the simplicity of solving sets of linear equations, and an obvious solution is to introduce some form of *non-linearity*. This can be done in several ways:

1. By replacing the modulo-2 feedback with an arbitrary Boolean function (Figure 4.26(a)). The function results in a *non-linear* feedback shift register, and it could be implemented using a ROM. The complete circuit could be used in the configuration of a self-synchronizing scrambler system (Figure 4.28(a)) to give a simple encryption system with modest security. In effect the function acts as a security key. Unfortunately, anyone with access to a decryption (descrambler) system can infer the key by examining the response to known input data.
2. By applying a Boolean function at the output of the linear feedback shift register. An example of this method is shown in Figure 4.26(b), the output repeating with a period of 7 in this case.
3. By combining the outputs of several linear feedback shift register generators using a nonlinear combining function.

Of the first two approaches, the second is perhaps more realistic. For method 1 it is known that there are a large number of functions that generate a cycle length of 2^n, but the determination of such functions which simultaneously give sequences with good, random-like

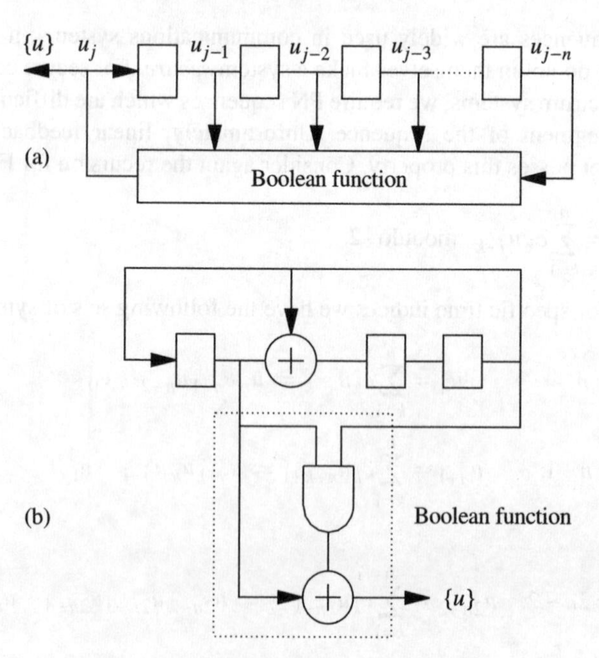

Figure 4.26 *Use of nonlinearity: (a) nonlinear feedback shift register; (b) nonlinear output logic.*

properties is not straightforward. On the other hand, the design of the nonlinear output logic in method 2 to give good random-like properties is discussed in the literature (Key, 1976).

4.4.3 Scrambling systems

As discussed, scrambling is often applied prior to data transmission in order to provide adequate timing content, or reduce clock jitter in PCM repeaters, or as part of a conditional access system. An elementary scrambling system is shown in Figure 4.27, and the scrambling action can be demonstrated by considering the probability of a 1 at the scrambler output:

$$P_C(1) = P_A(1)\,P_B(0) + P_A(0)\,P_B(1) \tag{4.29}$$

Since $P_B(0) \approx P_B(1) \approx 0.5$ for large n, then $P_C(1) \approx 0.5$, which implies that the transmitted bitstream appears random and is independent of the data. Also note that scrambling does not increase the bit rate, which means there is no bandwidth overhead. The same PN sequence is subtracted modulo-2 at the receiver, so that, provided there are no channel errors, the descrambled signal is

$$D = C + B = A + B + B = A \text{ over GF(2)} \tag{4.30}$$

In practice the receiver PN sequence is generated without the use of a second transmission channel, and one approach is to use a *self-synchronising* scrambler system. Figure 4.28(a) shows the self-synchronising system specified for the CCITT V29 modem, the shift register taps being defined by the polynomial

$$\phi(x) = 1 + x^{18} + x^{23} \tag{4.31}$$

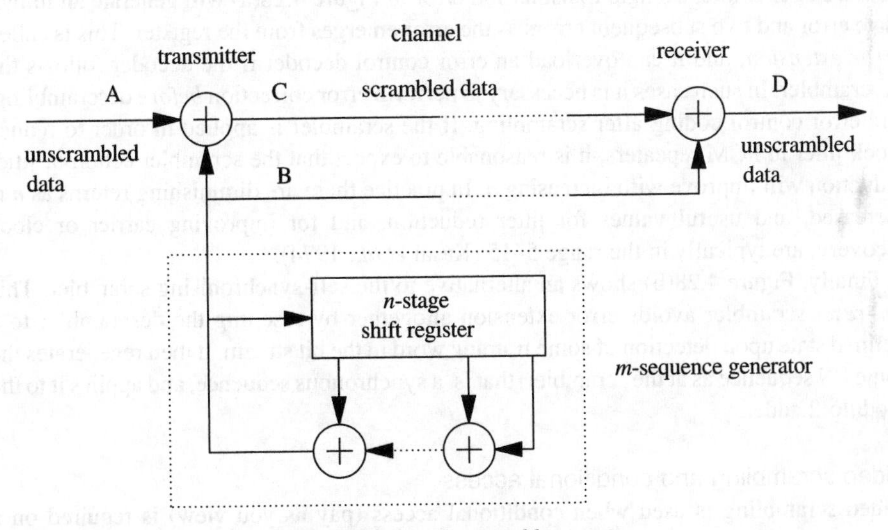

Figure 4.27 *Simple scrambler system.*

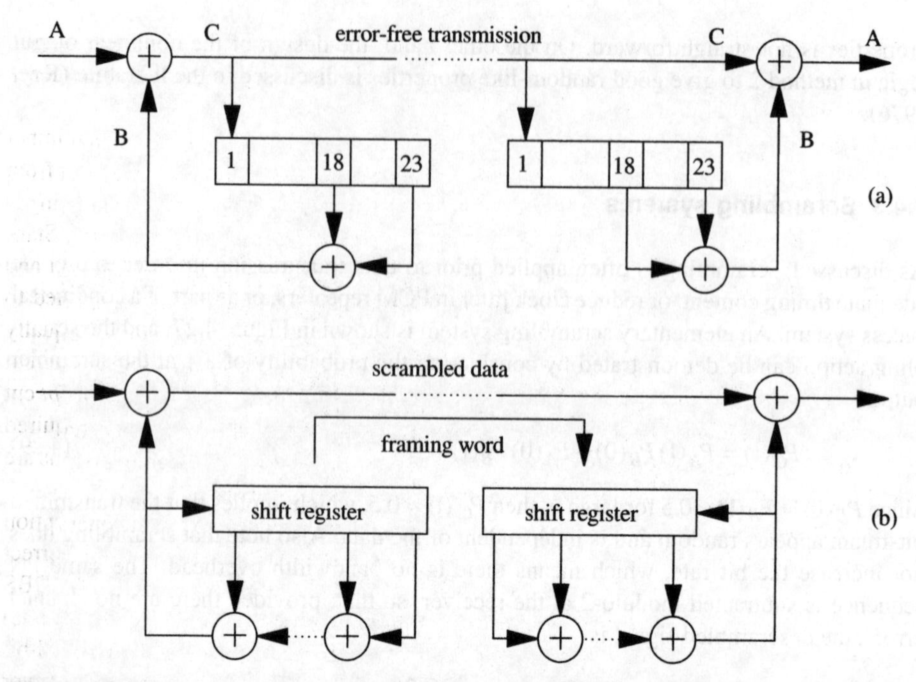

Figure 4.28 *Practical scrambler systems: (a) basic self-synchronising system for CCITT V29; (b) basic set–reset system.*

As before, provided that transmission is error-free, the descrambler output is identical to the scrambler input. Generally speaking, for a specified value of n, the optimal polynomial for a self-synchronising scrambler will have a minimal number of terms (minimal number of shift register taps) since this minimises both hardware and the effect of transmission errors. For example, a single transmission error in Figure 4.28(a) will generate an immediate error and two subsequent errors as the error emerges from the register. This is called *error extension*, and it can overload an error control decoder if the decoder follows the descrambler. In such cases it is necessary to perform error correction *before* descrambling, and error control coding after scrambling. If the scrambler is applied in order to reduce clock jitter in PCM repeaters, it is reasonable to expect that the scrambler action or jitter reduction will improve with increasing n. In practice there are diminishing returns as n is increased, and useful values for jitter reduction, and for improving carrier or clock recovery, are typically in the range 5–15 (Kasai *et al.*, 1974).

Finally, Figure 4.28(b) shows an alternative to the self-synchronising scrambler. This *set–reset* scrambler avoids error extension altogether by resetting the descrambler to a defined state upon detection of some framing word in the bit stream. It then regenerates the same PN sequence as at the scrambler, that is, a synchronous sequence, and applies it to the modulo-2 adder.

Video scrambling and conditional access

Video scrambling is used when conditional access (pay as you view) is required on a commercial video system. Usually the active part of each video line is digitised and stored to enable some form of *time scrambling* to be applied. For example, lines could be reversed at random, or lines within a large block of lines could be transmitted in random sequence (line dispersal).

Another common technique might be referred to as *segment swapping*. Here, a number of digital video lines are split into variable-sized segments and segments are read out from the store in shuffled order. To achieve a *secure* system, control of the segment shuffling is usually based upon some encryption algorithm. For example, the Data Encryption Standard (DES) algorithm has been used to control line dispersal scrambling (Kupnicki and Moote, 1984). In its simplest form the segment swapping technique may be applied to single lines. Typically, the active part of a line is randomly cut at one of 256 (8-bit) equally spaced positions and the two segments are interchanged prior to transmission. The cut position also varies in a pseudorandom manner down the picture. Note that an *abrupt* cut will generate rapid, wideband transitions, so some 'cross fading' between cuts is required to avoid any bandwidth increase. On the other hand, no signal level shifts or inversions are involved, so no artefacts will be generated by channel non-linearities.

A typical conditional access system achieves the required security by using encryption to 'lock' the scrambling process. In other words, de-scrambling is only possible if a correct 'unlocking key' is received. Figure 4.29 shows a basic conditional access system for DBS television based upon this principle. Here, a long control word, CW (typically 60 bits) seeds the PN generator (which has a long cycle time), which, in turn, could provide 8-bit numbers for segment swapping. An encrypted form of the CW is also sent to the receiver using an additional channel and it can only be decrypted by a secret 'session key' SK (which is transmitted on a third channel). In turn, decryption of SK is only possible by a properly authorised user.

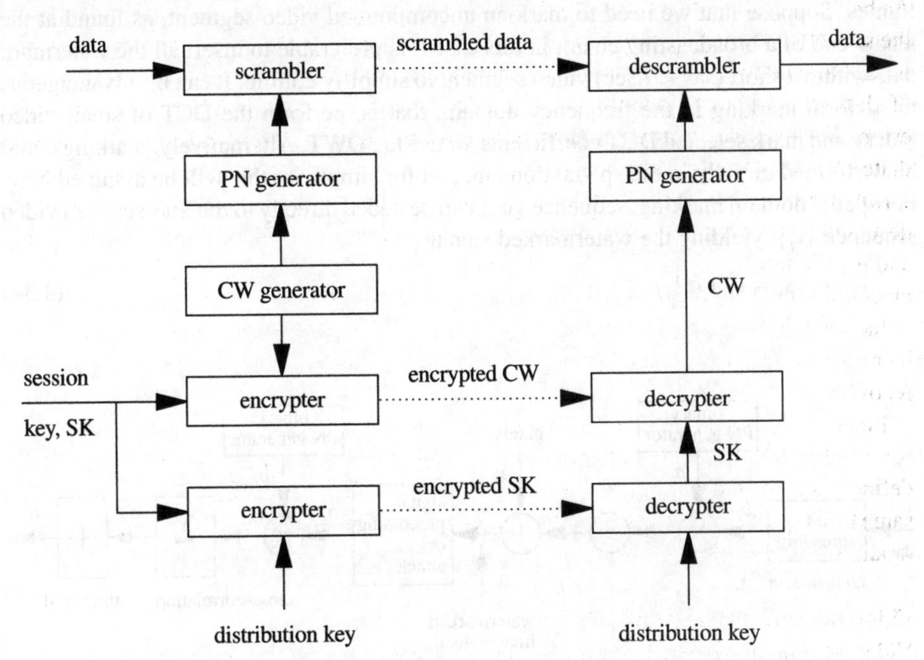

Figure 4.29 *Use of scrambling in a conditional access system.*

4.4.4 Watermarking using PN sequences

Watermarking is particularly applicable to audio, video and image material, and can be applied when the material is in either compressed or uncompressed form. Here we will consider image/video data, as found in networked multimedia systems, video discs, digital video broadcasting, and encrypted video-on-demand (VOD) systems. The idea is to embed a short sequence of binary data (either 'copyright' or 'fingerprint' data) in such a way that it is below the threshold of perception and yet is retrievable by the legal copyright holder. This 'watermark' must be retrievable even after considerable signal processing (such as MPEG compression), or after a determined attack to remove the watermark.

A common approach is based upon direct-sequence spread spectrum communications, Figure 4.30. Here, copyright data bits u_j, $u_j \in \{-1, +1\}$ are spread by a large factor, m (the *chip rate*), giving a spread data sequence $\{b_i\}$ where

$$b_i = u_j; \qquad jm \le i < (j+1)m \tag{4.32}$$

In the form of $\{b_i\}$, each data bit will now span m binary PN bits p_i, $p_i \in \{-1, +1\}$, as in spread spectrum systems. Sequence $\{b_i\}$ is then modulated with binary PN sequence $\{p_i\}$ to give a watermark signal

$$w_i = \alpha b_i p_i \tag{4.33}$$

Here α is a constant scaling factor, but in practice it can be advantageous to make it dependent upon the local image/video content in order to exploit the spatial and temporal masking effects of the eye. Sequence $\{w_i\}$ is now ready to watermark (modulate) the video frames. Suppose that we need to mark an uncompressed video segment, as found at the studio end of a broadcasting chain; in this case it is preferable to insert all the watermark data within a short (say < 1 sec) video segment to simplify editing. It can be advantageous to perform marking in the frequency domain, that is, perform the DCT of small video blocks and mark selected DCT coefficients, or use the DWT. Alternatively, marking could be performed directly in the spatial domain, and for simplicity this will be assumed here. For spatial domain marking, sequence $\{w_i\}$ can be added directly to the line-scanned video sequence $\{v_i\}$, yielding the watermarked signal

$$s_i = v_i + \alpha b_i p_i \tag{4.34}$$

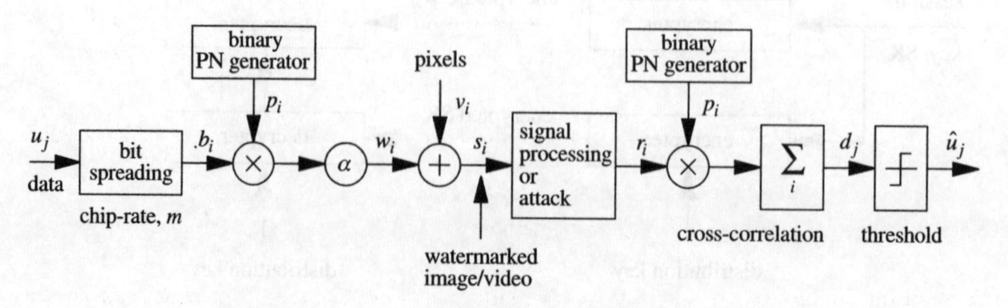

Figure 4.30 *Watermarking and retrieval (spatial domain).*

In fact, several watermark sequences could be added to $\{v_i\}$ providing different PN sequences are used. In this case the PN sequences would be selected to be mutually orthogonal (see Section 4.5.3).

Watermark retrieval follows basic spread spectrum receiver theory and uses a correlation detector (or matched filter) to extract the watermark bits from the received pixel sequence; see Figure 4.30. Each data bit is extracted by cross-correlation of received sequence $\{r_i\}$ with the same PN sequence $\{p_i\}$ over a window of m bits. Assuming $r_i = s_i$ (no intentional or unintentional attack), then

$$\begin{aligned} d_j &= \sum_{i=jm}^{(j+1)m-1} p_i s_i \\ &= \sum_{i=jm}^{(j+1)m-1} p_i v_i + \sum_{i=jm}^{(j+1)m-1} \alpha b_i p_i^2 \end{aligned} \tag{4.35}$$

Assuming the PN sequence and the received pixel sequence are uncorrelated over a large window, m, then

$$d_j \approx \sum_{i=jm}^{(j+1)m-1} \alpha b_i p_i^2 \tag{4.36}$$

Over the window, spread bits b_i simply take on the value of data bit u_j, so

$$d_j \approx \alpha m u_j \tag{4.37}$$

and u_j could be detected as $\hat{u}_j = \text{sign}(d_j)$.

There are several practical aspects worth mentioning here. First, the foregoing analysis makes the implicit assumption that the two PN sequences are correctly *synchronised* (in which case we are computing the peak of the cross-correlation function). Unfortunately, loss of synchronisation can occur in several ways, as for example when a determined attack is made to destroy the watermark by removing pixels or video frames. In practice it is therefore necessary to use a 'sliding correlator', which essentially searches for the cross-correlation peak over a limited search area. Secondly, the first term in (4.35) may not be negligible if sequence $\{p_i\}$ does not have an equal number of +1s and −1s. If this is the case it may be necessary to add a correction term to (4.36). More importantly, for spatial domain marking it is advantageous to remove the low-frequency content of the video sequence prior to cross-correlation (this can be done with a simple 3 × 3 Laplacian filter). Finally, it is not advisable to use a linearly generated sequence (*m*-sequence) for the PN sequence due to lack of security (Section 4.4.2). An aperiodic, non-shift register generated sequence could be preferable.

4.5 Correlation of sequences

4.5.1 Aperiodic and periodic correlation

Correlation can be used as a measure of the similarity between two data sequences and in this section we examine various correlation measures, with particular emphasis upon PN sequences. At the outset, we need to distinguish between *aperiodic* correlation and *periodic* correlation. Essentially, aperiodic correlation applies when a sequence is non-repetitive, with nothing before it or after it.

Example 4.18

Suppose we are given N samples $\{x_0 ... x_{N-1}\}$ of a random signal (formally we say we are given a sample function of a random process). In this case the symmetric autocorrelation function (ACF) of the signal could be estimated as

$$R_i = \frac{1}{N-i}\sum_{n=0}^{N-i-1} x_n x_{n+i}; \qquad i \geq 0 \tag{4.38}$$

This is an example of aperiodic correlation since, clearly, the sequence is non-repetitive. Also, when used in the context of a random signal, (4.38) is applicable to source coding (Section 3.1.3). Equation (4.38) is an *unbiased* estimate in the sense that the expected value of (4.38) is identical to the true ACF. On the other hand, as $i \Rightarrow n$, there are progressively fewer samples used in the estimate and the variance of the estimate increases. For this reason it is more usual to use the *biased* estimate

$$R_i = \frac{1}{N}\sum_{n=0}^{N-i-1} x_n x_{n+i}; \qquad i \geq 0 \tag{4.39}$$

This is biased in the sense that the expected value of (4.39) is $(N-i)/N$ times the true ACF, that is, it biases R_i towards zero for $i \Rightarrow N$. On the other hand, the bias disappears if $N >> i$, in which case (4.39) will give a good estimate of R_i. Usually the normalizing factor $1/N$ is omitted.

In this section we are more concerned with random binary sequences (PN sequences) rather than random signals. However, the same basic equations apply. For instance, Figure 4.31 illustrates the use of (4.39) for a simple bipolar sequence with $N = 3$. Occasionally we need to compute the ACF of a *complex* sequence $\{x_k\}$ of length N, in which case the aperiodic ACF is given by

```
... 0 0 0 + + − 0 0 0 ...      lag, i   R_i
+ + −                            −3      0
  + + −                          −2     −1
    + + −                        −1      0
      + + −                       0     +3
        + + −                     1      0
          + + −                   2     −1
            + + −                 3      0
```

Figure 4.31 *Determination of the ACF for an aperiodic sequence (N = 3).*

$$R_i = \sum_{n=0}^{N-i-1} x_n x_{n+i}^* \tag{4.40}$$

where * denotes complex conjugate.

Unique words

The ACF is of particular interest in radar, spread-spectrum, frame synchronisation and user identification applications. For instance, we could use an aperiodic random binary sequence or *unique word* (UW) to form a *preamble* for a data sequence in order to achieve frame synchronisation. In this case, the UW is usually optimised to make R_i, $i > 0$, as small as possible (although, with this optimisation criterion, there is no constraint on the number of different time displacements that can have the same maximum value of out of phase correlation).

Example 4.19

Where a short aperiodic sequence is permissible, we could use a *Barker code* (Barker, 1953). An Nth-order binary Barker code is a sequence of N positive or negative ones and its aperiodic ACF has a 'peak-to-maximum-sidelobe ratio' or PSR of N:1. The four highest-order Barker codes are given in Table 4.5.

Table 4.5 *Barker codes.*

Order, N	*PSR*	*Code*
5	5:1	+ + + − +
7	7:1	+ + + − − + −
11	11:1	+ + + − − − + − − + −
13	13:1	+ + + + + − − + + − + − +

Figure 4.32 shows a matched filter for detection of the 7th-order Barker code. In practice, if the matched filter output exceeds the required threshold (which will be set to allow for symbol errors), it is assumed that correct synchronisation has been established. Looked at

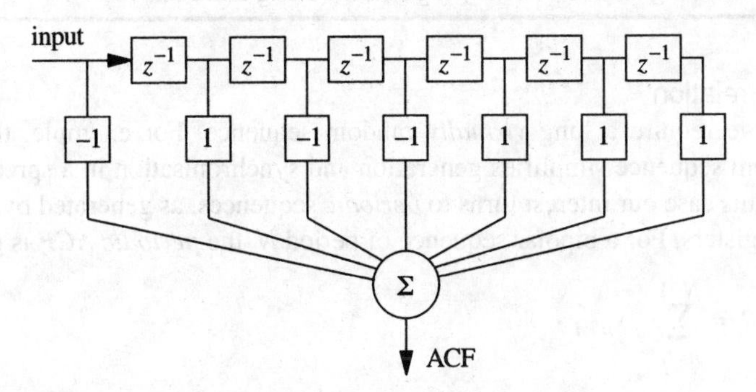

Figure 4.32 *Matched filter correlator for a 7th-order Barker code.*

another way, the matched filter *compresses* the energy of the PN sequence to a single large, short-duration output. It is generally believed that Barker sequences only exist up to $N = 13$, but generalised *Barker-like* sequences up to $N = 31$ have been generated by computer search (Chan and Lam, 1995).

Example 4.20

Very often synchronisation applications require longer aperiodic sequences than those in Example 4.19 in order to enhance the noise immunity or *processing gain*, and these are usually found by computer search. For instance, one search method (not restricted to particular sequence lengths) commences with a random sequence of +1s and –1s and first changes one bit of the sequence to see if there is any improvement in autocorrelation sidelobe level. If there is an improvement the new sequence is retained and the process is repeated. If there is no improvement, then two bits of the sequence are changed (all four combinations being tried). If there is still no improvement, then all combinations of three bits of the sequence are sequentially changed, and so on. As an example, a good 64-bit binary sequence, optimised against aperiodic function sidelobe level is (Tomlinson, 1997):

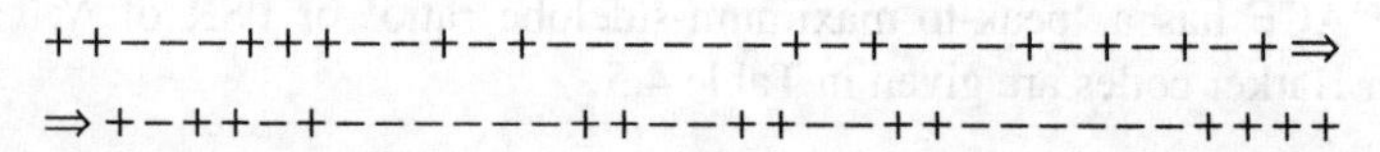

This sequence has a peak ACF value of 64 and a peak side lobe magnitude of 5.
Rather than minimise the peak sidelobe level, another optimisation criterion maximises the *merit factor*, *MF*. This is the ratio of the main lobe energy to the energy of the sidelobes and can be expressed as

$$MF = \frac{N^2}{2\Sigma_{i=1}^{N-1} R_i^2} \tag{4.41}$$

where N is the sequence length. The *MF* for the above 64-bit sequence is 3.68. Another optimised 64-bit sequence with a peak sidelobe of 5, but this time with $MF = 5.07$ is:

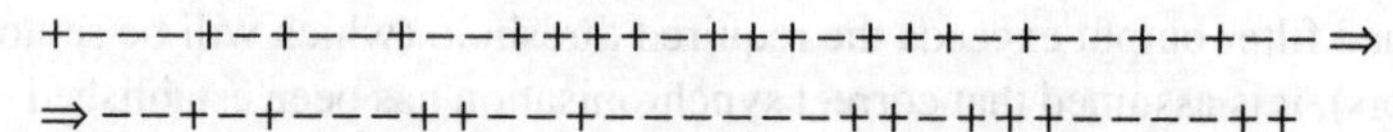

This was found using an evolutionary algorithm (Deng and Fan, 1999).

Periodic correlation

Sometimes we require a long *virtually* random sequence. For example, the use of a pseudorandom sequence simplifies generation and synchronisation in a spread spectrum receiver. In this case our interest turns to *periodic* sequences, as generated by linear feedback shift registers. For a bipolar sequence of period N, the *periodic ACF* is given by

$$R_i = \sum_{n=0}^{N-1} x_n x_{n+i} \tag{4.42}$$

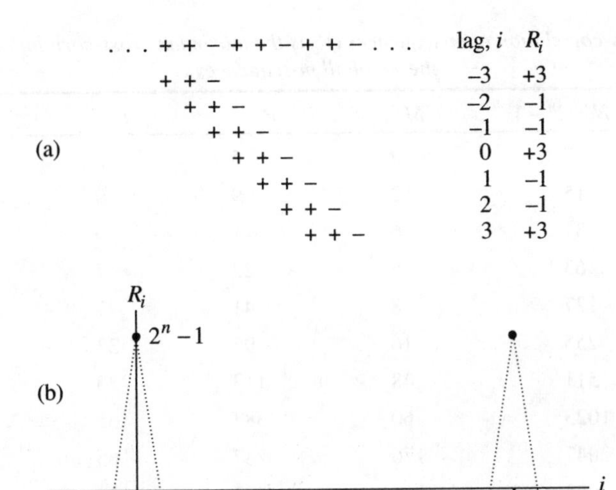

Figure 4.33 *Periodic ACF: (a) N = 3; (b) for an m-sequence period N = 2^n – 1.*

Figure 4.33(a) illustrates the use of (4.42) for the sequence {... + + − + + − + + − ...}, and Figure 4.33(b) shows the periodic ACF for an *m*-sequence generated by an *n*-stage register. It is apparent that *m*-sequences have an excellent periodic ACF, with a high peak-to-sidelobe ratio.

For some applications, such as code-division multiple access (CDMA), we are also interested in the *cross-correlation* properties of periodic sequences. In particular, for sequences $\{x\}$ and $\{y\}$, each of period N, the periodic cross-correlation function is defined as

$$R_i = \sum_{n=0}^{N-1} x_n y_{n+i}, \qquad i = 0, 1, \ldots, N-1 \tag{4.43}$$

where $y_N = y_0$, $y_{N+1} = y_1$, and so on. Usually we require $R_i / N \ll 1 \; \forall i$ in (4.43). Under some circumstances, (4.43) could also be applied to aperiodic sequences.

4.5.2 Preferred sequences

Generally speaking, two randomly selected *m*-sequences can have a significant cross-correlation function, although it *is* possible to select pairs of *m*-sequences that have a favourable (relatively small) cross-correlation function. Consider two *m*-sequences *u* and *v* each of length (period) $N = 2^n - 1$ where *n* is the shift register length, and let *u* and *v* be generated by degree-*n* primitive polynomials $\phi_u(x)$ and $\phi_v(x)$ respectively. In particular, let $\phi_u(x)$ be the *minimal polynomial* of primitive element α, $\alpha \in \mathrm{GF}(2^n)$ (that is, α is a root of $\phi_u(x)$) and let $\phi_v(x)$ be the minimal polynomial of α^t, where

$$t = 2^{\lfloor (n+2)/2 \rfloor} + 1 \tag{4.44}$$

Table 4.6 *Cross-correlation of m-sequences (R_c is the maximum cross-correlation magnitude for the set of all m-sequences).*

n	$N = 2^n - 1$	M	R_c	t	M_p
3	7	2	5	5	2
4	15	2	9	9	0
5	31	6	11	9	3
6	63	6	23	17	2
7	127	18	41	17	6
8	255	16	95	33	0
9	511	48	113	33	2
10	1023	60	383	65	3
11	2047	176	287	65	4
12	4095	144	1407	129	0

This being the case, it has been shown (Gold, 1967) that sequences u and v will have a periodic cross-correlation function with a magnitude which is upper bounded by t. In fact, providing $n \neq 0$ modulo-4, their exist pairs of m-sequences with three-valued cross-correlation functions, the values being $\{-1, -t, t-2\}$. Pairs with this cross-correlation are called *preferred pairs*, and for a specified value of n there are sets of m-sequences for which each pair in the set form a preferred pair. Unfortunately, the size, M_p, of such a set is generally small relative to the total number, M, of m-sequences (see Table 4.6). Also note that preferred pairs (that is, pairs with a *three*-valued cross-correlation function) do not exist when n is a multiple of 4. It is clear from Table 4.6 that random selection of m-sequence pairs can result in high cross-correlation values R_c compared with that achievable by preferred pairs, for example, 113 compared with 33 for $n = 9$.

Example 4.21

Consider $n = 5$ and the primitive polynomial

$$\phi_u(x) = x^5 + x^2 + 1 \equiv 100101 \equiv 45_8$$

(Note that it is usual to use compact octal notation). If α is a root we can write the identity $\alpha^5 = \alpha^2 + 1$, and since α is a primitive element of GF(2^5), that is, α, α^2, α^3, ... are all distinct, we can use the identity to generate GF(2^5). For example, $\alpha^6 = \alpha\alpha^5 = \alpha^3 + \alpha$, and $\alpha^{31} = 1$ (see (6.81)). According to Table 4.6, $\phi_v(x)$ will be the minimal polynomial of α^9, that is, α^9 will be a root. However, since over GF(2) we have $[\phi_v(x)]^2 = \phi_v(x^2)$, $[\phi_v(x)]^4 = \phi_v(x^4)$, and so on, then $(\alpha^9)^2$, $(\alpha^9)^4$, $(\alpha^9)^8$ and $(\alpha^9)^{16}$ will also be roots. Thus, the required 5th–order polynomial could be written

$$\phi_v(x) = (x + \alpha^9)(x + \alpha^{18})(x + \alpha^{36})(x + \alpha^{72})(x + \alpha^{144})$$

We can simplify $\phi_v(x)$ by reducing it using the rules of GF(2^5). As discussed in Section 6.6.1 these rules can be defined by setting $\phi_u(\alpha) = 0$ and deriving sets of simplifying identities. However, the simplest approach here is to use a Galois field table for GF(2^5) generated by $\phi_u(x)$ (Lin and Costello, 1983). Using this table

$$\phi_v(x) = x^5 + x^4 + x^2 + x + 1 \quad \equiv \ 110111 \equiv 67_8$$

Polynomials $\phi_u(x) = 45$ and $\phi_v(x) = 67$ therefore form a preferred pair for $n = 5$.

The use of decimation

Manipulation of minimal polynomials as in Example 4.21 is tedious, and there is a more straightforward way to identify preferred pairs based on sequence decimation or *sampling*. Suppose we deduce sequence v by taking every qth bit of sequence u (denoted by $v = u[q]$) where u is generated by $\phi_u(x)$. If $u[q]$ is not identically equal to zero, then v has period $N/\gcd(N,q)$ (where gcd denotes greatest common divisor) and the roots of $\phi_v(x)$ are the qth powers of the roots of $\phi_u(x)$. Thus, given $\phi_u(x)$, we can consult a table of minimal polynomials corresponding to various values of q (Peterson and Weldon, 1972) in order to directly identify $\phi_v(x)$. To help us do this we can invoke the following theorem (Sarwate and Pursley, 1980):

Theorem 4.1 Let u and v be m-sequences of period $N = 2^n - 1$, with $v = u[q]$ where

$$q = 2^k + 1 \quad \text{or} \quad q = 2^{2k} - 2^k + 1 \tag{4.45}$$

and k is a small integer. The periodic cross-correlation function for u and v is then three-valued with values

$$\{-1, -(1 + 2^{(n+e)/2}), -1 + 2^{(n+e)/2}\} \tag{4.46}$$

where $e = \gcd(n,k)$ such that n/e is odd. It is important to recognise that the cross-correlation values depend only upon q, and not upon the individual m-sequences.

Clearly, e should be small, and if $e = 1$ we must consider n odd only. In order to obtain $e = 1$ when n is odd we can let $k = 1$ or $k = 2$, for example. It is now apparent that, for n odd and $e = 1$, the following pairs will be three-valued with the cross-correlation magnitude bounded by (4.44):

$$u, u[3];\ u, u[5];\ u, u[13] \tag{4.47}$$

In other words, decimation of u by 3, 5 or 13 will give a preferred pair (although other decimations are possible). Cases for n odd and $v = u[3]$ have been extracted from Peterson and Weldon (1972) and are presented in Table 4.7. The table also gives *reciprocal pairs* of polynomials for $n = 6$ and $n = 10$, for which the peak magnitude cross-correlation is $t - 2$.

Example 4.22

According to Table 4.7, a preferred pair for $n = 9$ is

$$u \equiv 1021 \equiv 1000010001 \equiv x^9 + x^4 + 1 = \phi_u(x)$$

$$v = u[3] \equiv 1131 \equiv 1001011001 \equiv x^9 + x^6 + x^4 + x^3 + 1 = \phi_v(x)$$

Table 4.7 *Pairs of m-sequences with low cross-correlation. Polynomials are in octal form.*

n	$\phi_u(x)$	$\phi_v(x)$
5	45	75
6	103	141
7	211	217
9	1021	1131
10	2011	2201
11	4005	4445
13	20033	23261

The cross-correlation function for u and v will have values $\{-1, -33, 31\}$.

Example 4.23

According to (4.47) a preferred pair for n odd is simply $\{u, u[5]\}$. Thus, if for $n = 5$ we take

$$u \equiv 45 \equiv 100101 \equiv x^5 + x^2 + 1 = \phi_u(x)$$

then, from tables (Peterson and Weldon, 1972)

$$v = u[5] \equiv 67 \equiv 110111 \equiv x^5 + x^4 + x^2 + x + 1 = \phi_v(x)$$

Cross-correlation of u and v will give a cross-correlation function with values $\{-1, -9, 7\}$. Note that $\phi_v(x)$ is the same polynomial we deduced in Example 4.21, and we showed that α^9 is a root. However, from the above discussion on decimation we also expect α^5 to be a root of $\phi_v(x)$ over GF(2^5) as defined by $\phi_u(x)$. The reader should check that this is indeed the case. An MSRG for realising $\phi_v(x)$ (as defined in (4.20)) is given in Figure 4.34.

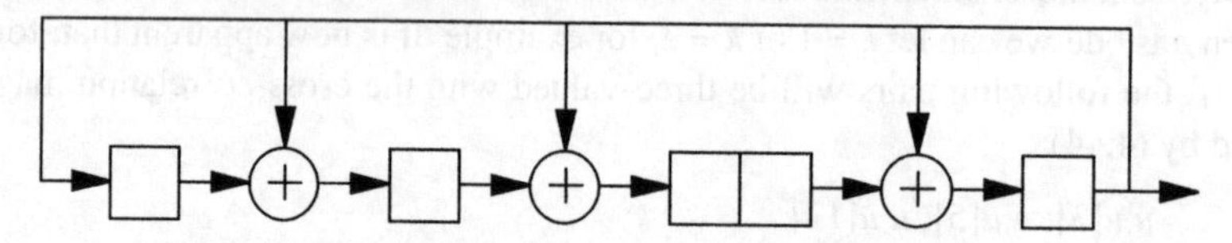

Figure 4.34 MSRG realisation for Example 4.23.

4.5.3 Gold sequences

We have seen that only relatively small sets of *m*-sequences have good periodic cross-correlation. This may be satisfactory in applications where only a few PN sequences are required. On the other hand, for code-division multiple access systems (CDMA), it is necessary to have *many* sequences with low cross-correlation, and a common solution is to use Gold codes.

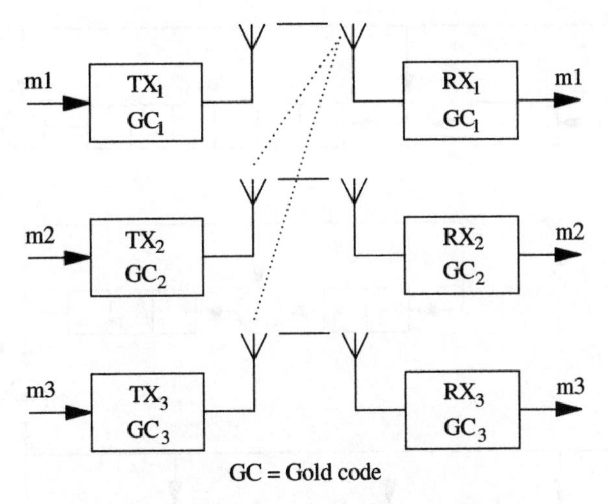

Figure 4.35 *Simple CDMA system using Gold codes.*

In Figure 4.35, for example, multiple users operate on the same carrier frequency and spread over the same bandwidth, but suffer minimal interference due to the use of different Gold codes. For instance, receiver RX_1 will only de-spread the signal from transmitter TX_1 even though it is receiving signals from other transmitters. The generation of Gold sequences is based upon the following theorem (Gold, 1967):

Theorem 4.2 Let $\phi_u(x)$ and $\phi_v(x)$ be a preferred pair of primitive polynomials, each of degree n. Then the $2n$-stage shift register corresponding to the product polynomial $\phi(x) = \phi_u(x)\phi_v(x)$ generates $2^n + 1$ different sequences each with period $2^n - 1$. The cross-correlation function of any pair of these sequences will be three-valued with values $\{-1, -t, t-2\}$, where t is given by (4.44).

Example 4.24

Consider the preferred pair $u = 45$, $v = 67$ in Example 4.23, corresponding to $n = 5$ and $t = 9$. We have

$$\phi(x) = (1 + x^2 + x^5)(1 + x + x^2 + x^4 + x^5) = 1 + x + x^3 + x^9 + x^{10} \text{ modulo-2}$$

and $\phi(x)$ could be implemented as shown in Figure 4.36(a). The 10-stage register generates sequences of period 31. Since there are $2^{10} - 1$ possible non-zero initial states, and each state corresponds to one bit of the 31 bit output sequence, then there are $(2^{10} - 1)/31 = 33$ distinct initial states which each result in a unique Gold sequence.

The more usual implementation is to modulo-2 add the two m-sequences u and v, as shown in Figure 4.36(b). This can generate $2^n - 1 = 31$ new periodic sequences (each of length 31) by changing the relative phase of the summed sequences. These, plus the two preferred sequences, constitute the $2^n + 1 = 33$ Gold sequences.

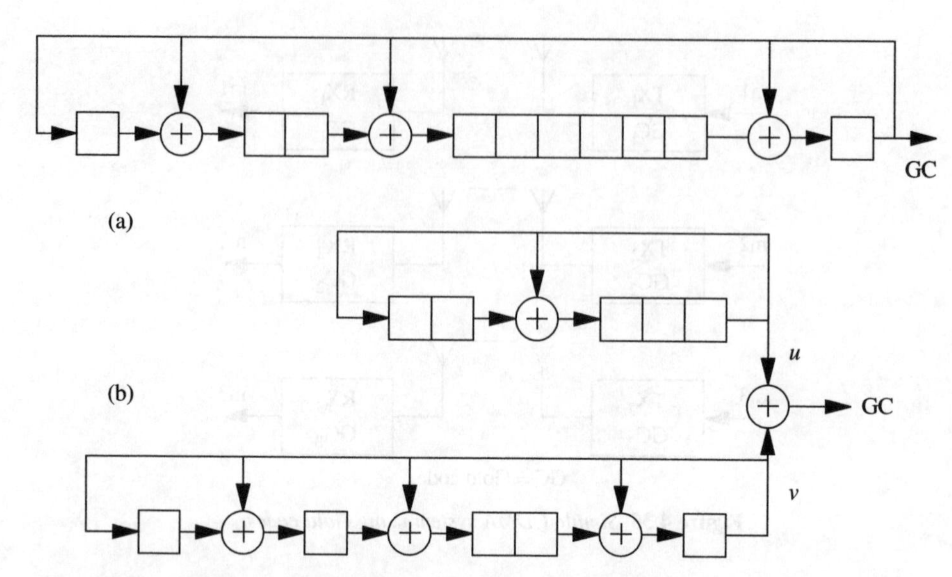

Figure 4.36 *Generation of 33 Gold codes, each of length 31: (a) direct generation; (b) generation by addition of two preferred sequences.*

Bibliography

Armour, S., Nix, A. and Bull, D. (2000) A pre-FFT equalizer design for OFDM, *Electronics Letts.* (in press).

Barker, R. (1953) Group synchronizing of binary digital systems, in *Communication Theory*, Butterworth, London, pp. 273–287.

Bhatoolaul, D. and Wade, G. (1995) Spectrum shaping in *N*-channel QPSK-OFDM systems, *IEE Proc.-Vis. Image Signal Process.*, **142**(5), 333–338.

Byrne, C. *et al.* (1963) Systematic jitter in a chain of digital repeaters, *Bell Syst. Tech. J.*, **42**, 2679–2714.

CENELEC (1990) *Specification of the radio data system (RDS)*, European Committee for Electrotechnical Standardisation, Rue de Stassart 35, B-1050 Brussels, EN 50067.

Chan, C. and Lam, W. (1995) Generalised Barker-like PN sequences for quasisynchronous spread-spectrum multiple-access communication systems, *IEE Proc.-Communications*, **142**(2), 91–98.

Deng, X. and Fan, P. (1999) New binary sequences with good aperiodic autocorrelations obtained by evolutionary algorithm, *IEEE Communications Lett.*, October.

Dolivo, F., Hermann, R. and Olcer, S. (1989) Performance and sensitivity analysis of maximum-likelihood sequence detection on magnetic recording channels, *IEEE Trans. Magnetics*, **25**(5), 4072–4074.

Franaszek, P. (1972) Run-length-limited variable length coding with error propagation limitation, *US Patent 3 689 899*.

Gold, R. (1967) Optimal binary sequences for spread spectrum multiplexing, *IEEE Trans. on Inform. Theory*, October, 619–621.

Justesen, J. (1982) Information rates and power spectra of digital codes, *IEEE Trans. Inform. Theory*, **IT-28**(3).

Kabal, P. and Pasupathy, S. (1975) Partial-response signalling, *IEEE Trans. Communications*, **COM-23**(9), 921–934.

Kallaway, M. and Mahadeva, W. (1977) CEEFAX: Optimum transmitted pulse-shape, *Report* **1977/15**, British Broadcasting Corporation Research Department.

Kasai, H. *et al.* (1974) PCM jitter suppression by scrambling, *IEEE Trans. Communications,* **COM-22**(8), 1114–1122.

Key, E. (1976) An analysis of the structure and complexity of nonlinear binary sequence generators, *IEEE Trans. Inform. Theory*, **IT-22**, 732–736.

Kupnicki, R. and Moote, S. (1984) High security television transmission using digital processing, *Milcom'84, IEEE Military Comms. Conf.*, Los Angeles, CA, pp. 284–289.

Lender, A. (1966) Correlative level coding for binary-data transmission, *IEEE Spectrum*, February, 104–115.

Lin, S. and Costello, Jr, D. J. (1983) *Error Control Coding*, Prentice-Hall, Englewood Cliffs, NJ.

Lucky, R., Salz, J. and Weldon, E. (1968) *Principles of Data Communication*, McGraw-Hill, New York.

Peterson, W. and Weldon, E. (1972) *Error Correcting Codes*, 2nd edn, MIT Press, Cambridge, MA.

Sarwate, D. and Pursley, M. (1980) Crosscorrelation properties of pseudorandom and related sequences, *Proc. IEEE*, **68**(5), 593–619.

Tomlinson, M. (1997) A state-of-the-art burst satellite communication system, *Satellite Centre Report No. 97102*, University of Plymouth, UK.

CHAPTER 5

Error control systems

Error control coding is a practical way of achieving very low *bit error rate* (BER) after transmission over a noisy, bandlimited channel. The fact that this is theoretically possible is stated by Shannon's noisy coding theorem (Theorem 1.1). In this chapter we first overview the error control scene. We then examine the implications of adding error control to a communications system, together with any theoretical limitations. The chapter concludes with a look at ideal or optimal decoding concepts, such as MAP and Maximum Likelihood (ML) decoding. The detailed design of error control encoders and decoders is discussed in Chapters 6 and 7.

5.1 Overview

Two main classes of error control are used to achieve an acceptable error rate at the receiver, namely, *automatic repeat request* (ARQ) and *forward error control* (FEC) (Figure 5.1(a)). Pure ARQ uses an error *detecting* code together with a feedback channel to initiate retransmission of any block or packet received in error. It can be used where constant throughput is not essential and variable delay is permissible. More generally, pure ARQ is combined with a moderate degree of error correction coding (*hybrid ARQ*) in order to increase the system throughput by reducing the number of retransmissions required.

There are two basic types of hybrid ARQ. Type-I incorporates error detection and error correction parity bits in every packet and requests retransmission of the original packet if an uncorrectable error pattern is detected. The uncorrectable packet is discarded even though it may contain useful information. The code power (specifically the code *rate*) can also be adjusted upon retransmission in order to match varying channel conditions. Type-II hybrid ARQ keeps the erroneous packet. It commences with a high rate (low power, low redundancy) code and if this fails the erroneous packet is used together with a transmission of additional parity bits to create a lower rate (higher power) code. If this fails, an even lower rate code could be tried. The variable code rate is usually achieved by using *punctured* convolutional codes (Section 7.1.2). Clearly, both Type-I and Type-II schemes offer forms of *adaptive error control* where the degree of error correction varies according to channel conditions. This adaptability makes hybrid ARQ attractive for time-varying channels. For example, it has been used on land mobile to satellite links, and for low bit rate video transmission over wireless networks.

As the name suggests, FEC controls the received error rate via forward transmission only. It is particularly applicable where a bounded delay and constant throughput is required, as in real-time systems. Here, error control codes are usually selected to guarantee a prescribed BER for the worst channel conditions, but clearly this will reduce the

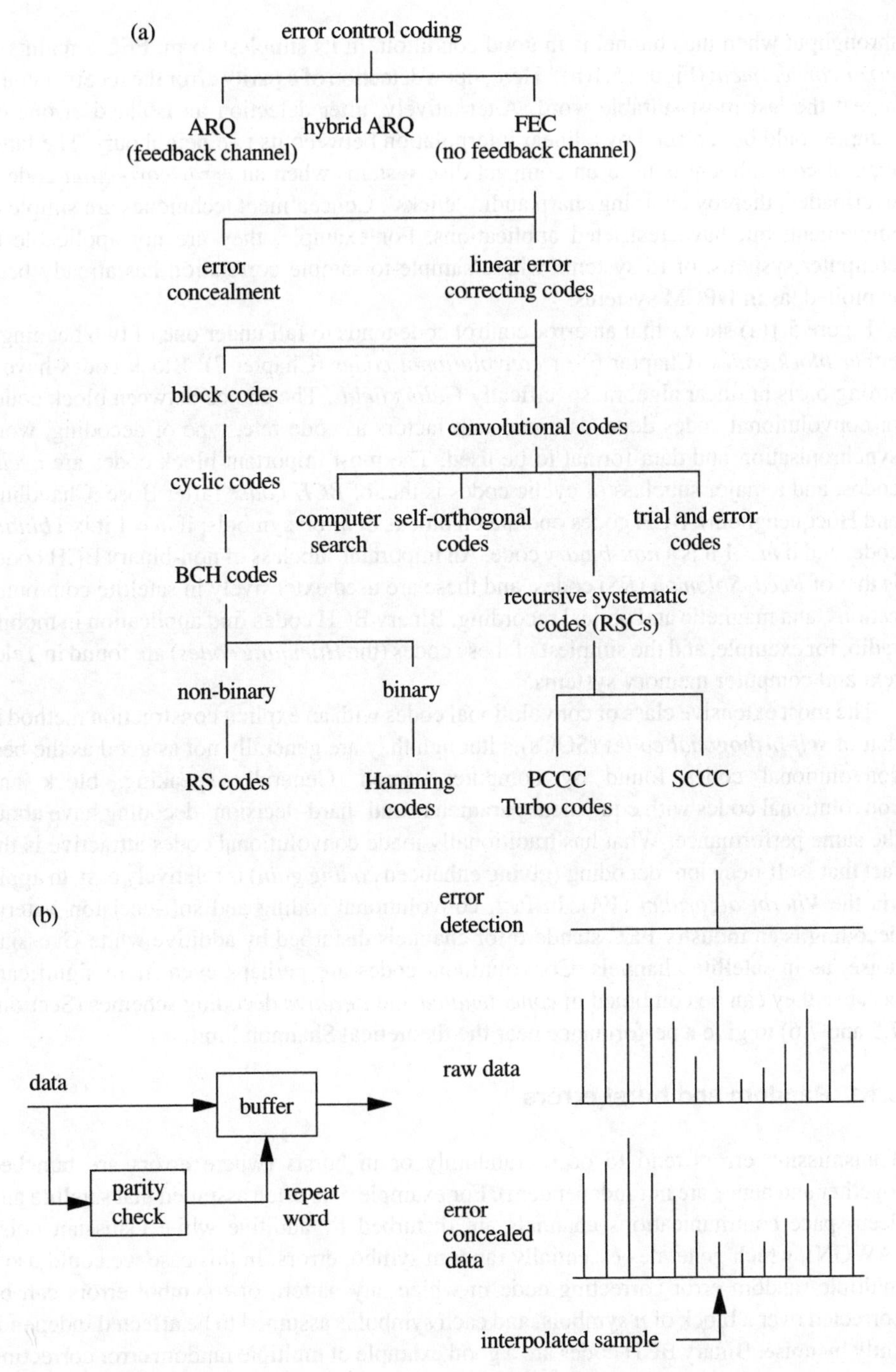

Figure 5.1 *(a) Overview of error control coding; (b) error concealment techniques using word repeat and sample interpolation.*

throughput when the channel is in good condition. In its simplest form, FEC amounts to *error concealment* (Figure 5.1(b)). Here, upon detection of a parity error the receiver could repeat the last most suitable word. Alternatively, after detection an isolated erroneous sample could be replaced by a linear interpolation between its two neighbours. The latter type of concealment is used on compact disc systems when an *error correction* code is overloaded, thereby avoiding sharp audio 'clicks'. Concealment techniques are simple to implement, but have restricted applications. For example, they are not applicable to computer systems, or to systems where sample-to-sample correlation has already been exploited, as in DPCM systems.

Figure 5.1(a) shows that an error control code tends to fall under one of two headings: either *block codes* (Chapter 6) or *convolutional codes* (Chapter 7). Block codes have a strong basis in linear algebra, specifically *Galois fields*. The choice between block codes or convolutional codes depends upon such factors as code rate, type of decoding, word synchronisation and data format to be used. The most important block codes are *cyclic* codes, and a major subclass of cyclic codes is that of *BCH codes* (after Bose, Chaudhuri and Hocquenghem). BCH codes operate on blocks of m-bit symbols; if $m = 1$ it is a *binary* code, and if $m > 1$ it is a *non-binary* code. An important subclass of non-binary BCH codes is that of *Reed–Solomon* (*RS*) codes, and these are used extensively in satellite communications, and magnetic and optical recording. Binary BCH codes find application in mobile radio, for example, and the simplest of these codes (the *Hamming codes*) are found in Teletext and computer memory systems.

The most extensive class of convolutional codes with an explicit construction method is that of *self-orthogonal codes* (SOCs), although they are generally not as good as the best convolutional codes found by computer search. Generally speaking, block and convolutional codes with equivalent parameters and 'hard-decision' decoding have about the same performance. What has traditionally made convolutional codes attractive is the fact that 'soft-decision' decoding (giving enhanced *coding gain*) is relatively easy to apply via the *Viterbi algorithm* (*VA*). In fact, convolutional coding and soft-decision Viterbi decoding is an industry FEC standard for channels disturbed by additive white Gaussian noise, as in satellite channels. Convolutional codes are perhaps even more significant because they can be combined in *concatenated* and *iterative* decoding schemes (Sections 7.5 and 7.6) to give a performance near the theoretical Shannon limit.

5.1.1 Random and burst errors

Transmission errors tend to occur randomly or in bursts (where errors are bunched together and hence are not independent). For example, it is often assumed that satellite and deep-space communications channels are disturbed by additive white Gaussian noise (AWGN), which generates essentially random symbol errors. In this case we could use a multiple random error-correcting code in which any pattern of t symbol errors can be corrected over a block of n symbols, and each symbol is assumed to be affected independently by noise. Binary BCH codes are a good example of multiple random error correcting codes.

In contrast, land mobile radio transmissions suffer from multipath propagation and a varying interference pattern as the vehicle moves. In turn, this gives amplitude variations (Rayleigh or Rician fading) and random phase variations in the received signal. Usually the fades occur at a much slower rate than the transmission rate, which means that errors occur in bursts. Magnetic and optical digital recording channels are also characterised as 'bursty channels'. One solution to the multipath propagation problem is to use some form of diversity. Another solution is to use FEC and a code specifically designed to combat burst errors, such as an RS code or a *Fire code*. In fact, it is interesting to note that the use of FEC on Rayleigh fading channels can be particularly effective in the sense that the coding gain (Section 5.4.4) can be considerably higher than that for Gaussian channels. Alternatively, instead of using a burst error-correcting code we could use a random error-correcting code together with a technique called *interleaving* (Section 6.4.8). Interleaving is particularly useful since it permits both random and burst errors to be corrected with one type of code, so it is tolerant of varying channel error statistics. In the context of a Rayleigh fading channel, interleaving can disperse error bursts to make it look similar to a Gaussian channel.

5.1.2 Concatenated codes

Shannon's noisy coding theorem implies that for any information rate less than channel capacity, it is theoretically possible to find a code such that the probability of incorrect decoding is arbitrarily small. In particular, Shannon showed that block codes exist for which the error probability decreases exponentially with block length n. Unfortunately, the complexity of the corresponding encoder and decoder will increase exponentially with n and ideal coding/decoding becomes impractical. Concatenated codes, and the *product codes* discussed in Section 6.8, are a practical way of obtaining very long codes (giving an exponential decrease in error probability with block length) *while keeping decoder complexity relatively low.*

Figure 5.2 shows a classical concatenated scheme, where, for simplicity, we assume block codes are used for both inner and outer codes (a concatenation of block and convolutional codes is more usual). In this case, an input of kK information bits could be split into K symbols each of k bits. This requires a *non-binary* outer code, as indicated, and the outer encoder would generate N k-bit symbols. Each k-bit symbol is then encoded into an n-bit binary codeword by the inner encoder, so that, for an input of kK bits, the overall coding process generates a stream of Nn bits. Overall, this results in an (Nn,Kk) code of long block length and rate equal to the product of the two code rates (which could be low). In general the concatenated code will not be as powerful as the best single stage code with the same rate and block length. However, a very long block length is now possible while the total complexity of the two decoders is low compared with the decoder complexity for a single block code. In general a store is necessary between the inner and outer decoders in order to redistribute the errors from the inner decoder, and so reduce the probability of failure of the outer decoder.

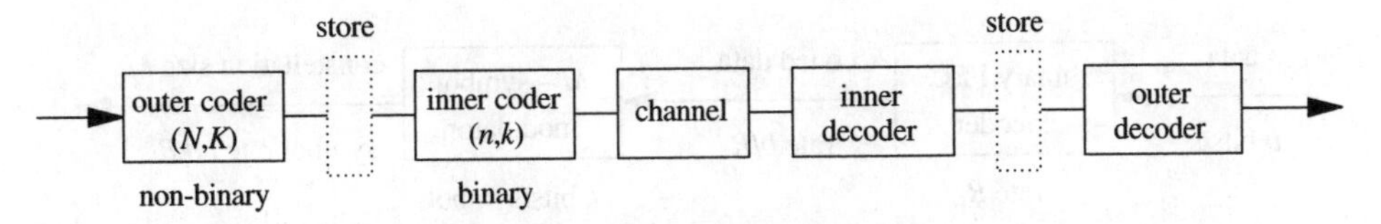

Figure 5.2 *A classical concatenated coding scheme.*

Usually the inner code is used to correct most random errors, and the outer code is used to correct those burst errors which overload the inner code. The outer code is then usually a high-rate RS code, while the inner code can be either block or convolutional. For example, the US Voyager space mission used a (255,223) RS outer code (8-bit symbols) and a rate 1/2 constraint length 7 convolutional inner code. A similar scheme is used for the European DVB system. Concatenation is also used in digital recording to combat dropout errors, and for iterative decoding (feedback between the decoders) in order to achieve near Shannon limit performance (Sections 7.5 and 7.6). The codes used in iterative concatenated schemes are usually simple convolutional codes. An advanced, concatenated, iterative decoding scheme is described by Burket and Hagenauer (1997).

5.2 Power and bandlimited channels

5.2.1 Classical error control systems

Here we examine some implications and limitations of adding error control to a communications system. A classical error control system is shown in Figure 5.3 (paradoxically, this is sometimes referred to as *uncoded* modulation, as opposed to *coded* modulation outlined in Section 5.2.3). In Figure 5.3 conventional FEC is used to achieve a *coding gain* and this is usually followed by a multilevel (M-ary) amplitude and/or phase modulator, such as M-PSK or M-QAM.

Assume a binary block or convolutional encoder with rate R ($R < 1$). The data rate at the encoder output is b/R bits/s and the modulator maps k bits at the encoder output into one of the $M = 2^k$ possible transmit signals that constitute an M-symbol *constellation.* The demodulator recovers the k bits by making an independent M-ary nearest-neighbour decision on each received signal. The symbol rate at the modulator output is now $r_s = b/(kR)$ baud and the theoretical minimum system bandwidth required to transmit such a signal is $B = r_s = b/(kR)$ Hz (a practical M-PSK system employing shaped data pulses with an overall raised-cosine rolloff, α, would have an occupied bandwidth $B = r_s(1+\alpha)$). The corresponding *spectral* or *bandwidth* efficiency (data rate/bandwidth) is

$$\eta = \frac{b}{r_s} = kR = R\log_2 M \quad \text{bits/s/Hz} \tag{5.1}$$

so that a rate 1/2 code followed by QPSK modulation corresponds to only 1 bit/s/Hz. Without FEC, $\eta = \log_2 M$, corresponding to 2 bits/s/Hz for QPSK and 3 bits/s/Hz for 8-PSK. Clearly, *the addition of FEC reduces the bandwidth efficiency of the modulation system by a factor R.*

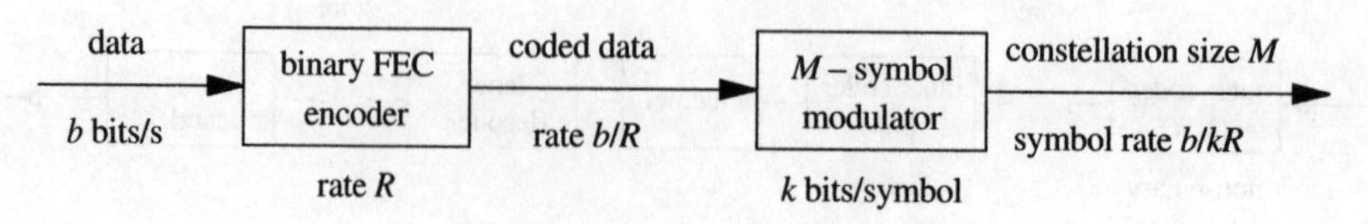

Figure 5.3 *Classical error control.*

The effect of FEC when applied to a modulation system depends upon the type of channel. For *bandwidth-limited* channels (those with a fixed bandwidth, such as telephone lines), the addition of FEC results in reduced data rate, b, by a factor R in order to maintain a fixed symbol rate (recall that bandwidth is proportional to symbol rate). Alternatively, the channel may be *power-limited*. Satellite channels have traditionally been power limited in the sense of low on-board transmitter power but plentiful bandwidth, although bandwidth limitation is becoming increasingly important. In this case the data rate can be maintained constant since the symbol rate and hence the bandwidth can be increased to accommodate the FEC (Figure 5.3). In turn, this gives increased noise at the receiver, for which the code has to more than compensate.

5.2.2 The Shannon limit

In Chapter 1 we introduced Shannon's capacity equation for a hypothetical coding scheme. We can write (1.1) in the more useful form

$$C = B\log_2\left(1+\frac{P}{N_0 B}\right) \quad \text{bits/s} \tag{5.2}$$

where P is the average signal power and B is the channel bandwidth (in Hz). The channel is assumed to be disturbed by additive white Gaussian noise having a one-sided power spectral density, N_0. Suppose we use Shannon's ideal coding scheme to transmit binary information at a rate $R_b = 1/T_b$ bits/s, where T_b is the bit duration. Since the energy/bit is just the received power multiplied by the bit duration, that is, $E_b = PT_b$, then (5.2) can be written

$$\frac{C}{B} = \log_2\left[1+\frac{E_b}{N_0}\frac{R_b}{B}\right] \quad \text{bits/s/Hz} \tag{5.3}$$

Taking the limiting case (letting $R_b = C$) then

$$\frac{C}{B} = \log_2\left[1+\frac{E_b}{N_0}\frac{C}{B}\right] \quad \text{bits/s/Hz} \tag{5.4}$$

and rearranging

$$\frac{E_b}{N_0} = \frac{1}{C/B}[2^{C/B}-1] \tag{5.5}$$

This relationship is useful in that it provides an E_b/N_0 target for practical coding/modulation schemes which have a specified bandwidth efficiency C/B.

Example 5.1
Consider Figure 5.3 with $R = 1/4$ and $M = 4$ (QPSK). From (5.1) the overall bandwidth efficiency is just 0.5 bits/s/Hz. According to (5.5), the *ideal* coding/modulation scheme having the same efficiency requires $E_b/N_0 > -0.82$ dB for error-free decoding.

Similarly, a rate 1/2 QPSK system has a bandwidth efficiency of 1 bit/s/Hz, and for this value the ideal scheme requires $E_b/N_0 > 0$ dB for error-free decoding. Rate 1/2 Turbo codes

(Section 7.5) with QPSK modulation can closely approach this theoretical limit, for example, a 10^{-5} bit error rate for $E_b/N_0 = 1$ dB.

A plot of (5.5) shows that E_b/N_0 increases exponentially with bandwidth efficiency C/B (corresponding to bandwidth-limited channels). Conversely, power-limited channels correspond to a reduction in C/B, and it is found that E_b/N_0 reduces to an asymptotic value below 0 dB as B is increased. As discussed, this latter case is particularly attractive for satellite systems since they are essentially power-limited. In practice, B could be increased by decreasing the code rate R since this increases the symbol rate (Figure 5.3). Suppose we let $B \rightarrow \infty$ in (5.2):

$$C_\infty = \lim_{B\rightarrow\infty}\left[B\log_2\left(1+\frac{P}{N_0 B}\right)\right] \tag{5.6}$$

$$= \lim_{B\rightarrow\infty}\left[\frac{B}{\ln 2}\ln\left(1+\frac{P}{N_0 B}\right)\right] \tag{5.7}$$

Noting that $\ln(1+x) \rightarrow x,\ x << 1$, then

$$C_\infty = \frac{1}{\ln 2}\frac{P}{N_0} \tag{5.8}$$

$$= \frac{E_b}{N_0}\frac{R_b}{\ln 2} \tag{5.9}$$

Letting $R_b = C_\infty$ gives $E_b/N_0 = -1.6$ dB. This means that some hypothetical coding/modulation scheme with infinite bandwidth and infinite coding (and decoding) delay will give arbitrarily small error rate providing $E_b/N_0 > -1.6$ dB. As an example, reference to the standard bit-error rate curve for QPSK (Figure 5.11(b)) shows that the Shannon limit of –1.6 dB offers a potential coding gain of some 11.2 dB at an error rate of 10^{-5}. It is interesting to note that, in simulation, iterative decoding schemes can achieve coding gains within 1 dB of the Shannon theoretical maximum for a specified code rate, for example, 9 dB coding gain for $R = 1/2$ (Berrou *et al.*, 1993).

5.2.3 Concept of Trellis Coded Modulation (TCM)

When coding gain is required *without* data rate reduction or bandwidth expansion (as in telephone channels), we can use *coded modulation*. Here, coding gain is achieved by increasing the size of the signal space constellation, rather than by increasing the symbol rate as in conventional FEC schemes. Increasing the constellation size introduces the redundancy necessary to accommodate the FEC. Figure 5.4 illustrates the concept. The modulator or mapper in Figure 5.4(a) is modified as in Figure 5.4(b) such that the k input bits generate a constellation of 2^{k+1} symbols rather than 2^k symbols. Thus, by doubling the constellation size we can accommodate a convolutional code of rate $(n-1)/n$. The significant point is that one output symbol is still generated for every k data bits, *so the symbol rate and hence bandwidth are unchanged, even though we have added FEC!* It is also

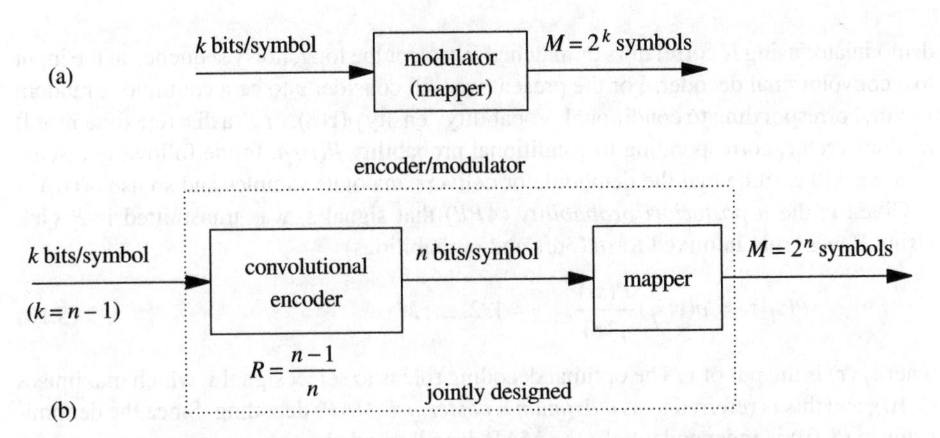

Figure 5.4 *Concept of TCM: (a) basic modulator; (b) coded modulation.*

important to note that in TCM systems the coding and mapping functions are designed *jointly,* rather than as the separate designs, as indicated in Figure 5.4(b).

Example 5.2
CCITT Recommendation V.32 for data modems specifies QAM and transmission at 2400 baud. At 9600 bits/s it specifies two possible modulation schemes: one uncoded (Figure 5.4(a)) and using 16-state QAM (4 bits/symbol); the other using TCM with 32 carrier states and 5 bits/symbol.

5.3 Optimal decoding

5.3.1 MAP and ML decoding

Figure 5.5 shows a generalised communications system. It is applicable to baseband or carrier systems, with or without FEC. Let s_i, $i = 1, 2, ..., M$ be one of M possible signals and let $P(s_i)$ be the *a priori probability* of s_i. The probabilistic transmission system could model a modulator, noisy analogue channel, and demodulator. The receiver will usually base its decision upon a received *vector* or *sequence* $\mathbf{r} = (r_1 r_2 ... r_N)$, rather than upon a single received signal value. For instance, $\mathbf{r}$ could be the output from an optimal

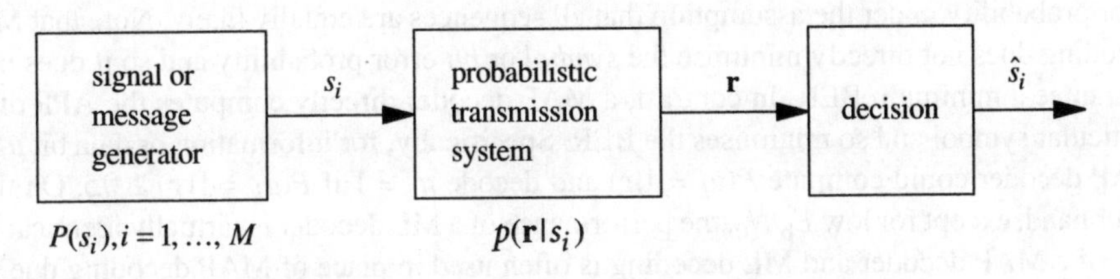

Figure 5.5 *A generalised communications system.*

demodulator using N correlators or matched filters, or the long, noisy sequence at the input to a convolutional decoder. For the present we will consider $\mathbf{r}$ to be a continuous random vector, corresponding to conditional probability density $p(\mathbf{r}|s_i)$, or as a discrete (quantised) random vector, corresponding to conditional probability $P(\mathbf{r}|s_i)$. In the following discussion we will assume that the demodulator delivers analogue samples and so use $p(\mathbf{r}|s_i)$.

Given $\mathbf{r}$, the *a posteriori probability* (*APP*) that signal s_i was transmitted is $P(s_i|\mathbf{r})$. Using Bayes' rule in mixed form (pdfs and probabilities)

$$P(s_i|\mathbf{r}) = p(\mathbf{r}|s_i)\frac{P(s_i)}{p(\mathbf{r})}, \quad i = 1, 2, \ldots, M \tag{5.10}$$

where $p(\mathbf{r})$ is the pdf of $\mathbf{r}$. The optimal decoding rule is to select signal s_i which maximises (5.10), and this is referred to as *maximum a posteriori* (*MAP*) decoding. Since the denominator in (5.10) is independent of i the MAP decoding rule becomes

$$\text{MAP: maximise } \{p(\mathbf{r}|s_i) \cdot P(s_i)\} \tag{5.11}$$

An apparent difficulty with MAP decoding is that it requires knowledge of the *a priori probabilities* $P(s_i)$. If these are unknown we can assume that they are equal, that is, $P(s_i) = 1/M$, so the optimum decision rule reduces to selecting s_i such that $p(\mathbf{r}|s_i)$ is maximised. In other words, we select s_i when

$$\text{ML:} \quad p(\mathbf{r}|s_i) \geq p(\mathbf{r}|s_j) \quad \forall j \neq i \tag{5.12}$$

The conditional pdf $p(\mathbf{r}|s_i)$ is the *likelihood function* of $\mathbf{r}$ given s_i and (5.12) represents *maximum likelihood (ML)* decoding. For iterative decoding it is also helpful to define the *log-likelihood ratio*, one form being

$$\Lambda_i(\mathbf{r}) = \log\left[\frac{p(\mathbf{r}|s_i)}{p(\mathbf{r}|s_j)}\right] \tag{5.13}$$

Signal s_i is then selected in preference to s_j if $\Lambda_i(\mathbf{r})$ is positive, and the magnitude of $\Lambda_i(\mathbf{r})$ is a measure of the confidence of the decision. The probability of a decision error is defined as $P(\varepsilon) = P(\hat{s}_i \neq s_i)$, and it is apparent from (5.11) and (5.12) that both MAP and ML decoders will make the same decision and so have the same $P(\varepsilon)$ when $P(s_i) = 1/M$. When the *a priori probabilities* are equal, ML decoding is therefore an optimal decision rule. When the probabilities are not equal, a MAP decoder will have a lower probability of error than an ML decoder.

In practice, ML decoding can be carried out by the *Viterbi algorithm* (*VA*) and the ith transmitted signal in Figure 5.5 is assumed to be a vector or sequence $\mathbf{s}_i = (s_{i,1}\ s_{i,2} \ldots s_{i,k} \ldots s_{i,N})$, that is, a codeword. The ML decoder then minimises the *sequence* or *word* error probability under the assumption that all sequences are equally likely. Note that ML decoding does not directly minimise the *symbol* or *bit* error probability and so it does not guarantee a minimum BER. In contrast, a MAP decoder directly computes the APP of a particular symbol and so minimises the BER. Specifically, for information or data bit u_k a MAP decoder could compute $P(u_k = 1|\mathbf{r})$ and decode $u_k = 1$ if $P(u_k = 1|\mathbf{r}) \geq 0.5$. On the other hand, except for low E_b/N_0, the performance of a ML decoder is virtually identical to that of a MAP decoder and ML decoding is often used in place of MAP decoding due to decoder complexity, unknown prior probabilities, and uncertainty in the noise variance (see (7.54)). The main attraction of MAP decoding lies in the fact that the APP of each symbol can be used for iterative decoding schemes (Sections 7.5 and 7.6).

5.3.2 Minimum distance decoding

Here we deduce a practical way of implementing the ML rule in (5.12). Let the transmitted signal in Figure 5.5 be the vector $\mathbf{s}_i = (s_{i,1}\, s_{i,2}...s_{i,k}...s_{i,N})$, corresponding to a demodulated signal $\mathbf{r} = (r_1 r_2...r_k...r_N)$ where $r_k = s_{i,k} + n_k$ and n_k is a noise component. If the channel noise is zero-mean AWGN of double-sided power spectral density $N_0/2$, then noise components n_k will be zero-mean Gaussian random variables with variance $\sigma^2 = N_0/2$. This means that components r_k will be Gaussian random variables with mean $s_{i,k}$ and variance $\sigma^2 = N_0/2$. Now, a Gaussian random variable x of mean μ and variance σ^2 has pdf

$$N(x,\mu,\sigma) = \frac{1}{\sqrt{2\pi\sigma^2}} \exp\left[-(x-\mu)^2/2\sigma^2\right] \tag{5.14}$$

so that

$$p(r_k | s_{i,k}) = \frac{1}{\sqrt{\pi N_0}} \exp\left[-(r_k - s_{i,k})^2/N_0\right] \tag{5.15}$$

The likelihood function can then be written

$$p(\mathbf{r}|\mathbf{s}_i) = \prod_k p(r_k | s_{i,k}) \qquad \forall i \tag{5.16}$$

$$= \frac{1}{\left[\pi N_0\right]^{N/2}} \exp\left[\sum_k \left[-(r_k - s_{i,k})^2/N_0\right]\right], \qquad \forall i \tag{5.17}$$

Taking logarithms

$$\ln\left[p(\mathbf{r}|\mathbf{s}_i)\right] = -\frac{N}{2}\ln\left[\pi N_0\right] - \frac{1}{N_0}\sum_k (r_k - s_{i,k})^2 \tag{5.18}$$

It is clear from (5.18) that maximising the log-likelihood function is equivalent to *minimising the (squared) Euclidean distance* $\Sigma(r_k - s_{i,k})^2$. In other words, *for AWGN channels the ML decoder reduces to a minimum distance decoder*. This is the approach adopted in Viterbi decoding (Section 7.3).

5.3.3 Application to error control systems

In this section we relate the optimal decoding concept to practical error control systems. Very often the transmitted signal $\mathbf{s}_i$ has *memory* in the sense that successive symbols are interdependent. The generation of such a signal can be modelled by the shift register circuit in Figure 5.6. We assume that symbol u_j in information sequence $\mathbf{u}_i = \{u_1 u_2...u_j...\}$ can have any one of m values and that it is generated independently according to probability distribution $P(u_j)$. Each symbol $s_{i,k}$ in the transmitted signal $\mathbf{s}_i = (s_{i,1} s_{i,2}...s_{i,k}...s_{i,N})$ is some deterministic function $f(\cdot)$ of the present and previous n inputs, giving the symbol interdependence (note that this models an nth-order m-ary Markov process). In contrast, it

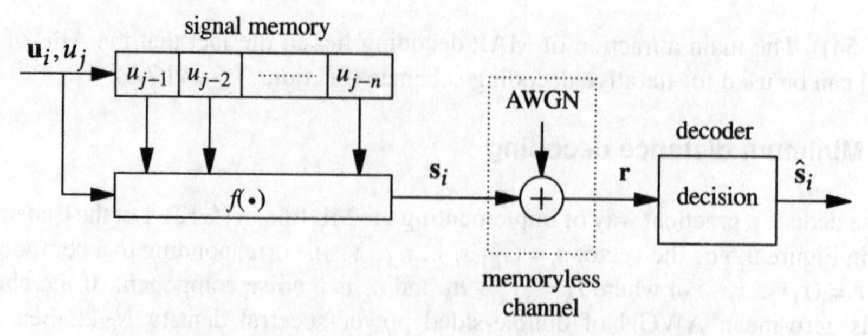

Figure 5.6 *Modelling a signal sequence which has memory.*

is usual to assume that the analogue channel is *memoryless* in that the noise affects each transmitted symbol independently of all others. The significant point here is that, when the transmitted signal has memory, the optimum decoder/detector bases its decisions on a received *sequence*, **r** (whereas, if there was no signal memory, the optimal detector works *symbol-by-symbol*). In other words, although the MAP decoder directly computes the APP of each symbol, each symbol decision is optimally based on sequence **r** when the signal has memory (see (5.20)).

Figure 5.6 applies directly to convolutional coding and decoding. The transmitted signal $\mathbf{s}_i$ becomes the code sequence or word, $\mathbf{v}_i$, $f(\cdot)$ becomes modulo-2 addition of selected register outputs, and each symbol $s_{i,k}$ becomes a block of bits. A practical system is shown in Figure 5.7. The modem and channel can be regarded as a *discrete memoryless channel* (*DMC*) in the sense that it has a discrete-time binary input, memoryless noise, and a *quantised* discrete-time output sequence **r**. In this case the ML decoding rule in (5.12) uses probabilities and can be written

$$P(\mathbf{r}|\mathbf{v}_i) \geq P(\mathbf{r}|\mathbf{v}_j), \quad \forall i \neq j \tag{5.19}$$

Expressed in words, the Viterbi decoder selects code sequence $\mathbf{v}_i$ (which may or may not be the transmitted sequence), when (5.19) is satisfied and then it uses the known *code trellis* (Section 7.2) to derive an estimate of the information sequence $\mathbf{u}_i$. The rule in (5.19) minimises the sequence error probability if all code or information sequences are equally likely. As discussed, in practice (5.19) is usually implemented by measuring Euclidean distance, and this is explained in Section 7.3.3. Also, **r** is quantised to typically 8 levels (3 bits) which means that the Viterbi decoder performs *soft-decision* decoding. On Gaussian

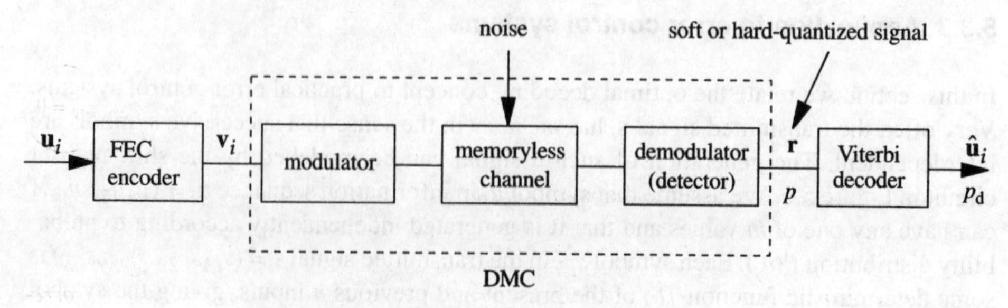

Figure 5.7 *ML decoding in an error control system.*

channels, soft-decision decoding gives some 2 dB improvement in E_b/N_0 compared to *hard-decision* decoding (just two output levels) since more information is fed to the decoder (see Figure 5.11(b)).

5.3.4 MAP decoding for a rate 1/2 convolutional code

This section provides the theoretical background for the practical MAP decoders discussed in Chapter 7. The theory uses the concept of forward and backward recursion (Bahl, 1974), and the objective is to obtain an expression for the log-likelihood ratio of each information or data bit. A simple (but useful) rate 1/2 convolutional code is shown in Figure 5.8(a), and we assume that data bit u_i generates code symbols $v1_i$ and $v2_i$, which are translated to the set $\{+1,-1\}$ before transmission. These symbols are then corrupted by a DMC disturbed by AWGN of variance σ^2, and received as $\mathbf{r}_i = (r1_i, r2_i)$. Block mode processing is used, whereby a complete block of data must be received before decoding commences. The inherent assumptions are that the decoding delay and memory requirements are acceptable. A block-mode MAP decoder operates on the set of received symbols $\mathbf{r}_1$ to $\mathbf{r}_N$, denoted $\mathbf{r}_1^N$, and the corresponding coded system becomes a form of block code. An alternative to block-mode MAP decoding is the *sliding window* MAP decoding algorithm (Benedetto, 1996).

A state transition diagram for the problem is shown in Figure 5.8(b). Here it is assumed that the encoder is in state S_{i-1} for data bit u_i, and that the encoder commences and ends in state zero (through the use of a *data tail*). The log-likelihood ratio of data bit u_i is

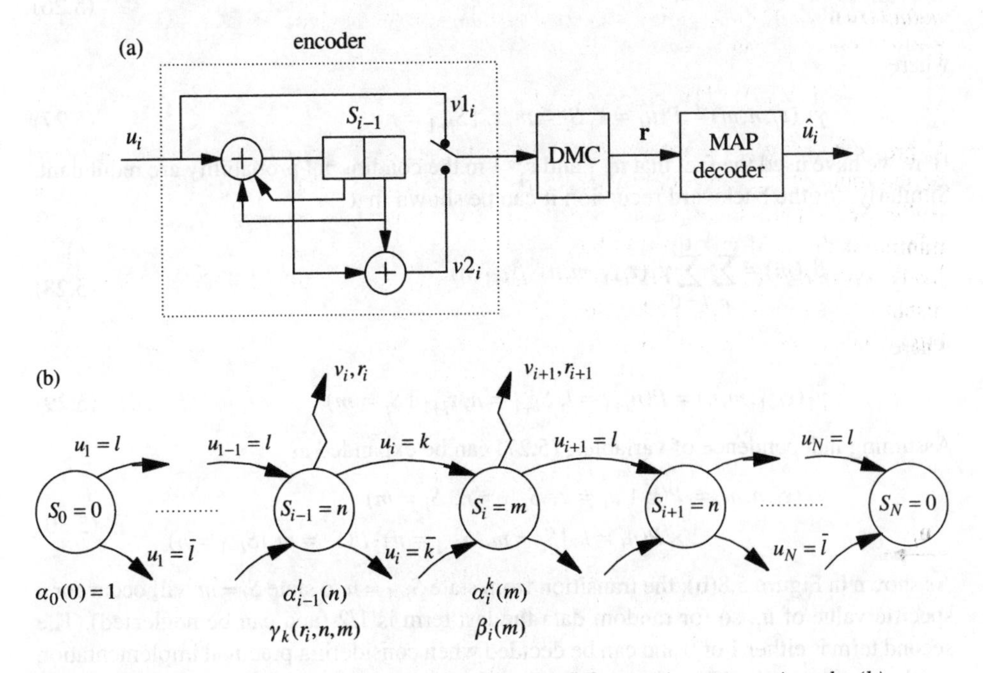

Figure 5.8 *MAP decoding: (a) overall system for a rate 1/2 recursive systematic code; (b) corresponding state diagram.*

$$\Lambda(u_i) = \log\left[\frac{P(u_i = 1 \mid \mathbf{r}_1^N)}{P(u_i = 0 \mid \mathbf{r}_1^N)}\right] \tag{5.20}$$

Using forward and backward parameters α and β respectively, this can be written as a summation over all possible states m

$$\Lambda(u_i) = \log\frac{\Sigma_m \alpha_i^1(m)\beta_i(m)}{\Sigma_m \alpha_i^0(m)\beta_i(m)} \tag{5.21}$$

where

$$\alpha_i^k(m) = P(u_i = k, S_i = m, \mathbf{r}_1^i) \tag{5.22}$$

$$\beta_i(m) = P(\mathbf{r}_{i+1}^N \mid S_i = m) \tag{5.23}$$

Expressing (5.22) as a summation over all possible states, n, the forward recursion is

$$\alpha_i^k(m) = \sum_n \sum_{l=0}^{1} P(u_i = k, u_{i-1} = l, S_i = m, S_{i-1} = n, \mathbf{r}_1^{i-1}, \mathbf{r}_i) \tag{5.24}$$

$$= \sum_n \sum_{l=0}^{1} P(u_i = k, S_i = m, \mathbf{r}_i \mid u_{i-1} = l, S_{i-1} = n, \mathbf{r}_1^{i-1}) P(u_{i-1} = l, S_{i-1} = n, \mathbf{r}_1^{i-1}) \tag{5.25}$$

$$= \sum_n \sum_{l=0}^{1} \gamma_k(\mathbf{r}_i, n, m)\alpha_{i-1}^l(n) \tag{5.26}$$

where

$$\gamma_k(\mathbf{r}_i, n, m) = P(u_i = k, S_i = m, \mathbf{r}_i \mid S_{i-1} = n) \tag{5.27}$$

Here we have used the fact that u_{i-1} and $\mathbf{r}_1^{i-1}$ in the conditional probability are redundant. Similarly, for the backward recursion it can be shown that

$$\beta_i(m) = \sum_n \sum_{l=0}^{1} \gamma_l(\mathbf{r}_{i+1}, m, n) \cdot \beta_{i+1}(n) \tag{5.28}$$

where

$$\gamma_l(\mathbf{r}_{i+1}, m, n) = P(u_{i+1} = l, S_{i+1} = n, \mathbf{r}_{i+1} \mid S_i = m) \tag{5.29}$$

Assuming independence of variables, (5.27) can be expanded as

$$\begin{aligned}\gamma_k(\mathbf{r}_i, n, m) &= P(\mathbf{r}_i \mid u_i = k,\ S_{i-1} = n,\ S_i = m) \\ &\times P(u_i = k \mid S_i = m,\ S_{i-1} = n) \cdot P(S_i = m \mid S_{i-1} = n)\end{aligned} \tag{5.30}$$

As shown in Figure 5.8(b), the transition from state $S_{i-1} = n$ to state $S_i = m$ will occur for a specific value of u_i, so for random data the last term is 1/2 (and can be neglected). The second term is either 1 or 0 and can be decided when considering practical implementation via the encoder trellis. Therefore, exploiting the systematic property of the code, (5.30) can be written

$$\begin{aligned}\gamma_k(\mathbf{r}_i,n,m) &= P(r1_i \mid u_i = k)\cdot P(r2_i|u_i = k,\ S_{i-1} = n,\ S_i = m)\\ &= P(r1_i \mid u_i = k)\cdot \gamma_k(r2_i,n,m)\end{aligned} \tag{5.31}$$

Using (5.31) and (5.26), the log-likelihood ratio can now be written

$$\Lambda(u_i) = \log\left[\frac{P(r1_i| u_i = 1)\cdot \Sigma_m \Sigma_n \Sigma_{l=0}^1 \gamma_1(r2_i,n,m)\alpha_{i-1}^l(n)\beta_i(m)}{P(r1_i|u_i = 0)\cdot \Sigma_m \Sigma_n \Sigma_{l=0}^1 \gamma_0(r2_i,n,m)\alpha_{i-1}^l(n)\beta_i(m)}\right] \tag{5.32}$$

Finally, in the analysis of iterative, concatenated, convolutional coding schemes (Sections 7.5 and 7.6), it is helpful to split (5.32) into two *independent* terms

$$\Lambda(u_i) = \log\left[\frac{P(r1_i|u_i = 1)}{P(r1_i|u_i = 0)}\right] + \log\left[\frac{\Sigma_m \Sigma_n \Sigma_{l=0}^1 \gamma_1(r2_i,n,m)\alpha_{i-1}^l(n)\beta_i(m)}{\Sigma_m \Sigma_n \Sigma_{l=0}^1 \gamma_0(r2_i,n,m)\alpha_{i-1}^l(n)\beta_i(m)}\right] \tag{5.33}$$

$$= \Lambda_{\mathrm{s}}(u_i) + \Lambda_{\mathrm{e}}(u_i) \tag{5.34}$$

The first term in (5.34) provides information about u_i from the systematic component of the code. The second term is particularly important since it acts as additional or *extrinsic* information in iterative decoding schemes. It can also be seen from Sections 7.5.5 and 7.6.3 that (5.33) requires an estimate of the noise variance.

5.3.5 Alternative APP decoding

So far we have computed the APP of information bit u_i but sometimes it is convenient to compute the APP of the encoder *output* bits, as for example in the iterative decoding of *product codes* (Section 6.8.1). These codes are serious contenders to Turbo codes when high coding gain is required at high code rate ($R > 0.7$).

Let the ith transmitted codeword be $\mathbf{v}_i = (v_1 v_2 ... v_k ... v_N)$, where $v_k \in \{-1, +1\}$, and assume $\mathbf{r}$ is the (unquantised) demodulator output after $\mathbf{v}_i$ is transmitted over a Gaussian channel. The APP of v_k can be written

$$P(v_k = 1|\mathbf{r}) = \sum_{i\in S(k,1)} P(\mathbf{v}_i|\mathbf{r}) \tag{5.35}$$

where the summation is over the set $S(k,1)$ of codewords corresponding to $v_k = 1$. Similarly

$$P(v_k = -1|\mathbf{r}) = \sum_{i\in S(k,-1)} P(\mathbf{v}_i|\mathbf{r}) \tag{5.36}$$

and using Bayes' rule on (5.35) gives

$$P(v_k = 1|\mathbf{r}) = \sum_{i\in S(k,1)} p(\mathbf{r}|\mathbf{v}_i)\cdot \frac{P(\mathbf{v}_i)}{p(\mathbf{r})} \tag{5.37}$$

Assuming all codewords are equiprobable, the log-likelihood ratio for v_k is

$$\Lambda(v_k) = \log\left[\frac{P(v_k = 1|\mathbf{r})}{P(v_k = -1|\mathbf{r})}\right] = \log\left[\frac{\Sigma_{i\in S(k,1)}\, p(\mathbf{r}|\mathbf{v}_i)}{\Sigma_{i\in S(k,-1)}\, p(\mathbf{r}|\mathbf{v}_i)}\right] \tag{5.38}$$

For Gaussian channels the pdf in (5.38) is a product of Gaussian pdfs and is given by

$$p(\mathbf{r}|\mathbf{v}_i) = \left(\frac{1}{\sqrt{2\pi}\sigma}\right)^N \exp\left(-\frac{|\mathbf{r}-\mathbf{v}_i|^2}{2\sigma^2}\right) \tag{5.39}$$

where

$$|\mathbf{r}-\mathbf{v}_i|^2 = \sum_{k=1}^{N}(r_k - v_k)^2 \tag{5.40}$$

Finally, substituting (5.39) into (5.38) gives

$$\Lambda(v_k) = \log\left[\frac{\sum_{i \in S(k,1)} \exp(-|\mathbf{r}-\mathbf{v}_i|^2/2\sigma^2)}{\sum_{i \in S(k,-1)} \exp(-|\mathbf{r}-\mathbf{v}_i|^2/2\sigma^2)}\right], \quad k = 1,\ldots,N \tag{5.41}$$

where $v_k \in \mathbf{v}_i$. As before, we note that APP decoding requires an estimate of the noise variance. Also, in practice some simplification of (5.41) would probably be necessary due to the large search space; for an (n,k) binary block code there are 2^k codewords, which could be a large number for high rate codes. In particular, only those codewords which are within a prescribed Euclidean distance could be used in the summation, as in the Chase algorithm (Chase, 1972). An extreme but practical simplification would be to find the codeword $\mathbf{v}^+$ which is at minimum Euclidean distance from $\mathbf{r}$ for which symbol $v_k = +1$. Similarly, we find the codeword $\mathbf{v}^-$ at minimum distance from $\mathbf{r}$ for which $v_k = -1$. Ignoring all other codewords, (5.41) then reduces to

$$\Lambda(v_k) \approx \frac{1}{2\sigma^2}[|\mathbf{r}-\mathbf{v}^-|^2 - |\mathbf{r}-\mathbf{v}^+|^2] \tag{5.42}$$

We will utilise this equation in the iterative decoding of product codes (Section 6.8.1).

5.4 Decoded error rate

Practical FEC systems can be overloaded, so there is always a finite bit error probability, p_d (or BER) at the FEC decoder output, as indicated in Figure 5.7. System design therefore usually specifies a maximum decoded BER.

Example 5.3

A particular satellite communications system required an output (decoded) 40 character message to be completely error-free for at least 99% of the time. Assuming 6-bit characters we have

$$((1-p_d)^6)^{40} \geq 0.99$$

giving $p_d \leq 4.2 \times 10^{-5}$. If we define a channel bit error rate, p, at the input to the decoder, the latter must then reduce this relatively high rate to the target specification, p_d (see Figure 5.7).

5.4.1 Hard and soft-decision decoding

Before estimating p_d and related error probabilities for block codes, it is helpful to expand on the idea of a discrete memoryless channel (DMC), introduced in Section 5.3.3. As shown in Figure 5.7, it is usual to model the modulator, analogue channel, and demodulator as a single block, or DMC. As indicated, the DMC has a quantised input and a quantised output. The general form of the DMC has an alphabet of L input symbols and an alphabet of J output symbols, and channel noise results in a set of conditional probabilities, as shown in Figure 5.9(a).

The model is 'memoryless' in the sense that the current output symbol Y_j depends only upon the current input symbol X_i. A typical practical system may have a binary input ($L = 2$) and an 8 level (3-bit) quantised output ($J = 8$). Given a binary input, if $J = 2$ we say the channel is *hard-quantised*, and if $J > 2$ we say the channel is *soft-quantised*. Soft

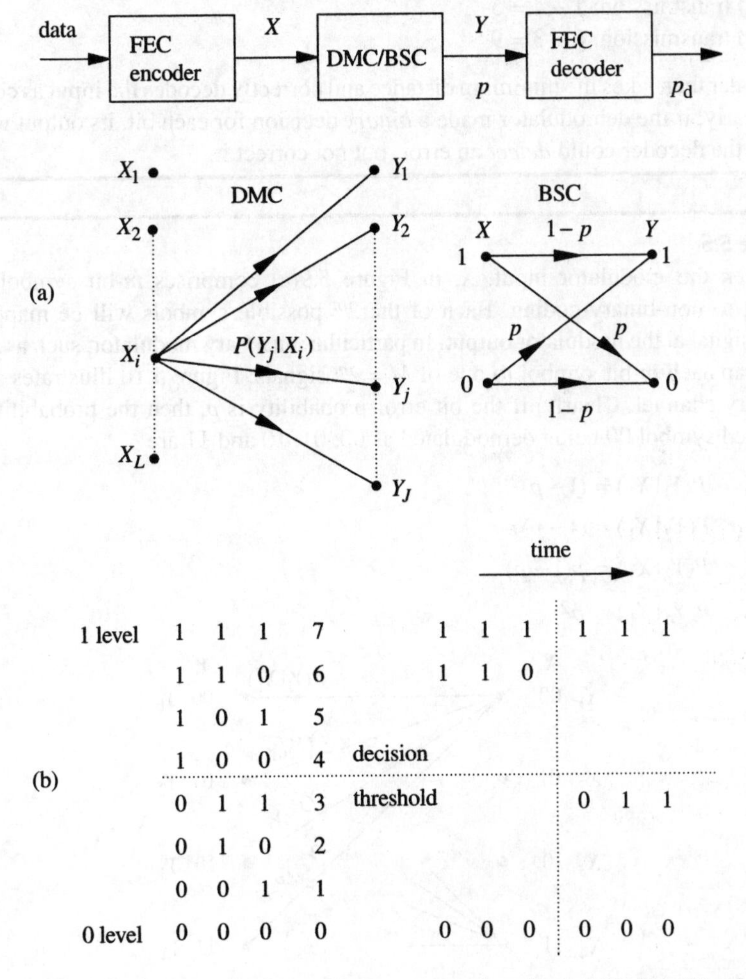

Figure 5.9 *(a) System modelling using a discrete memoryless channel (DMC); (b) soft-decision decoding, L = 2, J = 8.*

quantisation permits soft-decision decoding, whereby the J inputs to the decoder provide it with more information compared with that from a hard-quantised channel. For example, referring to Figure 5.9(b), a (111) input to the decoder denotes a transmitted 1 with high confidence and a (000) denotes a transmitted 0 with high confidence. In contrast, a (100) input would be decoded as a 1, but with low confidence. The decoder can use these inputs to measure 'soft distances', which in turn can give improved decoding.

Example 5.4

Consider the simple repetition code comprising codewords (11) and (00). Suppose (11) is transmitted (one bit at a time) and that channel noise results in the two bits being demodulated at levels 6 and 3 (Figure 5.9(b)). Note that 6 is a distance 1 from the 1 level (7), and 3 is a distance 4 from the 1 level. The soft distances are then

for a (11) transmission: $1 + 4 = 5$
for a (00) transmission: $6 + 3 = 9$

The decoder then takes the minimum distance and correctly decodes the input as codeword (11). Clearly, if the demodulator made a *binary* decision for each bit, its output would be (10) and the decoder could *detect* an error, but not correct it.

Example 5.5

Very often the modulator input, X, in Figure 5.9(a) comprises m-bit symbols corresponding to non-binary coding. Each of the 2^m possible symbols will be mapped to a specific signal at the modulator output. In particular, an M-ary modulator, such as M-PSK, would map each m-bit symbol to one of $M = 2^m$ signals. Figure 5.10 illustrates a simple non-binary channel. Clearly, if the bit error probability is p, then the probabilities of a transmitted symbol 00 being demodulated as 00, 01, 10 and 11 are

$$P(Y_1|X_1) = (1-p)^2$$
$$P(Y_2|X_1) = (1-p)p$$
$$P(Y_3|X_1) = p(1-p)$$
$$P(Y_4|X_1) = p^2$$

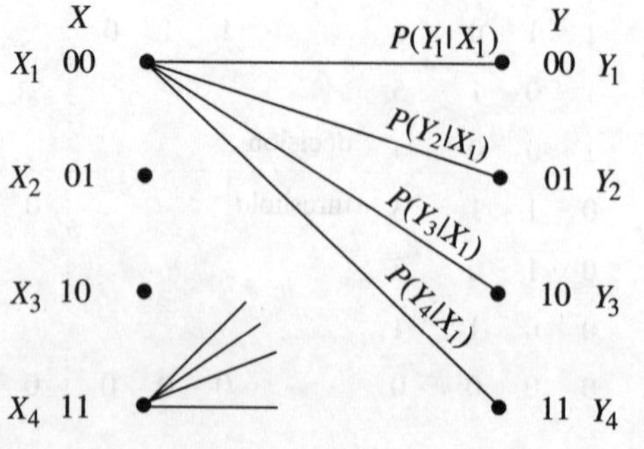

Figure 5.10 *Modelling a simple non-binary channel (m = 2).*

The performance advantage of soft-decision decoding depends on a number of factors, such as the transmission channel characteristics, the number and spacing of the quantisation levels, and the decoding algorithm. In practice there is little advantage in using more than 3-bit quantisation, and performance advantage tends to be smallest on a (non-fading) Gaussian channel and better on a Rayleigh fading channel. For a Gaussian channel and constant BER, 3-bit soft-decision decoding enables E_b/N_0 (as in Figure 5.11) to be reduced by approximately 2 dB compared with that for hard-decision decoding. On the other hand, soft decoding of block codes (Section 6.7.6) is relatively difficult to implement (compared with soft decoding of convolutional codes), and for non-binary codes (RS codes) there exist efficient hard-decision decoding algorithms. From now on we will therefore restrict discussion to hard-decision decoding.

For hard decoding we force the demodulator to make a definite decision at its output by incorporating a threshold device. Because of noise this will inevitably generate an error probability at the demodulator output, and in the case of non-binary codes we can associate the output with a *symbol* error probability, P_s. In the case of binary codes ($L = 2, J = 2$) we associate the output with a *bit* error probability or BER, p, and the DMC reduces to the special case of a *binary symmetric channel* (*BSC*); see Figure 5.9(a). That is, for any transmitted bit (1 or 0) there is a probability $1 - p$ that the bit is correctly detected, and a probability p of a detection error. Also, bit errors are assumed to be independent of one another. We can determine p by noting that $p = P(1|0) = P(0|1)$, where $P(1|0)$ is the probability that a transmitted 0 is detected as a 1, that is, $P(1|0)$ is the area under the Gaussian noise pdf above the binary threshold. This theory can be extended to bandpass systems (such as BPSK and QPSK) in order to express p in terms of E_b/N_0 (see Example 5.8).

5.4.2 Decoded block error rate (hard-decision decoding)

A word of caution before we compute some decoded error rates. Consider a t-error correcting binary block code. The decoder is certain to correct t or less errors but, in general, it will also correct some error patterns of *more* than t errors (exceptions are the so-called 'perfect codes', such as the Hamming codes and the Golay (23,12) code). This makes the calculation of decoded error rate complex, and so we will *assume* that only t or fewer errors can be corrected. In general, our calculations for decoded error rate will then simply be upper bounds.

Considering Figure 5.9(a), suppose an n-bit codeword of an (n,k,t) block code is received at the FEC decoder input with a single error in a particular position. The probability of this particular word occurring is simply

$$P(\text{specific word}) = p(1-p)^{n-1} \tag{5.43}$$

Since there are n possible error locations the probability of a codeword with a single error *somewhere* is the sum of the probabilities of n independent events

$$P(1\text{ error}) = np\,(1-p)^{n-1} = \binom{n}{1} p(1-p)^{n-1} \tag{5.44}$$

where

$$\binom{n}{r} = \frac{n!}{r!(n-r)!} \tag{5.45}$$

Similarly, the probability of a specific 2-error codeword is

$$P(\text{specific word}) = p^2(1-p)^{n-2} \tag{5.46}$$

and the number of different codewords having exactly two errors is given by

$$\binom{n}{2} = \frac{n(n-1)}{2}$$

$$\therefore P(2\text{ errors}) = \binom{n}{2} p^2(1-p)^{n-2} \tag{5.47}$$

Finally, generalising (5.47) gives

$$P(m\text{ errors}) = \binom{n}{m} p^m(1-p)^{n-m} \tag{5.48}$$

Assuming that a maximum of t errors can be corrected for each n-symbol block, then a block decoding error is sure to occur if $m > t$. Therefore

$$P(\text{block error}) = \sum_{m=t+1}^{n} P(m\text{ errors}) = 1 - \sum_{m=0}^{t} P(m\text{ errors}) \tag{5.49}$$

Equivalently,

$$P(\text{correct block}) = \sum_{m=0}^{t} \binom{n}{m} p^m(1-p)^{n-m} \tag{5.50}$$

As previously noted, most t-error correcting codes will perform better than this, in which case (5.50) is simply an upper bound.

Example 5.6

Consider the decoded block error rate for the SEC Hamming codes (Section 6.3.2). From (5.48) and (5.49) we have

$$P(\text{block error}) = 1-(1-p)^n - np(1-p)^{n-1} \tag{5.51}$$

We could expand this using the binomial theorem to give an approximate expression:

$$P(\text{block error}) \approx \frac{n}{2}(n-1)p^2, \qquad p \ll 1 \tag{5.52}$$

As an example, if $p = 10^{-3}$, the probability of an H(7,4) codeword being in error at the decoder output is approximately 2×10^{-5}.

Example 5.7

A similar approximate expression can be found for DEC block codes. We have

$$P(\text{correct decoding}) \geq \sum_{m=0}^{2} P(m \text{ errors})$$
$$\geq (1-p)^n + np(1-p)^{n-1} + \frac{n}{2}(n-1)p^2(1-p)^{n-2} \tag{5.53}$$

which reduces to

$$P(\text{correct decoding}) \geq 1 - \frac{n}{6}(n-1)(n-2)p^3, \qquad p \ll 1 \tag{5.54}$$

$$\therefore P(\text{decoding error}) = 1 - P(\text{correct decoding}) \approx \frac{1}{6}n^3p^3, \quad p \ll 1 \tag{5.55}$$

5.4.3 Decoded BER

Let the probability of a block error in the decoder output be x^{-1}. This means that, on average, one in x n-bit words is in error, or that there is a decoding failure every nx bits. The decoded BER, p_d, is simply the proportion of erroneous bits in the decoder output. Suppose that, on average, γ bits are in error at the decoder output for each decoding failure. Then

$$p_d = \frac{\gamma}{nx} = \frac{\gamma}{n} \cdot P(\text{block error}) \tag{5.56}$$

For SEC codes (5.52) gives

$$p_d \approx \frac{\gamma}{2}(n-1)p^2, \quad p \ll 1 \tag{5.57}$$

where, in general, $2 < \gamma < 3$. Therefore, when $p \ll 1$,

$$p_d \approx (n-1)p^2; \quad t = 1 \tag{5.58}$$

For DEC codes, γ is somewhat larger (typically 4 for large block length).

5.4.4 Coding gain

In the last section we examined the improvement in BER provided by the FEC decoder. In practice, such improvements are usually measured in terms of *coding gain,* as defined in Figure 5.11(a). Here, E_b is the mean received energy per information (or message) bit and N_0 is the single-sided noise power spectral density at the receiver input (it is usually assumed that the channel is disturbed by AWGN). The ratio E_b/N_0 is therefore a signal-to-noise measure and it acts as a figure of merit for different combinations of coding and modulation schemes. Figure 5.11(a) shows that coding gain, G, is measured relative to the channel error rate curve for a particular uncoded modulation scheme, such as BPSK or QPSK. Clearly, G varies, and so must be quoted at a specific output bit error rate, p_d

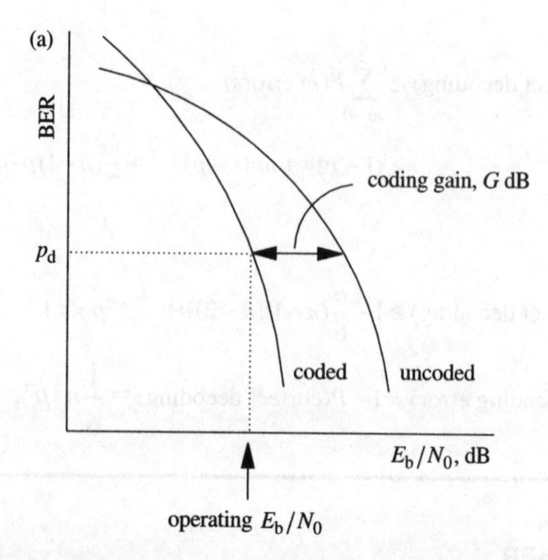

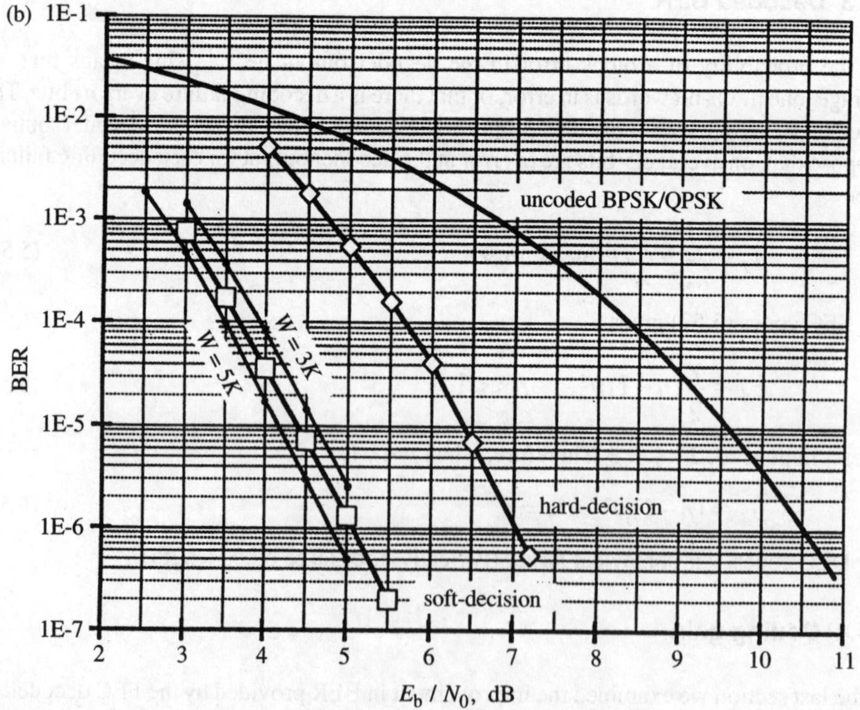

Figure 5.11 *Coding gain: (a) definition; (b) practical coding gains for Viterbi decoding of the rate 1/2, K = 7, (133,171) convolutional code (see text).*

(usually 10^{-5}). Also note that G can become negative for high error rates (typically around 10^{-1} or 10^{-2}), in which case FEC is ineffective.

As shown in Figure 5.3, for a fixed information rate, b, the addition of classical FEC of rate R to an uncoded modulation scheme will result in an increased symbol rate and hence increased bandwidth (by a factor $1/R$). In turn this leads to a larger BSC error probability, p. Looked at another way, assuming an (n,k) block code, the $n-k$ parity bits require some energy to transmit them, so the energy per *coded* bit will be only $E_c = RE_b$. This means that for a fixed noise density, N_0, more bits will be received in error at the demodulator output compared to a system without FEC. *Consequently, the code must have enough error control capability to more than compensate for the increased channel errors and so provide useful gain.* Finally, since $E_b = E_c/R$, we have

$$\left(\frac{E_b}{N_0}\right)\text{dB} = \left(\frac{E_c}{N_0}\right)\text{dB} + 10\log\frac{1}{R}\text{dB} \tag{5.59}$$

In other words, for a fair comparison between systems with and without FEC, the ratio E_c/N_0 measured on the FEC system must be corrected by a factor $10\log(1/R)$ dB to account for the added parity bits.

Example 5.8

A common design objective is to select a suitable combination of FEC and modulation schemes so as to minimise E_b/N_0 for a target output error rate, p_d. Consider a BPSK modulation scheme disturbed by AWGN. It can be shown that for coherent demodulation and binary thresholding, the channel BER is given by

$$p = Q\left[\left(2\frac{E_b}{N_0}\right)^{1/2}\right] \tag{5.60}$$

Here, $Q(x)$ is simply the integral of the zero-mean, unit variance Gaussian density function above x, and is given by

$$Q(x) = \frac{1}{\sqrt{(2\pi)}}\int_x^\infty \exp(-y^2/2)\,\mathrm{d}y \tag{5.61}$$

A table of $Q(x)$ is provided in Appendix A. Note that (5.60) also applies to QPSK (the BER is unchanged even though the transmission bandwidth is halved). Equation (5.60) therefore provides a good reference curve for practical uncoded systems (no FEC), as in Figure 5.11(a). Assuming the BPSK/QPSK receiver is to give an output BER of 10^{-5}, the reader should use (5.60) to show that the uncoded system requires $E_b/N_0 = 9.6$ dB. Using (5.56) with an appropriate block error probability, and accounting for the code rate R, it is readily shown that the DEC BCH(15,7) code will reduce the operating E_b/N_0 to 8.7 dB (a gain of 0.9 dB), while the higher rate BCH(127,113) code will give some 2 dB gain at $p_d = 10^{-5}$.

Figure 5.11(b) shows measured coding gains (denoted as boxed points) for both hard and soft-decision Viterbi decoding of the rate 1/2, $K = 7$, (133,171) convolutional code (QUALCOM, 1991). It is apparent from the graphs that soft-decision Viterbi decoding (Section 7.3.3) yields over 5 dB coding gain at $p_d = 10^{-5}$, which means that the transmitter power could be reduced by 5 dB relative to that for the uncoded system for the same information rate. As discussed in Section 7.3.2, the performance of the Viterbi decoder depends somewhat upon the size of the memory or decoding window, W, in the decoder. For the practical Viterbi decoder in Figure 5.11(b) (boxed points) the decoding window corresponds to about 7 times the constraint length ($W = 7K$ frames). Figure 5.11(b) also shows the performance of the soft-decision Viterbi decoder simulation in Appendix C for $W = 3K$ and $W = 5K$ frames.

5.4.4 Error rate for RS codes

Example 5.5 shows that M-ary modulation is well suited to the transmission of non-binary codes with m-bit symbols. In particular, we shall see in Section 6.7 that the non-binary RS codes transmit m-bit symbols where each symbol corresponds to one of the elements in Galois field GF(2^m). For an (n,k,t) code, n of these symbols would be transmitted in serial fashion for each codeword.

We will assume hard-decision decoding and associate a bit error probability, p, with the channel output. Clearly, an RS symbol will be in error when there are i bits ($i = 1, ..., m$) of the symbol in error. It follows from (5.48) that the probability of receiving a symbol with i bits in error is

$$P(i \text{ errors}) = \binom{m}{i} p^i (1-p)^{m-i} \tag{5.62}$$

so the symbol error probability is

$$P_s = \sum_{i=1}^{m} \binom{m}{i} p^i (1-p)^{m-i} = 1 - (1-p)^m \tag{5.63}$$

Note that P_s depends upon m. The error probability for an (n,k,t) RS code then follows from (5.50) and can be written

$$P(\text{correct word}) = \sum_{i=0}^{t} \binom{n}{i} P_s^i (1-P_s)^{n-i} \tag{5.64}$$

Put another way, the probability of an RS codeword error in hard-decision decoding is upper bounded by

$$P(\text{codeword error}) = \sum_{i=t+1}^{n} \binom{n}{i} P_s^i (1-P_s)^{n-i} \tag{5.65}$$

Bibliography

Bahl, L., Cocke, J., Jelinek, F. and Raviv, J. (1974) Optimal decoding of linear codes for minimising symbol error rate, *IEEE Trans. on Inform. Theory*, March, 284–287.

Benedetto, S., Divsalar, D., Montorsi, G. and Pollara, F. (1996) Algorithm for continuous decoding of Turbo codes, *IEE Electron. Lett.* **32**(4), 314–315.

Berrou, C., Glavieux, A. and Thitimajshima, P. (1993) Near Shannon limit error-correcting coding and decoding: Turbo codes, *Proc. ICC'93*, Geneva, Switzerland, pp. 1064–1070.

Burket, F. and Hagenauer, J. (1997), Improving channel coding of ETSI- and MPEG-satellite transmission standards, *IEEE GLOBECOM'97*, **2**, Phoenix, AZ, pp. 1539–1542.

Chase, D. (1972) A class of algorithms for decoding block codes with channel measurement information, *IEEE Trans. on Inform. Theory*, **IT-18**, 170–182.
QUALCOM (1991) Q1650 $K = 7$ multi-code rate Viterbi decoder, *Technical Data Sheet*, QUALCOM, Inc.

CHAPTER 6

Error control block codes

6.1 Introduction

In general coding theory a block code always codes a particular message into the same fixed sequence of code symbols (this is not true of a convolutional code). For instance, decimal number '3' is always coded into *codeword* or *codevector* (011) in a 3-bit binary number code (this code is represented by the vertices of a cube in Figure 6.1(a)). Also, whatever the message, the codeword length is always fixed at the *block length*, n, of the code. In order to convert Figure 6.1(a) to a simple error control block code, we could add an overall parity check symbol (or bit), as in Figure 6.1(b). Here we have added an even parity symbol (modulo-2 sum of the 1s in the codeword is zero) and the vertices of the cube represent the eight 'legal' code vectors. Odd parity vectors, such as (1000), are not part of the code and provide the redundancy necessary to generate an error control code.

In a general (n,k) error control block code, k message or information symbols are coded into a block of n code symbols by adding $n - k$ parity symbols. If the original k symbol block can be clearly identified within the codeword (usually at one end of the word) it is called a *systematic* code, otherwise it is *non-systematic*. Note that we used the term 'symbols' rather than 'bits'. This is because, in practice, each of the message symbols could be m bits deep (they could be ASCII characters), corresponding to 2^{km} different messages. Since each message has its own codeword, it follows that there are 2^{km} codewords, each

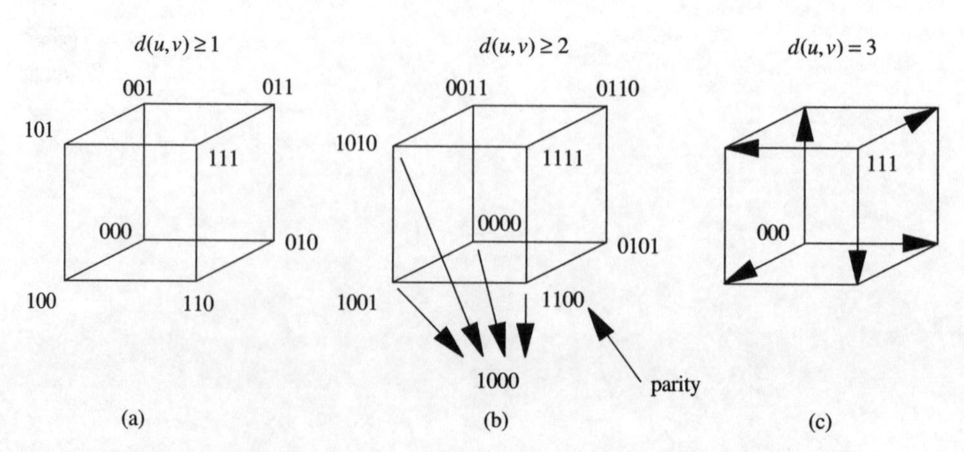

Figure 6.1 *Simple block codes: (a) no error control property; (b) SED; (c) SEC.*

one comprising n m-bit symbols. A binary (n,k) code will encode one of 2^k messages into one of 2^k codewords.

6.1.1 Code rate and code distance

The *rate* (or *efficiency*) of a block code is defined as $R = k/n$, and for a fixed n it will decrease as more parity symbols are added to achieve greater error control capability. As a general rule, short block codes (and unpunctured convolutional codes) are used for low and medium rates, and long block codes (and punctured convolutional codes) are used for high rates, for example, $R > 0.95$. As discussed in Section 5.2.1, a low-rate code will either significantly increase the required channel bandwidth (by a factor $1/R$), or force a significant reduction in the information rate (by a factor R).

The concept of code 'distance' is fundamental to both block and convolutional codes since it determines their error detection and correction capability. Distance is measured between two code vectors (or code *sequences* for a convolutional code), and the metric used is usually the *Hamming distance*. This is defined as the number of positions for which two codevectors have different *symbols*, and it is immaterial how many corresponding bits differ from each other within corresponding symbols. For binary codewords $\mathbf{u}$ = (10011010) and $\mathbf{v}$ = (01111000) the Hamming distance is simply $d_H(\mathbf{u},\mathbf{v}) = 4$. The significant point is that the *minimum* Hamming distance, d, can be simply related to the number of correctable errors, t, that can occur within the block length, n.

Suppose that a codevector $\mathbf{v}_1$ is received as vector $\mathbf{r}$ with t random errors (Figure 6.2). If the nearest codevector $\mathbf{v}_2$ is at least $2t + 1$ positions away from $\mathbf{v}_1$, then $\mathbf{r}$ must differ from $\mathbf{v}_2$ by at least $t + 1$ positions. *Assuming that a maximum of t errors have occurred*, the receiver can then associate $\mathbf{r}$ with a legal vector t or fewer errors away, that is, with $\mathbf{v}_1$. This can be formally expressed as

$$d \geq 2t + 1 \tag{6.1}$$

where d is the minimum Hamming distance between all possible pairs of codewords. We may of course simply wish to *detect* errors, in which case e errors can be detected provided $d \geq e + 1$. More generally, for an (n,k,t) code,

$$d = e + t + 1 \tag{6.2}$$

where $t \leq e$. In summary we can say that an (n,k,t) block code will correct all patterns of t or fewer errors in a block of n symbols, but there will be at least one pattern of $t + 1$ errors that cannot be corrected.

Consider the simple block codes in Figure 6.1. In Figure 6.1(a) $d = 1$ and the code has no error control property. In Figure 6.1(b) we assume that an error has occurred, giving

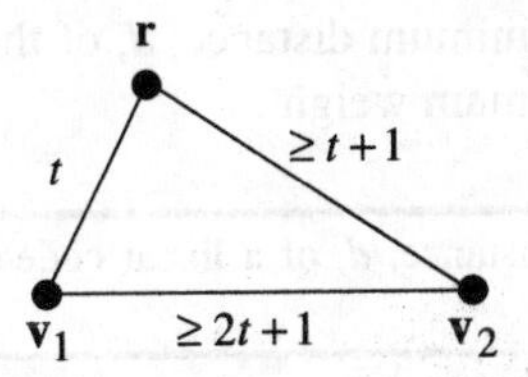

Figure 6.2 *Illustrating the code distance concept.*

received vector $\mathbf{r} = (1000)$. Clearly, an even parity check detects the error, but it cannot be corrected since any one of four codevectors could have been corrupted. Since $d = 2$ for Figure 6.1(b), we conclude that the code is simply single error detecting (SED), as suggested by (6.2). In Figure 6.1(c), $d = 3$ and (6.2) states that the code can detect *and* correct a single error. This is true providing only a single error occurs, since then $\mathbf{r}$ will be closest to the transmitted vector. Similarly, if $d = 4$, (6.2) states that the code is guaranteed to correct any single error and is also capable of detecting double errors (SEC + DED). Codes with $d = 4$ are used to protect UK teletext transmissions, for example.

6.1.2 Some algebraic concepts

The theory of error control block codes is heavily based upon algebraic systems such as *groups* and *fields*. In particular, the codevector can be written $\mathbf{v} = (v_1 v_2 ... v_n)$ where, in general, symbol v_i could be drawn from a finite mathematical field of q elements. In this case the field is called a *Galois* field and is denoted GF(q). For example, we might have $\mathbf{v} =$ (3410) where each symbol is drawn from the element range 0–4 and the relevant field is GF(5). Any multiplication and addition would then be performed modulo-5. Similar non-binary ($q > 2$) error control codes are discussed in Section 6.7, but for now we will concentrate on binary codes ($q = 2$). For instance, we could have $\mathbf{v} = (1001101)$ where symbols 0 and 1 are the 0 and 1 elements of GF(2). The significant point is that GF(2) defines the basic rules for any arithmetic operation performed by the encoder or decoder. These rules are simply those of modulo-2 arithmetic, and in particular $1 + 1 = 0$ modulo-2 (the exclusive-OR operation). Note that subtraction over GF(2) is identical to addition, and so is replaced by addition when it occurs. It is also worth noting that an *extension* of GF(2), to GF(2^m), m integer, is important when performing *algebraic* decoding of block codes (Section 6.6).

A linear binary block code is sometimes called a *group* code because its binary vectors $\mathbf{v}_1$, $\mathbf{v}_2$, $\mathbf{v}_3$, and so on, can be regarded as elements of a mathematical group G that is, $\mathbf{v}_1, \mathbf{v}_2, \mathbf{v}_3, \ldots \in G$ modulo-2. This is true if the group operation is defined as modulo-2 addition, so that any two elements $\mathbf{v}_i$, $\mathbf{v}_j$ could be added bit-by-bit over GF(2). It then follows from the closure postulate of G under addition that the resulting vector $\mathbf{v}_k$ also belongs to G

$$\mathbf{v}_i + \mathbf{v}_j = \mathbf{v}_k \in G \text{ modulo-2} \tag{6.3}$$

Put simply, the bit-by-bit modulo-2 addition of any two binary codewords yields another codeword (the reader should check this for the simple code in Figure 6.1(b)). In fact, *(6.3) is a necessary condition for the binary code to be linear, and it essentially embraces what we mean by a linear code.* Clearly, every linear code will also contain the all-zeros codeword since $\mathbf{v}_i + \mathbf{v}_i = \mathbf{0}$. Now, the number of 1s or non-zero elements of $\mathbf{v}_k$ is its *Hamming weight*, so this will be identical to the Hamming distance between $\mathbf{v}_i$ and $\mathbf{v}_j$. If, say, $\mathbf{v}_i$ is the all-zeros codeword, then the minimum distance, d, of the code is simply defined by the (non-zero) codeword $\mathbf{v}_j$ of minimum weight.

Theorem 6.1 The minimum distance, d, of a linear code equals the minimum Hamming weight of its non-zero vectors.

The following section expands upon the foregoing algebraic concepts.

6.2 The generator matrix of a linear block code

The linearity property in (6.3) enables a linear block to be easily specified. We do not need a long list of codewords. For instance, the triple error correcting binary BCH(31,16) code has 2^{16} codewords, but linearity means that it can be compactly specified by a 16×31 matrix! We call this the *generator matrix*, **G**.

Suppose, for an (n,k) linear block code, that we find a set of *k linearly independent* codewords or vectors $\mathbf{v}_1, \mathbf{v}_2, ..., \mathbf{v}_k$. For a binary code, linear independence simply means that the bit-by-bit modulo-2 sum of two or more vectors must not be $\mathbf{0}$. Vectors meeting this requirement form a *basis* for the code, and it is usual to put them into a $k \times n$ matrix

$$\mathbf{G} = \begin{bmatrix} \mathbf{v}_1 \\ \mathbf{v}_2 \\ \vdots \\ \mathbf{v}_k \end{bmatrix} = \begin{bmatrix} v_{11} & v_{12} & \cdots & v_{1n} \\ v_{21} & v_{22} & \cdots & v_{2n} \\ & & \vdots & \\ v_{k1} & v_{k2} & \cdots & v_{kn} \end{bmatrix} \tag{6.4}$$

The significant point is that all the 2^k linear combinations (vector sums) of the k row vectors of **G** *(the row space of* **G***) will generate all the 2^k vectors of the code,* **V**. Put another way, each codeword is generated as the row vector

$$\mathbf{v} = u_1\mathbf{v}_1 + u_2\mathbf{v}_2 + \ldots u_k\mathbf{v}_k \tag{6.5}$$

where, for a binary code, $u_i \in \mathrm{GF}(2)$ and the vectors are summed bit-by-bit modulo-2. Note that (6.5) is simply a generalisation of (6.3).

Example 6.1

The SED code in Figure 6.1(b) could be described as the row space of, say,

$$\mathbf{G} = \begin{bmatrix} 1001 \\ 1100 \\ 1010 \end{bmatrix} \tag{6.6}$$

where three linearly independent vectors have been selected from the eight codevectors (the choice of **G** is not unique, and the same code can be formed from a different set of basis vectors). Thus, a linear combination of two or three of the row vectors in (6.6) will give the code in Figure 6.1(b), the all-zeros vector being generated by setting $u_i = 0 \;\; \forall i$ in (6.5).

Finally, for message $\mathbf{u} = (u_1 u_2 ... u_k)$ we could write (6.5) as

$$\mathbf{v} = \mathbf{u}\mathbf{G} \tag{6.7}$$

which can be expanded as

$$\left.\begin{aligned} v_1 &= u_1 v_{11} + u_2 v_{21} + \ldots + u_k v_{k1} \\ v_2 &= u_1 v_{12} + u_2 v_{22} + \ldots + u_k v_{k2} \\ &\vdots \\ v_n &= u_1 v_{1n} + u_2 v_{2n} + \ldots + u_k v_{kn} \end{aligned}\right\} \tag{6.8}$$

where $\mathbf{v} = (v_1 v_2 \ldots v_n)$. Given message $\mathbf{u}$, (6.7) is the fundamental coding algorithm, that is, the encoder forms a particular linear combination of the row vectors in $\mathbf{G}$ to generate the codeword, as in (6.5).

6.2.1 Systematic form of G

Any $\mathbf{G}$ matrix can be put into a form which will generate a systematic block code. In general this is done by row operations (replacing rows by row additions), and possibly column reordering. For example, putting the columns in (6.6) in the order (4231), gives

$$\mathbf{G} = \begin{bmatrix} 1001 \\ 0101 \\ 0011 \end{bmatrix} \tag{6.9}$$

Using (6.9) in (6.7) we see that $v_1 = u_1$, $v_2 = u_2$, $v_3 = u_3$ and $v_4 = u_1 + u_2 + u_3$ modulo-2, that is, the first part of the codevector is the message, and v_4 is the overall parity check. In this particular case, despite column reordering, the row space of (6.9) (all the linear combinations of the rows) results in the same codevectors (Figure 6.1(b)). In general, column reordering changes the row space and results in an equivalent code having the same performance on a memoryless channel. On the other hand, elementary row operations on $\mathbf{G}$ (as in (6.5)) will result in the *same* row space, that is, the same code.

Equation (6.9) can be generalised to the partitioned matrix

$$\mathbf{G} = [\mathbf{I}_k \; \mathbf{P}_{k,n-k}] \tag{6.10}$$

where $\mathbf{I}_k$ is a $k \times k$ identity matrix and $\mathbf{P}$ is a $k \times (n-k)$ parity check matrix. This is sometimes referred to as the *reduced-echelon form.*

Example 6.2
A systematic (7,4) code can be generated from

$$\mathbf{G} = \begin{bmatrix} \mathbf{v}_1 \\ \mathbf{v}_2 \\ \mathbf{v}_3 \\ \mathbf{v}_4 \end{bmatrix} = \begin{bmatrix} 1000101 \\ 0100111 \\ 0010110 \\ 0001011 \end{bmatrix} \tag{6.11}$$

Using (6.5), the first part of the row space is

$u_1\ u_2\ u_3\ u_4$	$v_1\ v_2\ v_3\ v_4\ v_5\ v_6\ v_7$
0 0 0 0	0 0 0 0 0 0 0
1 0 0 0	1 0 0 0 1 0 1
0 1 0 0	0 1 0 0 1 1 1
1 1 0 0	1 1 0 0 0 1 0
0 0 1 0	0 0 1 0 1 1 0
⋮	⋮

We will return to the generator matrix when we look at cyclic codes (Section 6.4). There we will examine an algebraic approach to reducing a non-systematic matrix **G** to the systematic form of (6.10), and illustrate the use of the systematic form by looking at the coding for the European Radio Data System (RDS).

6.3 The parity check matrix of a linear block code

Rather than encode a message using the **G** matrix (equation (6.7)), an alternative matrix description of (n,k) linear block codes arises from the following definition:

Definition 6.1 A linear block code **V** is a set of vectors of length n where each vector **v** in **V** is orthogonal to every row of a matrix **H**.

Using the conventional definition of orthogonality this statement can be expressed as

$$\mathbf{H}\,\mathbf{v}^{\mathrm{T}} = \mathbf{0} \tag{6.12}$$

where **0** is a column vector, T denotes transpose, $\mathbf{v} = (v_1 v_2 \ldots v_n)$, and evaluation is over GF(2) for a binary code. Alternatively, we could define the code as $\mathbf{v}\mathbf{H}^{\mathrm{T}} = \mathbf{0}$, where **0** is a row vector. The $(n-k) \times n$ *parity check matrix* **H** defines the code. Expanding (6.12)

$$\begin{bmatrix} h_{11} & h_{12} & \cdots & h_{1n} \\ \vdots & & & \vdots \\ h_{i1} & h_{i}2 & \cdots & h_{in} \\ \vdots & & & \vdots \\ h_{(n-k)1} & h_{(n-k)2} & \cdots & h_{(n-k)n} \end{bmatrix} \begin{bmatrix} v_1 \\ \vdots \\ v_i \\ \vdots \\ v_n \end{bmatrix} = \begin{bmatrix} 0 \\ \vdots \\ 0 \\ \vdots \\ 0 \end{bmatrix} \tag{6.13}$$

This states that each codevector **v** must satisfy a set of $n-k$ independent equations each of the form

$$\sum_{j=1}^{n} h_{ij} v_j = 0 \ \text{ modulo-2}; \quad i = 1, 2, \ldots, n-k \tag{6.14}$$

For a given message, each equation is simply solved for one of the $n-k$ parity symbols. For instance, for a particular code, the first equation might give $P_1 + 1 + 0 + 1 = 0$ modulo-2, which means that parity symbol $P_1 = 0$. Since the number of 1s in the equation is now even, this results in an even parity code. An odd parity code would be defined as $\mathbf{H}\mathbf{v}^T = \mathbf{1}$. Clearly, (6.12) is the fundamental code generation mechanism.

Since we can generate a complete binary code of 2^k codevectors from **H**, it is reasonable to assume that we can also deduce the minimum Hamming distance, d, of the code from **H**. We simply invoke the following theorem:

Theorem 6.2 There exists at least one subset of d columns of **H** that sum to **0** and no subsets of $(d-1)$ or fewer columns will sum to **0**.

This theorem is somewhat more powerful than Theorem 6.1 since all we need is **H** (rather than the complete code) in order to find d. Conversely, the theorem could be used to *construct* **H** in order to provide a code with a specific value of d.

6.3.1 Decoding mechanism

Now consider the use of **H** for decoding. Suppose we transmit a vector **v** over a BSC and assume it is detected as **r** (see Figure 5.7). The role of the FEC decoder is then to correct **r**, if possible. The detection errors can be modelled as

$$\mathbf{r} = \mathbf{v} + \mathbf{e} \tag{6.15}$$

where $\mathbf{r} = (r_1 r_2 ... r_n)$, and $\mathbf{e} = (e_1 e_2 ... e_n)$ is the *error vector*. Symbol $e_i = 1$ if v_i is received in error, otherwise it is zero. The decoder performs the same parity checks on **r** as were carried out at the encoder, so that

$$\begin{aligned} \mathbf{H}\mathbf{r}^T &= \mathbf{H}(\mathbf{v}^T + \mathbf{e}^T) \\ &= \mathbf{H}\mathbf{v}^T + \mathbf{H}\mathbf{e}^T \\ &= \mathbf{H}\mathbf{e}^T \\ &= \mathbf{s}^T \end{aligned} \tag{6.16}$$

for an even parity code. Here, $\mathbf{s} = (s_1, s_2, ..., s_{n-k})$ is the *syndrome* vector and we use it to locate the error(s). As discussed in Section 6.1.1, a t-error correcting block code will correct all error patterns (error vectors) of weight t or less, but there will be at least one pattern of $t+1$ errors that cannot be corrected. In terms of Figure 6.2 (and assuming $d = 2t + 1$) if codevector $\mathbf{v}_1$ is received with $t+1$ errors it is apparent that it could be wrongly 'corrected' to codevector $\mathbf{v}_2$.

Suppose we have a random error pattern of t or fewer errors. In this case they can be corrected provided they result in a unique (distinct) and non-zero syndrome. In order to achieve this, we need certain restrictions on the columns of **H**. For example, if $t = 2$ and two errors occur at positions i and j respectively, that is, $e_i = e_j = 1$, then (6.16) gives

$$\mathbf{s}^T = \begin{bmatrix} h_{1i}e_i + h_{1j}e_j \\ h_{2i}e_i + h_{2j}e_j \\ \vdots \\ h_{(n-k)i}e_i + h_{(n-k)j}e_j \end{bmatrix} = \mathbf{h}_i + \mathbf{h}_j \quad i, j = 1, ..., n;\ i \neq j \tag{6.17}$$

where $\mathbf{h}_i$ and $\mathbf{h}_j$ correspond to columns i and j of $\mathbf{H}$. This means that the sum of any two columns of $\mathbf{H}$ must be unique and non-zero in order to give a unique and non-zero syndrome. Generalising, for a t-error correcting code, the sum of any t or fewer columns of $\mathbf{H}$ must be unique and non-zero (note that this is simply a weaker re-statement of Theorem 6.2).

6.3.2 Hamming codes

These are a class of SEC ($t = 1$) codes, and so from (6.17) a sufficient condition is that all columns of $\mathbf{H}$ must be unique and non-zero. For $n - k$ rows in $\mathbf{H}$ we can have up to $2^{n-k} - 1$ unique and non-zero columns, corresponding to a block length $n \le 2^{n-k} - 1$. The code is optimum or *perfect* when

$$2^{n-k} = n+1 \tag{6.18}$$

This equation defines a set of perfect SEC codes called *Hamming* codes. Clearly,

$$(n,k) = (3,1),(7,4),(15,11),(31,26),(63,57),\ldots$$

Note that R improves as n increases, but for a fixed channel error rate this is balanced by the increased probability of more than t errors in a block, that is, decoder failure. The SEC property can be proved by finding the minimum distance d of the code or by examining upper *coding bounds*. The so-called *Hamming bound* (a necessary upper bound) for a t-error correcting code defines the maximum value of k for a specified value of n (it is a plot of k against n). In fact, points defined by (6.18) lie on the Hamming bound for $t = 1$, thereby justifying the claim that Hamming codes are optimal codes. Also, Section 6.4 shows that Hamming codes (in their cyclic form) are really only a form of t-error correcting binary BCH code, with $t = 1$. The foregoing discussion can be expressed formally as follows:

Definition 6.2 A linear code defined by a matrix $\mathbf{H}$ comprised of $2^r - 1$ non-zero r-tuples arranged in any order is called a Hamming code.

A particular Hamming code is therefore constructed using a particular ordering of the columns in $\mathbf{H}$. For example, we might select the following 'binary-ordered' matrix for the H(7,4) code

$$\mathbf{H} = \begin{bmatrix} 0\,0\,0\,1\,1\,1\,1 \\ 0\,1\,1\,0\,0\,1\,1 \\ 1\,0\,1\,0\,1\,0\,1 \end{bmatrix} \tag{6.19}$$

According to (6.17) a single error at position i will give a syndrome

$$\mathbf{s}^{\mathrm{T}} = \mathbf{h}_i, \quad i = 1, \ldots, 7 \tag{6.20}$$

where $\mathbf{h}_i$ corresponds to the ith column of $\mathbf{H}$. Therefore, for the special case of a binary ordered matrix, $\mathbf{s}$ will have a binary value which directly gives the position of the error, for example, if $\mathbf{s} = (110)$ then $e_6 = 1$ and correction is performed by complementing r_6.

The *format* of the codevector follows from $\mathbf{H}$. If the matrix in (6.19) is used then (6.12) gives the following three equations, each one supplying one parity symbol:

$$\begin{aligned} v_4 + v_5 + v_6 + v_7 &= 0 \text{ giving } P_3 \\ v_2 + v_3 + v_6 + v_7 &= 0 \text{ giving } P_2 \\ v_1 + v_3 + v_5 + v_7 &= 0 \text{ giving } P_1 \end{aligned}$$

Recall that these equations are evaluated modulo-2, and would be equated to unity (not zero) for an odd-parity code. Since each equation yields one parity symbol, it is apparent that these symbols should be in positions one, two and four, corresponding to a codevector format $\mathbf{v} = (P_1P_2M_1P_3M_2M_3M_4)$. Message symbols, M, can be inserted in any order. Generalising, parity symbols are placed in positions corresponding to columns in **H** which have a single 1.

Example 6.3
Using the matrix in (6.19), the even parity H(7,4) codevector for message 13_{10} could be of the form $\mathbf{v} = (P_1P_21P_3101)$, where M_1 is the MSB. The parity symbols are given by

$$\begin{aligned} P_3 + 1 + 0 + 1 &= 0, \quad P_3 = 0 \\ P_2 + 1 + 0 + 1 &= 0, \quad P_2 = 0 \\ P_1 + 1 + 1 + 1 &= 0, \quad P_1 = 1 \end{aligned}$$

so $\mathbf{v} = (1010101)$. The reader should show that, using the same matrix and message format, the odd-parity codevector for message 8_{10} is $\mathbf{v} = (0011000)$.

Example 6.4
Suppose that an odd-parity H(7,4) code defined by

$$\mathbf{H} = \begin{bmatrix} 0011011 \\ 0110101 \\ 1100011 \end{bmatrix}$$

is used to transmit a binary number in the range 0–15_{10}. The codevector format is therefore $\mathbf{v} = (P_1M_1M_2P_2P_3M_3M_4)$ where, as before, we assume that M_1 is the MSB. If the detected vector is $\mathbf{r} = (0111101)$, what is the probable value of the number?
Since odd-parity is used

$$\mathbf{H}\mathbf{r}^{\mathrm{T}} = \mathbf{H}\mathbf{v}^{\mathrm{T}} + \mathbf{H}\mathbf{e}^{\mathrm{T}} = \mathbf{1} + \mathbf{s}^{\mathrm{T}} \tag{6.21}$$

The three modulo-2 sums defined by (6.21) can be represented as follows:

$$\begin{array}{ccccccclll} 0 & 1 & 1 & 1 & 1 & 0 & 1 & & & \\ & & \times & \times & & \times & \times & = 1 = 1 + s_1; & s_1 = 0 \\ & \times & \times & & \times & & \times & = 0 = 1 + s_2; & s_2 = 1 \\ \times & \times & & & & \times & \times & = 0 = 1 + s_3; & s_3 = 1 \end{array}$$

giving $\mathbf{s} = (011)$. Using (6.20) we conclude that the error is in position 2 ($e_2 = 1$), and the message would be decoded as (0101). This will be correct *provided that only a single error has occurred in the codeword.* Multiple errors in the block will cause decoder failure and, in general, the finite syndrome could lead to even more errors after 'correction'.

6.3.3 Extended Hamming codes

Basic codes are often modified to match them to a particular problem, or to enhance their error control capability. For instance, we may wish to protect a 32-bit data bus in a computer system ($k = 32$), or add double error detection capability to a Hamming code.

A Hamming code can be extended by adding an overall parity check symbol. Figure 6.3 shows the basic even-parity H(7,4) code generated from (6.19), and from Theorem 6.1 we deduce that $d = 3$. More fundamentally, it is apparent that no two columns in **H** sum to zero (since each column is unique), while several sets of three columns sum to zero. Therefore, according to Theorem 6.2, $d = 3$. Clearly, the minimum Hamming weight, and therefore d, can be increased to 4 by adding an overall even parity symbol P_4 such that the modulo-2 sum of all eight code symbols is zero. For example, message 5_{10} would be coded as (01001011). Since d is now four, we deduce from (6.2) that the extended code must have SEC and DED properties. The overall parity check is represented by adding a further column and a row of 1s to **H**, giving an H(8,4) code. Thus, from (6.16), $\mathbf{s}^{\mathrm{T}} = \mathbf{He}^{\mathrm{T}}$, which is expanded as

$$\begin{bmatrix} s_1 \\ s_2 \\ s_3 \\ s_4 \end{bmatrix} = \begin{bmatrix} 0\,0\,0\,0\,1\,1\,1\,1 \\ 0\,0\,1\,1\,0\,0\,1\,1 \\ 0\,1\,0\,1\,0\,1\,0\,1 \\ 1\,1\,1\,1\,1\,1\,1\,1 \end{bmatrix} \begin{bmatrix} e_1 \\ e_2 \\ e_3 \\ e_4 \\ e_5 \\ e_6 \\ e_7 \\ e_8 \end{bmatrix} \tag{6.22}$$

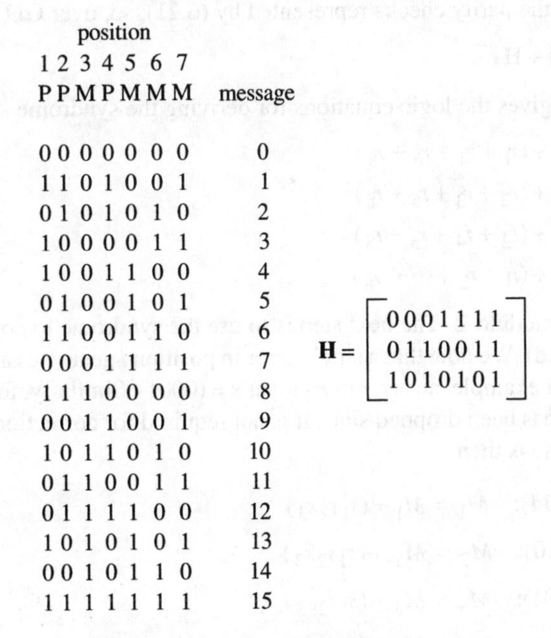

position 1 2 3 4 5 6 7 P P M P M M M	message
0 0 0 0 0 0 0	0
1 1 0 1 0 0 1	1
0 1 0 1 0 1 0	2
1 0 0 0 0 1 1	3
1 0 0 1 1 0 0	4
0 1 0 0 1 0 1	5
1 1 0 0 1 1 0	6
0 0 0 1 1 1 1	7
1 1 1 0 0 0 0	8
0 0 1 1 0 0 1	9
1 0 1 1 0 1 0	10
0 1 1 0 0 1 1	11
0 1 1 1 1 0 0	12
1 0 1 0 1 0 1	13
0 0 1 0 1 1 0	14
1 1 1 1 1 1 1	15

$$\mathbf{H} = \begin{bmatrix} 0\,0\,0\,1\,1\,1\,1 \\ 0\,1\,1\,0\,0\,1\,1 \\ 1\,0\,1\,0\,1\,0\,1 \end{bmatrix}$$

Figure 6.3 *An even parity H(7,4) code.*

Suppose there is a single error in position four ($e_4 = 1$). This will be flagged by $s_4 = 1$ and the error can be located from $\mathbf{s} = (s_1 s_2 s_3) = (011)$. On the other hand, suppose we have a double error, say, $e_4 = e_6 = 1$. Now $s_4 = 0$ and the basic syndrome is $\mathbf{s} = (110)$. The significant point is that at least one of the symbols s_1, s_2, s_3 in the basic syndrome will always be set to 1 by a double error. This means that any pattern of two errors in **r** could be flagged by the logical expression

$$\Omega = \overline{s_4} \cdot (s_1 + s_2 + s_3) \tag{6.23}$$

where, in this case, '+' denotes logical-OR. For any double error $\Omega = 1$ and **r** could be rejected. For three errors, $\Omega = 0$ and, in general, the non-zero syndrome will result in even more errors and decoding fails.

Example 6.5

An odd-parity H(8,4) code is useful for teletext systems. In this application it is helpful to interleave the message and parity symbols so that simple decoders can extract the message by simply clocking a shift register at half data rate. A suitable H(8,4) matrix is therefore

$$\mathbf{H} = \begin{bmatrix} 10010101 \\ 01100101 \\ 01011001 \\ 11111111 \end{bmatrix}$$

corresponding to a codeword format $\mathbf{v} = (P_1 M_4 P_2 M_3 P_3 M_2 P_4 M_1)$, where P_4 is the overall check bit and M_1 could be the MSB. Suppose that we require the decoder logic. The decoder performs the parity checks represented by (6.21), so, over GF(2), we can write

$$\mathbf{s}^{\mathrm{T}} = \mathbf{1} + \mathbf{H}\mathbf{r}^{\mathrm{T}} \tag{6.24}$$

Expanding (6.24) gives the logic equations for deriving the syndrome symbols:

$$s_1 = 1 + (r_1 + r_4 + r_6 + r_8)$$
$$s_2 = 1 + (r_2 + r_3 + r_6 + r_8)$$
$$s_3 = 1 + (r_2 + r_4 + r_5 + r_8)$$
$$s_4 = 1 + (r_1 + r_2 + \cdots + r_8)$$

where addition is modulo-2. The next step is to use the syndrome to correct the message symbols (if required). We note that a single error in position i generates a syndrome $\mathbf{s}^{\mathrm{T}} = \mathbf{h}_i$ (equation 6.20), for example, an error in P_4 gives $\mathbf{s} = (0001)$. Similarly, for an error in M_1, $\mathbf{s} = (111)$, where s_4 has been dropped since it is not required for correction of message bits. The correction logic is then

$$\mathbf{s} = (111): \quad \hat{M}_1 = M_1 + (s_1 s_2 s_3)$$
$$\mathbf{s} = (110): \quad \hat{M}_2 = M_2 + (s_1 s_2 \overline{s}_3)$$
$$\mathbf{s} = (101): \quad \hat{M}_3 = M_3 + (s_1 \overline{s}_2 s_3)$$
$$\mathbf{s} = (011): \quad \hat{M}_4 = M_4 + (\overline{s}_1 s_2 s_3)$$

Finally, (6.23) could be used to flag double errors.

6.3.4 Shortened Hamming codes

We may delete any q columns of a Hamming matrix $\mathbf{H}$ in order to match the code to a particular problem. A simple approach is given in the following example.

Example 6.6

Suppose we have extended the basic H(15,11) code to a H(16,11) code by adding an overall parity check. This ensures $d = 4$ (SEC+DED). We could then delete three arbitrarily selected 'message' columns in the 5×16 matrix to give an odd parity H(13,8) code suitable for protecting 8-bit data bytes. For instance, we might obtain the following matrix (corresponding to a systematic code)

$$\mathbf{H} = \begin{bmatrix} 0100010101011 \\ 0010001100111 \\ 0001000011111 \\ 0000111111110 \\ 1111111111111 \end{bmatrix} \tag{6.25}$$

Clearly, due to the overall parity check, it is apparent that no three columns sum to zero. On the other hand, it is readily seen that sets of four columns sum to zero, so $d = 4$ (Theorem 6.2), as expected. Now assume that several odd-parity codewords are generated using (6.25) and that the corresponding detector outputs are $\mathbf{r}_1$ = (1001001100110), $\mathbf{r}_2$ = (1011000000110). Since odd parity is used we can apply (6.21), and for $\mathbf{r}_1$ the parity equations can be written as

$$\begin{array}{ccccccccccccc l}
1&0&0&1&0&0&1&1&0&0&1&1&0 & \\
&\times&&&&\times&&\times&&\times&&\times&\times & =0=1+s_1; \quad s_1=1 \\
&&\times&&&&\times&\times&&&\times&\times&\times & =0=1+s_2; \quad s_2=1 \\
&&&\times&&&&&\times&\times&\times&\times&\times & =1=1+s_3; \quad s_3=0 \\
&&&&\times&\times&\times&\times&\times&\times&\times&\times& & =0=1+s_4; \quad s_4=1
\end{array}$$

The overall check gives $s_5 = 1$, flagging a probable single error. Applying (6.20) to (6.25) the syndrome indicates that $e_8 = 1$, so the probable transmitted message byte is (01000110). Similarly, for $\mathbf{r}_2$,

$$\begin{array}{ccccccccccccc l}
1&0&1&1&0&0&0&0&0&0&1&1&0 & \\
&\times&&&&\times&&\times&&\times&&\times&\times & =1=1+s_1; \quad s_1=0 \\
&&\times&&&&\times&\times&&&\times&\times&\times & =1=1+s_2; \quad s_2=0 \\
&&&\times&&&&&\times&\times&\times&\times&\times & =1=1+s_3; \quad s_3=0 \\
&&&&\times&\times&\times&\times&\times&\times&\times&\times& & =0=1+s_4; \quad s_4=1
\end{array}$$

and $s_5 = 0$. Since $\Omega = \bar{s}_5 \cdot (s_1 + s_2 + s_3 + s_4) = 1$ we assume a double error and reject $\mathbf{r}_2$.

We now consider a more general shortening approach. A shortened H($n-q$, $k-q$) Hamming code is generated by deleting q 'message' columns from the matrix. This corresponds to deleting q message symbols, and since there are now only 2^{k-q} codewords the minimum distance of the code may increase, that is, $d \geq 3$. In fact, to ensure d is at least 4, it is necessary that each column has an odd number of 1s. Deleting the 'even 1s' column in (6.19) for instance will give an H(4,1) code with $d = 4$.

Example 6.7

Consider the H(63,57) code. We commence shortening this by deleting the 31 columns in **H** having an even number of 1s. This leaves a 6 × 32 matrix, where six of the 32 columns have a single 1 (corresponding to the six parity symbols) and the remaining 26 columns correspond to message symbols. If this code is required to protect, say, 16-bit computer memory systems then a further ten 'message' columns must be deleted. This can be done in an optimal way in order to minimise hardware (Hsiao, 1970), and in this respect one criterion is to select columns such that the total number of 1s in **H** is minimised. In this example, 20 of the remaining 'message' columns will have just three 1s, and 16 of these can be selected for the final code. For example, an optimal matrix for the resulting H(22,16) SEC+DED code is

$$\mathbf{H} = \begin{bmatrix} 1\,0\,0\,0\,0\,0\,1\,0\,0\,1\,1\,0\,0\,1\,0\,0\,1\,1\,1\,1\,0\,0 \\ 0\,1\,0\,0\,0\,0\,0\,0\,1\,1\,1\,1\,1\,0\,1\,0\,0\,0\,1\,0\,1\,0 \\ 0\,0\,1\,0\,0\,0\,1\,1\,1\,0\,1\,1\,1\,0\,0\,1\,1\,0\,0\,0\,0\,0 \\ 0\,0\,0\,1\,0\,0\,1\,1\,1\,0\,0\,0\,0\,1\,1\,1\,0\,1\,0\,0\,0\,1 \\ 0\,0\,0\,0\,1\,0\,0\,0\,0\,1\,0\,0\,1\,1\,1\,1\,0\,0\,0\,1\,1\,1 \\ 0\,0\,0\,0\,0\,1\,0\,1\,0\,0\,0\,1\,0\,0\,0\,0\,1\,1\,1\,1\,1\,1 \end{bmatrix}$$

Figure 6.4 illustrates how this code can be used to increase computer memory reliability. On a memory-write cycle, six parity bits are generated and stored in a check memory. On a

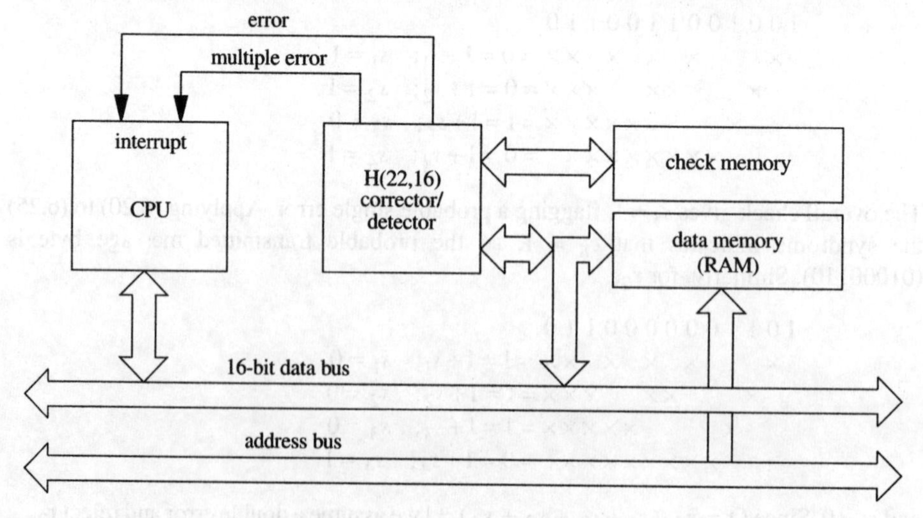

Figure 6.4 *Use of the H(22,16) code to enhance computer memory reliability.*

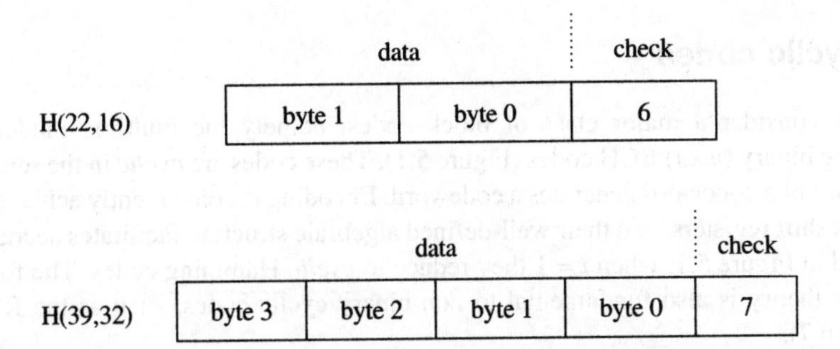

Figure 6.5 *Modified Hamming codes for 16- and 32-bit computer memory systems.*

read cycle new parity (or check) bits are generated from the data and these are compared with those in the check memory. The H(22,16) device interrupts the CPU only when an error is detected, so there is zero memory slow-down under zero error conditions. This arrangement will correct all single errors and detect all double errors. Note that a 32-bit data bus could be protected in a similar way using an H(39,32) code (Figure 6.5).

6.3.5 Systematic form of H

In practice, a systematic form of **H** tends to be preferred since this can simplify coding and decoding. Any binary matrix **H** can always be transformed into a systematic form to give an 'equivalent' code. For instance, the columns of (6.19) could be reordered to give

$$\mathbf{H} = \begin{bmatrix} 0111100 \\ 1011010 \\ 1101001 \end{bmatrix} = [\mathbf{Q}_{n-k,k}\ \mathbf{I}_{n-k}] \tag{6.26}$$

where $\mathbf{Q}$ is an $n - k \times k$ matrix. Using (6.12) then gives a systematic code with the format $\mathbf{v} = (M_1M_2M_3M_4P_1P_2P_3)$. More formally, for a code generated as the row space of (6.10), the same systematic code is generated using (6.12) with **H** given by

$$\mathbf{H} = [-\mathbf{P}^T_{n-k,k}\ \mathbf{I}_{n-k}] \tag{6.27}$$

The minus sign can be ignored over GF(2). As an example, the systematic matrix in (6.11) (Example 6.2) corresponds to the parity check matrix

$$\mathbf{H} = \begin{bmatrix} 1110100 \\ 0111010 \\ 1101001 \end{bmatrix} \tag{6.28}$$

The reader should check that using this in (6.12) gives the same code as given in Example 6.2. Also note that, using (6.10) and (6.27) we have

$$\mathbf{G}\mathbf{H}^{\mathrm{T}} = [\mathbf{I}_k\ \mathbf{P}_{k,\,n-k}]\begin{bmatrix} -\mathbf{P}_{k,\,n-k} \\ \mathbf{I}_{n-k} \end{bmatrix} = -\mathbf{P} + \mathbf{P} = \mathbf{0}_{k,\,n-k} \tag{6.29}$$

6.4 Cyclic codes

We now consider a major class of block codes, namely the multiple random error correcting binary (n,k,t) BCH codes (Figure 5.1). These codes are *cyclic* in the sense that a cyclic shift of a codeword generates a codeword. Encoding is conveniently achieved using feedback shift registers, and their well-defined algebraic structure facilitates decoding. As indicated in Figure 5.1, when $t = 1$ they reduce to *cyclic* Hamming codes. The following algebraic theory is also fundamental to non-binary cyclic codes, such as the RS codes (Section 6.7).

For an algebraic description of block codes we redefine a codeword of an (n,k,t) binary block code as $\mathbf{v} = (v_0 v_1 ... v_{n-1})$ and associate it with the polynomial

$$v(x) = v_0 + v_1 x + v_2 x^2 + ... + v_{n-1} x^{n-1} \tag{6.30}$$

Similarly, message $\mathbf{u} = (u_0 u_1 ... u_{k-1})$ is associated with polynomial

$$u(x) = u_0 + u_1 x + u_2 x^2 + ... + u_{k-1} x^{k-1} \tag{6.31}$$

An advantage of this algebraic representation can be seen from the following theorem:

Theorem 6.3 Every code polynomial in an (n,k,t) cyclic code can be derived as $v(x) = u(x)g(x)$.

---

where $g(x)$ is a *generator polynomial* of the form

$$g(x) = g_0 + g_1 x + g_2 x^2 + ... + g_{n-k} x^{n-k}; \quad g_0 = g_{n-k} = 1 \tag{6.32}$$

Theorem 6.3 states that a codeword for a particular message can be generated by simply multiplying polynomials! The code itself is defined by $g(x)$:

Theorem 6.4 An (n,k,t) cyclic code is completely specified by a generator polynomial $g(x)$ of degree $n - k$ that divides $x^n + 1$. Put another way, the basic block length, n, corresponds to the lowest degree polynomial $x^n + 1$ that is exactly divisible by $g(x)$.

Example 6.8
Over GF(2)

$$x^7 + 1 = (1 + x)(1 + x^2 + x^3)(1 + x + x^3)$$

Factors $(1 + x^2 + x^3)$ and $(1 + x + x^3)$ *each* generate a (7,4) cyclic code. For large n, $x^n + 1$ could have many factors of degree $n - k$ and only some of these will generate good codes. We shall see later that $g(x)$ can be selected from a table of polynomials.

It should be noted that, given the basic block length n as defined in Theorem 6.4, $g(x)$ will generate cyclic codes of block length $n' = in$, $i = 1, 2, 3, ...$. Also, since $g(x)$ completely specifies the code, we might hope to relate it to the error control property of the code, that is, to its minimum distance d. Unfortunately, in general this turns out to be difficult, although we can state a few simple rules which can be useful when determining the error control property of practical codes:

1. If $g(x)$ has at least two terms, then $d \geq 2$.
2. If the general block length of an error correcting code is $n' \leq n$, where n satisfies $g(x)$ as in Theorem 6.4, then $d \geq 3$. In other words, creating a shortened cyclic code (Section 6.4.6) sometimes leads to a more powerful code. Conversely, if $n' = in,\ i = 2, 3, 4, \ldots$, then in almost all cases $d = 2$, corresponding to an error detecting code.
3. If $g(x) = (1 + x)g_1(x)$ and $n' \leq n_1$, where $g_1(x)$ 'belongs' to the polynomial $x^{n_1} + 1$, as in Theorem 6.4, then $d \geq 4$.

Example 6.9
The ETSI-GSM channel coding specification for digital cellular radio (ETSI, 1991) protects speech bits by using a (systematic) cyclic code followed by a convolutional code. The cyclic code protects blocks of 50 speech bits using three parity bits and is based on the generator $g(x) = 1 + x + x^3$. According to Theorem 6.4, or Example 6.8, we see that $g(x)$ 'belongs' to polynomial $x^7 + 1$, so the basic block length is $n = 7$. On the other hand, $g(x)$ also defines the cyclic code (56,53), corresponding to the block length $n' = 8n$, and since $n' > n$ this code will have $d = 2$. The GSM specification therefore simply uses the code for error *detection*, and in fact the code will reliably detect any burst error of length 3 bits or fewer and 87.5% of all error patterns (see Section 6.5.3). Clearly, in order to protect the 50 speech bits, the code must be shortened to the (53,50) code, which is the code defined in the specification.

6.4.1 Generator matrix for a cyclic code

According to Theorem 6.3 we can construct the following cyclic code:

$u(x)$	$v(x)$
0	0
1	$g(x)$
x	$xg(x)$
$1+x$	$g(x)+xg(x)$
x^2	$x^2g(x)$
$1+x^2$	$g(x)+x^2g(x)$
$\vdots$	$\vdots$
$1+x+x^2+\ldots+x^{k-1}$	$g(x)+xg(x)+x^2g(x)+\ldots+x^{k-1}g(x)$

It will be apparent that code polynomials $v(x)$ could be obtained by taking all linear combinations (as in (6.5)) of the k rows of matrix

$$\mathbf{G}(x) = \begin{bmatrix} g(x) \\ xg(x) \\ x^2g(x) \\ \vdots \\ x^{k-1}g(x) \end{bmatrix} \tag{6.33}$$

Example 6.10

From Example 6.8 we see that a (7,4) cyclic code can be generated from $g(x) = 1 + x + x^3$. Inserting this into (6.33) gives the $k \times n$ generator matrix

$$\mathbf{G} = \begin{bmatrix} 1101000 \\ 0110100 \\ 0011010 \\ 0001101 \end{bmatrix} = \begin{bmatrix} \mathbf{v}_1 \\ \mathbf{v}_2 \\ \mathbf{v}_3 \\ \mathbf{v}_4 \end{bmatrix} \tag{6.34}$$

According to (6.7), a codevector is then generated as the linear combination

$$\mathbf{v} = u_0\mathbf{v}_1 + u_1\mathbf{v}_2 + u_2\mathbf{v}_3 + u_3\mathbf{v}_4 \tag{6.35}$$

and the first part of the code is given below:

u_0 u_1 u_2 u_3	v_0 v_1 v_2 v_3 v_4 v_5 v_6
0 0 0 0	0 0 0 0 0 0 0
0 0 0 1	0 0 0 1 1 0 1
0 0 1 0	0 0 1 1 0 1 0
0 0 1 1	0 0 1 0 1 1 1
0 1 0 0	0 1 1 0 1 0 0
0 1 0 1	0 1 1 1 0 0 1
0 1 1 0	0 1 0 1 1 1 0
0 1 1 1	0 1 0 0 0 1 1
1 0 0 0	1 1 0 1 0 0 0
⋮	⋮

Note that the code is non-systematic, and that a cyclic shift of the third codeword gives the second codeword, and so on.

Example 6.11

Protection for the European radio data system or RDS (CENELEC, 1992) is based upon a (341,331) cyclic block code defined by

$$g(x) = x^{10} + x^8 + x^7 + x^5 + x^4 + x^3 + 1 \tag{6.36}$$

Following (6.33), we could write the polynomial form of the generator matrix as

$$\mathbf{G}(x) = \begin{bmatrix} x^{330}g(x) \\ x^{329}g(x) \\ \vdots \\ xg(x) \\ g(x) \end{bmatrix} \tag{6.37}$$

The *basic* 331 × 341 generator matrix for the RDS code could therefore be written

$$\mathbf{G} = \begin{bmatrix} 1\,0\,1\,1\,0\,1\,1\,1\,0\,0\,1 & \cdots & 0 \\ 0\,1\,0\,1\,1\,0\,1\,1\,1\,0\,0\,1 & \cdots & 0 \\ & \vdots & \\ 0 \cdots & 1\,0\,1\,1\,0\,1\,1\,1\,0\,0\,1 & 0 \\ 0 \cdots & 1\,0\,1\,1\,0\,1\,1\,1\,0\,0 & 1 \end{bmatrix} \tag{6.38}$$

In practice, the RDS system uses a shortened version of this code, and a *systematic* form of G (Example 6.19).

6.4.2 Polynomial multiplication and division

The multiplication and division of polynomials are fundamental to the coding and decoding of cyclic codes. Polynomial division tends to be of most practical significance since it leads to systematic cyclic codes, although sometimes both multiplication and division are required in a single system. Either way, hardware implementation is conveniently based upon shift register circuits.

Consider the multiplication of a message $u(x)$ by $g(x)$. A general circuit for achieving this is shown in Figure 6.6(a) and is based upon an $n - k$ stage shift register. The assumptions are that the message symbols enter high-order first (u_{k-1} first), followed by $n - k$ zeros, and that the register is initially cleared. The output of the circuit will then correspond to the product

$$u(x)\,g(x) = g_0u_0 + (g_0u_1 + g_1u_0)x + \ldots + g_{n-k}u_{k-1}x^{n-1} \tag{6.39}$$

where the $g_{n-k}u_{k-1}$ term is generated first. Note that an equivalent 'transposed' form of Figure 6.6(a) is possible, but this is less attractive in practice, especially when both multiplication and division are performed simultaneously by a single shift register.

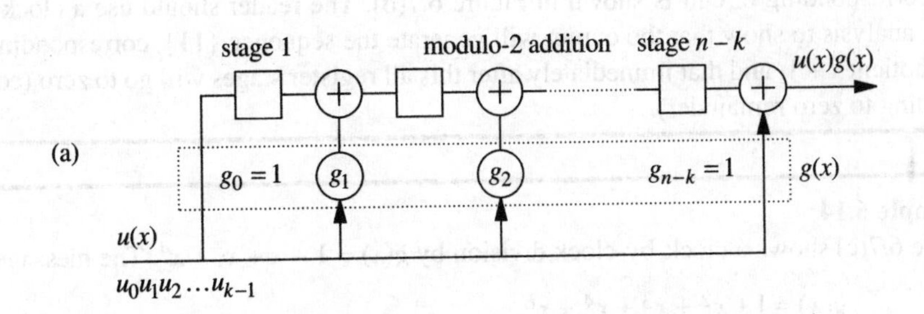

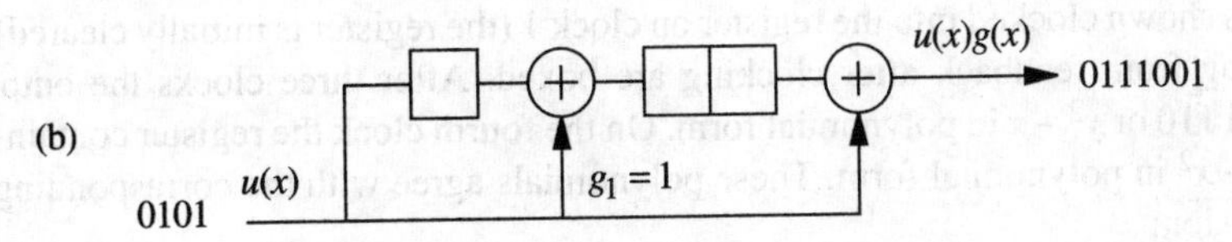

Figure 6.6 *Polynomial multiplication: (a) general shift register-based circuit; (b) multiplication of $u(x) = x + x^3$ by $g(x) = 1 + x + x^3$.*

Example 6.12

For message $\mathbf{u} = (u_0u_1u_2u_3) = (0101)$ and $g(x) = 1 + x + x^3$, Theorem 6.3 gives the codeword for the cyclic (7,4) code as

$$v(x) = (x + x^3)(1 + x + x^3) = x + x^2 + x^3 + x^6$$

corresponding to $\mathbf{v} = (0111001)$. The encoder circuit is shown in Figure 6.6(b) and the reader should use a clock-by-clock analysis to confirm that a (0101) input yields $\mathbf{v}$.

Example 6.12 and Theorem 6.3 show that cyclic codes are easily generated using polynomial multiplication. On the other hand, the resulting code is non-systematic and, in practice, it is more usual to generate cyclic codes in systematic form. This requires *division* by $g(x)$ rather than multiplication, and a general linear feedback shift register circuit for performing this over GF(2) is shown in Figure 6.7(a). Again, it is assumed that symbol u_{k-1} enters first and that the register is initially cleared. The circuit output is the quotient of the division and when all the quotient coefficients have appeared at the output the shift register contains the remainder.

Example 6.13

Assume the message $\mathbf{u} = (10111)$, corresponding to $u(x) = 1 + x^2 + x^3 + x^4$. Noting that + and – are equivalent over GF(2), the division of $u(x)$ by $g(x) = 1 + x + x^3$ is as follows:

$$\begin{array}{r} x+1 \\ x^3+x+1 \overline{\big) x^4+x^3+x^2+1} \\ \underline{x^4+x^2+x} \\ x^3+x+1 \\ \underline{x^3+x+1} \\ \text{zero} \end{array}$$

The corresponding circuit is shown in Figure 6.7(b). The reader should use a clock-by-clock analysis to show that the output will generate the sequence {11}, corresponding to the quotient $x + 1$, and that immediately after this all register stages will go to zero (corresponding to zero remainder).

Example 6.14

Figure 6.7(c) shows a clock-by-clock division by $g(x) = 1 + x + x^3 + x^4$. The message is

$$u(x) = 1 + x^2 + x^3 + x^4 + x^6$$

and this is shown clocked into the register on clock 1 (the register is initially cleared). Data bits arising from feedback after clocking are boxed. After three clocks the output has generated 110 or $x^2 + x$ in polynomial form. On the fourth clock the register contains 1110 or $1 + x + x^2$ in polynomial form. These polynomials agree with the corresponding algebraic division:

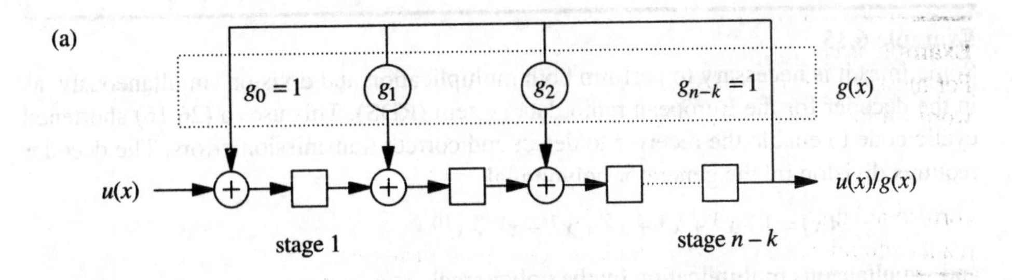

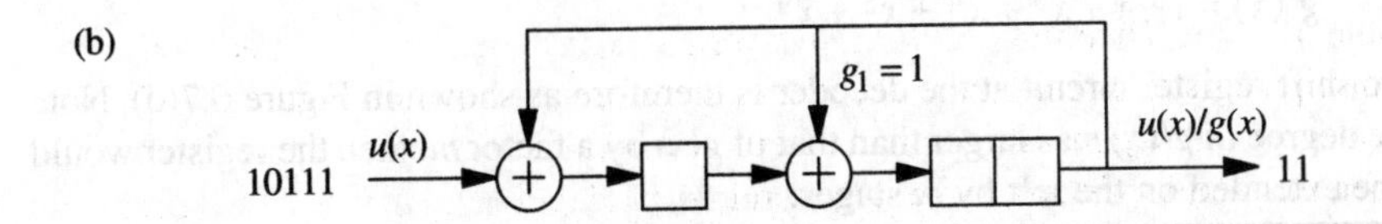

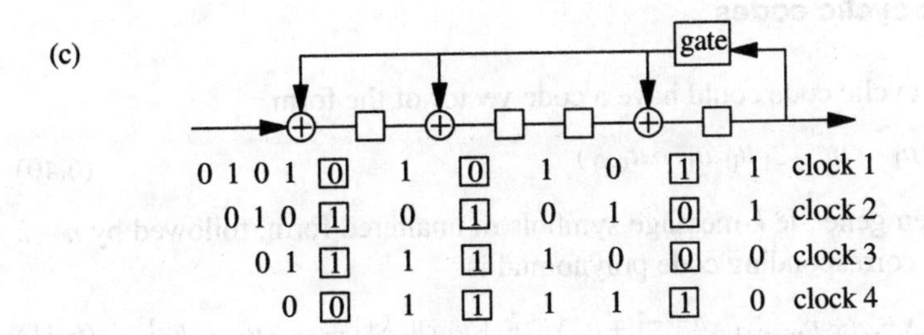

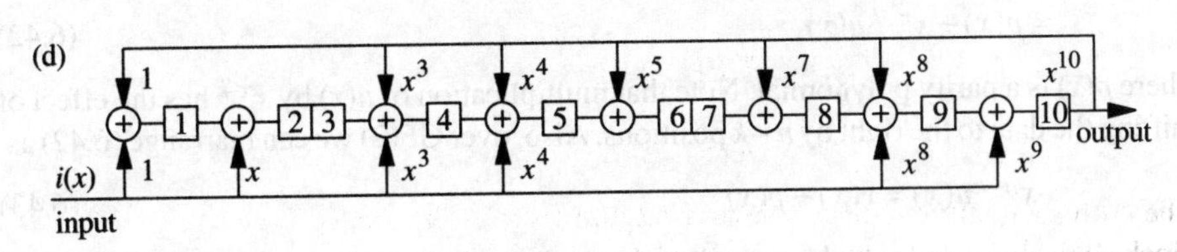

Figure 6.7 *Polynomial division: (a) division by a polynomial $g(x)$ of degree $n - k$; (b) division of $u(x) = 1 + x^2 + x^3 + x^4$ by $g(x) = 1 + x + x^3$; (c) clock-by-clock division; (d) simultaneous division by $1 + x^3 + x^4 + x^5 + x^7 + x^8 + x^{10}$ and multiplication by $1 + x + x^3 + x^4 + x^8 + x^9$.*

$$
\begin{array}{r}
x^2 + x \\
x^4 + x^3 + x + 1 \,\overline{\big)\, x^6 + x^4 + x^3 + x^2 + 1} \\
\underline{x^6 + x^5 + x^3 + x^2} \\
x^5 + x^4 + 1 \\
\underline{x^5 + x^4 + x^2 + x} \\
x^2 + x + 1 \equiv \text{remainder}
\end{array}
$$

Example 6.15

Sometimes it is necessary to perform both multiplication and division simultaneously, as in the decoder for the European radio data system (RDS). This uses a (26,16) shortened cyclic code to enable the receiver to detect and correct transmission errors. The decoder requires division by the generator polynomial

$$g(x) = 1 + x^3 + x^4 + x^5 + x^7 + x^8 + x^{10}$$

and simultaneous multiplication by the polynomial

$$g'(x) = 1 + x + x^3 + x^4 + x^8 + x^9$$

The basic shift register circuit at the decoder is therefore as shown in Figure 6.7(d). Note that if the degree of $g'(x)$ was larger than that of $g(x)$ by a factor m, then the register would need to be extended on the left by m stages.

6.4.3 Systematic cyclic codes

A systematic (n,k,t) cyclic code could have a code vector of the form

$$\mathbf{v} = (p_0\ p_1 \cdots p_{n-k-1}\ u_0\ u_1 \cdots u_{k-1}) \tag{6.40}$$

The encoder will then generate k message symbols in unaltered form, followed by $n-k$ parity symbols. The corresponding code polynomial is

$$v(x) = p_0 + p_1x + \ldots + p_{n-k-1}x^{n-k-1} + u_0x^{n-k} + u_1x^{n-k+1} + \ldots + u_{k-1}x^{n-1} \tag{6.41}$$

$$= p(x) + x^{n-k}u(x) \tag{6.42}$$

where $p(x)$ is a parity polynomial. Note that multiplication of $u(x)$ by x^{n-k} has the effect of shifting the data to the right by $n-k$ positions. Also, over GF(2) we can rearrange (6.42) as

$$x^{n-k}u(x) = v(x) + p(x) \tag{6.43}$$

Now, *since every codeword in a cyclic code is divisible by* $g(x)$, we can write (6.43) as

$$\frac{x^{n-k}u(x)}{g(x)} = a(x) + \frac{p(x)}{g(x)} \tag{6.44}$$

where $a(x)$ is the quotient (in this case of little interest) and $p(x)$ is the remainder, since its degree is less than that of $g(x)$. Equation (6.44) fully defines the following coding algorithm for a systematic cyclic code:

1. Multiply $u(x)$ by x^{n-k}
2. Divide $x^{n-k}u(x)$ by $g(x)$; the remainder gives the required parity symbols
3. Combine message and parity symbols

Example 6.16

Assuming $\mathbf{u} = (1011)$, find the codevector for the (7,4) systematic cyclic code defined by $g(x) = 1 + x + x^3$. We have

$$x^3u(x) = x^3 + x^5 + x^6$$

Dividing this by $g(x)$ gives remainder one, that is, $p(x) = 1$ and $\mathbf{p} = (100)$. Therefore

$$v(x) = p(x) + x^{n-k}u(x) = 1 + x^3 + x^5 + x^6$$

and $\mathbf{v} = (1001011)$. Similarly, the reader should show that if $\mathbf{u} = (0101)$ then $\mathbf{v} = (1100101)$.

The foregoing theory can be adapted to derive a systematic form for the generator matrix of a cyclic code. First note that $\mathbf{G}(x)$ in (6.33) is composed of codewords of the form $u_i(x)g(x)$, where $u_i(x) = x^i$, $i = 0, 1, \ldots, k-1$. Inserting message polynomial $u_i(x)$ in (6.44) we have

$$\frac{x^{n-k+i}}{g(x)} = a(x) + \frac{p(x)}{g(x)} \tag{6.45}$$

and the corresponding codeword for the systematic code would be, from (6.42)

$$v_i(x) = p(x) + x^{n-k+i}, \quad i = 0, 1, \ldots, k-1 \tag{6.46}$$

It follows that the k basis codewords for a matrix $\mathbf{G}$ that will generate a systematic code can be formed by finding remainder $p(x)$ from (6.45), and then using (6.46) to generate the codeword.

Example 6.17
The generator matrix of a non-systematic (7,4) cyclic code defined by $g(x) = 1 + x + x^3$ is given in (6.34). We can put this into systematic form as follows:

$$i = 3: \quad x^3 + x + 1 \overline{\big) x^6} \quad \text{quotient } x^3 + x + 1$$

$$\begin{array}{l} x^6 + x^4 + x^3 \\ \hline x^4 + x^3 \\ x^4 + x^2 + x \\ \hline x^3 + x^2 + x \\ x^3 + x + 1 \\ \hline p(x) = x^2 + 1 \end{array} \qquad \therefore v_3(x) = x^6 + x^2 + 1$$

Similarly, for

$$i = 2: \quad v_2(x) = x^5 + x^2 + x + 1$$

$$i = 1: \quad v_1(x) = x^4 + x^2 + x$$

$$i = 0: \quad v_0(x) = x^3 + x + 1$$

Finally, writing the matrix of basis polynomials as

$$\mathbf{G}(x) = \begin{bmatrix} v_3(x) \\ v_2(x) \\ v_1(x) \\ v_0(x) \end{bmatrix} \tag{6.47}$$

gives the systematic form of **G** (see also Section 6.2.1):

$$\mathbf{G} = \begin{bmatrix} 1000101 \\ 0100111 \\ 0010110 \\ 0001011 \end{bmatrix} = [\mathbf{I}_k \ \mathbf{P}_{k,\, n-k}] \tag{6.48}$$

Using $\mathbf{v} = \mathbf{uG}$ then gives a codeword of the form

$$\mathbf{v} = [u_0\, u_1\, u_2\, u_3\, (u_0 + u_1 + u_2)(u_1 + u_2 + u_3)(u_0 + u_1 + u_3)] \tag{6.49}$$

In this case the k message bits are placed at the start of the codeword.

6.4.4 Practical systematic encoders

The previous section described a simple algorithm for generating a systematic (n,k,t) cyclic code. In practice, multiplication of $u(x)$ by x^{n-k} can be achieved in a very simple way. Referring to Figure 6.7(d), we note that if we had simply multiplied by x^{10} then the input $u(x)$ would be applied at the output of stage ten, via a modulo-2 adder. The adder output would then be $x^{10}u(x)/g(x)$. Clearly then, multiplication by x^{n-k} can be achieved by applying $u(x)$ to one input of a modulo-2 adder, while the other input is fed from the output of stage $n - k$. The adder output is then fed back in order to perform division by $g(x)$.

Figure 6.8(a) shows a practical systematic encoder for a cyclic (7,4) code. Initially the gate is open (a short circuit) and the switch is in position A. The k message symbols are then shifted into the register (u_{k-1} first) and simultaneously to the channel. As soon as the k message symbols have entered the register the $n - k$ symbols in the register are the remainder, $p(x)$. The gate is then closed (with its output forced to '0'), the switch is placed

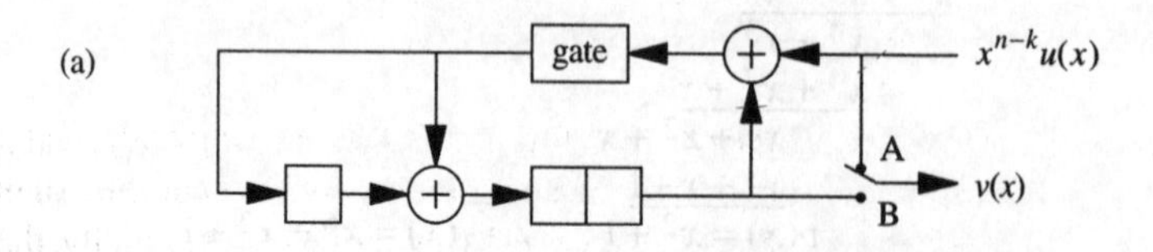

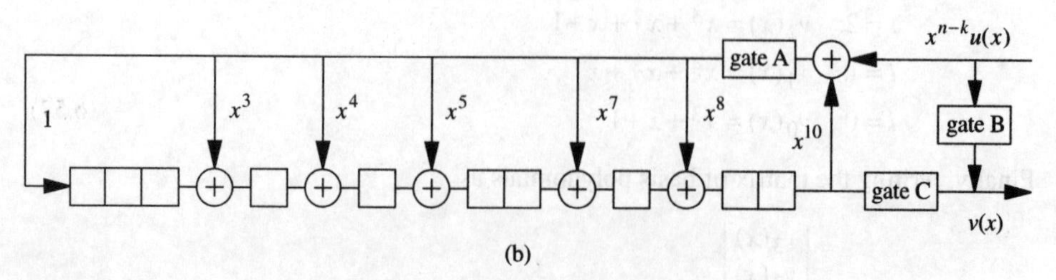

Figure 6.8 *Systematic coding: (a) (7,4) encoder for $g(x) = 1 + x + x^3$, (b) (26,16) encoder for $g(x) = 1 + x^3 + x^4 + x^5 + x^7 + x^8 + x^{10}$.*

in position B, and the parity symbols are clocked out to the channel (symbol p_{n-k-1} first). Similarly, Figure 6.8(b) shows a (26,16) encoder meeting the error protection requirements of the European radio data system (RDS). At the start of each 26-symbol block the shift register is cleared to the 'all 0s' state. After 16 message symbols have been clocked into the encoder and simultaneously to the channel, gates A and B are closed and gate C is opened. The register is then clocked a further ten times to shift the parity symbols to the channel.

The systematic parity check matrix associated with Figure 6.8(a) can be obtained as follows:

$$x^3u(x) = u_0x^3 + u_1x^4 + u_2x^5 + u_3x^6 \tag{6.50}$$

and dividing by $g(x)$ gives the remainder

$$p(x) = (u_0 + u_2 + u_3) + (u_0 + u_1 + u_2)x + (u_1 + u_2 + u_3)x^2 \tag{6.51}$$

$$= p_0 + p_1x + p_2x^2 \tag{6.52}$$

The three (even) parity check equations are therefore

$$p_0 + u_0 + u_2 + u_3 = 0 \tag{6.53}$$

$$p_1 + u_0 + u_1 + u_2 = 0 \tag{6.54}$$

$$p_2 + u_1 + u_2 + u_3 = 0 \tag{6.55}$$

which can be expressed as

$$\begin{bmatrix} 1\,0\,0\,1\,0\,1\,1 \\ 0\,1\,0\,1\,1\,1\,0 \\ 0\,0\,1\,0\,1\,1\,1 \end{bmatrix} \begin{bmatrix} p_0 \\ p_1 \\ p_2 \\ u_0 \\ u_1 \\ u_2 \\ u_3 \end{bmatrix} = [\mathbf{I}_{n-k}\,\mathbf{Q}_{n-k,k}]\mathbf{v}^{\mathrm{T}} = \begin{bmatrix} 0 \\ 0 \\ 0 \end{bmatrix} \tag{6.56}$$

Comparing (6.56) with (6.12) it is clear that (6.56) is a form of (7,4) Hamming code, and it is apparent that cyclic encoders (as in Figure 6.8(a)) are an efficient way of generating such codes. When this is done they are usually referred to as *cyclic* Hamming codes. Finally, the **H** matrix in (6.56) resulted from a particular $g(x)$. If we generalise Figure 6.8(a), the corresponding parity check matrix becomes

$$\mathbf{H} = \begin{bmatrix} 1 & 0 & 0 & \cdots & 0 & s_{00} & s_{10} & \cdots & s_{k-1,0} \\ 0 & 1 & 0 & & 0 & s_{01} & s_{11} & & s_{k-1,1} \\ & & & & \vdots & & & & \\ 0 & 0 & 0 & & 1 & s_{0,n-k-1} & & \cdots & s_{k-1,n-k-1} \end{bmatrix} \tag{6.57}$$

where

$$s_i(x) = \mathrm{Rem}\left[\frac{x^{n-k+i}}{g(x)}\right], \quad i = 0, 1, \ldots, k-1 \tag{6.58}$$

$$= s_{i0} + s_{i1}x + \ldots + s_{i,\,n-k-1}\,x^{n-k-1} \tag{6.59}$$

Example 6.18

Equation (6.57) suggests an alternative and flexible realisation for a systematic encoder (Bota, 1992). For a general systematic (7,4) code, (6.56) and (6.57) give

$$\begin{bmatrix} 1\,0\,0\; s_{00} & s_{10} & s_{20} & s_{30} \\ 0\,1\,0\; s_{01} & s_{11} & s_{21} & s_{31} \\ 0\,0\,1\; s_{02} & s_{12} & s_{22} & s_{32} \end{bmatrix} \begin{bmatrix} p_0 \\ \vdots \\ u_3 \end{bmatrix} = \begin{bmatrix} 0 \\ 0 \\ 0 \end{bmatrix}$$

$$p_0 + u_0 s_{00} + u_1 s_{10} + u_2 s_{20} + u_3 s_{30} = 0$$
$$p_1 + u_0 s_{01} + u_1 s_{11} + u_2 s_{21} + u_3 s_{31} = 0$$
$$p_3 + u_0 s_{02} + u_1 s_{12} + u_2 s_{22} + u_3 s_{32} = 0$$

$$\begin{bmatrix} p_0 \\ p_1 \\ p_2 \end{bmatrix} = u_0 \begin{bmatrix} s_{00} \\ s_{01} \\ s_{02} \end{bmatrix} + u_1 \begin{bmatrix} s_{10} \\ s_{11} \\ s_{12} \end{bmatrix} + u_2 \begin{bmatrix} s_{20} \\ s_{21} \\ s_{22} \end{bmatrix} + u_3 \begin{bmatrix} s_{30} \\ s_{31} \\ s_{32} \end{bmatrix} \tag{6.60}$$

Taking the message **u** = (1011) in Example 6.16, and the corresponding **H** matrix in (6.56), gives

$$\begin{bmatrix} p_0 \\ p_1 \\ p_2 \end{bmatrix} = \begin{bmatrix} 1 \\ 1 \\ 0 \end{bmatrix} + \begin{bmatrix} 1 \\ 1 \\ 1 \end{bmatrix} + \begin{bmatrix} 1 \\ 0 \\ 1 \end{bmatrix} = \begin{bmatrix} 1 \\ 0 \\ 0 \end{bmatrix}$$

that is, **p** = (100), as before. The significant point here is that the parity computation in (6.60) is attractive for *adaptive* FEC. For example, the 's' columns in **H** could be stored in EPROM and different sections of the EPROM used for different codes.

6.4.5 Binary BCH codes

We have used one particular generator polynomial ($g(x) = 1 + x + x^3$) to explain the concept of cyclic code generation. For applications requiring a standard t-error correction code we could refer to a list of generators, as in Table 6.1. This table lists some of the most popular cyclic codes, called BCH codes, after Bose and Chaudhuri (1960) and Hocquenghem (1959). Each generator $g(x)$ is given in octal form for compact representation, and its binary equivalent is interpreted from the right, as in the following examples:

(7,4) code: $g(x) = 13 \equiv 001011 \equiv 1 + x + x^3$

(15,7) code: $g(x) = 721 \equiv 111010001 \equiv 1 + x^4 + x^6 + x^7 + x^8$

Definition 6.3 For any positive integer m ($m \geq 3$) and t, $t < 2^{m-1}$ there exists a binary (n,k,t) BCH code having

Table 6.1 *Generators of primitive BCH codes. (Reprinted with permission from IEEE Trans. Inf. Theory,* ***IT-10****(4), October 1964, p. 391.)*

n	k	t	$g(x)$
7	4	1	13
15	11	1	23
	7	2	721
	5	3	2467
31	26	1	45
	21	2	3551
	16	3	107657
	11	5	5423325
	6	7	313365047
63	57	1	103
	51	2	12471
	45	3	1701317
	39	4	166623567
	36	5	1033500423
	30	6	157464165547
	24	7	17323260404441
	18	10	1363026512351725
	16	11	6331141367235453
	10	13	472622305527250155
	7	15	5231045543503271737
127	120	1	211
	113	2	41567
	106	3	11554743
	99	4	3447023271
	92	5	624730022327
	85	6	130704476322273
	78	7	26230002166130115
	71	9	6255010713253127753
	64	10	1206534025570773100045
	57	11	335265252505705053517721
	50	13	54446512523314012421501421
	43	14	17721772213651227521220574343
	36	15	3146074666522075044764574721735
	29	21	403114461367670603667530141176155
	22	23	123376070404722522435445626637647043
	15	27	22057042445604554770523013762217604353
	8	31	7047264052751030651476224271567733130217
255	247	1	435
	239	2	267543
	231	3	156720665
	223	4	75626641375
	215	5	23157564726421
	207	6	16176560567636227
	199	7	7633031270420722341
	191	8	2663470176115333714567
	187	9	52755313540001322236351

n	k	t	$g(x)$
255	179	10	22624710717340432416300455
	171	11	1541621421234235607706163 0637
	163	12	7500415510075602551574724514601
	155	13	3757513005407665015722506464677633
	147	14	1642130173537165525304165305441011711
	139	15	461401732060175561570722730247453567445
	131	18	215713331471510151261250277442142024165471
	123	19	120614052242066003717210326516141226272506267
	115	21	60526665572100247263636404600276352556313472737
	107	22	22205772322066256312417300235347420176574750154441
	99	23	10656667253473174222741416201574332252411076432303431
	91	25	6750265030327444172723631724732511075 55076272072434456 1
	87	26	110136763414743236435231634307172046 2 06722545273311721317
	79	27	6670003563765750002027034420736617462 101532671176654134235 5
	71	29	24024710520644321515554172112331163205444250362557643221706035
	63	30	1075447505516354432531521735770700366 611172645526761365670254330 1
	55	31	7315425203501100133015275306032054325 414326755010557044426035473617
	47	42	2533542017062646563033041377406233175 123334145446045005066024552543173
	45	43	1520205605523416113110134637642370156 3670024470762313033202157025051541
	37	45	5136330255067007414177447245437530420 735706174323432347644354737403044003
	29	47	3025715536673071465527064012361377115 342242324201174114060254757410403565037
	21	55	1256215257060332656001773153607612103 2273414056530745425211531216144665134737 25
	13	59	4641732005052564544426573714250066004 330677445476561403174677213570261344605005 47
	9	63	1572602521747246320103104325535513461 416236721204407454511276611554770556167751605 7

$$
\left.\begin{aligned} n &= 2^m - 1 \\ n-k &\le mt \\ d &\ge 2t+1 \end{aligned}\right\} \tag{6.61}
$$

In (6.61) t is the designed error correction capability and $2t + 1$ is referred to as the *design distance* of the code. In other words, (6.61) states that these codes will correct any pattern of t or less errors occurring in the block length n. Note, however, that the true or actual minimum distance, d, may be larger than the design distance. This implies that sometimes it is possible to correct *more* than t errors per block using specialised decoding techniques (Blahut, 1983). Also note that BCH codes with $t = 1$ correspond to cyclic Hamming codes.

The parameter m has important algebraic significance since it denotes the particular Galois field associated with a particular code (the reader unfamiliar with Galois fields is referred to Section 6.6.1). Here we are interested in *extension* fields of GF(2), that is, GF(2^m) and it is shown in Section 6.6.1 that

$$
\mathrm{GF}(2^m) = \{0, 1, \alpha, \alpha^2, \alpha^3, \ldots, \alpha^{2^m - 2}\} \tag{6.62}
$$

In words, GF(2^m) comprises a set of 2^m elements and α is the primitive element used to extend the basic field GF(2). The significant point is that a class of (primitive) BCH codes is defined by specifying that $2t$ consecutive powers of α are roots of $g(x)$, as stated by the following theorem:

Theorem 6.5 The generator polynomial $g(x)$ of a t-error correcting BCH code is the lowest degree polynomial over GF(2) which has $\alpha, \alpha^2, \alpha^3, \ldots, \alpha^{2t}$ as its roots, that is, $g(\alpha^i) = 0,\ i = 1, 2, \ldots, 2t$.

Consider the BCH (7,4) code defined by

$$g(x) = 1 + x + x^3; \quad m = 3, \quad t = 1$$

This code is associated with GF(2^3), so any arithmetic must obey the rules of this field. In particular, for GF(2^3) it is shown in Example 6.26 that

$$\alpha^3 = \alpha + 1; \qquad \alpha^6 = \alpha^2 + 1; \qquad \alpha^i + \alpha^i = 0$$

and we can use these relationships to demonstrate Theorem 6.5 for this code. Quite simply, according to the theorem, α and α^2 are roots of $g(x)$, so that

$$\text{for } \alpha: \quad 1 + \alpha + \alpha^3 = 1 + \alpha + (\alpha + 1) = 0$$
$$\text{for } \alpha^2: \quad 1 + \alpha^2 + \alpha^6 = 1 + \alpha^2 + (\alpha^2 + 1) = 0$$

In passing it is interesting to note that, over GF(2), the square of a root is also a root. Thus, since α^2 is a root then α^4 is also a root. Theorem 6.5 is fundamental to both code generation, (the derivation of $g(x)$), and to the decoding of BCH codes. For instance, since each code polynomial is divisible by $g(x)$, that is, $v(x) = a(x)g(x)$, then

$$v(\alpha^i) = a(\alpha^i)g(\alpha^i) = 0, \qquad i = 1, 2, \ldots, 2t \tag{6.63}$$

In words, the elements α^i are also the roots of each code polynomial. This fact is particularly useful in the *algebraic* decoding of BCH codes (Section 6.6).

6.4.6 Shortened cyclic codes

In practice a BCH code is often shortened to give either improved error probability performance or a wordlength suitable for a particular application. As in Section 6.3.4, it is shortened by deleting q message symbols from the basic (n,k,t) codeword. Essentially, we select the 2^{k-q} codewords that have their q leading high-order message digits identical to zero, and then shorten them by removing these leading, zero-valued message digits. This gives an $(n - q, k - q)$ shortened cyclic code comprising 2^{k-q} codewords and having at least the error control capability of the original code.

Example 6.19
Error protection for the European radio data system (RDS) is derived from the basic (341,331) code, and the corresponding 331 × 341 generator matrix is shown in (6.38). According to (6.7), codewords for the basic (n,k) code would be derived as

$$[v_0 v_1, \ldots, v_{n-1}] = [u_0\ u_1, \ldots, u_{k-1}] \begin{bmatrix} g_{11} & g_{12} & \cdots & g_{1n} \\ g_{21} & g_{22} & & g_{2n} \\ & & \vdots & \\ g_{k1} & g_{k2} & \cdots & g_{kn} \end{bmatrix}, \ k = 331, \ n = 341$$

$$v_0 = u_0 g_{11} + u_1 g_{21} +, \ldots, + u_{k-1} g_{k1}$$

$$v_1 = u_0 g_{12} + u_1 g_{22} +, \ldots, + u_{k-1} g_{k2}$$

$$\vdots$$

$$v_{n-1} = u_0 g_{1n} + u_1 g_{2n} +, \ldots, + u_{k-1} g_{kn}$$

Clearly, setting leading symbols $u_0, u_1, \ldots, u_{q-1}$ to zero corresponds to deleting the first q rows of **G**. Also, removing the first q columns in **G** gives the required block length of $n - q$ symbols. For the RDS system we would delete the first $q = 315$ rows and columns in **G** to obtain the required (26,16) shortened cyclic code, and the corresponding non-systematic generator matrix would be of the form (6.38), but with $k = 16$ rows and $n = 26$ columns. According to Section 6.4.3, **G** can be obtained in systematic form by first dividing x^{n-k+i}, $i = 0, 1, \ldots, k-1$ by $g(x)$ (see (6.36)), and then adding x^{n-k+i} to the remainder (see (6.46)). For example, setting $i = 15$, we divide x^{25} by $g(x)$ to give remainder $x^6 + x^5 + x^4 + x^2 + x + 1$ and code polynomial $x^{25} + x^6 + x^5 + x^4 + x^2 + x + 1$ (corresponding to the first row of **G**). Similarly, letting $i = 14$ gives code polynomial $x^{24} + x^9 + x^7 + x^6 + x^5 + x^2 + x + 1$, corresponding to the second row of **G**. Thus, from Section 6.4.3, the systematic form of **G** is

$$\mathbf{G} = \left[\begin{array}{ll}
1000000000000000 & 0001110111 \\
01 & 0101110011 \\
0\ \ 1 & 0111010111 \\
0\ \ \ \ 1 & 0110000101 1 \\
0\ \ \ \ \ \ 1 & 0110101100 1 \\
0\ \ \ \ \ \ \ \ 1 & 0110111000 0 \\
0\ \ \ \ \ \ \ \ \ \ 1 & 0011011100 0 \\
0\ \ \ \ \ \ \ \ \ \ \ \ 1 & 0001101110 0 \\
0\ \ \ \ \ \ \ \ \ \ \ \ \ \ 1 & 0000110111 0 \\
0\ \ \ \ \ \ \ \ \ \ \ \ \ \ \ \ 1 & 0000011011 1 \\
0\ \ \ \ \ \ \ \ \ \ \ \ \ \ \ \ \ \ 1 & 0101100011 1 \\
0\ 1 & 0111011111 1 \\
0\ 1 & 0110000001 1 \\
0\ 1 & 0110101110 1 \\
0 & 101101110010 \\
000000000000000 & 10110111001
\end{array}\right] = [\mathbf{I}_{16}\ \mathbf{P}_{16,10}]$$

The resulting (26,16) code is an optimal burst-error correcting code (Kasami, 1963, 1964) capable of correcting any single error burst spanning 5 bits or less. Finally, we note from (6.35) that a codevector **v** is a linear combination of the rows of **G**, so in this case $\mathbf{v} = u_0\mathbf{v}_1 + u_1\mathbf{v}_2 + \ldots + u_{15}\mathbf{v}_{16}$. If, for example, $\mathbf{u} = (0\ldots011)$ we modulo-2 add rows 15 and 16 to obtain $\mathbf{v} = (0\ldots0111011001011)$, the ten parity bits being at the end of the vector. Similarly, for the all 1s message we would modulo-2 add all rows in **G** to obtain **v**.

Example 6.20
A particular application required a BCH code to protect television lines during 34 Mbit/s DPCM video transmission. For a 15.625 kHz line rate, a digitised television line comprises approximately 2199 bits, so a block length of this order was required. The nearest BCH code corresponds to $m = 12$, that is

$$n = 2^{12} - 1 = 4095$$

and this had to be shortened to approximately 2199. The actual codeword length was restricted by counters to any length less than 4095 which is divisible by 4, so a value of 2196 was selected.
In order to meet the required decoded error rate (Section 5.4) of 10^{-8} for a channel error rate of 10^{-4} it was found necessary to select a code with $t \geq 4$, and $t = 5$ was finally chosen. The corresponding generator polynomial then had a degree

$$n - k = mt = 60$$

so the encoder required a 60-stage linear feedback register. Apart from the larger register, the systematic encoder was similar to those in Figure 6.8. The final (2196,2136) code was found by extensive search of possible codes and was defined by the polynomial $g(x)$ = 101467071546246223445 (octal).

Generally speaking, for any application, one objective is to maximise the code efficiency or rate R, or minimise the coding redundancy $1 - R$. In Example 6.20 the redundancy is an acceptable 2.7%. However, as the channel error rate increases, say to 10^{-2}, BCH coding becomes less efficient and other techniques should be used, such as convolutional codes, or concatenated codes.

6.4.7 Cyclic redundancy check (CRC) codes

Binary BCH codes are the basis of a very common error detection technique for checking the integrity of serial data transfers. The technique is used for example in data communications systems, computer networks, and computer disk systems.

The basic error detection system has identical linear feedback shift registers at the transmitter and receiver (Figure 6.9). The data $u(x)$ is coded as in conventional systematic coding, that is, $u(x)$ is transmitted first and is followed by the parity polynomial $p(x)$. As $p(x)$ enters the receiver it is compared symbol-by-symbol with the check polynomial $p'(x)$ generated at the receiver. Clearly, if there is no transmission error $p'(x) = p(x)$ and the final contents of the receive shift register will be zero. Looked at another way, for error-free transmission the receiver effectively divides the codeword by $g(x)$, and since all codewords are divisible by $g(x)$ the remainder will be zero. Conversely, any remainder indicates a transmission error, in which case an automatic repeat request (ARQ) could be made for a repeat transmission. The significant point is that, given a data transmission error, it is highly unlikely that $p'(x) = p(x)$ due to the fact that the state of the shift registers just after message transmission depends upon the feedback of two long, and different, data sequences. Mathematically, given a transmission error, the final state of the receiver

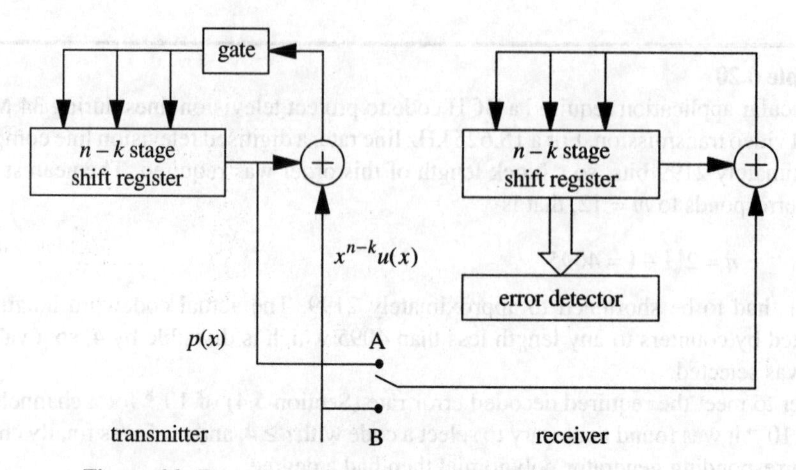

Figure 6.9 *Error detection using a cyclic redundancy check (CRC).*

Table 6.2 *Some standard CRC polynomials.*

Standard	$g(x)$
CRC-8	$1 + x^8$
CRC-12	$1 + x + x^2 + x^3 + x^{11} + x^{12}$
CRC-16	$1 + x^2 + x^{15} + x^{16}$
CRC-CCITT	$1 + x^5 + x^{12} + x^{16}$

register will be zero only when the effective error polynomial is a code polynomial (Section 6.5.3).

The transmitted check vector is called the *CRC character*, and CRC chips usually offer a range of standard polynomials (Table 6.2). The CRC-16 and CRC-CCITT standards, for example, operate with 8-bit or 16-bit characters and transmit a 16-bit CRC character. It is shown in Section 6.5.3 that a BCH (n,k) code can *reliably* detect a burst of errors of length $n - k$ or less, where $g(x)$ is of degree $n - k$. This means that the codes in Table 6.2 can reliably detect bursts up to and including 8, 12 or 16 bits.

6.4.8 Interleaving

Interleaving is an efficient way of using a t-random error correcting code to combat both random and burst errors. Essentially, it converts bursty channels (as in Rayleigh fading channels and compact disc channels) into nearly independent error or Gaussian channels by dispersing burst errors. For example, burst errors are common in mobile radio systems, since signal fades usually occur at a much slower rate than the transmission rate. However, with sufficient interleaving, a mobile radio channel could be assumed to be memoryless.

Interleaving can be implemented using either block or convolutional codes, and Figure 6.10(a) illustrates the principle for an (n,k) code (to simplify notation we assume a systematic codeword of the form $\mathbf{v} = (p_1p_2...p_{n-k}u_1u_2...u_k)$, where u_k is generated first). At the transmitter an $i \times n$ matrix is formed by reading in i codewords row-by-row. The full

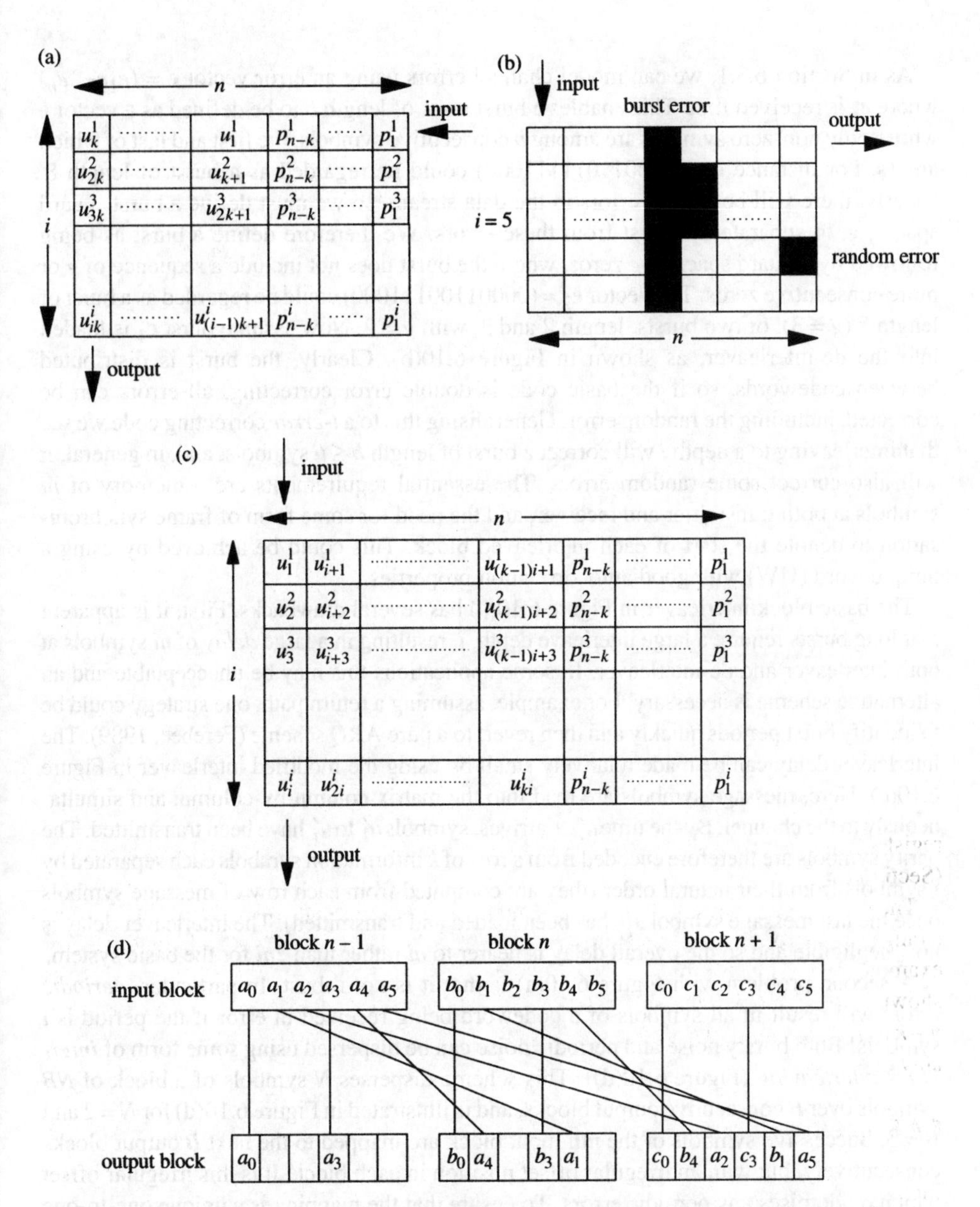

Figure 6.10 *Interleaving: (a) block interleaver for an (n,k) block code; (b) block de-interleaver for i = 5; (c) modified block interleaver; (d) inter-block interleaving, N = 2, B = 3.*

matrix is then transmitted column by column to give an *(in,ik) interleaved code.* At the receiver the data is loaded into a similar matrix column-by-column and then de-interleaved by reading the matrix a row at a time (Figure 6.10(b)). The de-interleaver output is then applied to the FEC decoder.

As in Section 6.3.1, we can model channel errors using an error vector $\mathbf{e} = (e_1 e_2 ... e_n)$ where e_n is received first. This enables a burst error of length b to be defined as a vector $\mathbf{e}$ whose only non-zero symbols are among b consecutive symbols, the first and last of which are 1s. For instance $\mathbf{e}_1 = (0001101111000)$ could be regarded as a burst of length 8. Clearly, there will be further errors in the data stream, so we must define a burst 'guard space', g, to separate the burst from these errors. We therefore define a burst as being followed by a guard space of g zeros, where the burst does not include a sequence of g or more consecutive zeros. The vector $\mathbf{e}_2 = (0000110011000)$ could be regarded as a burst of length 7 ($g = 3$), or two bursts, length 2 and 3, with $g = 2$. Now assume burst $\mathbf{e}_1$ is loaded into the de-interleaver, as shown in Figure 6.10(b). Clearly, the burst is distributed between codewords, so if the basic code is double error correcting, all errors can be corrected, including the random error. Generalising this to a *t-error* correcting code we see that interleaving to a depth i will correct a burst of length $b \leq ti$ symbols, and, in general, it will also correct some random errors. The essential requirements are a memory of ni symbols at both transmitter and receiver, and the need for some form of frame synchronisation to denote the start of each interleaved block. This could be achieved by using a unique word (UW) with good autocorrelation properties.

The basic block interleaver in Figure 6.10(a) has several drawbacks. First, it is apparent that long bursts require a large interleave depth, i, resulting in a large *delay* of ni symbols at both interleaver and de-interleaver. In some applications this may be unacceptable and an alternative scheme is necessary. For example, assuming a return path, one strategy could be to identify burst periods quickly and then revert to a pure ARQ scheme (Ferebee, 1989). The interleaver delay can be made relatively small by using the modified interleaver in Figure 6.10(c). Here, message symbols are read into the matrix column-by-column, and simultaneously to the channel. By the time u'_{i+1} arrives, symbols u'_1 to u'_i have been transmitted. The parity symbols are therefore encoded from a row of k information symbols each separated by i symbols from their natural order (they are computed from each row of message symbols once the last message symbol u^i_{ki} has been loaded and transmitted). The interleaver delay is now negligible and so the overall delay is nearer to ni rather than $2ni$ for the basic system.

A second problem with Figure 6.10(a) is that it is not robust. In particular, *periodic* errors will result in all symbols of a codeword being received in error if the period is i symbols! Both bursty noise and periodic noise can be dispersed using some form of *inter-block interleaving* (Figure 6.10(d)). This scheme disperses N symbols of a block of NB symbols over B consecutive output blocks, and is illustrated in Figure 6.10(d) for $N = 2$ and $B = 3$. Successive symbols of the nth input block are mapped to the next B output blocks consecutively, but with an irregular offset position in each block. It is this irregular offset which randomises any periodic errors. To ensure that the mapping is a unique one-to-one operation, B and N cannot have a common multiple. A similar technique is used for interleaving speech data in the GSM digital cellular radio system (ETSI-GSM, 1991).

Finally, it is worth noting that compact disc systems often use a powerful *cross-interleaving* system to combat burst errors. Typically, an RS encoder is followed by a 'convolutional interleave', formed by delays which are functions of the symbol position within the coded block. The interleaver is then followed by a second RS encoder before data recording. In this way, codewords are formed both before and after the interleave process. Upon detection of multiple errors, the first RS decoder passes on erasure flags to the second RS decoder via a de-interleaver.

6.5 Non-algebraic decoding of cyclic codes

Decoding techniques for (n,k,t) cyclic codes divide into two general categories: algebraic and non-algebraic. Algebraic techniques use linear algebra and Galois field arithmetic to solve for the *error locations* (and *error values* in the case of non-binary codes). The simpler non-algebraic techniques (such as *Meggitt* decoding and *threshold* decoding) use non-algebraic determination of the error pattern, and are suitable for relatively short-length BCH codes and for modest values of t, say $t \leq 3$. Simplifications of the basic Meggitt decoder are useful for burst detection and burst-error correction (error trapping).

6.5.1 Meggitt decoding

A general objective in the decoding of error control codes is that of *syndrome calculation.* This is true for the Hamming codes in Section 6.3.1, the algebraic techniques in Section 6.6, and in some forms of convolutional decoder (threshold decoding). It is also true for non-algebraic decoding. Following the approach in Section 6.3.1, the received codeword in polynomial form is

$$\begin{aligned} r(x) &= r_0 + r_1 x + r_2 x^2 + \ldots + r_{n-1} x^{n-1} \\ &= v(x) + e(x) \end{aligned} \tag{6.64}$$

Here, code polynomial $v(x)$ is defined by (6.30) and

$$e(x) = e_0 + e_1 x + e_2 x^2 + \ldots + e_{n-1} x^{n-1} \tag{6.65}$$

is the *error polynomial* such that $e_i = 1$ iff symbol r_i is received in error. The syndrome polynomial

$$s(x) = s_0 + s_1 x + \ldots + s_{n-k-1} x^{n-k-1} \tag{6.66}$$

is generated by dividing $r(x)$ by $g(x)$, and since all codewords are divisible by $g(x)$ we can write

$$\frac{r(x)}{g(x)} = a(x) + \frac{e(x)}{g(x)} \tag{6.67}$$

The required syndrome $s(x)$ is the remainder resulting from the division $e(x)/g(x)$:

$$s(x) = \text{Rem}\left[\frac{e(x)}{g(x)}\right] \equiv e(x) \text{ modulo } g(x) \tag{6.68}$$

and clearly, $s(x) = 0$ if $e(x) = 0$. As in the encoding of cyclic codes, we use a linear feedback $n - k$ stage shift register to perform the division by $g(x)$, and Figure 6.11 shows the basic circuit.

The syndrome register (SR) is initially reset and $r(x)$ is applied to the input (symbol r_{n-1} first). After the entire polynomial $r(x)$ has been clocked in, the SR contains the syndrome $s(x)$ of $r(x)$. Since $r(x)$ may have up to t correctable errors, we might reasonably wonder how a single syndrome $s(x)$ can correct all t errors! In fact, error correction is performed over a number of clock cycles using a number of different syndromes, and is based upon the following cyclic decoding theorem:

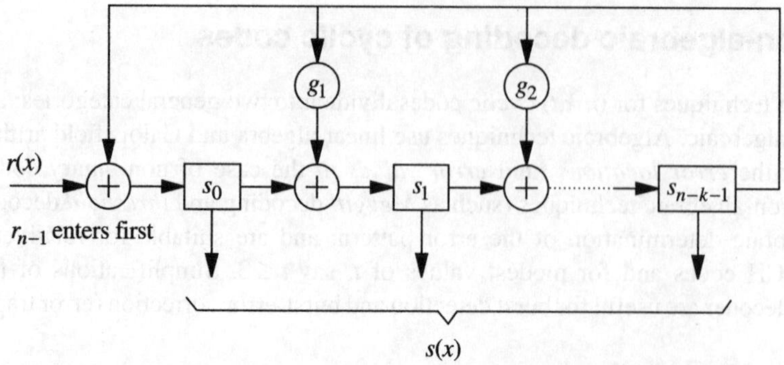

Figure 6.11 *Syndrome calculation for the Meggitt decoder.*

Theorem 6.6 If $s(x)$ is the syndrome of $r(x)$, then the remainder $s'(x)$ resulting from dividing $xs(x)$ by $g(x)$ is the syndrome of $r'(x)$, which is a cyclic shift of $r(x)$ (the division of $xs(x)$ by $g(x)$ is achieved by shifting the SR once with $s(x)$ as the initial contents).

This theorem states that, given the initial syndrome $s(x)$ of $r(x)$, then repeated clocking of the SR with the input open circuit will generate syndromes corresponding to cyclically shifted versions of $r(x)$. If then $r(x)$ is held in an n-symbol (circulating) buffer, we can decode those syndromes corresponding to an error at the end of the buffer and so perform *symbol-by-symbol* decoding. The basic Meggitt decoder for a binary BCH code therefore requires an n-stage buffer, an $n - k$ stage SR, and syndrome detection logic, as shown in Figure 6.12. Note that in practice it is not essential to shift the buffer *cyclically*.

The decoder operates as follows:

1. With gate 1 open (a short circuit) and gate 2 closed the entire $r(x)$ is clocked into the buffer and simultaneously into the SR. When symbol r_{n-1} occupies the last stage of the buffer (as shown), the SR contains $s(x)$ corresponding to $r(x)$.

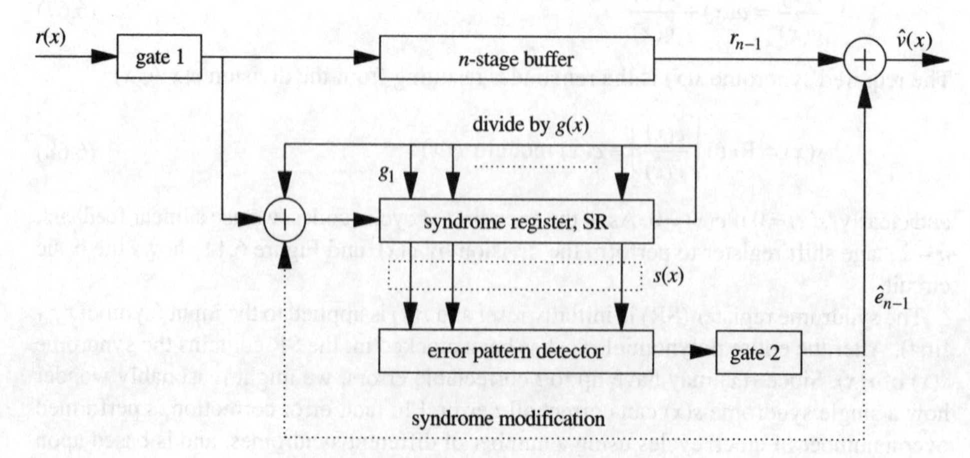

Figure 6.12 *A Meggitt decoder for cyclic codes.*

2. Symbol-by-symbol error correction is now performed with gate 1 closed and gate 2 open. If r_{n-1} is not in error, then both the SR and the buffer are shifted once simultaneously giving $s'(x)$ corresponding to $r'(x)$ (assuming the theoretical *cyclic* shift of $r(x)$). If $s(x)$ corresponds to a correctable error pattern (t or fewer errors in $r(x)$), with an error in symbol r_{n-1}, then the pattern detector generates $\hat{e}_{n-1} = 1$ to complement (correct) r_{n-1}. The corrected polynomial is

$$r_1(x) = r_0 + r_1 x + \ldots + r_{n-2}\, x^{n-2} + (r_{n-1} + \hat{e}_{n-1})\, x^{n-1} \tag{6.69}$$

and a cyclic shift right would give

$$r_1'(x) = (r_{n-1} + \hat{e}_{n-1}) + r_0 x + \ldots + r_{n-2}\, x^{n-1} \tag{6.70}$$

3. Theoretically, if r_{n-1} is in error, the true syndrome corresponding to $r'(x)$ is obtained as $s_1'(x) = s'(x) + 1$, that is, we must add a 1 to the SR input while it is shifted. Modifying the SR in this way removes the effect of the error in r_{n-1} from $s(x)$.
4. Symbol r_{n-2} is now at the end of the buffer and can be corrected if necessary using the new $s(x)$, i.e. $s'(x)$ or $s_1'(x)$. Decoding stops when $r(x)$ has been clocked out of the buffer.

In practice syndrome modification is not essential and is therefore drawn dotted in Figure 6.12. When it is used, the SR will be zero at the end of decoding, and when it is omitted the SR will generally be non-zero after decoding, although decoding will still be correct. Also note that the basic circuit in Figure 6.12 is non-real-time since only alternate words are decoded. A real-time decoder could be realised using two n-stage buffers, or dual decoders.

Example 6.21

Suppose we require a Meggitt decoder for the (15,11) cyclic Hamming code defined by $g(x) = 1 + x + x^4$. The single error can occur in any one of 15 positions, corresponding to $e(x) = 1, x, x^2, \ldots, x^{14}$. As the buffer is clocked the error eventually ends up at the end of the buffer, corresponding to $e(x) = e^{14}$, and in this position it can be corrected. Since $s(x) = e(x)$ modulo $g(x)$ we perform the following division over GF(2):

$$\begin{array}{r} x^{10} + x^7 + x^6 + x^4 + x^2 + x + 1 \\ x^4 + x + 1 \overline{) \; x^{14} \qquad\qquad\qquad\qquad\qquad\qquad} \\ \underline{x^{14} + x^{11} + x^{10}} \qquad\qquad\qquad \\ x^{11} + x^{10} \qquad\qquad\qquad \\ \underline{x^{11} + x^8 + x^7} \qquad\qquad \\ x^{10} + x^8 + x^7 \qquad \\ \underline{x^{10} + x^7 + x^6} \qquad \\ x^8 + x^6 \qquad \\ \vdots \qquad\qquad \\ x^3 + 1 = s(x) \end{array}$$

$$\therefore\; \mathbf{s} = (s_0\, s_1\, s_2\, s_3) = (1001)$$

The pattern detector must therefore generate $\hat{e}_{n-1} = 1$ for this particular vector, and the basic circuit is shown in Figure 6.13. Note that the cyclic structure of the code results in a simpler decoder circuit when compared with a (15,11) decoder derived using the approach

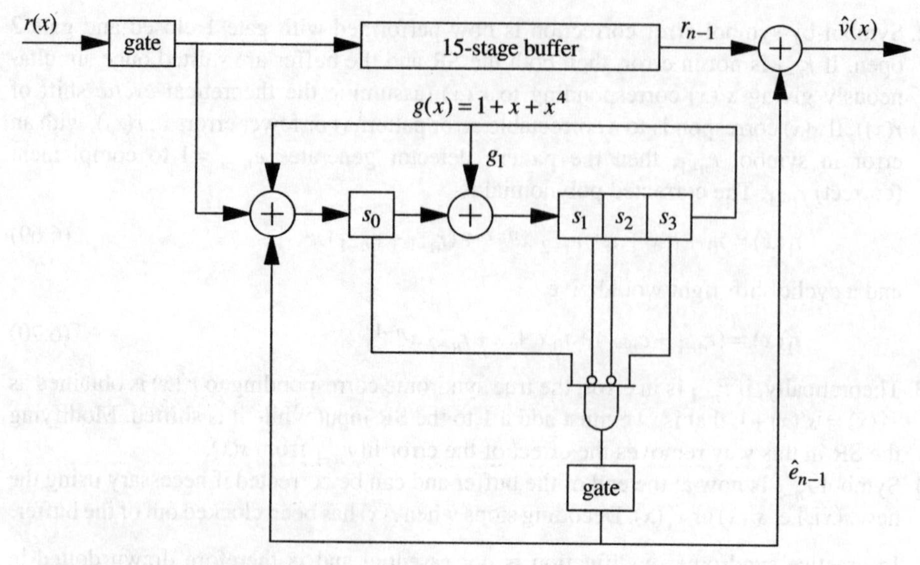

Figure 6.13 *Meggitt decoder for the BCH(15,11) code.*

in Example 6.5. On the other hand, decoding now takes longer, since it is carried out in a serial rather than a parallel manner.

Example 6.22

The decoder for the DEC BCH(15,7) code is more complex since there are significantly more correctable error patterns. The simplest case corresponds to a single error located at r_{n-1}, in which case $s(x) = x^{14}$ modulo $g(x)$, where

$$g(x) = 1 + x^4 + x^6 + x^7 + x^8$$

However, there are 14 other correctable error patterns, namely

$$e(x) = \begin{cases} x^{13} + x^{14} \\ x^{12} + x^{14} \\ \vdots \\ 1 + x^{14} \end{cases}$$

For example, when adjacent errors occur with one error in symbol r_{n-1} the syndrome is

$$\begin{aligned} s(x) &= (x^{13} + x^{14}) \text{ modulo } g(x) \\ &= x^2 + x^3 + x^4 + x^7 \\ \mathbf{s} &= (00111001) \end{aligned}$$

This is one of 15 possible syndrome vectors that must generate $\hat{e}_{n-1} = 1$, and so a small PROM could be used for error pattern detection, as shown in Figure 6.14. Clearly, the complexity of the pattern detector will rapidly increase with t, and this is probably the main

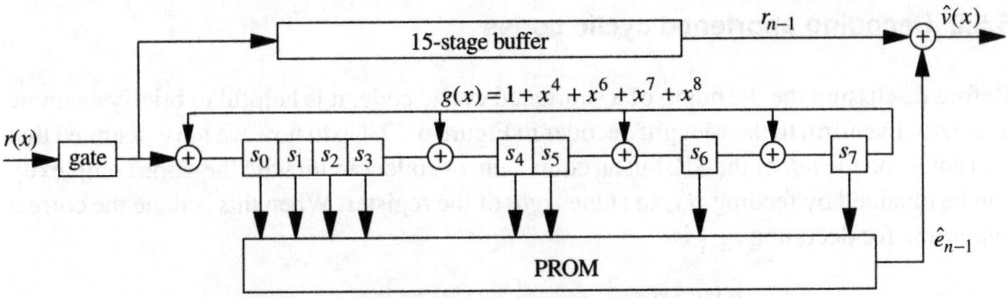

Figure 6.14 *Meggitt decoder for the BCH(15,7) code.*

limitation of the Meggitt decoder. The simulation in Appendix B is useful for examining the operation of the Meggitt decoder under random and burst errors.

Example 6.23
Assume a BCH(31,21) codeword is generated in systematic form for the message $u(x) = 1 + x^5$, and let the hard-decision input to the decoder correspond to

$$r(x) = x^2 + x^4 + x^6 + x^9 + x^{15}$$

From Table 6.1 we have

$$g(x) = 1 + x^3 + x^5 + x^6 + x^8 + x^9 + x^{10}$$

Using (6.44) to obtain the parity polynomial requires the following division:

$$\begin{array}{r} x^5 + x^4 + x^2 \\ x^{10} + x^9 + x^8 + x^6 + x^5 + x^3 + 1 \,\overline{\big)\, x^{15} + x^{10}} \\ \underline{x^{15} + x^{14} + x^{13} + x^{11} + x^{10} + x^8 + x^5} \\ x^{14} + x^{13} + x^{11} + x^8 + x^5 \\ \vdots \end{array}$$

giving remainder $p(x) = x^2 + x^4 + x^9$. The code polynomial is therefore

$$v(x) = x^2 + x^4 + x^9 + x^{10} + x^{15}$$

It follows that

$$\begin{aligned} e(x) &= r(x) + v(x) \quad \text{modulo-2} \\ &= x^6 + x^{10} \end{aligned}$$

and positions 7 and 11 of the decoder must be in error. The first syndrome polynomial generated by the Meggitt decoder is

$$s(x) = \text{Rem}\left[\frac{x^6 + x^{10}}{g(x)}\right] = 1 + x^3 + x^5 + x^8 + x^9$$

6.5.2 Decoding shortened cyclic codes

Before discussing the decoding of a shortened cyclic code, it is helpful to briefly examine an alternative form to the Meggitt decoder in Figure 6.12. Up to now we have assumed that $r(x)$ enters on the *left* of the SR, but an equivalent decoder circuit with the same complexity can be obtained by feeding $r(x)$ in at the *right* of the register. When this is done the correct syndrome for decoding r_{n-1} is

$$s(x) = \text{Rem}\left[\frac{x^{n-k}r(x)}{g(x)}\right] = \text{Rem}\left[\frac{x^{n-k}e(x)}{g(x)}\right] \tag{6.71}$$

that is, feeding $r(x)$ into the right of the SR premultiplies $e(x)$ by x^{n-k}.

Example 6.24

Suppose we require the decoder for the (7,4) cyclic code defined by $g(x) = 1 + x + x^3$. As before, an error will be correctable when it is at the end of the buffer register (when $e_{n-1} = 1$ or $e(x) = x^{n-1}$). In this case, due to entry from the right, such an error corresponds to $e(x) = x^9$ and the required syndrome is $s(x) = x^9$ modulo $g(x)$. Long division gives $s(x) = x^2$ or $\mathbf{s} = (001)$, and the corresponding circuit (without gating) is shown in Figure 6.15(a). Note that

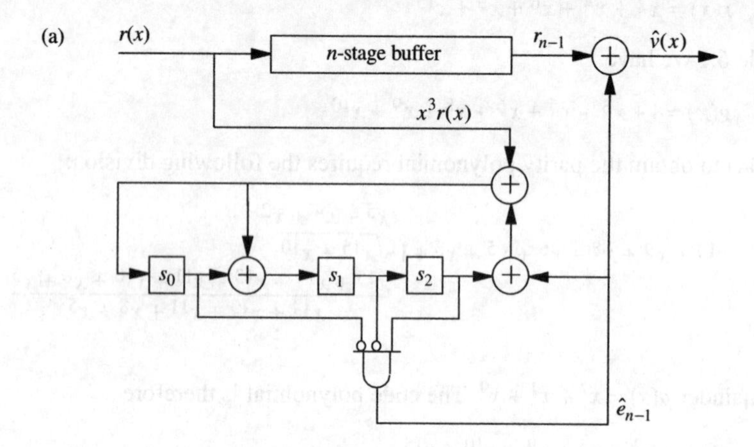

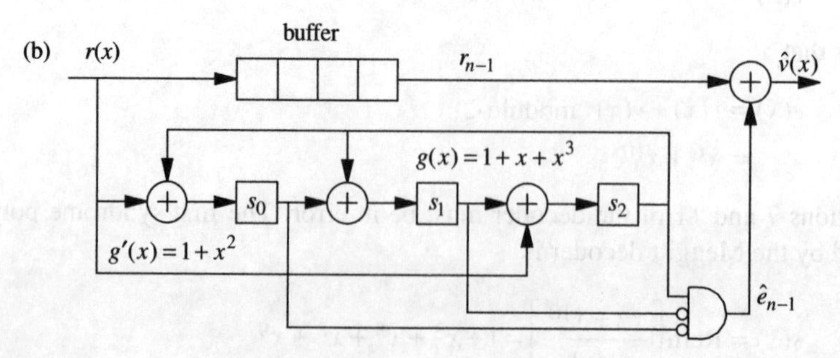

Figure 6.15 *(a) Meggitt decoder for a (7,4) cyclic code, with r(x) entered at the right of the SR; (b) decoder for the shortened (4,1) cyclic code.*

syndrome modification is achieved by feeding $\hat{e}_{n-1} = 1$ into the right-hand side of the register via modulo-2 addition.

Now suppose we have designed the pattern detector logic for a particular (n,k) cyclic code, with $r(x)$ entering at the right of the SR. If this code is now shortened to an $(n-q, k-q)$ cyclic code, that is,

$$r(x) = r_0 + r_1 x + \ldots + r_{n-q-1} x^{n-q-1}$$

then the correct syndrome for decoding r_{n-q-1} using the same detector is obtained by effectively restoring the degree of $r(x)$ to $n-1$. Mathematically

$$s(x) = \text{Rem}\left[\frac{x^{n-k+q}\, r(x)}{g(x)}\right] \tag{6.72}$$

To achieve this, it would be necessary to fully enter $r(x)$ into the $n-k$ stage SR, and then to cyclically shift the SR a further q times. Unfortunately, when q is large (see Example 6.19) this extra clocking results in significant decoding delay. Fortunately, it can be shown that (6.72) can be expressed in the form

$$s(x) = \text{Rem}\left[\frac{g'(x)\, r(x)}{g(x)}\right] = \text{Rem}\left[\frac{g'(x) e(x)}{g(x)}\right] \tag{6.73}$$

where

$$g'(x) = \text{Rem}\left[\frac{x^{n-k+q}}{g(x)}\right] \tag{6.74}$$

This states that, using pattern detection logic based upon right-hand entry (equation (6.72)), the correct syndrome for r_{n-q-1} can be obtained by simply pre-multiplying $r(x)$ by $g'(x)$, and the q cyclic shifts of the SR are no longer required. Note that the buffer is shortened to length $n-q$, and, according to (6.73), $r(x)$ now enters the SR on the left.

For a simple illustration, suppose we shorten the (7,4) code in Example 6.24 to the (4,1) cyclic code by deleting $q = 3$ information symbols. In this case

$$g'(x) = \text{Rem}\left[\frac{x^6}{1+x+x^3}\right] = x^2 + 1$$

As usual, the decoder performs correction when the erroneous symbol is at the end of the $n-q$ stage buffer, that is, when $e(x) = x^{n-q-1} = x^3$. Therefore

$$s(x) = \text{Rem}\left[\frac{(x^2+1)x^3}{1+x+x^3}\right] = x^2$$

This means that the pattern detection logic should give a correcting '1' when s = (001). The corresponding decoder is shown in Figure 6.15(b).

Example 6.25

RDS decoding is based upon a shortened (26,16) cyclic code with $q = 315$ and $g(x) = 1 + x^3 + x^4 + x^5 + x^7 + x^8 + x^{10}$ (see Example 6.19). Therefore,

$$g'(x) = x^{325} \text{ modulo } g(x) = 1 + x + x^3 + x^4 + x^8 + x^9$$

The syndrome calculation in (6.73) is therefore performed by the 10-stage shift register circuit in Figure 6.7(d).

6.5.3 Burst detection

Besides random error correction, an (n,k) cyclic code can also detect *burst errors*, as defined in Section 6.4.8. More specifically, we now show that any cyclic code defined by a generator of degree $n - k$ detects any burst $b \le n - k$. According to (6.68), the syndrome $s(x)$ is guaranteed to be non-zero if the degree of $e(x)$ is less than that of $g(x)$. In other words, when $e(x)$ has degree $n - k - 1$ or less (corresponding to a burst length $b \le n - k$) the division of $r(x)$ by $g(x)$ will always give a non-zero syndrome. A burst of length $b \le n - k$ can therefore be detected by examining the SR (via an OR-gate) for at least one non-zero bit after division. For example, the BCH(15,11) code defined by $g(x) = 1 + x + x^4$ could detect the burst

$$e(x) = 1 + x + x^2 + x^3; \qquad \mathbf{e} = (11110000\ldots)$$

Similarly, a shifted version of $e(x)$ is also non-divisible by $g(x)$ and so is detectable. In fact, most bursts of length $b > n - k$ will be detected, and detection will only fail when $e(x)$ is of the form $e(x) = a(x)g(x)$, that is, when $e(x)$ is a codeword. For example

$$e(x) = (x + x^3)(1 + x + x^4) = x + x^2 + x^3 + x^4 + x^5 + x^7$$
$$\mathbf{e} = (01111101000\ldots)$$

will go undetected by the BCH(15,11) code since $e(x)$ is divisible by $g(x)$. On this analysis, since there are 2^k codewords and very nearly 2^n possible n-bit error patterns, the probability of an undetected error is very nearly $2^k/2^n = 2^{-(n-k)}$. In practice, efficient high-rate, long block length codes tend to be used, with enough parity digits to give an acceptably low probability of undetected error. For example, using the BCH(255,231) code this probability is approximately 2^{-24} or 6×10^{-8}. Once a burst has been detected a request can be made for retransmission (ARQ).

6.5.4 Burst correction (error trapping)

The Meggitt decoder can also be simplified to the form of an *error-trapping* decoder. This approach replaces the possibly complex error pattern detector with simple combinational logic and can be applied to both random error correction and burst error correction. The latter application will be considered here.

Suppose we wish to correct a single burst error of b bits, or fewer, using a form of Meggitt decoder. For an (n,k) code there exists a large class of burst-error correcting cyclic codes with the restriction that

$$b \le \left\lfloor \frac{n-k}{2} \right\rfloor \tag{6.75}$$

The upper limit is called the *Reiger bound* and codes meeting this are said to be optimum. Optimum shortened cyclic codes for burst-error correction are described by Kasami (1963). Referring to Kasami's results, the (15,9) shortened cyclic code has $g(x) = 1 + x + x^2 + x^3 + x^6$ and $b = 3$, while the (27,17) shortened cyclic code has $g(x) = 1 + x^3 + x^4 + x^5 + x^7 + x^8 + x^{10}$ and $b = 5$. The latter code is the basis of the burst-error correction used in the RDS system (see Figure 6.7(d)).

A form of Meggitt decoder suitable for single burst correction is shown in Figure 6.16. At the start of each n-bit block the k message bits of the systematic codeword are fed into the syndrome and buffer registers (gates G1 and G2 open). Gate G2 is then closed and the $n - k$ parity bits are fed into the SR. At this point, the syndrome $s(x)$ corresponds to the received polynomial $r(x)$, where

$$s(x) = s_0 + s_1 x + s_2 x^2 + \ldots + s_{n-k-1}\, x^{n-k-1}$$
$$r(x) = r_0 + r_1 x + r_2 x^2 + \ldots + r_{n-k-1} x^{n-k-1} + r_{n-k} x^{n-k} + \ldots + r_{n-1} x^{n-1}$$

and r_0 to r_{n-k-1} are parity bits. With G1 open, the SR is then clocked i times, corresponding to a cyclic shift of $r(x)$. The subsequent detailed operation is somewhat tedious, but we can cite several cases for illustration.

First, suppose that up to b errors are confined to the b high-order parity positions, that is

$$e(x) = x^{n-k-b} + \ldots + x^{n-k-1}$$

In this case, since the degree of $e(x)$ is less than that of $g(x)$, and since $s(x) = e(x)$ modulo $g(x)$, then $s(x) = e(x)$. In other words, when $r(x)$ has been clocked in, *the b high-order bits of the SR match the error pattern and bits* $s_0 s_1 \ldots s_{n-k-b-1}$ *must be zero.* Of course, in general, the b errors will not span $r_{n-k-b} \ldots r_{n-k-1}$ of $r(x)$, but will be located elsewhere. To illustrate

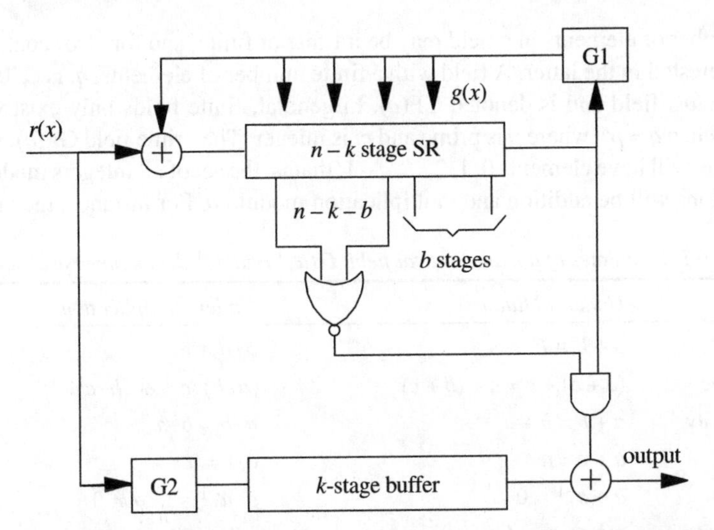

Figure 6.16 *Error trapping decoder for a b-burst-error correcting cyclic code.*

the point, suppose information bits $r_{n-b}...r_{n-2}r_{n-1}$ are in error. According to the cyclic decoding theorem (Theorem 6.6) we can clock the SR $i = n - k$ times to give a syndrome $s'(x)$ corresponding to a received polynomial $r'(x)$ which has errors spanning r'_{n-k-b} to r'_{n-k-1}. Now, as before, the b high-order bits of $s'(x)$ match the actual error pattern and the $n - k - b$ low-order bits of $s'(x)$ will be zero. In other words, *the b rightmost bits in the SR match the errors in* $r_{n-b}...r_{n-2}r_{n-1}$, so clocking both the SR (with G1 closed) and the buffer will correct the burst bit-by-bit.

As a second illustration, if the b errors are confined to the b low-order parity positions, corresponding to $e(x) = 1 + ... + x^{b-1}$, then $i = n - k - b$ clocks of the SR are required to ensure $s_0s_1...s_{n-k-b-1}$ are zero (recall that $s(x) = e(x)$). Thus, if after $0 \le i \le n - k - b$ clocks $s_0s_1...s_{n-k-b-1}$ become zero we conclude that the burst is in the parity bits and the k-bit buffer can be clocked out without correction.

6.6 Algebraic decoding of cyclic codes

In this section we discuss several powerful decoding techniques which are applicable to word-orientated codes (RS codes) as well as to binary cyclic codes. They rely heavily upon Galois field theory to find *algebraic* solutions for the *error locations* and also, in the case of RS codes, the *error values*. We therefore review aspects of Galois field theory which are relevant to algebraic decoding.

6.6.1 Basic Galois field theory

A mathematical field, F, is a system having two operations, both with inverses. Usually these are addition (with its inverse, subtraction) and multiplication (with its inverse, division). If a, b and c are elements of F, that is, $a, b, c \in F$, then the postulates or laws of a field are defined as in Table 6.3. The identity postulate indicates that every field must have at least two elements: the additive identity element, 0, and the multiplicative identity element, 1.

The number of elements in a field may be infinite or finite, and for error control coding we are interested in the latter. A field with a finite number of elements, q, is called a finite field or *Galois* field and is denoted GF(q). In general, finite fields only exist when q is prime, or when $q = p^m$ where p is prime and m is integer. The prime field GF(q), with $q > 1$ and q prime, will have elements 0, 1, 2, ..., $q - 1$, that is, the set of all integers modulo q, and the operations will be addition and multiplication modulo q. For instance, the prime field

Table 6.3 *Postulates of a mathematical field, F (a_+^{-1} and $a_\times^{-1}$ denote inverse elements).*

	Under addition	*Under multiplication*
Closure	$a + b \in F$	$a \cdot b \in F$
Associative	$(a + b) + c = a + (b + c)$	$(a \cdot b) \cdot c = a \cdot (b \cdot c)$
Commutivity	$a + b = b + a$	$a \cdot b = b \cdot a$
Identity	$a + 0 = a$	$a \cdot 1 = a$
Inverse	$a + a_+^{-1} = 0$	$a \cdot a_x^{-1} = 1, \; a \neq 0$
Distributivity	$a(b + c) = ab + ac$	$a(b + c) = ab + ac$

GF(5) will have elements {0,1,2,3,4} and 3 + 4 = 2 modulo-5. Similarly, over the same field, $2 \times 3 = 1$ modulo-5. Clearly, the simplest prime field, GF(2), has elements {0,1} and all arithmetic is modulo-2 (exclusive-OR). We used this arithmetic when discussing the coding and decoding of Hamming codes and cyclic codes.

Field GF(p^m), with p prime and integer $m > 1$, is called an *extension* field of GF(p), and p is the *characteristic*. Extension fields with characteristic 2 are particularly relevant to binary BCH codes and to the RS codes. One attraction is that data sources are usually binary and so can be neatly represented in GF(2^m); for example, an RS(n,k) code has n symbols and each will be m bits wide and correspond to one of 2^m field elements. Another attraction is that arithmetic is simplified; in particular, for GF(2^m) addition and subtraction are identical, so there are no 'carries' as in conventional arithmetic (note that addition and subtraction are identical only for fields of characteristic 2). In order to generate a field GF(2^m) we extend the elements 0, 1 using a primitive element α. If

$$\alpha \in GF(2^m)$$

then the closure postulate under multiplication means that $\alpha\alpha = \alpha^2, \alpha\alpha^2 = \alpha^3$ (and so on) are also elements of GF(2^m). Therefore,

$$GF(2^m) = \{0, 1, \alpha, \alpha^2, \alpha^3, \ldots\} \tag{6.76}$$

By definition, GF(2^m) must be finite, and this is achieved by associating it with a polynomial $p(x)$ of degree m which is *primitive* over GF(2) (see Definition 4.1). A list of primitive polynomials is given in Table 4.4. For instance, taking the first entry for $n = 8$ in this table, the polynomial $p(x) = x^8 + x^4 + x^3 + x^2 + 1$ could be used to construct GF(2^8) or GF(256). The field GF(2^m) is limited to just 2^m elements by setting

$$p(\alpha) = 0 \tag{6.77}$$

The following example illustrates how (6.77) limits the field and effectively defines the addition and multiplication rules for the elements of GF(2^m).

Example 6.26
Consider GF(2^3). From Table 4.4 we select a primitive polynomial of degree 3, such as $p(x) = 1 + x + x^3$. Using (6.77) this gives the identity

$$\alpha^3 + \alpha + 1 = 0 \tag{6.78}$$

To generate the field we also note that for GF(2^m) each element is its own additive inverse, so that from Table 6.3 we can state $\alpha^i + \alpha^i = 0$. This simply restates our previous discussion, namely, that addition and subtraction are identical over GF(2^m). Using (6.78) we can now write

$$\alpha^3 = \alpha + 1$$

$$\alpha^4 = \alpha\alpha^3 = \alpha^2 + \alpha$$

$$\alpha^5 = \alpha\alpha^4 = \alpha^3 + \alpha^2 = \alpha^2 + \alpha + 1$$

$$\alpha^6 = \alpha\alpha^5 = \alpha^3 + \alpha^2 + \alpha = \alpha^2 + 1$$

Note in particular that

Table 6.4 *Representations for GF(2^3) generated by p(x) = 1 + x + x^3.*

Power representation	*Polynomial representation*	*Vector representation*
0	0	(000)
1	1	(100)
α	α	(010)
α^2	α^2	(001)
α^3	$1 + \alpha$	(110)
α^4	$\alpha + \alpha^2$	(011)
α^5	$1 + \alpha + \alpha^2$	(111)
α^6	$1 + \alpha^2$	(101)

$$\alpha^7 = \alpha\alpha^6 = \alpha^3 + \alpha = 1 \tag{6.79}$$

This means that $\alpha^8 = \alpha,\ \alpha^9 = \alpha^2$ etc., and that the elements of GF(2^3) are restricted to

$$\mathrm{GF}(2^3) = \{0, 1, \alpha, \alpha^2, \alpha^3, \alpha^4, \alpha^5, \alpha^6\}$$

The full addition and multiplication rules for GF(2^3) are now easily derived. For example, $\alpha^5 + \alpha^6 = \alpha^2 + \alpha + 1 + \alpha^2 + 1 = \alpha$ and $\alpha^3\alpha^6 = \alpha^9 = \alpha^7\alpha^2 = \alpha^2$ modulo-$p(\alpha)$. In practice, addition is easily accomplished by using a vector representation for each element (Table 6.4). We see that GF(2^3) comprises all 3-bit binary numbers, and addition is simply component-wise addition over GF(2). Thus, $\alpha^3 + \alpha^5 \equiv (110) + (111)$ modulo-2 = (001) $\equiv \alpha^2$. Galois field addition could then be done by bit-wise modulo-2 addition of the contents of two binary registers. Also note from (6.68) that a polynomial $r(\alpha)$ can be reduced modulo-$p(\alpha)$ by division, that is

$$r(\alpha) \text{ modulo } p(\alpha) = \mathrm{Rem}\left[\frac{r(\alpha)}{p(\alpha)}\right] \tag{6.80}$$

For example, if $r(\alpha) = \alpha^6$ the remainder will be $\alpha^2 + 1$. Finally, generalising (6.79), and using (6.61)

$$\alpha^{2^m - 1} = \alpha^n = 1 \tag{6.81}$$

$$\mathrm{GF}(2^m) = \{0, 1, \alpha, \alpha^2, \ldots, \alpha^{2^m - 2}\} \tag{6.82}$$

Example 6.27

The BCH(15,7) code is defined on GF(2^4) (since, from (6.61), $15 = 2^4 - 1$). The field can therefore be constructed from primitive polynomial $p(x) = 1 + x + x^4$ (see Table 4.4), and various forms are shown in Table 6.5. Suppose the message is $u(x) = 1 + x^3 + x^4 + x^5$. Using Table 6.1 we have $g(x) = 1 + x^4 + x^6 + x^7 + x^8$ and the corresponding systematic code polynomial is

$$v(x) = x + x^7 + x^8 + x^{11} + x^{12} + x^{13}$$

Table 6.5 *Representations for GF(2^4) generated by p(x) = 1 + x + x^4.*

Power representation	*Polynomial representation*	*Vector representation*
0	0	(0000)
1	1	(1000)
α	α	(0100)
α^2	α^2	(0010)
α^3	α^3	(0001)
α^4	$1 + \alpha$	(1100)
α^5	$\alpha + \alpha^2$	(0110)
α^6	$\alpha^2 + \alpha^3$	(0011)
α^7	$1 + \alpha + \alpha^3$	(1101)
α^8	$1 + \alpha^2$	(1010)
α^9	$\alpha + \alpha^3$	(0101)
α^{10}	$1 + \alpha + \alpha^2$	(1110)
α^{11}	$\alpha + \alpha^2 + \alpha^3$	(0111)
α^{12}	$1 + \alpha + \alpha^2 + \alpha^3$	(1111)
α^{13}	$1 + \alpha^2 + \alpha^3$	(1011)
α^{14}	$1 + \alpha^3$	(1001)

Now, according to (6.63), $v(\alpha) = 0$ since α is a root of any code polynomial. We can check this for $v(x)$ using Galois field addition and Table 6.5:

$$\begin{array}{r} \alpha: 0\,1\,0\,0 \\ \alpha^7: 1\,1\,0\,1 \\ \alpha^8: 1\,0\,1\,0 \\ \alpha^{11}: 0\,1\,1\,1 \\ \alpha^{12}: 1\,1\,1\,1 \\ \alpha^{13}: 1\,0\,1\,1 \\ \hline 0\,0\,0\,0 \end{array} \quad \text{modulo-2}$$

The reader should show that the same result is obtained if we use (6.80) to reduce $v(\alpha)$ modulo-$p(\alpha)$, that is, the remainder will be zero.

6.6.2 Decoding binary BCH codes

BCH codes are attractive because they are easy to encode and especially because of the existence of efficient (fast) algebraic decoding algorithms based upon an *error location polynomial.* The following theory uses this concept to locate the errors in binary BCH codes, and is simply extended to find the error *values* for non-binary BCH codes, such as RS codes.

Consider a t-error correcting binary BCH(n,k) code defined on GF(2^m), that is, $n = 2^m - 1$. As for Meggitt decoding, we model the received polynomial as

$$r(x) = v(x) + e(x)$$

and we require the syndrome of $r(x)$. Since α^i, $i = 1, 2, ..., 2t$ is a root of $v(x)$ (see (6.63))

$$\begin{aligned} r(\alpha^i) &= v(\alpha^i) + e(\alpha^i) \\ &= e(\alpha^i) \\ &= s_i \end{aligned} \tag{6.83}$$

where s_i is a syndrome symbol. Restating this result, the syndrome symbols can be found from

$$s_i = r(\alpha^i), \qquad i = 1, 2, \ldots, 2t \tag{6.84}$$

and this is the first step in BCH decoding.

Example 6.28

A BCH(15,7) codeword is transmitted and received at the decoder input as

$$r(x) = x + x^7 + x^8 + x^{11}$$

Using the Galois field in Table 6.5, the corresponding syndrome symbols are as follows:

$$s_1 = r(\alpha) = \alpha + \alpha^7 + \alpha^8 + \alpha^{11} = \alpha + 1 + \alpha + \alpha^3 + 1 + \alpha^2 + \alpha + \alpha^2 + \alpha^3 = \alpha$$

$$s_3 = r(\alpha^3) = \alpha^3 + \alpha^{21} + \alpha^{24} + \alpha^{33} = \alpha^3 + \alpha^2 + \alpha^3 + \alpha + \alpha^3 + \alpha^3 = \alpha^5$$

According to (6.93), $s_2 = s_1^2$ and $s_4 = s_2^2$, giving $\mathbf{s} = (\alpha\ \alpha^2\ \alpha^5\ \alpha^4)$.

When decoding Hamming codes we saw that the syndrome, **s**, directly gave the error location, see (6.20). In algebraic decoding the syndrome is used in a less direct way to perform the same task. Essentially, it will be used to determine the coefficients of an *error location polynomial* $\sigma(x)$, the roots (or reciprocal roots) of which define the error location(s). Suppose that, for a particular $r(x)$

$$e(x) = x^8 + x^{11}$$

so that

$$e(\alpha^i) = (\alpha^8)^i + (\alpha^{11})^i = X_2^i + X_1^i \tag{6.85}$$

Here, $X_1 = \alpha^{11}$ is an *error locator*, denoting an error in received symbol r_{11}, and $X_2 = \alpha^8$ is a locator denoting an error in symbol r_8. Generalising, the kth error will have a locator of the form

$$X_k = \alpha^j, \quad \alpha^j \in \mathrm{GF}(2^m), \quad k = 1, 2, \ldots, t \tag{6.86}$$

and denotes an error in symbol r_j, $j = 0, 1, ..., n-1$. Since we are considering binary codes, r_j will then be complemented to perform error correction. Generalising (6.85),

$$s_i = e(\alpha^i) = \sum_{k=1}^{t} X_k^i, \quad i = 1, 2, \ldots, 2t \tag{6.87}$$

In principle we could solve (6.87) for the t values of X_k, but since it involves *non-linear* equations the usual approach is to use an error location polynomial, $\sigma(x)$. The significance of $\sigma(x)$ is that it translates the problem into a set of linear equations, and that its roots (or reciprocal roots) turn out to be the error locators, X_k. The following sections examine several algorithms for finding $\sigma(x)$.

6.6.3 Peterson's direct method and the Chien search

This is a straightforward approach for determining $\sigma(x)$, but it is computationally inefficient for large values of t when compared to the *iterative* techniques in Section 6.6.4. For instance, it has been shown that iterative techniques (specifically the Berlekamp–Massey frequency domain algorithm) can be preferable for t as low as three (Borda and Terebes, 1996). We commence by defining $\sigma(x)$ as

$$\sigma(x)=\prod_{k=1}^{t}(x+X_k)=x^t+\sigma_1 x^{t-1}+\sigma_2 x^{t-2}+\ldots+\sigma_t \tag{6.88}$$

so that locator X_k is a root of $\sigma(x)$. Since it is a root we can write (6.88) as

$$X_k^t+\sigma_1 X_k^{t-1}+\sigma_2 X_k^{t-2}+\ldots+\sigma_t=0, \quad k=1,2,\ldots,t \tag{6.89}$$

Multiplying by X_k^j, and then expanding

$$X_k^{t+j}+\sigma_1 X_k^{t+j-1}+\ldots+\sigma_t X_k^j=0, \quad k=1,2,\ldots,t \tag{6.90}$$

$$\begin{aligned}\therefore\ & X_1^{t+j}+\sigma_1 X_1^{t+j-1}+\ldots+\sigma_t X_1^j=0\\ & X_2^{t+j}+\sigma_1 X_2^{t+j-1}+\ldots+\sigma_t X_2^j=0\\ & \qquad\vdots\\ & X_t^{t+j}+\sigma_1 X_t^{t+j-1}+\ldots+\sigma_t X_t^j=0\end{aligned}$$

$$\therefore \sum_{k=1}^{t}X_k^{t+j}+\sigma_1\sum_{k=1}^{t}X_k^{t+j-1}+\ldots+\sigma_t\sum_{k=1}^{t}X_k^j=0 \tag{6.91}$$

Finally, substituting for the summation terms from (6.87) gives

$$s_{t+j}+\sigma_1 s_{t+j-1}+\ldots+\sigma_t s_j=0, \quad j=1,2,\ldots,t \tag{6.92}$$

Equation (6.92) represents a set of t *linear* equations which can be solved for the coefficients σ_i, $i=1,2,\ldots,t$, giving $\sigma(x)$. In addition, for *binary* codes,

$$s_{2k}=s_k^2, \quad \forall k \tag{6.93}$$

so (6.84) need only be computed for 'odd' values of i.

Example 6.29

For a binary code with $t=1$ we have $\sigma(x)=x+\sigma_1$, and the error locator is the root of $\sigma(x)$ i.e. $X_1+\sigma_1=0$, or $X_1=\sigma_1$. From (6.92) we have $s_2+\sigma_1 s_1=0$, and since $s_2=s_1^2$, we

Table 6.6 *Coefficients of $\sigma(x)$ for some t-error correcting binary BCH codes.*

t	Coefficients σ_i
1	$\sigma_1 = s_1$
2	$\sigma_1 = s_1$
	$\sigma_2 = \dfrac{s_3 + s_1^3}{s_1}$
3	$\sigma_1 = s_1$
	$\sigma_2 = \dfrac{s_1^2 s_3 + s_5}{s_1^3 + s_3}$
	$\sigma_3 = (s_1^3 + s_3) + s_1\sigma_2$

have $s_1 = \sigma_1$. Clearly, error location is simply $X_1 = s_1$, and if $s_1 = \alpha^j, j = 0, 1, ..., n-1$ then symbol r_j must be complemented. Similar manipulation of (6.92) and (6.93) for larger values of t gives the coefficients in Table 6.6.

Given the error location polynomial, $\sigma(x)$, the next step is to determine its roots. These will be the error locators X_k, $k = 1, 2, ..., t$, and each locator will be of the form given in (6.86). The obvious approach is therefore to try each possible root $\alpha^j, j = 0, 1, ..., n-1$ in (6.89). The substitution is usually carried out in the following way. First, rewrite (6.88) as

$$\frac{\sigma(x)}{x^t} = 1 + \sigma_1 x^{-1} + ... + \sigma_t x^{-t} \tag{6.94}$$

Over GF(2^m), equating $\sigma(x)$ to zero is equivalent to finding the roots of

$$\sigma_1 x^{-1} + \sigma_2 x^{-2} + ... + \sigma_t x^{-t} = 1$$

$$\sum_{i=1}^{t} \sigma_i x^{-i} = 1 \tag{6.95}$$

The roots of (6.95) are of the form α^j, $j = 0, 1, \ldots, n-1$, that is, root or error locator α^j corresponds to an error in symbol r_j. Suppose we first test for an error in symbol r_{n-1} by substituting α^{n-1} for x. Term x^{-i} then reduces to (see (6.81))

$$\alpha^{-i(n-1)} = \alpha^{-n\,i}\alpha^i = \left(\frac{1}{\alpha^n}\right)^i \alpha^i = \alpha^i \tag{6.96}$$

Similarly, x^{-i} reduces to α^{2i} when substituting α^{n-2}, and to α^{ji} when substituting α^{n-j}. Equation (6.95) can now be written as the search equation

$$\sum_{i=1}^{t} \sigma_i\, \alpha^{ji} \overset{?}{=} 1, \quad j = 1, 2, \ldots, n \tag{6.97}$$

A circuit for implementing (6.97) is commonly called the *Chien search* (Chien, 1964). The circuit sets a correction bit to 1 whenever (6.97) is satisfied, which then complements the appropriate bit in the received word. For instance, if (6.97) is satisfied for $j = 3$, then symbol r_{n-3} is complemented. Error correction starts on symbol r_{n-1} and proceeds bit by bit just as in the Meggitt decoder. Note that the Chien search is not required for $t = 1$, since solution of (6.89) is trivial (see Example (6.29)).

Example 6.30

Suppose the DEC BCH(15,7) code polynomial $v(x)$ in Example 6.27 is received as

$$r(x) = x + x^7 + x^{12} + x^{13}$$

In other words, r_8 and r_{11} are in error. Using (6.84) and reducing polynomials in α modulo-$p(\alpha)$ via Table 6.5 gives

$$s_1 = r(\alpha) = \alpha + \alpha^7 + \alpha^{12} + \alpha^{13} = \alpha^7$$

$$s_3 = r(\alpha^3) = \alpha^3 + \alpha^{21} + \alpha^{36} + \alpha^{39} = \alpha^3 + \alpha^6 + \alpha^6 + \alpha^9 = \alpha \quad \text{(note } \alpha^{15} = 1\text{)}$$

Using (6.93), the full syndrome vector is $\mathbf{s} = (\alpha^7\ \alpha^{14}\ \alpha\ \alpha^{13})$. From Table 6.6 we have

$$\sigma_1 = \alpha^7$$

$$\sigma_2 = (\alpha + \alpha^{21})(\alpha^{-7})$$

Here, α^{-7} is the multiplicative inverse of α^7 (Table 6.3), that is, $\alpha^7\alpha^{-7} = 1$. Clearly, since $\alpha^{15} = 1$, then $\alpha^{-7} = \alpha^8$ and

$$\sigma_2 = (\alpha + \alpha^6)\alpha^8 = \alpha^4 \text{ modulo } p(\alpha)$$

The error location polynomial is therefore

$$\sigma(x) = x^2 + \alpha^7 x + \alpha^4$$

Applying the Chien search in (6.97) for the roots of $\sigma(x)$, and reducing using Table 6.5, gives

$$j = 1: \quad \sigma_1\alpha + \sigma_2\alpha^2 = \alpha^7\alpha + \alpha^4\alpha^2 = \alpha^{14}$$

$$j = 2: \quad \sigma_1\alpha^2 + \sigma_2\alpha^4 = \alpha^7\alpha^2 + \alpha^4\alpha^4 = \alpha^{12}$$

$$j = 3: \quad \sigma_1\alpha^3 + \sigma_2\alpha^6 = \alpha^7\alpha^3 + \alpha^4\alpha^6 = 0$$

$$j = 4: \quad \sigma_1\alpha^4 + \sigma_2\alpha^8 = \alpha^7\alpha^4 + \alpha^4\alpha^8 = 1$$

Therefore, $\alpha^{n-4} = \alpha^{11}$ is a root of $\sigma(x)$, that is, $X_1 = \alpha^{11}$ satisfies (6.89), so symbol r_{11} should be complemented. The reader should show that (6.97) is also satisfied for $j = 7$, that is, $X_2 = \alpha^8$, and that r_8 should also be complemented.

Clearly, in practice the *actual* number of errors in $r(x)$ will be $\nu \le t$, which means that the degree of $\sigma(x)$ can be less than t. In other words, not all the coefficients may be required. For instance, if we assume that only r_{11} is in error in Example 6.30, we find $\sigma_1 = \alpha^{11}$, $\sigma_2 = 0$ and $\sigma(x) = x + \alpha^{11}$. The formal approach in Peterson's method is therefore to first estimate the actual number of errors, v, by performing a determinant test on the syndrome values. In particular, the matrix form of (6.92) over $GF(2^m)$ is

$$\begin{bmatrix} s_1 & s_2 & \cdots & s_t \\ s_2 & s_3 & \cdots & s_{t+1} \\ & \vdots & & \\ s_t & s_{t+1} & \cdots & s_{2t-1} \end{bmatrix} \begin{bmatrix} t \\ t-1 \\ \vdots \\ 1 \end{bmatrix} = \begin{bmatrix} s_{t+1} \\ s_{t+2} \\ \vdots \\ s_{2t} \end{bmatrix} \tag{6.98}$$

and solution for the coefficients is possible if the $t \times t$ matrix is invertible (non-singular). A general approach is as follows. If the determinant of this matrix is non-zero, assume t errors. If it is zero we reduce t by one, corresponding to $v = t - 1$, and test again. If now the determinant is non-zero then v is taken as the actual number of errors, otherwise we assume $v = t - 2$, and so on. Note that the number of computations required to invert a $v \times v$ matrix is proportional to v^3, and so, as already pointed out, Peterson's method is only suitable for low values of t. This being the case, a simple equivalent approach to determining v is given in Section 6.7.3.

6.6.4 Berlekamp's algorithm

Direct solution of (6.98) for the coefficients of the error location polynomial requires the inversion of a $v \times v$ matrix, where $v \leq t$. In fact, a similar inversion problem occurs in 'frequency domain' decoding (Section 6.7.4). Unfortunately, the number of computations required for inversion is proportional to v^3, so this direct approach becomes unattractive for large values of t. In contrast, the computational complexity of Berlekamp's iterative approach to finding $\sigma(x)$ grows more slowly with t. In practice, the iterative approach is usually referred to as the Berlekamp–Massey algorithm (since a similar algorithm was developed by Massey (1969)), and this technique is widely recognised as the most efficient decoding method for RS codes. As we shall see, the Berlekamp–Massey algorithm can be used either in the domain in which the data is received, or in the frequency domain (Section 6.7.4); either way, the algorithm delivers an error-location polynomial in an iterative way. The following section describes a table-based approach to Berlekamp's algorithm. The reader interested in a more formal matrix-based approach is referred to Blahut (1992).

The equations defined by (6.92) are one form of *Newton's identities*, and Berlekamp's algorithm essentially derives $\sigma(x)$ in an iterative way by satisfying successive identities. Initially the algorithm assumes $\sigma(x) = 1$ and at iteration r the polynomial is of the general form $\sigma^r(x) = 1 + \sigma_1^r x + \sigma_2^r x^2 + \ldots$. It finally builds an error location polynomial of the form

$$\sigma(x) = \prod_{k=1}^{t} (1 + X_k x) = 1 + \sigma_1 x + \sigma_2 x^2 + \ldots + \sigma_t x^t \tag{6.99}$$

Note that $\sigma(x)$ has roots at $x = X_k^{-1}$; that is, the roots are now the *inverse* of the error location numbers. At the rth iteration the polynomial will be denoted by $\sigma^r(x)$, and if the coefficients do not satisfy the $(r+1)$th Newton identity, a correction term is added to $\sigma^r(x)$ to give $\sigma^{r+1}(x)$, and so on. This procedure continues until enough identities are satisfied, that is, to $\sigma^{2t}(x)$ for a t-error correcting code, in which case $\sigma(x) = \sigma^{2t}(x)$. The full Berlekamp

Table 6.7 *Iteration table for the full Berlekamp algorithm.*

Iteration, r	$\sigma^r(x)$	c_r	k_r	$r - k_r$
–1	1	1	0	–1
0	$\sigma^0(x) = 1$	s_1	0	0
1	$\sigma^1(x) =$			
2	$\sigma^2(x) =$			
$\vdots$	$\vdots$			
$2t$	$\sigma^{2t}(x) = \sigma(x)$	–	–	–

algorithm is applicable to both binary and non-binary BCH codes and can be described in the context of Table 6.7.

Given iteration r in Table 6.7, the $(r + 1)$th row can be obtained using the following pseudocode:

```
for (r = 0, 1, ..., 2t – 1) {
   if (c_r = 0) {
      σ^{r+1}(x) = σ^r(x)
      k_{r+1} = k_r
   }
   else {
      find a row, i, in the table prior to row r such that c_i ≠ 0
      and the number (i – k_i) in the last column of the table has
      the largest value. Then add the correction to give
         σ^{r+1}(x) = σ^r(x) + c_r c_i^{-1} x^{r-i} σ^i(x)
         k_{r+1} = max[k_r, k_i + r – i]
   }
   c_{r+1} = s_{r+2} + σ_1^{r+1} s_{r+1} + σ_2^{r+1} s_r + ... + σ_{k_{r+1}}^{r+1} s_{r+2-k_{r+1}}
}
```

The formal Chien search (see (6.97)) can be applied once $\sigma(x)$ is determined, or we could simply try roots α^j, $j = 0, 1, ..., n-1$. Clearly, the actual number of errors, ν, in $r(x)$ can be less than t, but it is interesting to note that the Berlekamp algorithm does not need to know ν prior to decoding. For instance, a $t = 3$ code will yield a $\sigma(x)$ of degree three when $\nu = 3$ (giving three error locators), and a $\sigma(x)$ of degree two for $\nu = 2$. For $\nu < t$, $\sigma(x)$ is simply obtained in fewer iterations. If $\nu > t$ the algorithm will usually yield a polynomial of degree t or less, and so fail. Finally, it can be shown that the algorithm and Table 6.7 can be simplified from $2t$ to just t iterations for *binary* BCH codes (Peterson and Weldon, 1970).

Example 6.31

Consider the double error correcting BCH(15,7) code discussed in Example 6.30. For a given message and two specific errors (r_8 and r_{11} in error) it is shown that $\mathbf{s} = (\alpha^7\ \alpha^{14}\ \alpha\ \alpha^{13})$. Instead of using Peterson's method for direct determination of $\sigma(x)$ (as in Table 6.6), we will use the full Berlekamp algorithm. The iterations are summarised in Table 6.8. The error location polynomial differs from that in Example 6.30 (due to the

Table 6.8 *Iterative determination of $\sigma(x)$ using the full Berlekamp algorithm.*

Iteration, r	$\sigma^r(x)$	c_r	k_r	$r-k_r$
–1	1	1	0	–1
0	1	$s_1=\alpha^7$	0	0
1	$\sigma^1(x)=1+\alpha^7x$ $=1+\sigma_1^1x$	0	1	0
2	$\sigma^2(x)=1+\alpha^7x$ $=1+\sigma_1^2x$	α^{11}	1	1
3	$\sigma^3(x)=1+\alpha^7x+\alpha^4x^2$ $=1+\sigma_1^3x+\sigma_2^3x^2$	0	2	1
4	$\sigma(x)=1+\alpha^7x+\alpha^4x^2$	–	–	–

redefinition in (6.99)) but the final result is the same. In particular, simply trying $\alpha^j, j=0, 1, ..., n-1$ in $\sigma(x)=1+\alpha^7x+\alpha^4x^2$ we find that α^4 is a root, that is, $1+\alpha^{11}+\alpha^{12}=0$ over GF(2^4) (see Table 6.5). According to (6.99) the inverses of the roots are the error location numbers, so $X_1=\alpha^{-4}=\alpha^{11}$, corresponding to an error in r_{11}. Similarly, α^7 is found to be a root, giving $X_2=\alpha^{-7}=\alpha^8$, corresponding to an error in r_8.

The detailed derivation of Table 6.8 is as follows:

$r=0$: $c_0\neq 0$ so use row -1 ($i=-1$)

$$\sigma^1(x)=\sigma^0(x)+c_0c_{-1}^{-1}x^1\sigma^{-1}(x)=1+\alpha^7\cdot 1\cdot x\cdot 1=1+\alpha^7x$$

$$k_1=\max[k_0,k_{-1}+0+1]=\max[0,1]=1$$

$$c_1=s_2+\sigma_1^1s_1=\alpha^{14}+\alpha^7\alpha^7=0$$

$r=1$: $c_1=0$ so $\sigma^2(x)=\sigma^1(x)$, $k_2=k_1=1$

$$c_2=s_3+\sigma_1^2s_2 \quad (\text{note } k_{r+1}=k_2=1 \text{ so no } s_1 \text{ term})$$

$$=\alpha+\alpha^7\alpha^{14}=\alpha^{11}$$

$r=2$: $c_2\neq 0$ so use row 0 ($i=0$)

$$\sigma^3(x)=\sigma^2(x)+c_2c_0^{-1}x^2\sigma^0(x)=1+\alpha^7x+\alpha^{11}\alpha^{-7}x^2\cdot 1$$

$$=1+\alpha^7x+\alpha^4x^2$$

$$k_3=\max[k_2,k_0+2-0]=\max[1,2]=2$$

$$c_3=s_4+\sigma_1^3s_3+\sigma_2^3s_2 \quad (\text{note } k_{r+1}=k_3=2 \text{ so no } s_1 \text{ term})$$

$$=\alpha^{13}+\alpha^7\alpha+\alpha^4\alpha^{14}=0$$

$r=3$: $c_3=0$ $\therefore \sigma^4(x)=\sigma^3(x)=\sigma(x)$

Example 6.32

Example 6.27 derives a BCH(15,7) code polynomial for a specific message. Suppose this is received with just r_{12} in error, giving $r(x) = x + x^7 + x^8 + x^{11} + x^{13}$. The corresponding error location polynomial will be first-order (a single root). The reader should use (6.84) and (6.93) to show that the decoder generates syndrome $\mathbf{s} = (\alpha^{12}, \alpha^9, \alpha^6, \alpha^3)$. Using Table 6.7 (the full Berlekamp algorithm) we then have

$$r = 0: \quad c_0 \neq 0 \quad \text{so use row} -1 \ (i = -1)$$

$$\sigma^1(x) = \sigma^0(x) + c_0 c_{-1}^{-1} x^1 \sigma^{-1}(x) = 1 + \alpha^{12} \cdot 1 \cdot x \cdot 1 = 1 + \alpha^{12} x$$

$$k_1 = \max\left[k_0, k_{-1} + 0 + 1\right] = \max[0,1] = 1$$

$$c_1 = s_2 + \sigma_1^1 s_1 = \alpha^9 + \alpha^{12}\alpha^{12} = 0$$

$$r = 1: \quad c_1 = 0 \ \text{and so}\ \sigma^2(x) = \sigma^1(x),\ k_2 = k_1 = 1$$

$$c_2 = s_3 + \sigma_1^2 s_2 \quad (\text{note } k_{r+1} = k_2 = 1 \text{ so no } s_1 \text{ term})$$

$$= \alpha^6 + \alpha^{12}\alpha^9 = 0$$

$$r = 2: \quad c_2 = 0 \ \text{so}\ \sigma^3(x) = \sigma^2(x) = 1 + \alpha^{12} x, \quad k_3 = k_2 = 1$$

$$c_3 = s_4 + \sigma_1^3 s_3 \quad (\text{note } k_{r+1} = k_3 = 1 \text{ so no further terms})$$

$$= \alpha^3 + \alpha^{12}\alpha^6 = 0$$

$$r = 3: \quad c_3 = 0 \quad \therefore \sigma^4(x) = \sigma^3(x) = \sigma(x)$$

Solving for the root of $\sigma(x)$, we have $\alpha^{12} x = 1$ or $x = \alpha^3$. Therefore, $X_1 = \alpha^{-3} = \alpha^{12}$, denoting an error in r_{12}. The iterations are summarised in Table 6.9.

Table 6.9 *Iterations for Example 6.32.*

Iteration, r	$\sigma^r(x)$	c_r	k_r	$r - k_r$
–1	1	1	0	–1
0	1	$s_1 = \alpha^{12}$	0	0
1	$\sigma^1(x) = 1 + \alpha^{12}x$ $= 1 + \sigma_1^1 x$	0	1	0
2	$\sigma^2(x) = 1 + \alpha^{12}x$ $= 1 + \sigma_1^2 x$	0	1	1
3	$\sigma^3(x) = 1 + \alpha^{12}x$ $= 1 + \sigma_1^3 x$	0	1	2
4	$\sigma(x) = 1 + \alpha^{12}x$	–	–	–

Example 6.33

A particular message is coded by the triple error correcting BCH(15,5) code and is received with two errors as

$$r(x) = x + x^3 + x^4 + x^5 + x^6 + x^{14}$$

Using the Galois field in Table 6.5 the syndrome can be shown to be $\mathbf{s} = (\alpha^9, \alpha^3, \alpha^7, \alpha^6, 0, \alpha^{14})$. The reader should confirm the iterations in Table 6.10, and show that α^2 and α^5 are roots of $\sigma(x)$. The corresponding errors are in r_{10} and r_{13}.

Table 6.10 *Iterations for Example 6.33.*

Iteration, r	$\sigma^r(x)$	c_r	k_r	$r - k_r$
–1	1	1	0	–1
0	1	$s_1 = \alpha^9$	0	0
1	$\sigma^1(x) = 1 + \alpha^9 x$	0	1	0
	$= 1 + \sigma_1^1 x$			
2	$\sigma^2(x) = 1 + \alpha^9 x$	α^2	1	1
	$= 1 + \sigma_1^2 x$			
3	$\sigma^3(x) = 1 + \alpha^9 x + \alpha^8 x^2$	0	2	1
	$= 1 + \sigma_1^3 x + \sigma_2^3 x^2$			
4	$\sigma^4(x) = 1 + \alpha^9 x + \alpha^8 x^2$	0	2	2
	$= 1 + \sigma_1^4 x + \sigma_2^4 x^2$			
5	$\sigma^5(x) = 1 + \alpha^9 x + \alpha^8 x^2$	0	2	3
	$= 1 + \sigma_1^5 x + \sigma_2^5 x^2$			
6	$\sigma(x) = 1 + \alpha^9 x + \alpha^8 x^2$	–	–	–

We can summarise the algebraic decoding procedure for binary BCH codes as follows:

1. Compute the syndrome $\mathbf{s} = (s_1, s_2, ..., s_{2t})$ from $r(x)$. For software implementation $\mathbf{s}$ can be computed using $s_i = r(\alpha^i)$, $i = 1, 3, ..., 2t - 1$ and by using (6.93) if even terms are required.
2. If $\mathbf{s}$ is non-zero, find the error location polynomial $\sigma(x)$. For small values of t, Peterson's direct solution for the coefficients σ_1 to σ_t can be used by solving $s_{t+j} + \sigma_1 s_{t+j-1} + ... + \sigma_t s_j = 0$, $j = 1, 2, ..., t$, as in Table 6.6. Otherwise, the Berlekamp–Massey algorithm is usually used.
3. Determine the error locators X_k, $k = 1, 2, ..., t$ and perform error correction. If $t = 1$ correction is trivial (see Example 6.29); otherwise use trial substitution or the formal Chien search to find the roots of $\sigma(x)$.

6.7 Reed–Solomon (RS) codes

Binary BCH codes can be regarded as a subclass of the more general q-ary cyclic codes where $q = p^m$ (p prime, m integer). These q-ary codes ($q > 2$) are non-binary BCH codes, since the code symbols are drawn from GF(q) and so can be other than 0 or 1. As we shall see, since most data sources are in m-bit byte form, it is convenient to use field GF(2^m). This leads us to the RS codes, which can be regarded as a subclass of the q-ary BCH codes (see Figure 5.1(a)). Put simply, while RS codes can be defined on any field, the usual RS codes are non-binary BCH codes with symbols drawn from GF(2^m).

Definition 6.4 A t-error correcting RS code with symbols drawn from GF(2^m) has

$$\begin{aligned} n &= 2^m - 1 \\ n - k &= 2t \\ d &= 2t + 1 \end{aligned} \qquad (6.100)$$

From this definition it is evident that these codes are very efficient, that is, to correct t errors each codeword only requires $2t$ check symbols. The DEC RS(15,11) code, for example, needs only four check symbols compared with eight check symbols for the DEC binary BCH(15,7) code (see (6.61)).

An RS codeword $\mathbf{v} = (v_0 v_1 \ldots v_{n-1})$ will have symbols v_i drawn from GF(2^m) or, in other words, each symbol will be m-bits deep and corresponds to an m-bit binary word. As noted, this makes the codes attractive since data is often in m-bit byte form. Put the other way round, the number of bits in a data symbol determines both the Galois field, and the number of symbols in a codeword. Also note that the minimum distance d between codewords is the number of positions in which there are different *symbols*, and it is immaterial how many corresponding bits differ from each other within the corresponding symbols. Reed–Solomon codes are also attractive when burst errors span a whole symbol or a number of adjacent symbols, since, clearly, by correcting a single symbol we can correct many bit errors.

Several practical examples are useful at this point to establish the symbol concept.

1. A particular satellite paging system used an RS(30,18) code to protect text (ASCII characters). This is a shortened RS(31,19) code, which is capable of correcting up to 6 *symbol* errors over a block of 31 symbols. Since each symbol was drawn from GF(2^5) it was 5-bits deep, so it was necessary to shorten each ASCII character to 5 bits. Encoding was achieved by dividing the message into groups of 18 5-bit characters and each group was encoded into an RS(30,18) codeword. The paging system also used the DEC RS(7,3) code to protect the 'unique word' transmissions required for synchronisation.
2. The DEC RS(63,59) code has been used to protect 68-Mbit/s digital PAL television transmissions (Stott, 1984). The code used 6-bit symbols drawn from GF(2^6), and each symbols was actually a 6-bit DPCM sample. An interleave depth of six generated a matrix of 378 symbols (see Figure 6.10) and provided significant burst correction capability. The reader should show that this arrangement can correct a single burst of 67 bits or fewer, or two bursts each of length 31 bits or fewer.

3. RS codes are also used to protect MPEG-2 data packets in terrestrial and satellite DVB systems (see Section 1.7.2).

6.7.1 Encoding RS codes

RS codes are found in a similar way to codewords for binary BCH codes, and a coding algorithm for a systematic RS code will follow the same form as (6.44). First we need the generator polynomial $g(x)$ of the RS code. For binary BCH codes, Theorem 6.5 implies that $g(x)$ will be a product of the *minimal* polynomials of the elements α^i, $i = 1, 2, ..., 2t$. A similar argument applies for non-binary BCH codes, and for RS codes the minimal polynomial of α^i is simply $(x - \alpha^i)$. Therefore

$$g(x) = (x-\alpha)(x-\alpha^2)...(x-\alpha^{2t})$$
$$= (x+\alpha)(x+\alpha^2)...(x+\alpha^{2t}) \tag{6.101}$$

As before, α is a primitive element of GF(2^m) and, clearly, α, α^2, ..., α^{2t} are all roots of $g(x)$.

Example 6.34

Consider the DEC RS(7,3) code defined over GF(2^3). In this case

$$g(x) = (x+\alpha)(x+\alpha^2)(x+\alpha^3)(x+\alpha^4)$$

and using the field defined in Table 6.4 this reduces to

$$g(x) = x^4 + \alpha^3 x^3 + x^2 + \alpha x + \alpha^3$$

Note that the coefficients of $g(x)$ are now drawn from GF(2^3) rather than from GF(2). For a systematic code

$$\mathbf{v} = (p_0 p_1 p_2 p_3 u_0 u_1 u_2)$$
$$v(x) = p(x) + x^{2t} u(x)$$

where $u(x) = u_0 + u_1 x + u_2 x^2$ and $p(x)$ is the remainder from the division $x^{2t}u(x)/g(x)$. For the RS(7,3) code we therefore divide $x^6 u_2 + x^5 u_1 + x^4 u_0$ by $g(x)$, and after some manipulation based on Table 6.4 we obtain the following parity equations:

$$p_0 = \alpha^3 u_0 + \alpha^6 u_1 + \alpha^5 u_2$$
$$p_1 = \alpha u_0 + \alpha^6 u_1 + \alpha^4 u_2$$
$$p_2 = u_0 + u_1 + u_2$$
$$p_3 = \alpha^3 u_0 + \alpha^2 u_1 + \alpha^4 u_2$$

Since message symbol $u_i \in$ GF(2^3) (it is one of 8 elements) we could assume the arbitrary message

$$u(x) = \alpha^3 + \alpha x + \alpha^6 x^2$$

Use of the above parity equations then gives the complete RS(7,3) codeword

$$\mathbf{v} = (\alpha\ \alpha^2\alpha^2\alpha^6\alpha^3\alpha\ \alpha^6)$$

For DSP implementation it is convenient to simply number the elements of $GF(2^3)$ as follows:

$$v_i \in \begin{Bmatrix} 0 & 1 & \alpha & \alpha^2 & \alpha^3 & \alpha^4 & \alpha^5 & \alpha^6 \\ 0 & 1 & 2 & 3 & 4 & 5 & 6 & 7 \end{Bmatrix}$$

in which case the codeword is represented as $\mathbf{v} = (2337427)$, where each code symbol is 3-bits deep.

6.7.2 Decoding RS codes

For binary BCH codes the essential task is to solve (6.87) for the error locators X_k, $k = 1, 2, \dots, t$. For RS codes it is necessary to compute not only the locators (positions) but also the error *values*, Y_k, $k = 1, 2, \dots, t$ (assuming t erroneous symbols each m-bits deep). In this case, for t errors in the received codeword, (6.87) can be rewritten as

$$s_i = \sum_{k=1}^{t} Y_k X_k^i, \quad i = 1, 2, \dots, 2t \tag{6.102}$$

The RS decoding algorithm therefore includes all the steps for decoding binary BCH codes, plus a further step to compute the Y_k from (6.102). The significant point here is that, once the X_k have been determined (by deducing $\sigma(x)$ and then finding its roots), then (6.102) reduces to a set of *linear* simultaneous equations which are readily solved for the Y_k. Solving (6.102) for single and double errors, gives

$$t = 1: \quad Y_1 = \frac{s_1^2}{s_2} \tag{6.103}$$

$$t = 2: \quad Y_1 = \frac{s_1 X_2 + s_2}{X_1 X_2 + X_1^2} \tag{6.104}$$

$$Y_2 = \frac{s_1 X_1 + s_2}{X_1 X_2 + X_2^2} \tag{6.105}$$

Error correction is then performed by adding Y_k to the symbol identified by X_k, the addition being carried out using the rules of $GF(2^m)$.

Finally, note that (6.93) does not hold for RS codes and so we need to compute all syndrome symbols s_i. If we then use Peterson's direct solution for $\sigma(x)$ the expressions for its coefficients must be modified accordingly. For example, re-evaluating (6.92), Table 6.6 becomes Table 6.11.

6.7.3 A decoding algorithm for the RS(7,3) code

The following discussion is for the DEC RS(7,3) code, although it is easily generalised. We will assume that the code is generated as in Example 6.34. The received vector is $\mathbf{r} = (r_0 r_1 \dots r_6)$, corresponding to the polynomial $r(x) = r_0 + r_1 x + \dots + r_6 x^6$, where $r_i \in GF(2^3)$.

Table 6.11 *Coefficients of $\sigma(x)$ for single and double errors.*

t	Coefficients σ_i
1	$\sigma_1 = \dfrac{s_2}{s_1}$
2	$\sigma_1 = \dfrac{s_1 s_4 + s_3 s_2}{s_2^2 + s_1 s_3}$
	$\sigma_2 = \dfrac{s_2 s_4 + s_3^2}{s_2^2 + s_1 s_3}$

Step 1: The four syndrome symbols are computed using (6.84) and noting that $\alpha^7 = 1$:

$$s_1 = r_0 + r_1\alpha + r_2\alpha^2 + r_3\alpha^3 + r_4\alpha^4 + r_5\alpha^5 + r_6\alpha^6$$
$$s_2 = r_0 + r_1\alpha^2 + r_2\alpha^4 + r_3\alpha^6 + r_4\alpha + r_5\alpha^3 + r_6\alpha^5$$
$$s_3 = r_0 + r_1\alpha^3 + r_2\alpha^6 + r_3\alpha^2 + r_4\alpha^5 + r_5\alpha + r_6\alpha^4$$
$$s_4 = r_0 + r_1\alpha^4 + r_2\alpha + r_3\alpha^5 + r_4\alpha^2 + r_5\alpha^6 + r_6\alpha^3$$

For a DSP implementation these syndromes are easily computed in a recursive loop, and multiplication and addition over GF(2^3) can be implemented using look-up tables.

Step 2: Determine the number of errors present. Table 6.11 suggests the following rules:
if $s_1 = s_2 = s_3 = s_4 = 0$ no errors have occurred
if $s_2^2 + s_1 s_3 = 0$ and $s_1 \neq 0$ assume one error
if $s_2^2 + s_1 s_3 \neq 0$ assume two errors

Step 3: Compute the coefficients σ_i of the error location polynomial using Table 6.11. Assuming a single error, (6.88) gives

$$\sigma(x) = x + \sigma_1$$

$$\therefore X_1 = \sigma_1 = \frac{s_2}{s_1} \tag{6.106}$$

Step 4: For two errors, use the Chien search (either in hardware or as a software equivalent) to find the error locators X_1 and X_2. Here, $\sigma(x)$ is assumed to have degree two and if the Chien search fails to find two distinct roots, more than two errors have probably occurred.

Step 5: Determine the error value(s) using (6.103)–(6.105) and correct the appropriate symbol(s) by adding the value(s) to the symbol(s) over GF(2^3).

Step 6: Recompute the syndromes for the corrected vector. If these are non-zero then indicate a decoding failure.

When implemented in assembly language on a general purpose processor, the above algorithm took 2440 machine cycles to correct two errors and 1350 cycles to correct a single error.

Example 6.35

The RS(7,3) codeword for message $u(x) = \alpha^2 + x + \alpha^5 x^2$ is $\mathbf{v} = (1226316)$. Here we have used the practical notation discussed in Example 6.34, and GF(2^3) as defined in Table 6.4. Suppose this vector is received with an error in symbol r_0, that is, $\mathbf{r} = (5226316)$. In order to solve for the syndrome symbols (Step 1) it is helpful first to extract the following identities from Example 6.26:

$$\alpha^3 = \alpha + 1$$

$$\alpha^4 = \alpha^2 + \alpha$$

$$\alpha^5 = \alpha^2 + \alpha + 1$$

$$\alpha^6 = \alpha^2 + 1$$

$$\alpha^7 = 1$$

Using (6.84) gives

$$s_1 = \alpha^4 + \alpha^2 + \alpha^3 + \alpha^8 + \alpha^6 + \alpha^5 + \alpha^{11} = \alpha^2 + \alpha + 1 = \alpha^5 \equiv 6$$

Solving for all symbols we find

$$\mathbf{s} = (s_1 s_2 s_3 s_4) \equiv (6666)$$

Decoding step 2 indicates a single error, so the error location is simply

$$X_1 = \sigma_1 = \frac{s_2}{s_1} = 1$$

Using (6.86) we deduce that $j = 0$ and so the error is in symbol r_0. Finally, since we are assuming a single error the error value is obtained from (6.103) as

$$Y_1 = \frac{s_1^2}{s_2} = \alpha^5$$

Symbol r_0 is therefore corrected by adding its value (α^4) to Y_1, giving $\alpha^4 + \alpha^5 = 1$ (see Table 6.4).

Example 6.36

In Example 6.34 we derived the RS(7,3) codeword $\mathbf{v} = (2337427)$, using GF(2^m) as defined in Table 6.4. Suppose this is received as $\mathbf{r} = (2737426) \equiv (\alpha\ \alpha^6\ \alpha^2\ \alpha^6\ \alpha^3\ \alpha\ \alpha^5)$. Decoding step 1 gives $\mathbf{s} = (\alpha^3 1 \alpha^2\ 0) \equiv (4130)$ and step 2 gives

$$s_2^2 + s_1 s_3 = 1 + \alpha^3 \alpha^2 = \alpha^4$$

This indicates two errors, so the equations for σ_1 and σ_2 corresponding to $t = 2$ in Table 6.11 must be evaluated. The denominator in these equations is α^4 and requires the multiplicative inverse defined in Table 6.3. Since $\alpha^4 \alpha^{-4} = 1$ and $\alpha^7 = 1$ it follows that $\alpha^{-4} = \alpha^3$. Using this and the identities in Example 6.35, then gives

$$\sigma_1 = \alpha^5, \quad \sigma_2 = 1, \quad \sigma(x) = x^2 + \alpha^5 x + 1 \text{ (from (6.88))}$$

Now apply the Chien search. This starts by evaluating (6.97) starting at $j = 1$, that is, we first look for an error in symbol $r_{n-1} = r_6$ ($\equiv 6$):

$$j = 1: \quad \sigma_1\alpha + \sigma_2\alpha^2 = \alpha^5\alpha + \alpha^2 = 1$$

This means $\alpha^{n-1} = \alpha^6$ is a root. Since we are assuming Peterson's approach, that is, $\sigma(x)$ of the form given in (6.88), then the roots directly give the error locators and $X_1 = \alpha^6$. According to (6.86) this means that r_6 is in error. Proceeding with the search subsequently locates the second error:

$$j = 6: \quad \sigma_1\alpha^6 + \sigma_2\alpha^{12} = \alpha^4 + \alpha^5 = 1$$

Therefore, $\alpha^{n-6} = \alpha$ is a root, $X_2 = \alpha$ and r_1 is in error. Using (6.104) and (6.105) for the error values gives (see Table 6.4)

$$Y_1 = \frac{\alpha^3\alpha + 1}{\alpha^6\alpha + \alpha^{12}} = \frac{\alpha^2 + \alpha + 1}{1 + \alpha^5} = \frac{\alpha^5}{\alpha^4} = \alpha$$

Similarly, $Y_2 = 1$. Finally, we add the error values to the appropriate symbols:

$$r_6: \quad \alpha + \alpha^5 = \alpha^6 \equiv 7$$

$$r_1: \quad 1 + \alpha^6 = \alpha^2 \equiv 3$$

The reader should verify that the Berlekamp algorithm in Section 6.6.4 gives $\sigma(x) = 1 + \alpha^5 x + x^2$ (as defined in (6.99)) and that this also correctly locates the errors.

6.7.4 Discrete transforms in Galois fields

The RS decoding technique discussed in Section 6.7.2 is summarised in Figure 6.17(a), and a computationally efficient decoder based on this approach is given in (Wei *et al.*, 1993). While DSP implementation of this technique is straightforward for small values of t (see Section 6.7.3), it is not the only method. Another approach called 'frequency domain' decoding is also widely used, and one form of this is shown in Figure 6.17(b). The Berlekamp–Massey algorithm now operates on *discrete transforms* of the received symbols. Yet another approach to RS decoding operates in the 'time domain', whereby the Berlekamp–Massey algorithm operates directly on the received symbols, without any syndrome computation (Blahut, 1984). The time-domain algorithm is attractive for VLSI implementation since it can be implemented by repeating a single operation, but it can be computation intensive (Choomchuay and Arambepola, 1993). Here we will examine the frequency domain technique, which can be advantageous in certain implementation configurations. For example, a comparison of Figure 6.17(a) and (b) implies that frequency domain decoding could be advantageous when discrete transforms can be performed more efficiently than syndrome computation and error value computation (assuming that recursion takes about the same number of computations as the Chien search). Clearly, this might be the case if an FFT algorithm is used.

Consider a codevector $\mathbf{v} = (v_0 v_1 \ldots v_{n-1})$ where each code symbol v_i is drawn from GF(q). Let α be an element in GF(q) such that $\alpha^n = 1$ modulo-q, where $n = q - 1$. We can now define a discrete transform or 'spectrum' of $\mathbf{v}$ as

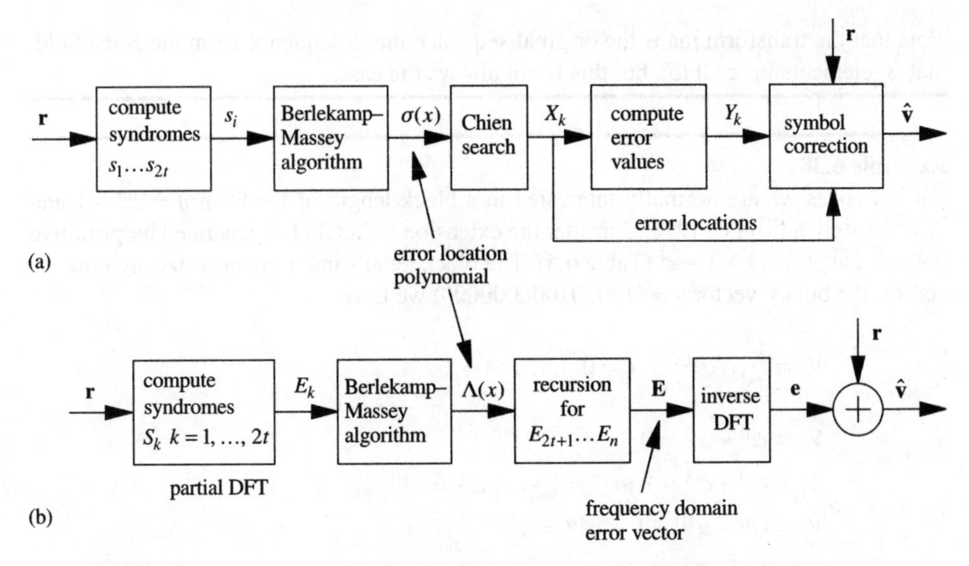

Figure 6.17 *Reed–Solomon decoding: (a) basic non-transform approach; (b) a frequency domain approach.*

$$V_k = \sum_{i=0}^{n-1} v_i \alpha^{ik}; \quad k = 0, 1, \ldots, n-1 \tag{6.107}$$

where the operations of addition and multiplication obey the rules of GF(q). For instance, given GF(25) we could perform a 24-point transform. In (6.107), α^{ik} has the role of a forward transform kernel and in the coding literature (6.107) is referred to as the discrete Fourier transform (DFT) of codevector **v**. Assuming the transform is over a field of characteristic two, any normalising factor is unity and the corresponding inverse DFT is simply

$$v_i = \sum_{k=0}^{n-1} V_k \alpha^{-ik}; \quad i = 0, 1, \ldots, n-1 \tag{6.108}$$

Example 6.37

Consider GF(5) and its element set {0,1,2,3,4}. Since $2^4 = 1$ modulo-5, that is, $\alpha = 2$, $n = 4$, we have the 4-point transform

$$V_k = \sum_{i=0}^{3} v_i 2^{ik}; \quad k = 0, \ldots, 3$$

Expanding this gives

$$\begin{bmatrix} V_0 \\ V_1 \\ V_2 \\ V_3 \end{bmatrix} = \begin{bmatrix} 1\,1\,1\,1 \\ 1\,2\,4\,3 \\ 1\,4\,1\,4 \\ 1\,3\,4\,2 \end{bmatrix} \begin{bmatrix} v_0 \\ v_1 \\ v_2 \\ v_3 \end{bmatrix}$$

Note that the transform maps the original sequence into a sequence from the *same* field, that is, elements $V_k \in \mathrm{GF}(5)$, but this is not always the case.

Example 6.38

For RS codes we are normally interested in a block length of the form $n = 2^m - 1$ and symbols drawn from $\mathrm{GF}(2^m)$. Consider the extension field GF(16) generated by primitive polynomial $p(x) = 1 + x + x^4$ (Table 6.5). The block length and transform size are now 15, and for the binary vector $\mathbf{v} = (100001000000000)$ we have

$$V_k = \sum_{i=0}^{14} v_i \alpha^{ik}; \qquad k = 0, 1, \ldots, 14$$

$$V_0 = \alpha^0 + \alpha^0 = 0$$

$$V_1 = \alpha^0 + \alpha^5 = 1 + \alpha^5 = 1 + \alpha + \alpha^2 = \alpha^{10}$$

$$V_2 = \alpha^0 + \alpha^{10} = 1 + \alpha^{10} = \alpha^5$$

$$\vdots$$

and so on, giving $V = (0, \alpha^{10}, \alpha^5, 0, \alpha^{10}, \alpha^5, 0, \ldots)$. In this case the DFT has mapped symbols in GF(2) and transformed them into symbols from GF(16).

Use of FFT algorithms

We note at the outset that the discrete Fourier transform (DFT) is valid in *any* field, although the signal processing engineer meets it most often in the field of complex numbers. Here, we are concerned with finite fields, and in general a DFT of block length n exists in $\mathrm{GF}(q)$ for every n that divides $q - 1$, as well as for $n = q - 1$. Thus, GF(25) could have DFT sizes of 24, 12, 8, 6, 4, 3 or 2, and GF(16) could have 15, 5 or 3-point transforms. It is worth noting that computationally efficient 3-point and 5-point DFTs exist (Nussbaumer, 1981), and that a 15-point DFT could be computed efficiently using a Prime Factor FFT. Similarly, a 256-point DFT could be performed over the prime field GF(769), which raises the question 'could we use the common radix-2 256-point FFT algorithm to perform the DFT?'. In fact, we can, since the structure of this and similar FFT algorithms is independent of the selected field.

Turning to RS codes, we note from Definition 6.4 that these are *usually* defined over $\mathrm{GF}(2^m)$ since each symbol is then exactly m bits deep. However, an RS code can be defined over any prime field, and so could be defined over GF(769). In this case a radix-2 256-point FFT could be used and the RS code would have a block length of 256 (this is somewhat inefficient in that each field symbol now represents some 9.6 bits, rather than 9 or 10 bits). In fact, one can always use a radix-2 FFT if the block length is of the form $n = 2^m$ (m integer). This could be achieved by adding an overall parity check symbol. Also note that a radix-2 FFT would be readily applicable over $\mathrm{GF}(2^m + 1)$.

6.7.5 Frequency domain RS decoding

Consider the codeword polynomial

$$v(x) = v_0 + v_1 x + v_2 x^2 + \cdots + v_{n-1} x^{n-1} = \sum_{i=0}^{n-1} v_i x^i \tag{6.109}$$

Now, any code polynomial is divisible by $g(x)$, that is

$$v(x) = a(x)g(x) \tag{6.110}$$

where the generator polynomial $g(x)$ has degree $n-k=2t$. By definition, for a general field characteristic, $g(x)$ for an RS code is given by

$$g(x) = (x-\alpha)(x-\alpha^2)\cdots(x-\alpha^{2t}) \tag{6.111}$$

so, clearly, α^k, $k = 1, 2, \ldots, 2t$ is a root of $v(x)$. We can therefore write

$$v(\alpha^k) = \sum_{i=0}^{n-1} v_i \alpha^{ik} = V_k = 0; \quad k = 1, 2, \ldots, 2t \tag{6.112}$$

which is fundamental to the frequency domain decoding of RS codes. As usual, the received codeword is modelled as $\mathbf{r} = \mathbf{v} + \mathbf{e}$, corresponding to $r(x) = r_0 + r_1 x + \ldots + r_{n-1}x^{n-1}$, and due to the linearity of the DFT we can write

$$\mathbf{R} = \mathbf{V} + \mathbf{E} \tag{6.113}$$

where

$$R_k = V_k + E_k = \sum_{i=0}^{n-1} r_i \alpha^{ik}; \quad k = 0, 1, \ldots, n-1 \tag{6.114}$$

Using (6.112) we then have

$$R_k = E_k = S_k; \quad k = 1, 2, \ldots, 2t \tag{6.115}$$

where S_k is a syndrome of $\mathbf{r}$. Expressing this another way

$$S_k = E_k = \sum_{i=0}^{n-1} r_i \alpha^{ik} = r(\alpha^k); \quad k = 1, 2, \ldots, 2t \tag{6.116}$$

The significant point here is that the syndromes can be computed as $2t$ *components of an n-point transform (which could be an FFT). This processing is sometimes referred to as a partial DFT (see Figure 6.17(b)), and is analogous to (6.84). Also, the transform delivers* $2t$ *components of the error vector transform* $\mathbf{E}$*, and the decoder can use the syndromes to help determine the remaining components of* $\mathbf{E}$*.*

As shown in Figure 6.17(b), once $\mathbf{E}$ has been determined it can be inverse transformed to give the time domain vector $\mathbf{e}$, and error correction is then performed as $\mathbf{v} = \mathbf{r} + \mathbf{e}$ (recall that addition and subtraction are identical over a field of characteristic two). It is interesting to see that, in this procedure, both error locations and error values are computed together! Finally, we note that error correction can give an invalid codeword when there are more than t errors, so it is generally important to recompute the syndromes, as discussed in Section 6.7.3.

Recursive computation of E

First represent $\mathbf{E}$ in polynomial form:

$$E(x) = E_0 + E_1x + \ldots + E_kx^k + \ldots + E_{n-1}x^{n-1} \tag{6.117}$$

Given E_1 to E_{2t}, the remaining coefficients can be found using an error location polynomial $\Lambda(x)$, where

$$\Lambda(x)E(x) = 0 \quad \text{modulo } x^n - 1 \tag{6.118}$$

The error location polynomial is defined in a similar way to (6.99), but in the frequency domain, and for $v \le t$ errors is of the form

$$\Lambda(x) = 1 + \Lambda_1 x + \Lambda_2 x^2 + \ldots + \Lambda_v x^v \tag{6.119}$$

Expanding (6.118) for $v = t$, and equating coefficients of powers of x to zero, gives

$$E_j + \sum_{k=1}^{t} \Lambda_k E_{j-k} = 0$$

$$E_j = -\sum_{k=1}^{t} \Lambda_k E_{j-k}; \quad j = 0, 1, \ldots, n-1 \tag{6.120}$$

(note that negative subscripts of E are meaningless). There are t unknown values of Λ_k in (6.120) but these can be found from t linear equations involving the known values $E_1 \ldots E_{2t}$. The required t equations are defined by the recursive relationship

$$E_j = -\sum_{k=1}^{t} \Lambda_k E_{j-k}; \quad j = t+1, \ldots, 2t \tag{6.121}$$

Once $\Lambda(x)$ has been determined, we can extend (6.121) to determine $E_{2t+1} \ldots E_n$, noting that $E_0 = E_n$ since $n = 0$ modulo n. For example,

$$E_{2t+1} = -\Lambda_1 E_{2t} - \Lambda_2 E_{2t-1} - \ldots - \Lambda_t E_{t+1} \tag{6.122}$$

Example 6.39

If $t = 2$ we know E_1 to E_4 from (6.116), and from (6.121) we have

$$E_3 = -\Lambda_1 E_2 - \Lambda_2 E_1$$

$$E_4 = -\Lambda_1 E_3 - \Lambda_2 E_2$$

giving

$$\Lambda_1 = \frac{E_2E_3 - E_1E_4}{E_1E_3 - E_2^2} \tag{6.123}$$

$$\Lambda_2 = \frac{E_2E_4 - E_3^2}{E_1E_3 - E_2^2} \tag{6.124}$$

As discussed previously, the negative signs can be dropped for field GF(2^m). More significantly, we note that the solution for $\Lambda(x)$ can be expressed formally as

$$\begin{bmatrix} \Lambda_2 \\ \Lambda_1 \end{bmatrix} = \begin{bmatrix} E_1 & E_2 \\ E_2 & E_3 \end{bmatrix}^{-1} \begin{bmatrix} -E_3 \\ -E_4 \end{bmatrix} \tag{6.125}$$

which highlights the main problem for general t-error correction. *As in (6.98), there is still a problem of matrix inversion despite the transform approach.* A basic approach would initially assume that there are t errors and perform a determinant test, as discussed in Section 6.6.3. However, the usual approach is to use the Berlekamp–Massey algorithm to solve for $\Lambda(x)$, as indicated in Figure 6.17(b).

As just discussed, the solution of (6.121) for coefficients Λ_k is the central (most difficult) problem in RS decoding. To this end we note that the recursion in (6.120) can be carried out using a linear feedback shift register, as indicated in Figure 6.18. Here, the Λ_k are the tap weights defining feedback. In fact, the use of such a register to provide an iterative solution for the error location polynomial in BCH/RS decoding was recognised by Massey (1969). Essentially, his algorithm adjusts the Λ_k and/or the register length in order to satisfy (6.120). Also, once $\Lambda(x)$ has been determined, the register can be used to recursively extend elements E_j, as in (6.122), in order to obtain **E**.

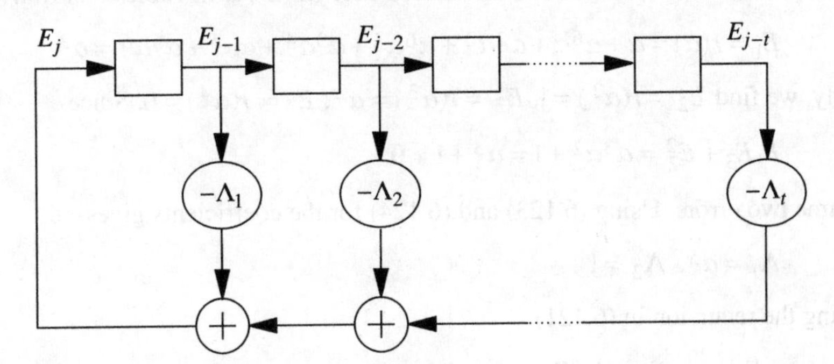

Figure 6.18 *Recursive generation of* **E** *using a linear feedback shift register.*

Example 6.40
Given received polynomial $r(x)$, the steps for the frequency domain decoding of the DEC RS(31,27) code defined over GF(2^5) could be as follows. From (6.116) we have

$$E_1 = r(\alpha) \quad E_2 = r(\alpha^2) \quad E_3 = r(\alpha^3) \quad E_4 = r(\alpha^4)$$

Now test the determinant of the square matrix in (6.125): if $E_1E_3 + E_2^2 \neq 0$ we assume two errors (as in Section 6.7.3) and compute Λ_1 and Λ_2 from (6.123) and (6.124). Otherwise we assume a single error and compute Λ_1 from (6.121):

$$E_2 = -\Lambda_1 E_1; \quad \Lambda_1 = \frac{E_2}{E_1} \text{ over GF}(2^5)$$

Assuming two errors we extend the recursion in (6.121) as follows:

$$-E_5 = \Lambda_1 E_4 + \Lambda_2 E_3$$
$$-E_6 = \Lambda_1 E_5 + \Lambda_2 E_4$$
$$\vdots$$
$$-E_{30} = \Lambda_1 E_{29} + \Lambda_2 E_{28}$$
$$-E_{31} = -E_0 = \Lambda_1 E_{30} + \Lambda_2 E_{29}$$

The time domain error vector **e** is found from the inverse DFT (see (6.108)):

$$e_i = \sum_{k=0}^{30} E_k \alpha^{-ik} = E(\alpha^{-i}); \qquad i = 0, 1, \ldots, 30$$

where $E(x)$ is the error spectrum. Note that both *error locations* and *error values* are contained in **e**, and this is illustrated in the following example.

Example 6.41
Here we decode the problem in Example 6.36 using the frequency domain approach. The DEC RS(7,3) codeword $\mathbf{v} = (\alpha\alpha^2\alpha^2\alpha^6\alpha^3\alpha\alpha^6)$ has been corrupted to $\mathbf{r} = (\alpha\alpha^6\alpha^2\alpha^6\alpha^3\alpha\alpha^5)$, and the code is defined over GF(2^3) as in Table 6.4. From (6.116)

$$E_1 = r(\alpha) = \alpha + \alpha^6\alpha + \alpha^2\alpha^2 + \alpha^6\alpha^3 + \alpha^3\alpha^4 + \alpha\alpha^5 + \alpha^5\alpha^6 = \alpha^3$$

Similarly, we find $E_2 = r(\alpha^2) = 1$, $E_3 = r(\alpha^3) = \alpha^2$, $E_4 = r(\alpha^4) = 0$. Since

$$E_1E_3 + E_2^2 = \alpha^3\alpha^2 + 1 = \alpha^5 + 1 \neq 0$$

we assume two errors. Using (6.123) and (6.124) for the coefficients gives

$$\Lambda_1 = \alpha^5, \; \Lambda_2 = 1$$

Extending the recursion in (6.121):

$$-E_5 = \Lambda_1 E_4 + \Lambda_2 E_3 = \alpha^2$$
$$-E_6 = \Lambda_1 E_5 + \Lambda_2 E_4 = \alpha^5\alpha^2 = 1$$
$$-E_7 = -E_0 = \Lambda_1 E_6 + \Lambda_2 E_5 = \alpha^5 + \alpha^2 = \alpha^3$$

Therefore, ignoring signs, the full error transform is $\mathbf{E} = (\alpha^3\alpha^3\,1\,\alpha^2\;0\,\alpha^2\;1)$. Applying the inverse transform as in Example 6.40, and noting that $\alpha^i\alpha^{-i} = \alpha^n = 1$:

$$e_i = E(\alpha^{-i}) = E(\alpha^{7-i})$$

In particular,

$$e_1 = E(\alpha^6) = \alpha^3 + \alpha^3\alpha^6 + \alpha^{12} + \alpha^2\alpha^{18} + \alpha^2\alpha^{30} + \alpha^{36} = 1$$

$$e_6 = E(\alpha) = \alpha^3 + \alpha^3\alpha + \alpha^2 + \alpha^2\alpha^3 + \alpha^2\alpha^5 + \alpha^6 = \alpha$$

and the complete error vector is found to be $\mathbf{e} = (0\,1\,0\,0\,0\,0\,\alpha)$. Over GF($2^3$) error correction is then performed as

$$\begin{aligned}\mathbf{v} &= \mathbf{r}+\mathbf{e}\\ &= (\alpha(\alpha^6+1)\alpha^2\alpha^6\alpha^3\alpha(\alpha^5+\alpha)) = (\alpha\alpha^2\alpha^2\alpha^6\alpha^3\alpha\alpha^6)\end{aligned}$$

In general it is possible that $\mathbf{v}$ is invalid due to the actual number of errors ν exceeding t. We should therefore check for a valid codeword by recomputing the syndromes for the corrected codevector $\mathbf{v}$. If these are non-zero then indicate a decoding failure.

6.7.6 Soft decoding of RS codes

The performance advantage of soft-decision decoding compared with hard-decision decoding is undisputed, but usually it is only applied to convolutional codes. This is because it is relatively easy to incorporate soft-decisions into the regular trellis of a convolutional code (soft-decision Viterbi decoding), but the trellises of block codes are generally more complex and irregular. On the other hand, soft *iterative* decoding of block codes is of interest since, in product code form, these can be a competitive alternative to Turbo codes when a high code rate ($R > 0.7$) is required (Pyndiah *et al.*, 1994). Here we will outline the soft decoding of block codes and extend it to RS codes. The complexity of MAP or symbol decoding (Section 5.3) will be avoided, and instead we will use Maximum Likelihood decoding theory to estimate the most likely transmitted code *sequence*.

Suppose codeword $\mathbf{v}_i = (v_1 \ldots v_k \ldots v_n)$, $v_k \in \{\pm 1\}$ is transmitted over a Gaussian channel using some form of digital modulation. For generality we assume an unquantised demodulator output, $\mathbf{r}$, corresponding to the addition of Gaussian noise samples of variance σ^2 to $\mathbf{v}_i$. Following the optimum decoding concept in Section 5.3.1, the optimal decoded sequence (codeword) is defined by the rule

$$\text{select } \mathbf{v}_i \text{ if: } \quad P(\mathbf{v}_i|\mathbf{r}) \geq P(\mathbf{v}_j|\mathbf{r}) \qquad \forall j \neq i \tag{6.126}$$

Section 5.3.2 showed that the practical form of (6.126) for an AWGN channel reduces to minimum Euclidean distance decoding, that is

$$\text{select } \mathbf{v}_i \text{ if: } \quad |\mathbf{r}-\mathbf{v}_i|^2 \leq |\mathbf{r}-\mathbf{v}_j|^2 \qquad \forall j \neq i \tag{6.127}$$

where the squared Euclidean distance is

$$|\mathbf{r}-\mathbf{v}_i|^2 = \sum_{k=1}^{n} (r_k - v_k)^2 \tag{6.128}$$

As noted in Section 5.3.5, the search through 2^k codewords for an (n,k) code can be prohibitive and a good simplification is provided by the Chase algorithm.

Suppose now that $\mathbf{v}_i$ is an RS codeword of block length $n = 2^m - 1$ with symbols drawn from $GF(2^m)$. Each symbol is m bits deep and the demodulator output (RS decoder input) can be written

$$\mathbf{r} = \begin{bmatrix} r_{11} \cdots r_{1k} \cdots r_{1n} \\ \vdots \qquad r_{jk} \qquad \vdots \\ r_{m1} \cdots r_{mk} \cdots r_{mn} \end{bmatrix} \tag{6.129}$$

corresponding to transmitted codeword

$$\mathbf{v}_i = \begin{bmatrix} v_{11} & \cdots & v_{1k} & \cdots & v_{1n} \\ \vdots & & v_{jk} & & \vdots \\ v_{m1} & \cdots & v_{mk} & \cdots & v_{mn} \end{bmatrix} \tag{6.130}$$

The minimum distance rule in (6.127) still applies, but now

$$|\mathbf{r} - \mathbf{v}_i|^2 = \sum_{k=1}^{n} \sum_{j=1}^{m} (r_{jk} - v_{jk})^2 \tag{6.131}$$

As stated, an important application of soft decoded block codes is in the area of iterative decoding. This is true whether we are dealing with binary or non-binary BCH codes. To extend the decoding to iterative decoding, we must first recognise that (6.127), evaluated by the Chase algorithm, accepts a soft input, $\mathbf{r}$, but yields a hard-decision output. In other words, a particular code sequence $\mathbf{v}_i$ is selected, but there is no side information giving the *reliability* of each code symbol. For iterative decoding we must therefore not only perform the ML decoding in (6.127), but also generate reliability information for each code symbol. This could be done by estimating the log-likelihood ratio for each code symbol using (5.42). The use of iterative decoding in the context of product codes is discussed in Section 6.8.1.

6.8 Product codes

In Section 5.1.2 we saw how a concatenated coding scheme enabled two (or more) block codes to be combined to effectively obtain a long block code. The objective was to obtain a powerful code while keeping the code rate high and overall decoder complexity relatively low. This is also the general objective in product decoding; in fact, a product code (also called an *array* code) can be regarded as a serial concatenated coding scheme.

Consider two linear block codes $\mathbf{V}_1 = (n_1 k_1 t_1)$, $\mathbf{V}_2 = (n_2 k_2 t_2)$ with minimal distances d_1, d_2 respectively. The product code $\mathbf{P} = \mathbf{V}_1 \otimes \mathbf{V}_2$ can be obtained by

1. Placing $k_1 \times k_2$ information symbols in an array of k_1 rows and k_2 columns.
2. Coding the k_1 rows using $\mathbf{V}_2$.
3. Coding the n_2 columns using $\mathbf{V}_1$.

This procedure is illustrated in Figure 6.19, and corresponds to a code rate $R = R_1 R_2 = k_1 k_2 / n_1 n_2$. Note that we could let $\mathbf{V}_1 = \mathbf{V}_2$. Also, the exact coding order is not important, for example, we could first encode the k_2 columns using $\mathbf{V}_1$ and then code the n_1 rows using $\mathbf{V}_2$, and the $(n_1 - k_1) \times (n_2 - k_2)$ parity symbols would still be the same.

As for many types of code, the error control capability of a product code depends upon how it is decoded. The most obvious approach is to apply simple hard-decision decoding first to the rows, and then to the columns. We will then have a block code of length $n_1 n_2$ decoded with a complexity of two relatively simple decoders. For the coding in Figure 6.19, a row decoding error can occur when the number of errors in any row is $t_2 + 1$. Also, we need at least $t_1 + 1$ such row errors (with careful alignment) for code $\mathbf{V}_1$ to fail, that is, we must have at least $(t_1 + 1)(t_2 + 1)$ errors for a possibility of overall failure. This implies that simple row then column hard-decision decoding will correct any error pattern with $t =$

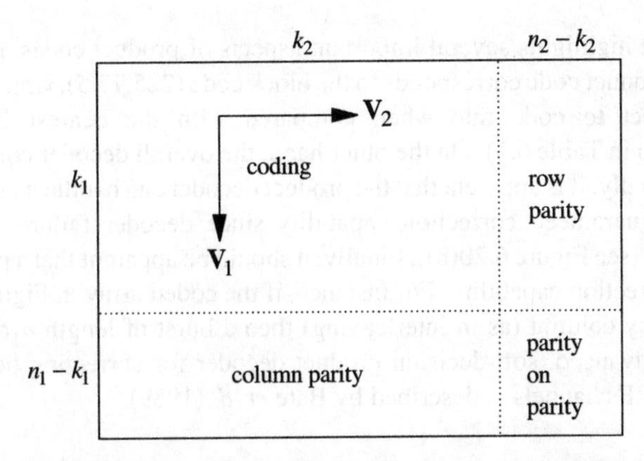

Figure 6.19 *A two-dimensional product code* $\mathbf{P} = \mathbf{V}_1 \otimes \mathbf{V}_2$.

$(t_1 + 1)(t_2 + 1) - 1$ errors or fewer. This guaranteed error control capability is, in fact, less than we might expect from a product code since its minimum distance is $d = d_1 d_2$ where d_i is the minimum distance of code $\mathbf{V}_i$. According to (6.1) we might then expect

$$t = \left\lfloor \frac{d_1 d_2 - 1}{2} \right\rfloor \tag{6.132}$$

Unfortunately, in general, simple row–column hard-decision decoding will not give the capability in (6.132), although this capability *is* achievable by more complex decoding.

Example 6.42

Let $\mathbf{V}_1 = (15,7,2)$, $\mathbf{V}_2 = (15,11,1)$, corresponding to $d_1 = 5$, $d_2 = 3$. Thus, the guaranteed number of correctable errors with simple decoding is 5, while (6.132) suggests 7. Suppose for instance that there are 6 carefully located errors as in Figure 6.20(a). Clearly, each row decoding will fail (and may generate even more errors), and the column of 3 errors will cause an overall decoding failure. However, removal of any one error will guarantee correct overall decoding.

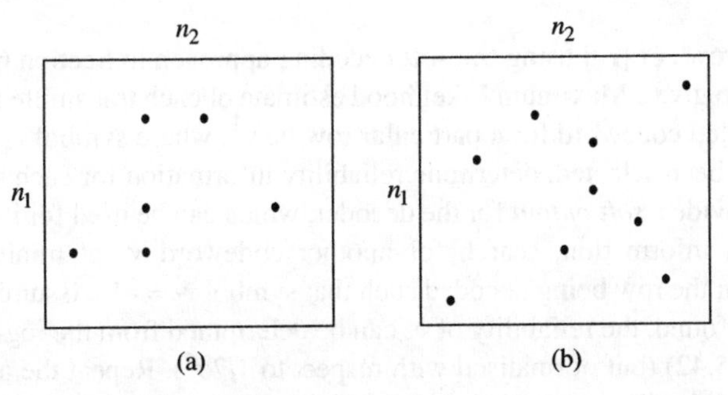

Figure 6.20 *Product codes: (a) decoder failure for 6 errors; (b) a correctable pattern of 9 errors.*

Example 6.42 highlights several important aspects of product codes. First, for simple decoding, the product code corresponds to the block code (225,77,5), which is clearly inferior with respect to code rate when compared with the nearest BCH code (the BCH(255,215,5) in Table 6.1). On the other hand, the overall decoder complexity is relatively low. Secondly, it is apparent that the product decoder can handle many error patterns *exceeding* the guaranteed correction capability since decoder failure requires careful placing of errors (see Figure 6.20(b)). Finally, it should be apparent that a product code has a burst error correction capability. For instance, if the coded array in Figure 6.19 is transmitted column by column (as in interleaving) then a burst of length n_1t_2 or less can be corrected. An advanced, soft-decision product decoder for correcting both random and burst errors on HF channels is described by Bate *et al.* (1989).

6.8.1 Iterative decoding of product codes

The power of product codes is sometimes exploited in data storage applications (recall that digital recording media, whether magnetic or optical, are characterised as bursty channels). In such an application it is usual to use RS product codes over GF(2^m) where the component codes have the same length, that is, $n_1 = n_2 = n$ in Figure 6.19. For instance, the two codes could be the RS(32,28) and RS(32,26) codes. This ensures that the codes have the same m-bit symbols. Despite such applications and the high potential of product codes, the overall performance can be disappointing. We have already hinted at this by the fact that the guaranteed error correction capability with simple decoding is generally lower than expected (Example 6.42). This disappointing performance is due to the use of suboptimal hard decoders for row and column decoding, and one solution is to employ an iterative approach based on soft-decision decoding (with *soft output*) of the component codes (albeit at the cost of increased complexity).

An iterative decoding algorithm (Pyndiah *et al.*, 1994)

We will assume a product code $\mathbf{P}$ comprised of $n_1 \times n_2$ binary symbols each selected from the set $\{\pm 1\}$. Given product matrix $[\mathbf{r}_1]$ at the demodulator output (the subscript corresponding to the first iteration), the essential steps in an iterative decoding procedure could be as follows:

1. Decode the rows of $[\mathbf{r}_1]$ using the soft decoding approach in Section 6.7.6, that is, we use (6.127) to give a Maximum Likelihood estimate of each transmitted row codeword. Let the selected codeword for a particular row be $\mathbf{v}^+$, where symbol $v_k = +1$.
2. Once $\mathbf{v}^+$ has been selected, determine reliability information for each symbol v_k in $\mathbf{v}^+$. This will provide a *soft output* for the decoder, which can be used for iteration. In order to find such information, search for another codeword $\mathbf{v}^-$ at minimum Euclidean distance from the row being decoded such that symbol $v_k = -1$. Assuming such a codeword can be found, the reliability of v_k can be determined from the log-likelihood ratio estimate in (5.42) (but normalised with respect to $1/2\sigma^2$). Repeat the above procedure for all symbols in all rows to obtain a matrix $[\mathbf{r}_1']$ of normalised log-likelihood estimates.
3. Generate the output of the first decoder as

$$[\mathbf{w}_2] = [\mathbf{r}_1'] - [\mathbf{r}_1] \tag{6.133}$$

Note that $[\mathbf{w}_2]$ can be regarded as additional or *extrinsic* information about $[\mathbf{r}_1]$ and it takes on the same role as the extrinsic information found in Turbo codes (Section 7.5).

4. The second decoder now performs the same operations on the columns, using as input

$$[\mathbf{r}_2]=[\mathbf{r}_1]+\alpha_2[\mathbf{w}_2] \tag{6.134}$$

where α_i is a constant which is used to reduce the influence of $[\mathbf{w}_i]$ in early iterations. This completes the elementary decoding process.

5. Repeat steps 1 to 4 for a number of iterations.

Using a similar approach, iterative decoding of an RS product code gave over 5 dB coding gain at a BER of 10^{-5} after just four iterations (Aitsab and Pyndiah, 1996).

Bibliography

Aitsab, O. and Pyndiah, R. (1996) Performance of Reed–Solomon block turbo code, *IEEE GLOBECOM '96*, **1**, London, 18-22 November, pp. 121–125.

Bate, S., Honary, B. and Farrell, P. (1989) Error control techniques applicable to HF channels, *IEE Proc.*, **136**, Pt. 1 (1), 57–63.

Blahut, R. (1983) *Theory and Practice of Error Control Codes*, Addison-Wesley, Reading, MA.

Blahut, R. (1984) A universal Reed–Solomon decoder, *IBM J. Res. Develop.*, **28**(2), 150–158.

Blahut, R. (1992) *Algebraic Methods for Signal Processing and Communications Coding*, Springer-Verlag, New York.

Borda, M. and Terebes, R. (1996) *Decoding algorithms for Reed–Solomon codes*, Internal report, Technical University of Cluj-Napoca, Romania.

Bose, R. and Chaudhuri, D. (1960) On a class of error correcting binary group codes, *Inf. Control*, **3**, 68–79.

Bota, V. (1992) University of Cluj-Napoca, Romania, personal communication.

CENELEC (1992) *Specification of the radio data system* (RDS), European Committee for Electrotechnical Standardisation, Rue de Stassart 35, B-1050 Brussels, EN 50067.

Chien, R. (1964) Cyclic decoding procedures for Bose–Chaudhuri–Hocquenghem codes, *IEEE Trans. Inf. Theory*, **IT-10**, 357–363.

Choomchuay, S. and Arambepola, B. (1993) Time domain algorithms and architectures for Reed–Solomon decoding, *IEE Proc.-I*, **140**(3), 189–196.

ETSI-GSM (1991) *GSM full rate speech transcoding*, European Telecommunications Standards Institute, Valbonne Cedex, France, **GSM 06.10** (Version 3.2.0).

Ferebee, I., Tait, D. and Taylor, D. (1989) An ARQ/FEC coding scheme for land mobile communication, *International Journal of Satellite Communication*, **7**(2), 19–224.

Hocquenghem, A. (1959), Codes correcteurs d'erreurs, *Chiffres*, **2**, 147–156.

Hsiao, M. T. (1970) A class of optimum odd-weight-column SEC-DED codes, *IBM J. Res. Dev.*, **14**, July.

Kasami, T. (1963) Optimum shortened cyclic codes for burst error correction, *IEEE Trans. on Information Theory*, **IT-9**(4), 105–109.

Kasami, T. (1964) Some efficient shortened cyclic codes for burst-error correction, *IEEE Trans. on Information Theory*, **IT-10**, 252–253.

Massey, J. (1969) Shift-register synthesis and BCH decoding, *IEEE Trans. on Information Theory*, **IT-15**(1), 122–127.

McEliece, R. (1977) *The Theory of Information and Coding*, Addison-Wesley, Reading, MA.

Nussbaumer, H. (1981) *Fast Fourier Transform and Convolution Algorithms*, Springer-Verlag, New York.

Peterson, W. and Weldon, E. (1970) *Error-Correcting Codes*, 2nd edn, MIT Press, Cambridge, MA.

Pyndiah, R., Glavieux, A., Picart, A. and Jacq, S. (1994) Near optimum decoding of product codes, *IEEE GLOBECOM '94*, **1**, San Francisco, 28 November–2 December, 339–343.

Pyndiah, R., Combelles, P. and Adde, P. (1996) A very low complexity block turbo decoder for product codes, *IEEE GLOBECOM '96*, **1**, London, 18–22 November, 101–105.
Stott, J. H. (1984) Digital transmission: 68 Mbit/s PAL field trial system, *Report 1984/5*, BBC Research Department.
Wei, C., Chen, C. and Liu, G. (1993) High-speed Reed–Solomon decoder for correcting errors and erasures', *IEE Proc.-I,* **140**(4), 246–254.

CHAPTER 7

Convolutional codes

As shown in Figure 5.1, convolutional codes are the main alternative to block codes when FEC is required. They are particularly attractive when soft-decision decoding is required, as in soft-decision Viterbi decoding. They are also used as the constituent codes in powerful *iterative* decoding schemes for concatenated codes (Sections 7.5 and 7.6).

7.1 Introduction

A generalised binary convolutional encoder is shown in Figure 7.1(a). The binary information sequence is shifted into the register or memory k bits at a time, corresponding to one input frame, and the register stores m_e frames. For each input frame the encoder generates a codeword frame of n bits, corresponding to a code rate $R = k/n$. For example, a rate 2/3 encoder would shift in 2 bits and generate 3 code bits by making parity checks (modulo-2 sums) on selected subsets of the $m_e k$ bits. The *constraint length*, K, of the convolutional code can be defined as the total span of information bits which influence the encoder output frame, and for the model in Figure 7.1(a),

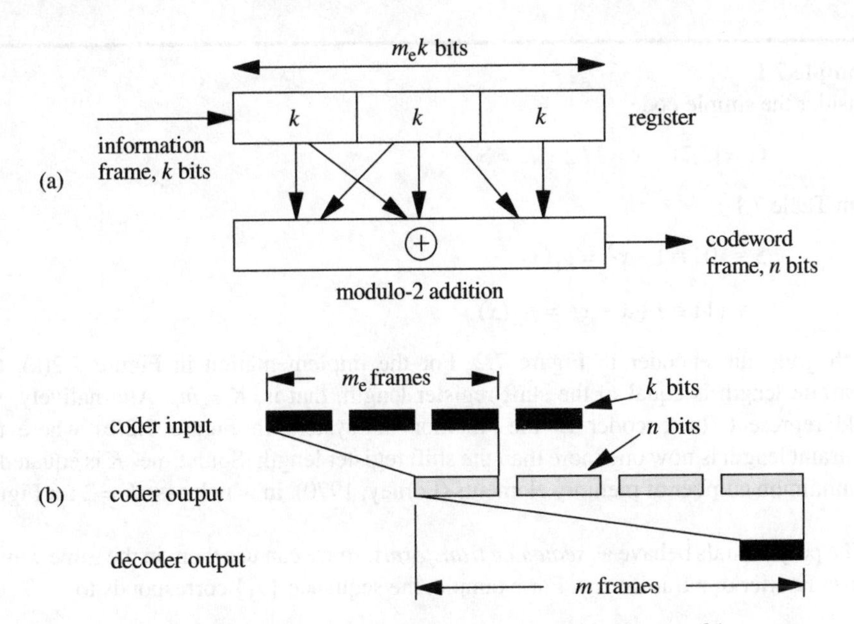

Figure 7.1 *Conventional binary convolutional encoder, rate k/n.*

$$K = m_e k \tag{7.1}$$

The corresponding decoding process (Figure 7.1(b)) uses a memory of m codeword frames to decode a k-bit information frame. It is important to note that, usually, $m > m_e$ and, ideally, m tends to infinity for probabilistic decoding methods, such as Viterbi decoding.

7.1.1 Generator polynomial and optimal codes

A block code can be defined by a single generator polynomial, $g(x)$, while a convolutional code can be defined by kn polynomials and a $k \times n$ generator-polynomial matrix. For implementation reasons we usually select $k = 1$ and this will be assumed from now on. This means, of course, that the *basic* convolutional code cannot have a rate close to one (unlike a block code). For $k = 1$, a convolutional code can be compactly defined by the generator-polynomial matrix

$$\mathbf{G}(x) = [g_1(x), g_2(x), \ldots g_n(x)] \tag{7.2}$$

where

$$g_j(x) = g_{j0} + g_{j1}x + g_{j2}x^2 + \ldots + g_{jK-1}x^{K-1} \tag{7.3}$$

As for block codes, the problem is to find generator polynomials which have good error control properties, and most convolutional codes have been found by computer search. Optimal codes (codes with maximum *free distance*, d_{free}) are given in the literature (Larsen, 1973; Paaske, 1974) and an abbreviated list is given in Table 7.1. Here the polynomials are in octal format and are interpreted in the following example.

Example 7.1

Consider the simple code

$$\mathbf{G} = [5,7] \quad R = 1/2 \quad K = 3$$

From Table 7.1

$$5 \equiv 101 \equiv 1 + x^2 = g_1(x)$$

$$7 \equiv 111 \equiv 1 + x + x^2 = g_2(x)$$

which gives the encoder in Figure 7.2. For the implementation in Figure 7.2(a), the constraint length is equal to the shift register length, that is, $K = m_e$. Alternatively, we could represent the encoder as the 'memory-2' system in Figure 7.2(b) where the constraint length is now one more than the shift register length. Sometimes K is equated to the minimum number of memory elements (Forney, 1970), in which case $K = 2$ for Figure 7.2.

The polynomials behave as *sequence transforms*, so we can use them in the same way as we use Fourier or z transforms. For example, the sequence $\{v_1\}$ corresponds to

$$v_1(x) = g_1(x)\,u(x) \tag{7.4}$$

Table 7.1 *Optimal convolutional codes.*

	K	d_{free}	$g_1(x)$	$g_2(x)$	$g_3(x)$	$g_4(x)$
$R = 1/2$	3	5	5	7		
	4	6	15	17		
	5	7	23	35		
	6	8	53	75		
	7	10	133	171		
	8	10	247	371		
	9	12	561	753		
	10	12	1167	1545		
$R = 1/3$	3	8	5	7	7	
	4	10	13	15	17	
	5	12	25	33	37	
	6	13	47	53	75	
	7	15	133	145	175	
	8	16	225	331	367	
	9	18	557	663	711	
	10	20	1117	1365	1633	
$R = 1/4$	3	10	5	7	7	7
	4	13	13	15	15	17
	5	16	25	33	35	37
	6	18	53	67	71	75
	7	20	135	135	147	163
	8	22	235	275	313	357
	9	24	463	535	733	745
	10	27	1117	1365	1633	1653

where multiplication is carried out over GF(2). Sequence $\{v_1\}$ can therefore be regarded as the result of a filtering or *convolution* operation upon input $u(x)$ – hence the term convolutional coding. To illustrate, consider the semi-infinite input sequence

$$\{u\} = 0110100\ldots$$

Therefore

$$g_1(x)u(x) = (1+x^2)(x+x^2+x^4) = x+x^2+x^3+x^6$$
$$\{v_1\} = 0111001\ldots$$

Similarly

$$g_2(x)u(x) = (1+x+x^2)(x+x^2+x^4) = x+x^5+x^6$$
$$\{v_2\} = 0100011\ldots$$

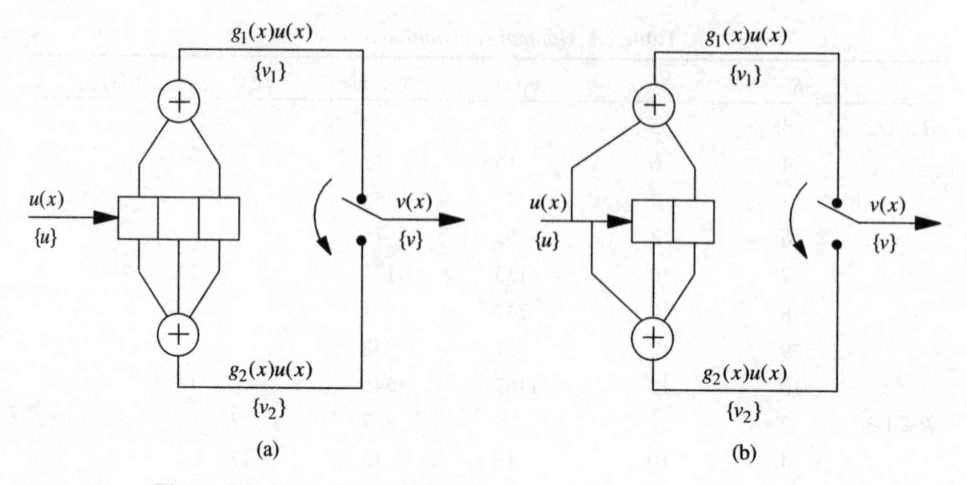

Figure 7.2 *Two representations of the [5,7], R = 1/2, K = 3 encoder.*

giving the semi-infinite code sequence

$$\{v\} = 00111010000111\ldots$$

Note that the information sequence cannot be identified in the code sequence, so the code is non-systematic. Non-systematic codes are usually preferred when Viterbi decoding is used since, in general, they offer the maximum possible d_{free} for a given rate and constraint length (Viterbi and Omura, 1979).

7.1.2 Puncturing

Rate $1/n$ codes can be modified via a process of 'puncturing' to achieve higher rates (at the expense of decreased performance). The advantage of puncturing over the classical way to achieve high rate is that it enables a *single* decoder designed for a low rate (such as 1/2) to also decode higher rate codes. In particular, we could use a rate-selectable encoder and a single Viterbi decoder in an adaptive FEC scheme. Consider the rate 1/2 [133,171] code in Table 7.1:

$$\mathbf{G} = [133, 171] \quad R = 1/2 \quad K = 7$$

$$133 \equiv 1011011 \equiv 1 + x^2 + x^3 + x^5 + x^6 = g_1(x)$$

$$171 \equiv 1111001 \equiv 1 + x + x^2 + x^3 + x^6 = g_2(x)$$

The [133,171] generator provides the best error-correcting performance of all rate 1/2, $K =$ 7 codes, and so is often used in satellite communications systems. Figure 7.3 shows the corresponding encoder, together with a puncturing (or *perforation*) matrix to increase the code rate. Consider the puncturing matrix

$$\mathbf{P} = \begin{bmatrix} 110 \\ 101 \end{bmatrix} \tag{7.5}$$

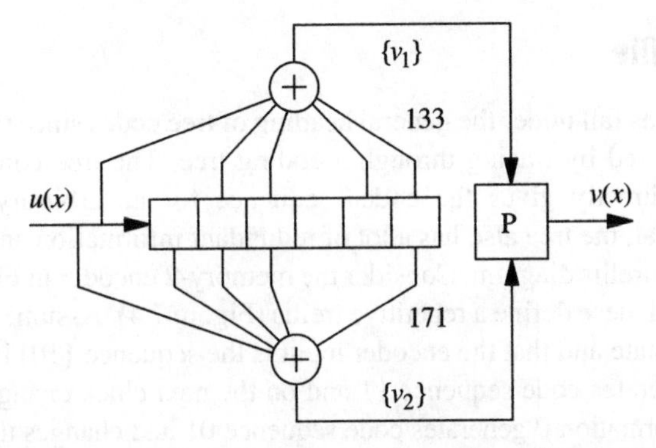

Figure 7.3 *[133,171] R = 1/2, K = 7 encoder with puncturing.*

where a zero corresponds to a deleted or punctured code bit (the bit is not transmitted). The rows of the matrix correspond to code sequences $\{v_1\}$ and $\{v_2\}$ and the columns correspond to a sequence of three information bits. For the first information bit, both v_1 and v_2 are transmitted, for the second bit only v_1 is transmitted, and for the third bit only v_2 is transmitted. Only four code bits are transmitted for every three information bits, corresponding to $R = 3/4$, and the cycle of puncturing is then repeated. Equation (7.5) can be generalised by deleting $l-1$ code bits from a total of $2l$ code bits for every l information bits. The code rate is then $l/(l+1)$, and is usually written in the form $R = (n-1)/n$, $n = 3, 4....$

Puncturing weakens the code by reducing its d_{free} and **P** should be chosen to give the best possible BER performance. For instance, the basic rate 1/2 code in Figure 7.3 has $d_{free} = 10$ (Table 7.1), while a properly punctured rate 2/3 code has $d_{free} = 6$. Table 7.2 gives optimum **P** matrices, and the corresponding free distances, for the basic rate 1/2 code. More general tables are given in the literature (Yasuda *et al.*, 1984; Lee, 1988).

Decoding (specifically Viterbi decoding) is performed as for the original rate 1/2 code after inserting dummy or null bits (erasures) into positions where coded bits are deleted. In coding theory, an erasure is simply an unknown (uncertain) value in a known location of the decoder input. The decoder therefore treats the null bits or erasures as being exactly midway between 1 and 0 (which of course conveys no information and so does not contribute to the trellis metric (Section 7.3.3)). Finally, it is important to note that a punctured code requires a larger 'decoding window' (Section 7.3.2) compared to the basic code, for example, the decoding window for the rate ¾ code should be approximately double that of the basic rate 1/2 code.

Table 7.2 *Optimal **P** matrices for the [133,171] R = 1/2, K = 7 code.*

Rate R	2/3	3/4	4/5	5/6	6/7	7/8
d_{free}	6	5	4	4	3	3
133	11	110	1111	11010	111010	1111010
171	10	101	1000	10101	100101	1000101

7.2 Code trellis

Convolutional codes fall under the general heading of tree codes since their coding operation can be visualised by tracing through a coding tree. The tree comprises nodes and branches and it directly gives the coded sequence for an arbitrary input sequence. However, in general, the tree also has a lot of redundant information and a more compact visualisation is the trellis diagram. Consider the memory-2 encoder in Figure 7.2(b). It has $2^{K-1} = 4$ states, and these define a repetitive trellis (Figure 7.4). Assume that the register is initially in the 00 state and that the encoder input is the sequence {10110001...}. The first information 1 generates code sequence 11 and on the next clock changes the state to 10. The following information 0 generates code sequence 01 and changes the state to 01. The third bit (a 1) generates code sequence 00 and changes the state back to 10, and so on. Continuing in this way gives the full coding path (dotted) and codeword $\mathbf{v}$ for the above sequence. Generalising, each branch will be labelled with n code symbols for a rate $R = 1/n$ code, and a *message tail* of $K-1$ zeros could be used to return the encoder (or decoder) to the all-zeros state. For example, the ETSI-GSM specification for cellular radio uses a $K = 5$ convolutional code and appends a tail of four zeros to force the memory to the all-zeros state.

The trellis concept is fundamental to Viterbi decoding, although the diagram itself is limited to low values of K due to the exponential growth in the number of states (2^{K-1} states). More importantly, this exponential growth means that Viterbi decoding is usually limited to low values of K, and typically $K = 7$.

Example 7.2
The [5,7,7], $R = 1/3$, $K = 3$ code is defined by

$$\mathbf{G}(x) = [1 + x^2, 1 + x + x^2, 1 + x + x^2]$$

and the corresponding encoder is shown in Figure 7.5(a). The reader should use this circuit to deduce the trellis in Figure 7.5(b).

7.2.1 Free distance

Just as for block codes, the minimum distance of a convolutional code is a fundamental parameter since it determines its error control capability when the average number of errors/codeword is small. However, compared with the distance measure for block codes, the measure for a convolutional code is rather more complex since it depends upon the type of decoding used. More precisely, it depends upon the number of frames, m, used by the decoder (see Figure 7.1(b)).

Definition 7.1 The mth-order minimum distance d_m of a convolutional code is the minimum Hamming distance between all possible pairs of coded sequences or codewords m frames (or branches) long which differ in their initial frame. Mathematically

$$d_m = \min d_{\mathrm{H}}(\mathbf{u}_m, \mathbf{v}_m), \quad \mathbf{u}_1 \neq \mathbf{v}_1 \tag{7.6}$$

where the first m frames of two codewords are denoted $\mathbf{u}_m$ and $\mathbf{v}_m$.

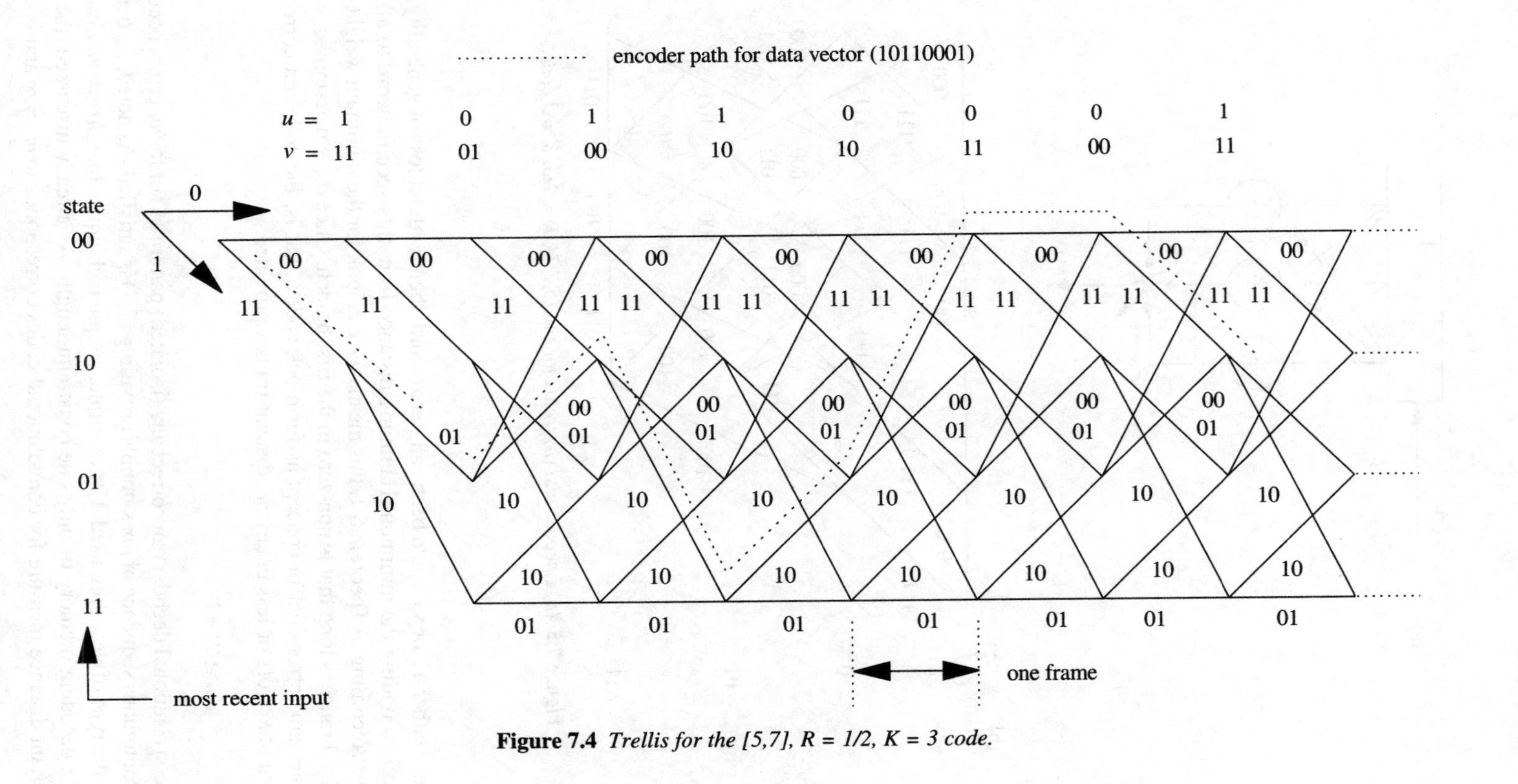

Figure 7.4 *Trellis for the [5,7], R = 1/2, K = 3 code.*

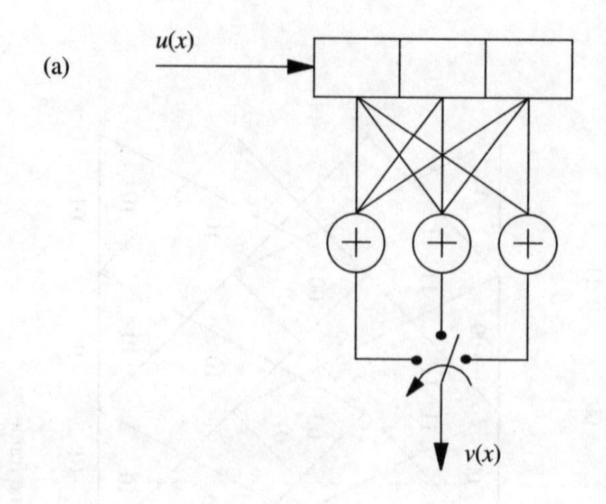

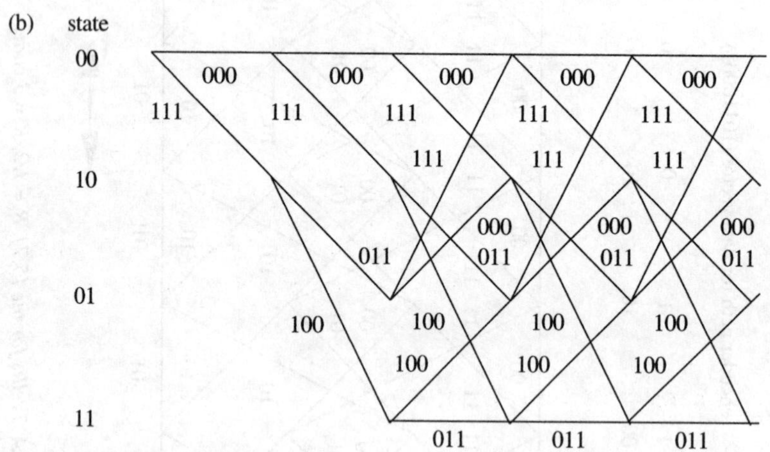

Figure 7.5 *(a) Encoder and (b) trellis for the [5,7,7], R = 1/3, K = 3 code.*

Since either $\mathbf{u}_m$ or $\mathbf{v}_m$ could be the all-zeros word, then, without loss of generality, (7.6) reduces to finding the minimum Hamming distance from this specific word to all other words. According to Theorem 6.1, this means we can look for the minimum weight codeword m branches long that is non-zero in the first branch. Once d_m is determined, we can use the same error control concept as for block codes, that is, the code can correct any pattern of t errors or less in any m adjacent frames provided

$$d_m \geq 2t + 1 \tag{7.7}$$

A convolutional decoder can correct any (isolated) pattern of t or fewer errors occurring in a contiguous sequence of mn symbols providing (7.7) is satisfied. As noted, d_m depends upon the type of decoding used. For example, simple techniques like *threshold decoding* have a decoding memory of only one constraint length. As already discussed, the most important distance measure for convolutional codes corresponds to $m \Rightarrow \infty$ and is called

d_{free}. This distance is appropriate to probabilistic decoding methods, such as Viterbi decoding, and sequential decoding, since here the decoding memory is, in principle, unlimited.

Example 7.3

Consider the trellis diagram in Figure 7.5. To find d_{free} we examine all paths leaving the zero state and returning to that state, and the path with the minimum Hamming weight gives d_{free}. For most codes the path corresponding to d_{free} is only a few constraint lengths long, and often is just one constraint length long. For Figure 7.5, $d_{free} = 8$, corresponding to codewords (111,011,111) and (111,100,100,111), and this is confirmed in Table 7.1. Similarly, d_{free} for the [5,7] code in Figure 7.4 is easily seen to be 5, corresponding to the codeword (11,01,11).

7.3 Viterbi decoding

As discussed in Section 5.3.2, ML decoding is normally accomplished via a minimum distance decoder, and this is usually referred to as the Viterbi algorithm (VA). The VA is therefore an optimal decoding technique for a memoryless channel, although we note that it minimises the *sequence* error probability and does not directly minimise the BER. The VA is available in VLSI form and finds application, for example, in satellite networks, spread spectrum systems, high-speed data communications and digital recording.

7.3.1 Hard-decision Viterbi decoding

As shown in Section 5.3.3, for a quantised channel output sequence, $\mathbf{r}$, ML decoding selects code sequence $\mathbf{v}_i$ when

$$P(\mathbf{r}|\mathbf{v}_i) \geq P(\mathbf{r}|\mathbf{v}_j), \quad \forall i \neq j \tag{7.8}$$

When $\mathbf{r}$ is quantised to just two levels the DMC in Figure 5.7 becomes a binary symmetric channel (BSC) and the Viterbi decoder is fed with a hard-decision input. In this case $\mathbf{r}$ is an N-dimensional binary vector corresponding to an N-dimensional code sequence $\mathbf{v}_i$, and errors in the hard-decision generate a finite Hamming distance between the two vectors. Put another way, there is a finite channel BER at the input of the Viterbi decoder. Inserting binary symbols into (5.18), we see that ML decoding now reduces to finding a code sequence $\mathbf{v}_i$ which minimises the Hamming distance between $\mathbf{v}_i$ and $\mathbf{r}$. Hard-decision Viterbi decoding therefore seeks a trellis path which has minimum Hamming distance from $\mathbf{r}$.

The VA operates frame by frame over a finite number of frames. At any frame the decoder does not know the node the encoder reached, so it labels the possible nodes with metrics – in this case, the running Hamming distance between the trellis path and the input sequence. In the next frame the decoder uses these metrics to deduce the most likely path and drop other paths.

Example 7.4

Consider the trellis in Figure 7.4 and assume the following information and code vectors:

$$\mathbf{u} = (1\ \ 0\ \ 1\ \ 1\ \ 0\ \ 0\ \ 0\ \ 1\ \ 0\ \ 0\ \ 1\ldots)$$

$$\mathbf{v} = (11\ 01\ 00\ 10\ 10\ 11\ 00\ 11\ 01\ 11\ 11\ldots)$$

$$\mathbf{r} = (11\ 1\underline{1}\ 00\ 10\ 10\ \underline{0}1\ 00\ 11\ 01\ 11\ 11\ldots)$$

Note that **r** is assumed to have two errors (underscored). Decoding starts by computing the metrics of encoder paths up to the first remerging node, that is, for the first K branches (Figure 7.6(a)). For ML decoding we retain only those paths with minimum Hamming distance, and the 'surviving' paths are extended by one frame (Figure 7.6(b)). The new survivors are then extended for the following frame, Figure 7.6(c), and so on. Where two metrics at a node are identical, as in Figure 7.6(f) we make a random choice. The significant point here is that, providing enough frames are examined, erroneous paths are eventually eliminated (compare Figures 7.6(h) and (i)), and a unique path covering both error bits in **r** is generated. Referring to Figure 7.4, we can see that this path corresponds to **v**, so the two errors are corrected by the decoder.

The code used in Example 7.4 has $d_{\text{free}} = 5$, so according to (7.7) any pattern of two channel errors can be corrected. In fact, for this code, (7.7) implies that any two errors are correctable, while three *closely spaced* (virtually adjacent) errors are not. This is illustrated in Example 7.5.

Example 7.5

The failure of the [5,7] $R = 1/2$, $K = 3$ code for three closely spaced errors can be demonstrated by assuming the following sequences

$$\mathbf{v} = (00\ \ 00\ \ 00\ \ 00\ \ 00\ \ 00\ldots)$$

$$\mathbf{r} = (00\ \ \underline{1}0\ \ 0\underline{1}\ \ \underline{1}0\ \ 00\ \ 00\ldots)$$

As before, the decoder keeps only the minimum distance paths at each new frame (Figure 7.7). Note, however, that the correct (all-zeros) path has been eliminated in Figure 7.7(c), implying an inevitable decoding error. The solution here is to use a code with a larger d_{free}, such as $d_{\text{free}} = 7$. On the other hand, the [5,7] $R = 1/2$, $K = 3$ code will correct three more widely spaced errors.

To summarise, in general we cannot state the error correction capability of a convolutional code as succinctly as we can for a block code since it depends upon the error distribution. In particular, Viterbi decoding does not perform well against burst errors due to error propagation. In such cases it is usual to use an outer block code to combat the error bursts at the output of the Viterbi decoder (Section 5.1.2), or to use interleaving.

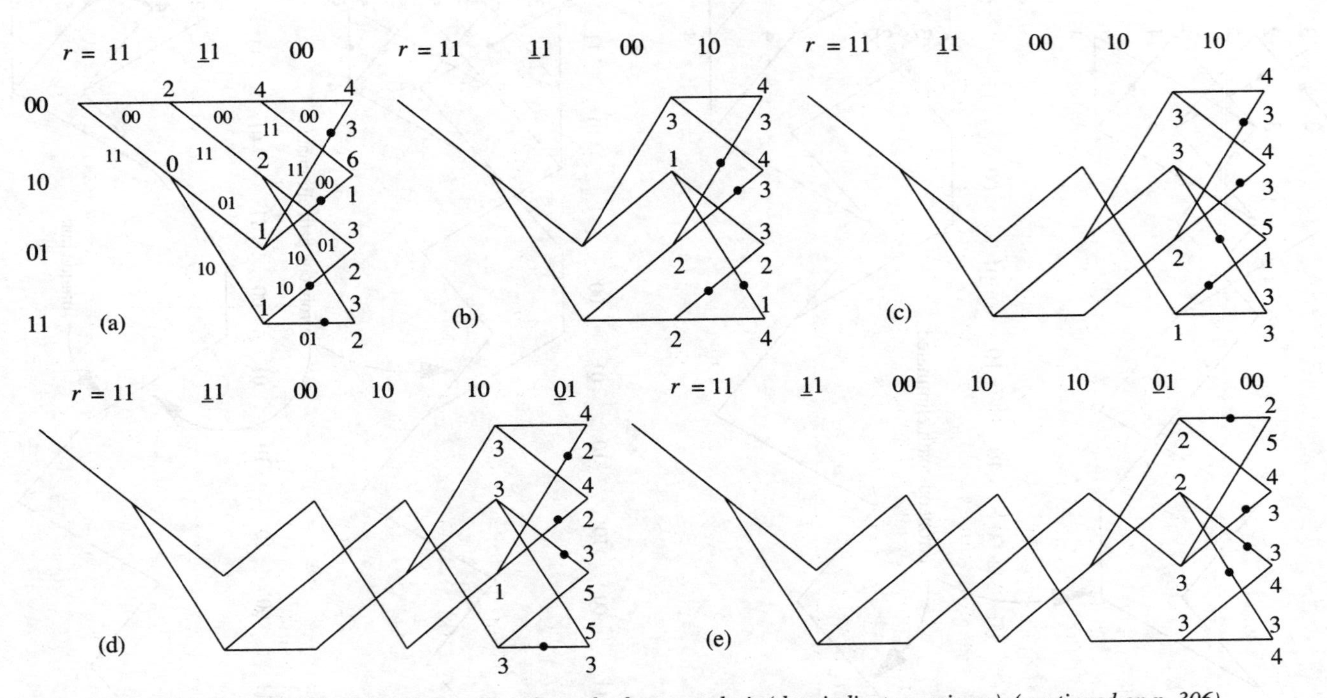

Figure 7.6 *Hard-decision Viterbi decoding: frame by frame analysis (dots indicate survivors).* *(continued on p. 306)*

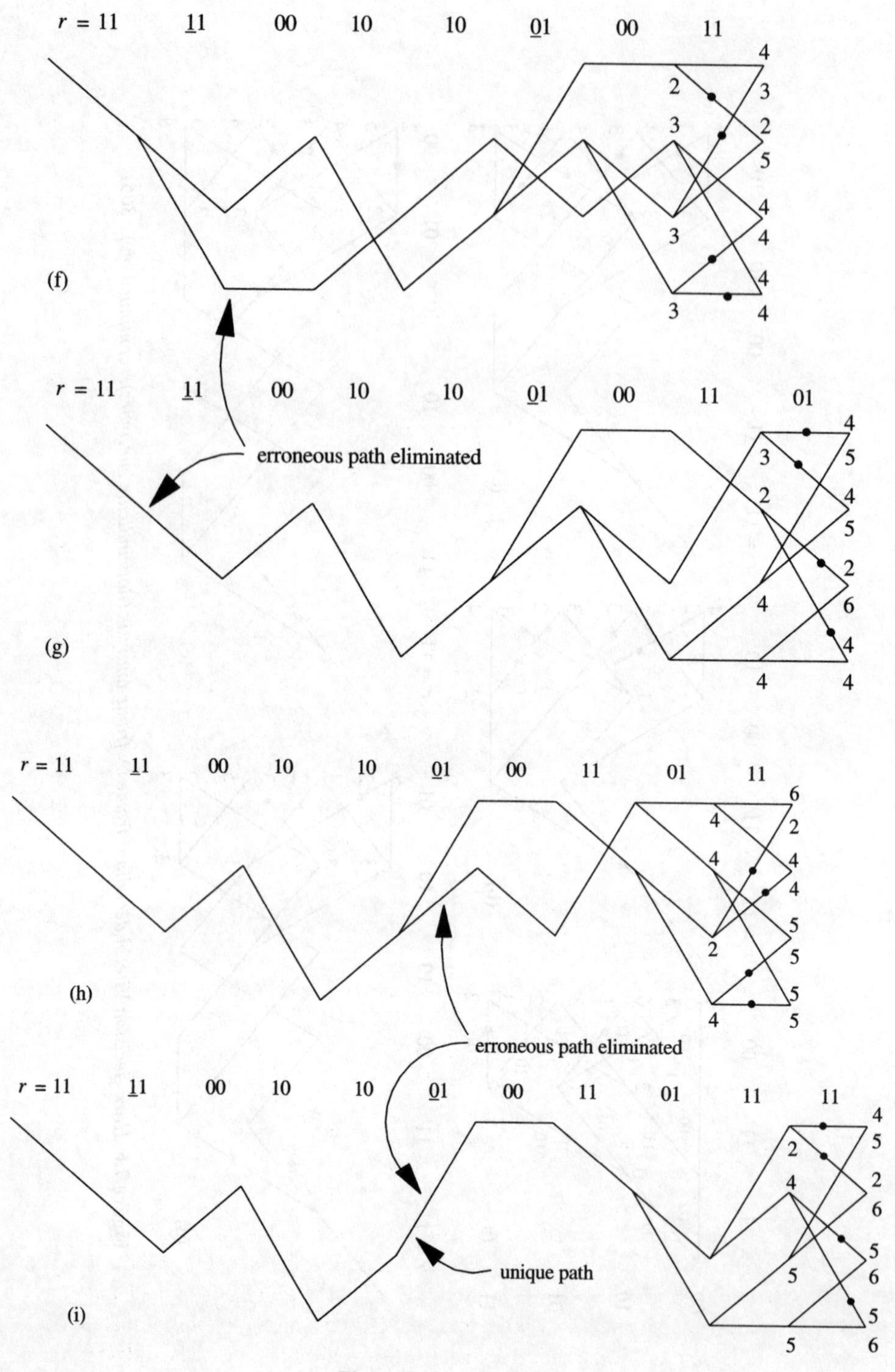

Figure 7.6 *(continued)*

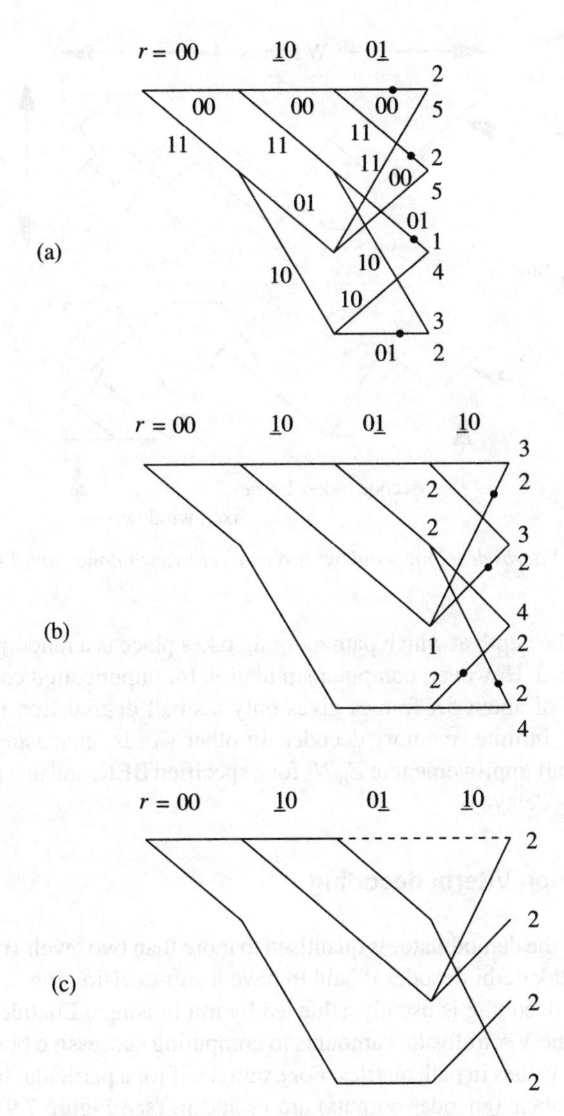

Figure 7.7 *Illustrating failure of Viterbi decoding.*

7.3.2 Decoding window

In practice, a Viterbi decoder has a finite memory which can be visualised as a fixed window, W, holding a finite portion of the trellis. For example, Figure 7.6(f) and (g) could be redrawn in a fixed 8-frame window as shown in Figure 7.8. The depth of the window or *chainback* memory must be large enough to ensure that a well-defined decision can be made in the oldest frame. In other words, there should be a high probability that all paths merge to a single branch for the oldest frame, as shown in Figure 7.8. The oldest frame is then decoded and effectively the trellis is shifted one frame to the left, giving a new frame on the right. A decoding failure occurs when paths fail to merge for the oldest frame.

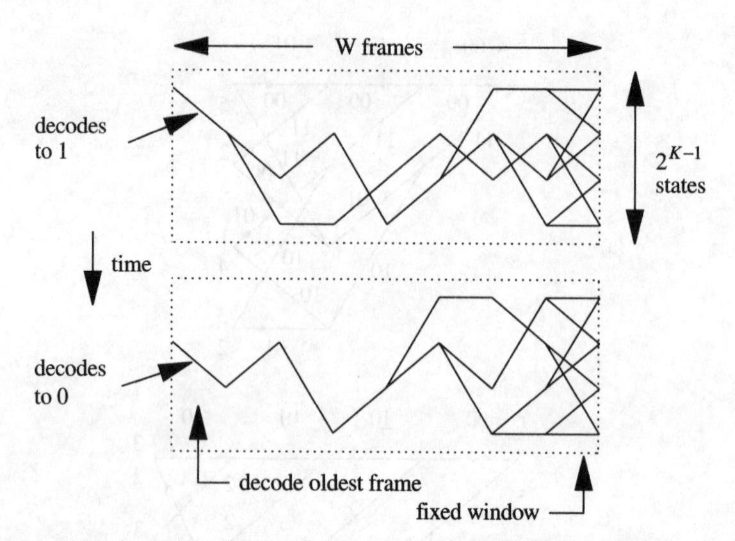

Figure 7.8 *Viterbi decoding window: two consecutive windows from Figure 7.6.*

Unfortunately, the depth at which path merging takes place is a random variable and so cannot be guaranteed. However, computer simulation for unpunctured codes shows that a decoding depth W of about $5K$ frames gives only a small degradation in decoded BER compared with an 'infinite' memory decoder. In other words, increasing W beyond this will give only a small improvement in E_b/N_0 for a specified BER, and this at the expense of increased decoding delay.

7.3.3 Soft-decision Viterbi decoding

Often the output of the demodulator is quantised to more than two levels (typically to 3 bits or 8 levels) and the Viterbi decoder is said to have a soft-decision input. As discussed in Section 5.3.2, ML decoding is usually achieved by minimising a Euclidean distance (see 5.18). In terms of the VA trellis this amounts to computing successive branch metrics and accumulating their values in path metrics. For example, if for a particular branch of the VA trellis the known labels (encoder outputs) are v_1 and v_2 (see Figure 7.9), and the corresponding decoder inputs are r_1 and r_2 then the branch metric is computed as

$$bm = \sum_{j=1}^{2}(r_j - v_j)^2 \tag{7.9}$$

We will outline the essential steps for soft-decision Viterbi decoding using the industry standard [133,171] $R = 1/2$, $K = 7$ code ($2^{K-1} = 64$ states). We assume that the encoder output is +1 or –1 to which is added an independent Gaussian noise sample. For simplicity the decoder inputs r_{2i} and r_{2i+1} on frame i are therefore taken as continuous valued. For $0 \le i \le K-2$ the trellis grows and we use the general state transition shown in Figure 7.9. A running path metric is assigned to each node, and for node p it is given by

$$pm[i+1][p] = pm[i][n] + (r_{2i} - v_1)^2 + (r_{2i+1} - v_2)^2 \tag{7.10}$$

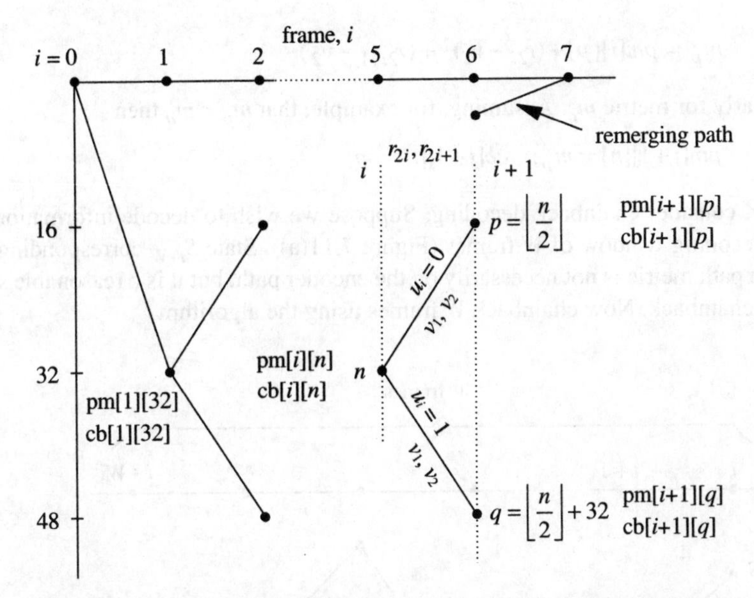

Figure 7.9 *Trellis growth for the [133,171] R = 1/2, K = 7 Viterbi decoder.*

Here, v_1 and v_2 are the branch labels for the case $u_i = 0$ given current state n. In addition to the two path metrics, we assign chainback information to each node in order to define a likely path back through the trellis. For the general transition in Figure 7.9 we assign $cb[i + 1][p] = n$ and $cb[i + 1][q] = n$ so that chainback points to the previous state.

The repetitive part of the trellis (path merging) commences on frame $K - 1$ and we now use the state transitions in Figure 7.10 in addition to the general transition in Figure 7.9. For branch pn a temporary path metric is computed as

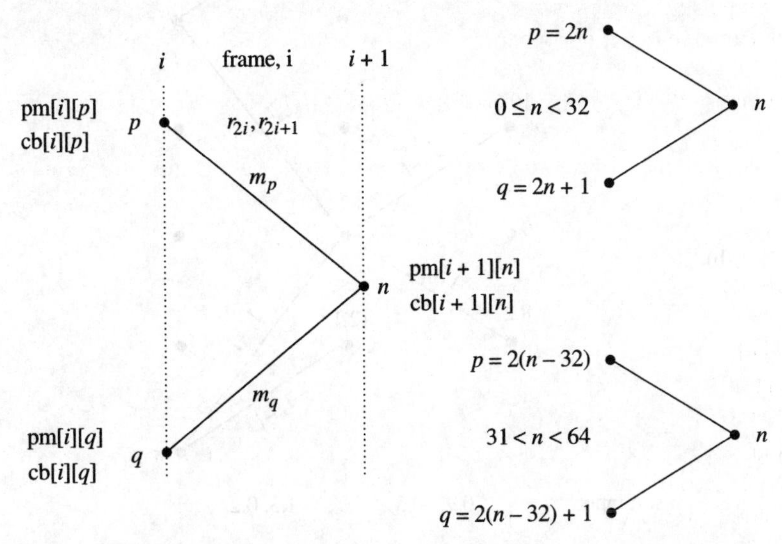

Figure 7.10 *Repetitive trellis for the [133,171] R = 1/2, K = 7 Viterbi decoder.*

$$m_p = pm[i][p] + (r_{2i} - v_1)^2 + (r_{2i+1} - v_2)^2 \tag{7.11}$$

and similarly for metric m_q. Assuming, for example, that $m_p < m_q$ then

$$pm[i+1][n] = m_p; \quad cb[i+1][n] = p \tag{7.12}$$

Finally, consider chainback decoding. Suppose we wish to decode information bit u_i given a decoding window of W frames (Figure 7.11(a)). State S_{i+W} corresponding to the minimum path metric is not necessarily on the encoder path, but it is a reasonable starting point for chainback. Now chainback W frames using the algorithm

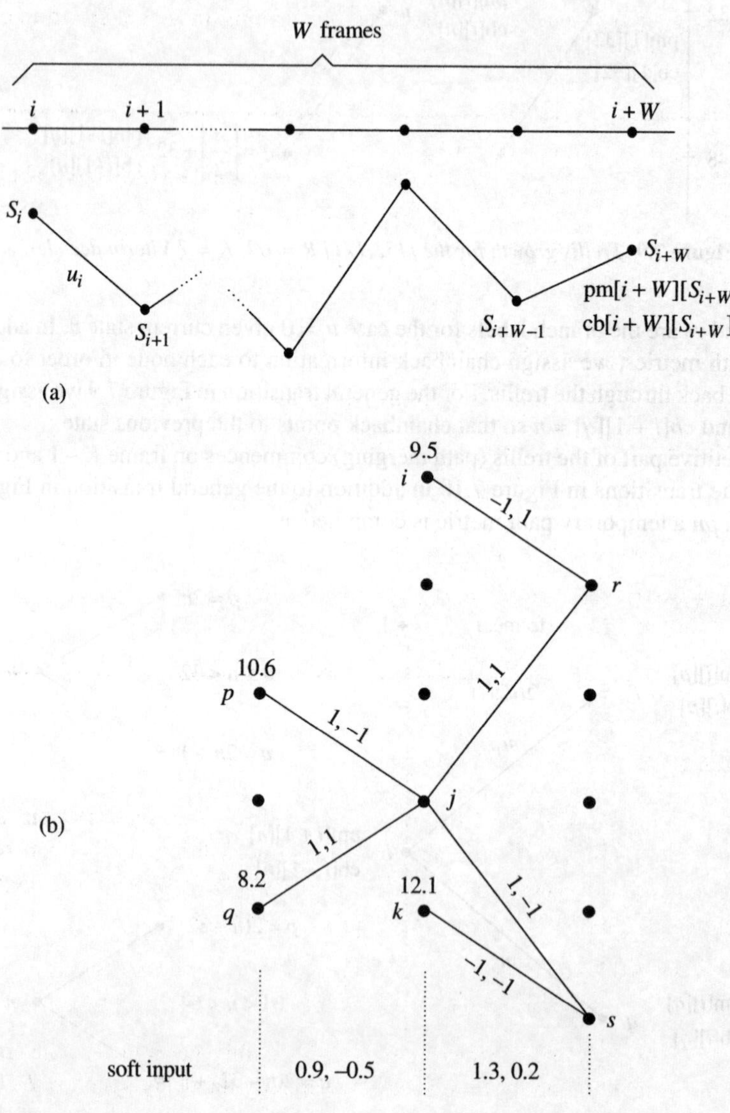

Figure 7.11 *(a) Chainback decoding for u_i; (b) illustrating soft-decision decoding.*

$$S[j] = cb[j+1][S_{j+1}], \quad j = (i+W-1), \ldots, i+1, i \tag{7.13}$$

Providing $W \approx 5K$ or more, then states S_i and S_{i+1} will most likely identify the minimum path metric for the corresponding frame, and therefore are most likely to identify the path used by the encoder for frame i. Simple rules are then used for decoding, for example, $u_i = 0$ if $S_{i+1} < S_i$. Appendix C provides a soft-decision Viterbi decoding simulation for the [133,171], rate 1/2, $K = 7$ code, and can be used to examine the effect of decoding window size (the exact BER will be slightly dependent upon the noise simulation algorithm).

Finally, it should be noted that a VA can also have a soft *output* (Hagenauer and Hoeher, 1989). The Soft Output VA (SOVA) is a VA which uses soft or hard decisions to compute path metrics, but also decides on the output bits in a soft way by providing reliability information together with the output bits. The SOVA algorithm finds application in concatenated VA schemes, for example.

Example 7.6

Figure 7.11(b) shows part of a trellis for a rate 1/2 convolutional code. Branch labels correspond to the code symbol pair v_1, v_2 and running path metrics are shown for nodes p, q, i and k. The path metric for node j is found as

$$m_p = 10.6 + (0.9-1)^2 + (-0.5-(-1))^2 = 10.86$$

$$m_q = 8.2 + (0.9-1)^2 + (-0.5-1)^2 = 10.46$$

so the metric at node j is 10.46. The path metric for node r is found as

$$m_j = 10.46 + (1.3-1)^2 + (0.2-1)^2 = 11.19$$

$$m_i = 9.5 + (1.3-(-1))^2 + (0.2-1)^2 = 15.43$$

so the metric at node r is 11.19. Finally, for node s we have

$$m_j = 10.46 + (1.3-1)^2 + (0.2-(-1))^2 = 11.99$$

It is now apparent that the optimal chain-back is $r \Rightarrow j \Rightarrow q$.

7.4 Code spectra

Here we take a fundamental look at convolutional codes with respect to their 'spectra', particularly their *weight spectrum*, and we will do this with emphasis upon *recursive systematic codes* (*RSCs*). These codes are the basic or *constituent* codes used in powerful concatenated coding schemes (Turbo encoders and decoders).

7.4.1 Recursive systematic codes

Consider the convolutional encoder in Figure 7.12; this generates a rate 1/2 RSC. Following the usual notation for these codes, we employ the delay operator, D, to create the D-transform of the associated sequences. As in Example 7.1, this enables the

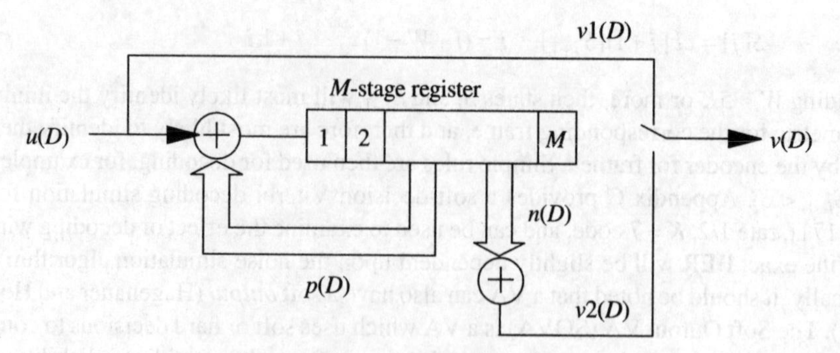

Figure 7.12 *General encoder for a rate 1/2 RSC.*

convolutional operation of the encoder to be represented as a product of transforms. For example

$$\{u_0, u_1, u_2, \ldots\} \Leftrightarrow u(D) = u_0 + u_1 D + u_2 D^2 + \ldots \tag{7.14}$$

The recursive loop in Figure 7.12 gives polynomial division, while the feedforward loop gives polynomial multiplication, so the encoder output is

$$v(D) = \left[u(D), u(D)\frac{n(D)}{p(D)} \right] = u(D)\mathbf{G}(D) \tag{7.15}$$

where the generator matrix is given by

$$\mathbf{G}(D) = \left[1, \frac{n(D)}{p(D)} \right] \tag{7.16}$$

A rate $1/n$ RSC would be generated by

$$\mathbf{G}(D) = \left[1, \frac{n_1(D)}{p(D)}, \frac{n_2(D)}{p(D)}, \ldots, \frac{n_{n-1}(D)}{p(D)} \right] \tag{7.17}$$

If polynomials $n(D)$ and $p(D)$ in (7.16) are *relatively prime* (no common factor) then we can write

$$\frac{n(D)}{p(D)} = a_0 + a_1 D + a_2 D^2 + \ldots \tag{7.18}$$

corresponding to a semi-infinite code sequence. In this case, encoding by (7.16) results in an infinite weight code sequence (infinite number of 1s), except when $u(D)$ is of the form $u(D) = w(D)p(D)$, where $w(D)$ is an arbitrary polynomial. For iterative decoding, $p(D)$ is often selected to be a primitive polynomial of degree M for a memory-M encoder.

7.4.2 Code transfer function

Consider the encoder in Figure 7.13, corresponding to

$$\mathbf{G}(D) = \left[1, \frac{1}{1 + D + D^3} \right] \tag{7.19}$$

From the trellis it is apparent that there is one path with Hamming distance $d = 4$, that is, $d_{free} = 4$, which leaves and then remerges with the all-zeros path, and several remerging paths with $d = 6$. In general there will be a_d paths of Hamming distance d from the all-zeros path, where the set a_d could be referred to as a *distance spectrum.* This spectrum is of interest (particularly the d_{free} term) since it directly determines the error probability when decoding (see (7.31)). The complete distance spectrum can be formally determined by evaluating a *transfer function* $T(D)$ for the code, although the complexity of this approach increases exponentially with constraint length K. For, say, $K > 7$, and for punctured codes, it may be easier to compute the first few (the most important) terms of $T(D)$ using an exhaustive search procedure, by applying test information sequences to a model of the encoder. Alternatively, we could exhaustively search all paths in the code trellis (Cedervall and Johannesson, 1970). For Figure 7.13 the first few non-zero terms of the spectrum are shown in Table 7.3.

The corresponding transfer function for Figure 7.13 is then

$$T(D) = \sum_{d=d_{free}}^{\infty} a_d D^d = D^4 + 6D^6 + 16D^8 + 69D^{10} + 232D^{12} + \ldots \qquad (7.20)$$

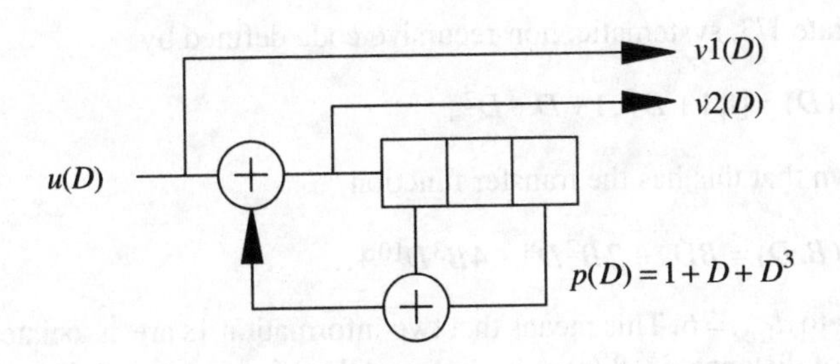

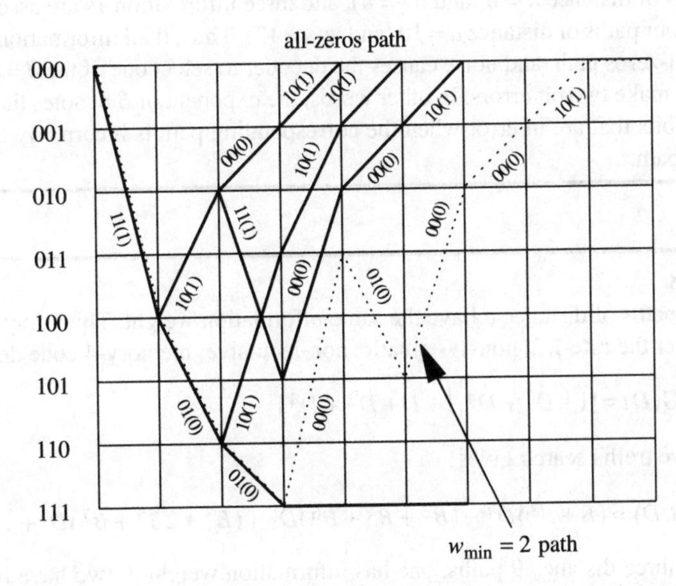

Figure 7.13 *Memory-3 RSC. The information bit is shown in brackets.*

Table 7.3 *Initial distance spectrum for Figure 7.13.*

d	4	6	8	10	12
a_d	1	6	16	69	232

In fact, the transfer function of any binary convolutional encoder not exhibiting catastrophic error propagation can always be expanded in a generalised form of (7.20).

For a full code spectrum we must also include the information *weight spectrum*, w_d, where w_d *is the total number of information 1s in all paths of distance d from the all-zeros path*. This spectrum is important since it directly determines the decoded BER; see (7.32). Note that Figure 7.13 shows the appropriate information bit for each branch (in brackets). This information can be included in $T(D)$ by associating a factor B with all branch transitions caused by an information 1.

Example 7.7

Consider the rate 1/3, systematic, non-recursive code defined by

$$\mathbf{G}(D) = [1, 1+D^2, 1+D+D^2] \tag{7.21}$$

It can be shown that this has the transfer function

$$T(B,D) = BD^6 + 2B^2D^8 + 4B^3D^{10} + \ldots \tag{7.22}$$

corresponding to $d_{\text{free}} = 6$. This means that two information 1s are associated with each of the two paths of distance $d = 8$ (and $w_8 = 4$), and three information 1s are associated with each of the four paths of distance $d = 10$ (and $w_{10} = 12$). Thus, if all information bits are 0s, giving the all-zeros path, and noise causes the decoder to select one of the $d = 8$ paths, the decoder will make two bit errors. In other words, the exponent of B denotes the number of information bits that are in error when the corresponding path is incorrectly chosen over the all-zero path.

Example 7.8

In (7.22) all paths of distance d have the same information weight. This is not always the case. Consider the rate 1/2, non-systematic, non-recursive, memory-4 code defined by

$$\mathbf{G}(D) = [1+D^3+D^4, 1+D+D^2+D^4] \tag{7.23}$$

An exhaustive trellis search gives

$$T(B,D) = (B+B^3)D^7 + (B^2+B^4+B^6)D^8 + (B^3+2B^5+B^7)D^9 + \ldots \tag{7.24}$$

Thus, of the three distance 9 paths, one has information weight 3, two have information weight 5, and one has information weight 7.

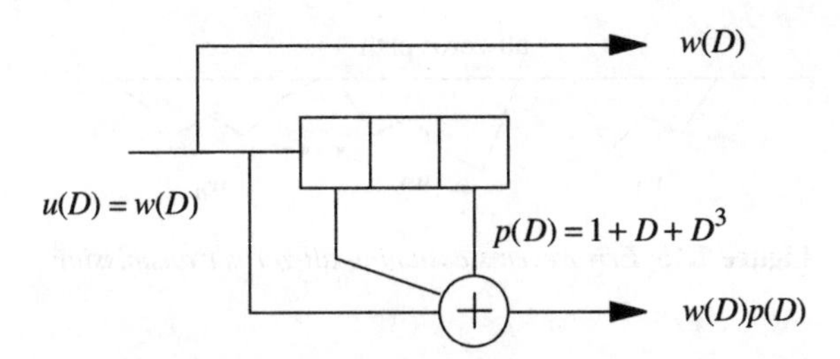

Figure 7.14 *Non-recursive form of Figure 7.13.*

7.4.3 A comparison of recursive and non-recursive coders

For the recursive encoder in Figure 7.13 we have $v2(D) = u(D)/p(D)$, so that if $u(D) = w(D)p(D)$ then $v2(D) = w(D)$. Now assume $w(D)$ is applied to a non-recursive encoder having the same polynomial, $p(D)$ (Figure 7.14). It is interesting to observe that we obtain the *same* code sequence as generated in Figure 7.13, and, since all input sequences are possible, it follows that the two encoders will have the same distance spectrum, a_d. A non-recursive encoder can therefore have the same coefficients a_d as the corresponding recursive encoder, and therefore the same 'first event' error probability (see (7.31)). Furthermore, non-recursive, *non-systematic* encoders have been shown to have a larger d_{free} than non-recursive, systematic encoders (Forney, 1970), and so for the same memory have a lower BER at high E_b/N_0. For these reasons, the classical approach has been to use non-recursive, non-systematic convolutional codes (NSCs), some of which are given in Table 7.1, and the critical parameter has been d_{free}.

The question then arises: why use RSCs? It turns out that, at *low* E_b/N_0, the d_{free} parameter tends to be an insufficient criterion for determining BER performance, and the weight spectrum w_d becomes important. Now, it is true that for *high rate* (punctured) RSCs this spectrum tends to be better than that for NSCs (Berrou *et al.*, 1992). On the other hand, for unpunctured codes, non-recursive codes have marginally better performance than recursive codes, even at low E_b/N_0. The advantage of RSCs (punctured or unpunctured) only really becomes apparent when they are incorporated into concatenated coding schemes, and a fundamental reason for selecting RSCs in preference to NSCs is discussed in Section 7.5.3.

7.4.4 Error events

Since convolutional codes are group codes we can exploit this property when computing code performance (error rates). In particular, it is convenient to assume that the all-zeros codeword is transmitted, without loss of generality. An error event is then defined as a path that leaves and joins (remerges) with the all-zeros path, as shown in Figure 7.15, and we could define d_{free} as the minimum Hamming weight of all error events. Also note that the nth error event will have an associated information weight, w_n, and, as suggested in Example 7.7, we can use this to deduce a decoded bit error rate. In particular, an incorrect path could be selected by the decoder because it is closer in distance terms to the received sequence than the all-zeros path, thereby generating bit errors equal to the information

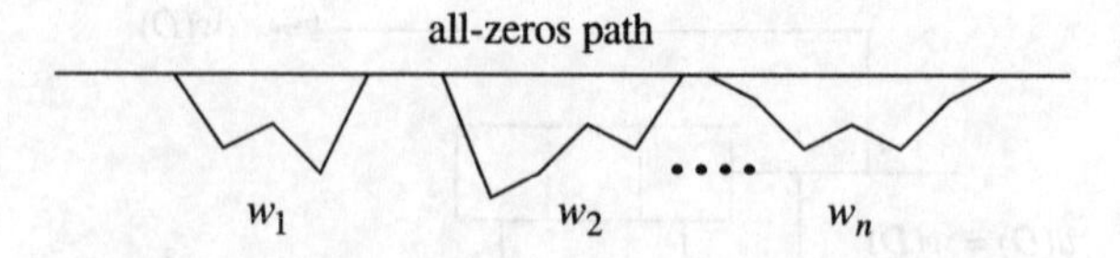

Figure 7.15 *Error events assuming all-zeros transmission.*

weight of the path. Clearly, the larger the Hamming distance of the path the more transmitted bits must be in error and the less likely that particular error event will be selected. In other words, the d_{free} path represents the most probable error event.

Example 7.9

The memory-2 encoder in Figure 7.16 is defined by

$$\mathbf{G}(D) = \left[1, \frac{1+D^2}{1+D+D^2}\right] \tag{7.25}$$

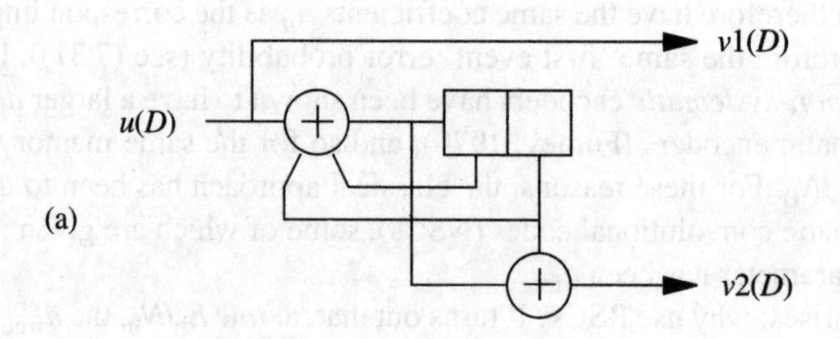

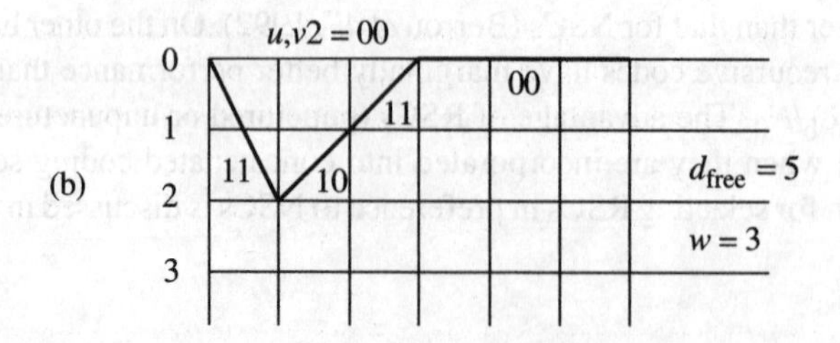

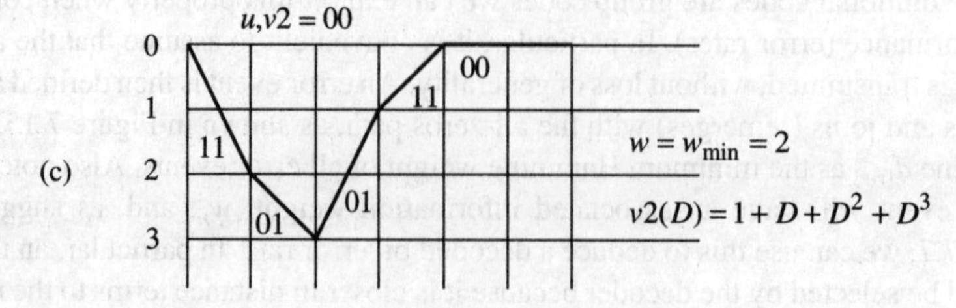

Figure 7.16 *Memory-2 encoder for a rate 1/2 RSC, and two error events.*

so

$$v2(D) = u(D)\frac{1+D^2}{1+D+D^2} \tag{7.26}$$

It is possible to select certain sequences $u(D)$ such that $v2(D)$ has finite weight (usually it will be semi-infinite). Clearly, if $u(D) = 1 + D + D^2$, then $v2(D) = 1 + D^2$ and the path has $w = 3$. This in fact corresponds to the most probable error event and the encoder traces the d_{free} path for the code (Figure 7.16(b)). As we shall see, the *minimum* information weight, w_{min}, associated with an error event is of particular interest since it effects the performance of iterative decoding systems.

Theorem 7.1 For a recursive convolutional code, the minimum information weight associated with an error event is $w_{min} = 2$.

Proof: Consider the general memory-M encoder in Figure 7.12, where

$$v2(D) = u(D)\frac{n(D)}{p(D)} \tag{7.27}$$

$$p(D) = 1 + \ldots + D^M \tag{7.28}$$

For a finite weight sequence $v2(D)$, corresponding to an error event, $p(D)$ should divide $u(D)n(D)$. Now, every polynomial $p(D)$ of degree M divides a polynomial of the form $1 + D^i$, where i is the period of the memory-M linear feedback shift register with connection polynomial $p(D)$. But since $p(D)$ has the form of (7.28), it does not divide D^i for any i. Therefore, if $u(D) = 1 + D^i$, then $p(D)$ divides it for some i and hence $v2(D)$ corresponds to a finite weight error event with information weight $w = 2$. However, $u(D) = D^i$ (corresponding to $w = 1$) will not generate a finite weight, so $w_{min} = 2$. Note that for a *non-recursive* code, the minimum information weight associated with an error event is $w_{min} = 1$. This is easily seen by letting $u(D) = 1$ in Figure 7.14.

Example 7.10
Consider the encoder in Figure 7.16. It is easily shown that $p(D)$ divides the numerator of (7.26) if we let $u(D) = 1 + D^3$, giving $v2(D) = (1 + D)(1 + D^2) = 1 + D + D^2 + D^3$. This corresponds to an error event with $w = w_{min} = 2$, as shown in Figure 7.16(c).

Theorem 7.1 can also be seen non-mathematically. It is apparent from Figure 7.16 that a recursive encoder can be forced into the zero state by applying a data tail involving a 1 (to cancel state one). A memory-M encoder would require a non-zero tail of M bits, and for Figure 7.16(c) the tail is 01. Since at least one non-zero bit is required to initiate the error event, then $w_{min} = 2$. Another example is shown in Figure 7.13. The reader should show that for this encoder, $v2(D)$ is a finite weight sequence (that is, $u(D)$ is divisible by $p(D)$) when $u(D) = 1 + D^7$. The corresponding minimum information weight path is shown dotted, and the data tail is 001.

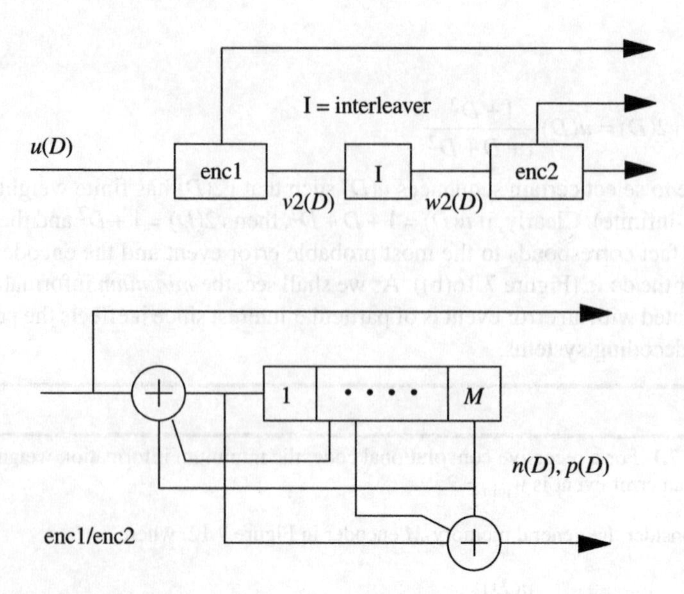

Figure 7.17 *Simple coding scheme for a rate 1/3 concatenated convolutional code (I is an interleaver).*

The value of w_{min} is important when designing iterative decoding systems based upon several convolutional decoders. For example, a simple concatenated coding scheme for an iterative system is shown in Figure 7.17, where

$$u(D) = v2(D)\frac{p(D)}{n(D)} \tag{7.29}$$

Since $n(D)$ has degree M, we could consider it to be the feedback polynomial of a linear feedback shift register of memory M. If this register has period i then $n(D)$ will divide $1 + D^i$ (Theorem 7.1). Now suppose $u(D)$ is an arbitrary finite weight information sequence; it is clear that this can be achieved if we make $v2(D) = 1 + D^i$, which is of weight 2. The fact that $v2(D)$ can have weight as low as 2 for a finite weight input sequence requires some modification of the basic coding scheme in Figure 7.17 in order to take full advantage of concatenation (Section 7.6).

7.4.5 Error rate

As discussed, the code spectrum (a_d and w_d) has a direct bearing upon the code performance, that is, on the error rate at the decoder output. The detailed error rate computation will depend, for example, upon the type of channel (hard or soft-decision decoder input) and upon the form of decoder (Viterbi decoding, or blockwise, iterative decoding).

Consider the error rate computation for the usual non-iterative decoding of convolutional codes (Viterbi decoding). We will assume soft-decision decoding and a memoryless AWGN channel. We also assume that the all-zeros sequence is transmitted, without loss of generality. As discussed in Section 7.3.3, the VA computes a running path metric for an arbitrary frame i and selects the most favourable branch at each node (see

(7.12)). Clearly, over a long period of time (we assume the VA is operating on a semi-infinite sequence) the decoder will diverge and remerge a number of times from the all-zeros path as noise causes errors in the computed metric. Thus, over a long period of time the decoder will exhibit a number of error events, as in Figure 7.15, and the probability of an error event, or *sequence* error, tends to unity. Since the VA decodes frame by frame, it is more meaningful to compute the error probability *per unit time*, that is, at time *i*. It is quite possible, for instance, that the decoder selects, *for the first time,* a path distance *d* from the all-zeros path which remerges with the all-zeros path at frame *i*. In other words, an error event (the first event) terminates on frame *i*, and for the first time a path distance *d* is selected in preference to the all-zeros path. This gives rise to the concept of *pairwise error probability*, P_d, which is the probability of the decoder selecting a path of distance *d* in preference to the all-zeros path at some arbitrary time *i*. For a memoryless Gaussian channel and code rate *R* this is given by

$$P_d = \frac{1}{2}\,\mathrm{erfc}\left(\sqrt{R\frac{E_b}{N_0}d}\right) \tag{7.30}$$

where erfc(x) is defined in Appendix A. By accounting for all possible paths (or error events) we can compute the probability that the all-zeros path is excluded for the first time at frame *i*. This probability is referred to as the *first event error probability*, $P(\varepsilon)$, and it can be the same for different encoders because they generate the same set of code sequences (see Section 7.4.3). By summing the error probabilities for all possible paths which merge with the all-zeros path at frame *i* we obtain the *union bound* for $P(\varepsilon)$ (Viterbi, 1971)

$$P(\varepsilon) < \sum_{d=d_{\mathrm{free}}}^{\infty} a_d P_d \tag{7.31}$$

Here, a_d is the number of paths of distance *d* from the all-zeros path which merge with the all-zeros path for the first time, that is, we are not considering a concatenation of error events. Equation (7.31) is an upper bound primarily because several paths may have a distance closer to the received sequence than the all-zeros path, but only one is selected.

More often we require the *bit error probability* (BER), which differs from $P(\varepsilon)$ by weighting the probability of each path by the number of information bits resulting in error along that path. Since the all-zeros path is the correct one, any branch of a path with an information 1 on it results in a bit error should the decoder select that path. It is therefore necessary to count the number of 1s on each remerging path. This leads to the union bound for the BER:

$$\begin{aligned} P_b(\varepsilon) &< \sum_{d=d_{\mathrm{free}}}^{\infty} w_d P_d \\ &< \frac{1}{2}\sum_{d=d_{\mathrm{free}}}^{\infty} w_d\,\mathrm{erfc}\left(\sqrt{R\frac{E_b}{N_0}d}\right) \end{aligned} \tag{7.32}$$

Recall that w_d is the total information weight associated with all paths of distance *d*. More specifically, terms w_d are the coefficients of the polynomial $\partial T(B,D)/\partial B$ with $B = 1$. For

Table 7.4 *Sensitivity of erfc(·) to E_b/N_0 (R = 1/2).*

E_b/N_0 dB	d	erfc(·)
4	5	0.0003942
4	10	0.00000054
1	5	0.01211
1	10	0.000388

example, referring to (7.24), $w_8 = 12$ and $w_9 = 20$. The d_{free} term dominates the expansion in (7.32) for channels with high E_b/N_0, although it may be necessary to include many more terms for low E_b/N_0. Table 7.4 illustrates this point for a rate 1/2 code. At 4 dB the $d = 10$ term is nearly three orders of magnitude down on the d_{free} term, but this is not true at 1 dB.

Blockwise decoding

For iterative decoding of convolutional codes (Section 7.5) it is convenient to regard the code as a block code. In this case, a block of N information bits can yield 2^N codewords, and a particular codeword of distance d to the all-zeros word could be considered to comprise a number of error events. In particular, w_d is now measured over the full length of the codeword (N branches), and so in general could span several error events. The elemental BER is therefore w_d/N, and accounting for all values of d the union bound becomes

$$P_b(\varepsilon) \le \frac{1}{2} \sum_{d=d_{\text{free}}}^{\infty} \frac{w_d}{N} \operatorname{erfc}\left(\sqrt{R\frac{E_b}{N_0}d}\right) \tag{7.33}$$

Clearly, for finite values of N and R the upper limit in the summation will also be finite.

7.5 Parallel concatenated convolutional codes

In general, the weight spectrum, w_d, of a punctured (high rate) RSC is superior to that of the corresponding non-recursive, non-systematic (NSC) punctured code (Section 7.4.3). According to (7.32) this implies that the BER performance of punctured RSCs will be somewhat superior to that of punctured NSCs at low E_b/N_0. This superior performance, coupled with the systematic structure of the code, made punctured RSCs look attractive for concatenated, iterative decoding schemes which work towards the Shannon limit (very low E_b/N_0) (Berrou *et al.*, 1992). However, as we shall see in Section 7.5.3, there is a much more fundamental reason why RSCs are used in concatenated schemes.

Concatenated convolutional coding schemes tend to fall into two categories: *parallel* concatenated convolutional codes (PCCCs, or *Turbo codes*), and *serial* concatenated convolutional codes (SCCCs). Both schemes use relatively simple (small memory) RSCs, several MAP decoders, and some mechanism of using information generated by one MAP decoder to aid decoding in a second MAP decoder. *This interchange of information between decoders is crucial in that it provides a mechanism for the iterative decoding of the current information bit* (albeit in a suboptimal way). An interleaving (or scrambling)

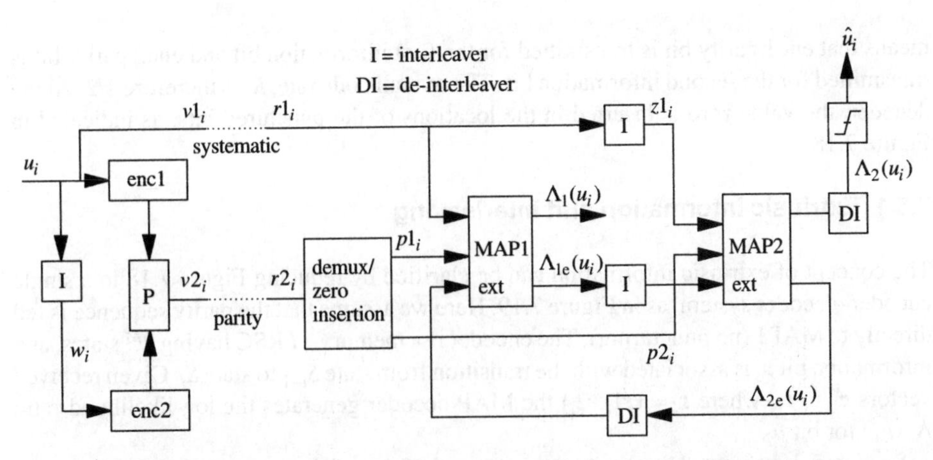

Figure 7.18 *A basic rate 1/2 PCCC scheme (Turbo encoder and decoder).*

technique is used to ensure that the additional information fed from one MAP decoder to another is approximately independent of the primary information used by the MAP decoder, at least in the early iterations. Since it is initially independent, this additional information is usually called *extrinsic* information, and it boosts decoder performance – hence the term Turbo code. Given sufficient iterations, the E_b/N_0 for very low BER can be within a fraction of a dB of the Shannon limit for the selected modulation scheme and code rate (Section 5.2.2). *The significant point is that this represents considerably better performance than that offered by classical large constraint length convolutional codes, and at much lower decoding complexity!*

Figure 7.18 shows the basic rate 1/2 Turbo code introduced by Berrou *et al.* (1993). It is also referred to as a PCCC because the two encoders operate on the same set of input bits, rather than one encoder operating on the output of another, as in Figure 7.17. The simplest approach is to treat the Turbo code as a block code by assuming a block of information bits $\mathbf{u} = (u_1, u_2, \ldots, u_i, \ldots, u_N)$. In practice these could first be stored in an input buffer, size N, and all bits in the buffer would then be interleaved and placed into a second N-bit buffer (as in Figure 7.20(a)). Each buffer would be loaded in serial fashion into enc1 and enc2, respectively, such that the N bits entering enc2 are simply a permuted or scrambled version of those entering enc1. There is therefore an inherent interleaver delay of N bits although, depending on design, this is not necessarily associated with the encoder. High encoder/decoder delay is to be expected anyway as system performance approaches that of Shannon's ideal coding scheme (Section 1.1). There is also delay due to iterative decoding, although this becomes irrelevant if hardware permits decoding to be performed well within the acquisition time for a data block. The *constituent codes* (CCs) generated by enc1 and enc2 are typically simple, small constraint length (K = 3 to 5) RSCs (the use of *systematic* codes is not strictly required, but it simplifies the extraction of extrinsic information in the decoder). The encoder parity sequences can be punctured to improve the overall code rate. In particular, setting

$$\mathbf{P} = \begin{bmatrix} 1 & 0 \\ 0 & 1 \end{bmatrix}$$

means that enc1 parity bit is transmitted for the first information bit and enc2 parity bit is transmitted for the second information bit. The overall code rate, R, is therefore 1/2. At the decoder, the value zero is inserted in the locations of the punctured bits, as indicated in Figure 7.18.

7.5.1 Extrinsic information and interleaving

The concept of extrinsic information can be clarified by reducing Figure 7.18 to a single encoder–decoder system, as in Figure 7.19. Here we assume that the parity sequence is fed directly to MAP1 (no puncturing). The encoder is a memory-M RSC having 2^M states, and information bit u_i is associated with the transition from state S_{i-1} to state S_i. Given received vectors $\mathbf{r}_1$ to $\mathbf{r}_N$ where $\mathbf{r}_i = (r1_i, r2_i)$ the MAP decoder generates the log-likelihood ratio $\Lambda_1(u_i)$ for bit u_i.

Section 5.3.4 shows that, due to the systematic form of the code, $\Lambda_1(u_i)$ readily splits into two *independent* terms, that is

$$\Lambda_1(u_i) = \Lambda_{1s}(u_i) + \Lambda_{1e}(u_i) \tag{7.34}$$

The first term arises from systematic information, $r1_i$, while the second term is independent of $r1_i$ in the initial iterations and is derived from parity information and the code structure. The important point is that only $\Lambda_{1e}(u_i)$ is passed on to MAP2 in Figure 7.18 since this component is essentially independent of the other two inputs to MAP2 and so can aid the decoder in its estimate of $\Lambda_2(u_i)$. As shown in Figure 7.18, MAP2 also generates a term $\Lambda_{2e}(u_i)$ which can be de-interleaved and fed back as extrinsic information for MAP1. Again, this extrinsic information should ideally be independent of the other two MAP1 inputs, and now MAP1 must avoid passing on information to MAP2 which has been generated by MAP2 in the first place!

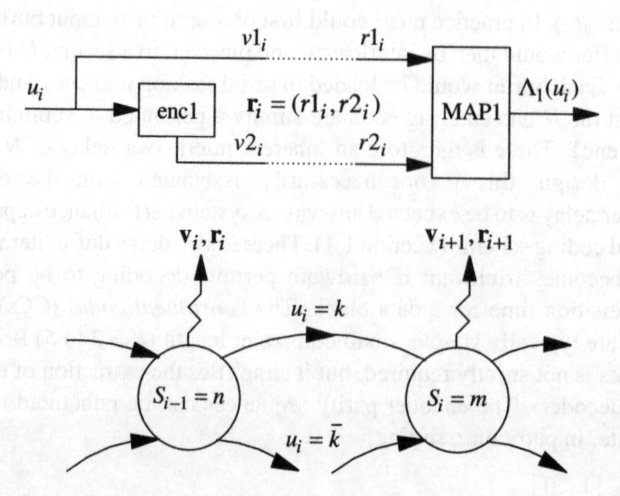

Figure 7.19 *Single encoder–decoder scheme corresponding to Figure 7.18, and its general state transition diagram.*

The need for independent information terms is met by the interleaver, I. As discussed, in the initial iterations the extrinsic input to MAP2 should be independent of the other two inputs, and this would not be the case if no interleaver was present during coding (bearing in mind that $\Lambda_{1e}(u_i)$ in Figure 7.18 depends upon $p1_i$, and this would be correlated with $p2_i$ in the absence of interleaving). Intuitively, the larger the interleaver size, N, the smaller the correlation between the interleaved form of $\Lambda_{1e}(u_i)$ and $p2_i$. Similarly, interleaving tries to ensure that the initial extrinsic input to MAP1 is only weakly correlated with inputs $r1_i$ and $p1_i$. Note that due to feedback, MAP1 output now has an additional term generated by MAP2:

$$\Lambda_1(u_i) = \Lambda_{1s}(u_i) + \Lambda_{12}(u_i) + \Lambda_{1e}(u_i) \tag{7.35}$$

and it is important to feed only the last term to MAP2, as shown in Figure 7.18.

Interleaver design

An objective in interleaver design is to spread low-weight input sequences so that the resulting codeword has high weight. Generally speaking, increased spreading will give a longer error event which will tend to increase the d_{free} of the code. In practice, a randomly selected interleaver can give satisfactory performance. One such approach is to generate random integers, n_i, using a linear feedback shift register and a randomly selected primitive polynomial of sufficient order to accommodate the interleaver size, N. In terms of Figure 7.18, and with the input bits u_i, $i = 0, 1, ..., N-1$, the interleaver function is simply $w_j = u_i$, $j = n_i$.

A somewhat better approach is to use a semi-random interleaver or *S-interleaver*. Here, each randomly selected integer is compared with S previously selected integers. If the current selection is equal to any S previous selections within a distance of $\pm S$, then the current selection is rejected. This process is repeated for all N values of n_i. The search time increases with S and can be computationally intensive. Typically, $S = 16$ for $N = 500$, $S = 21$ for $N = 1000$, and $S = 42$ for $N = 10\,000$.

7.5.2 Code tree

Normally the tree structure of a convolutional code is of little interest since it is highly redundant and can be represented more concisely by a trellis diagram. On the other hand, it is interesting to develop the tree for a Turbo code (PCCC) since it turns out to have a non-uniform branch structure due to the interleaver. Moreover, the tree can be used to determine the information weight distribution w_d and hence an estimate of the BER performance using the union bound in (7.33). It is important to note, however, that the union bound tends to be inaccurate at very low E_b/N_0 since higher weight (and therefore higher distance) events cannot be neglected (see Table 7.4).

Consider the encoder for a rate 1/3 PCCC system, Figure 7.20(a). Here, I denotes an interleaver of length N, and we assume the memory-2 encoder in Figure 7.16(a) is used for each CC ($C1$ and $C2$). Clearly, N information bits must first be stored and interleaved, generally resulting in a coding delay of N. Figure 7.20(b) illustrates part of the code tree for each CC, where the encoder state is shown in brackets and branch labels denote information bits. The tree redundancy is readily seen since there are two identical subtrees corresponding to state 1. The coding scheme in Figure 7.20(a) can be considered as an

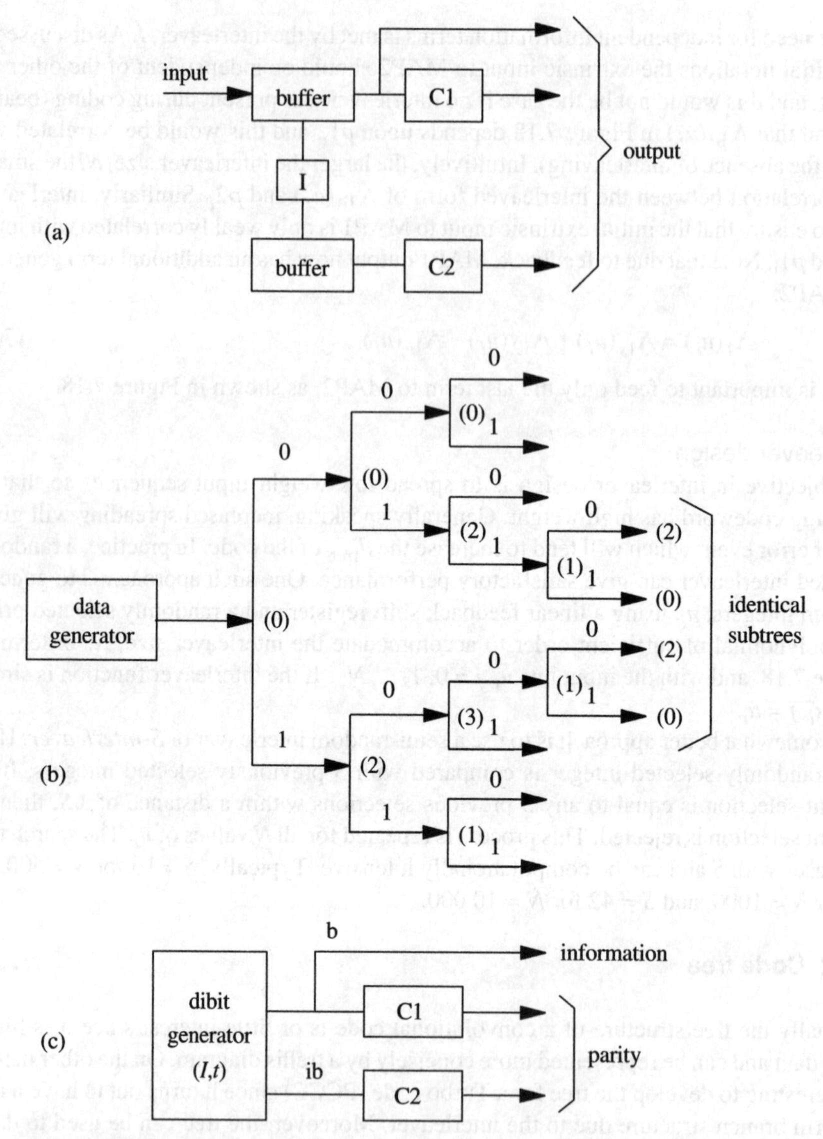

Figure 7.20 *Rate 1/3 PCCC: (a) encoder; (b) CC tree; (c) equivalent encoder for tree generation.*

equivalent rate 2/3 block code preceded by a bit-pair (dibit) generator, Figure 7.20(c), where *ib* denotes the interleaved bit-stream. Generalising, if $C1$ has n_1 states and $C2$ has n_2 states, the equivalent code has $n_1 n_2$ states, and the maximum depth of the corresponding tree will be N. In general, the dibit applied to $C1$ and $C2$ will be constrained by the interleaver in the sense that the generation of valid bit-pairs will be based on previous bit pairs.

To illustrate PCCC tree generation, assume $N = 7$ and the interleaver mapping (0123456) ⇒ (6142305), that is, $ib_0 = b_6$, $ib_1 = b_1$, $ib_2 = b_4$ and so on. Part of the resulting

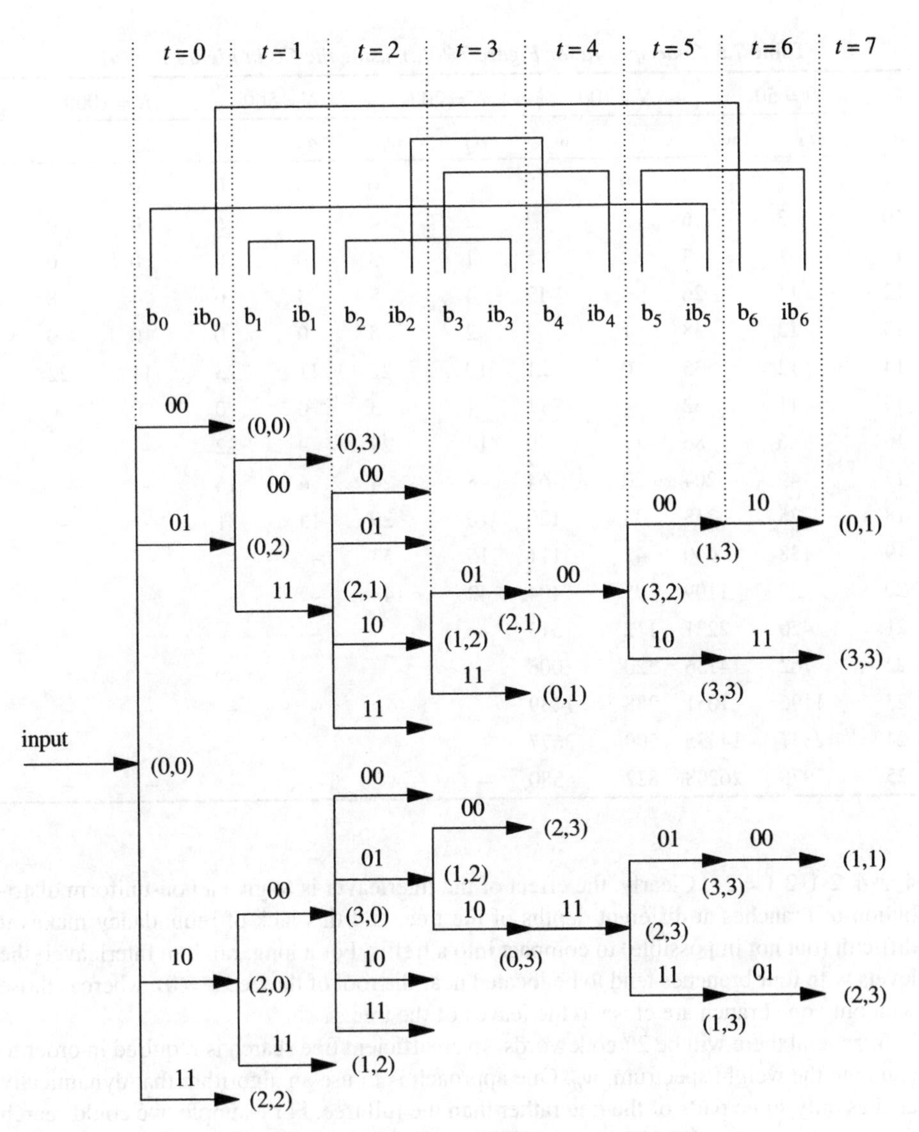

Figure 7.21 *Example of a PCCC tree for N = 7.*

tree is shown in Figure 7.21. At any node, the dibit generator checks to see whether a particular bit depends upon the previous bit. Clearly, at $t = 0$ all four bit pairs are possible at state (0,0), resulting in states (0,0)(0,2)(2,0) and (2,2). At $t = 1$ the interleaver mapping forces $ib_1 = b_1$, resulting in only two possible transitions from states (0,2) and (2,0). At $t = 2$, neither b_2 nor ib_2 have been constrained, so there are four possible transition regions for every state. At $t = 3$, b_3 is unconstrained but ib_3 is constrained to b_2, giving just two possible transitions for any state. At $t = 4$, both b_4 and ib_4 are constrained to ib_2 and b_3 respectively, so there is only one possible transition from each state. The tree is completed in a similar manner, and it is apparent that the total number of codewords is

Table 7.5 *Code spectra for Figure 7.20(a), using the CC in Figure 7.16(a).*

	$N = 50$		$N = 100$		$N = 200$		$N = 500$		$N = 1000$	
d	a_d	w_d	a_d	w_d	a_d	w_d	a_d	w_d	a_d	w_d
7	2	5	0	0	0	0	0	0	0	0
10	3	6	3	7	2	4	3	6	0	0
11	1	3	2	5	1	3	0	0	0	0
12	11	26	6	12	4	8	3	6	4	8
13	12	38	1	3	2	5	0	0	0	0
14	12	35	10	21	10	22	11	23	11	22
15	11	32	4	11	1	3	0	0	0	0
16	26	86	17	49	14	29	9	22	–	–
17	49	204	20	64	8	24	6	16	–	–
18	75	313	34	120	10	29	13	31	–	–
19	138	640	42	171	15	53	–	–	–	–
20	230	1109	95	404	42	148	–	–	–	–
21	420	2231	112	513	–	–	–	–	–	–
22	762	4156	220	1006	–	–	–	–	–	–
23	1196	7051	288	1439	–	–	–	–	–	–
24	2337	14435	509	2677	–	–	–	–	–	–
25	3978	26208	822	4580	–	–	–	–	–	–

$4 \cdot 2 \cdot 4 \cdot 2 \cdot 1 \cdot 2 \cdot 1 = 2^7$. Clearly, the effect of the interleaver is to give a non-uniform distribution of branches at different depths of the tree, and this lack of redundancy makes it difficult (but not impossible) to compact into a trellis. For a long, random interleaver, the levels with four branches tend to be located near the root of the tree ($t = 0$), whereas those with only one branch are close to the leaves of the tree.

In general there will be 2^N codewords, so an efficient tree search is required in order to compute the weight spectrum, w_d. One approach is to use an algorithm that dynamically creates only some parts of the tree rather than the full tree. For example, we could search for all codewords with a Hamming weight up to w_{max}, and if at any state the cumulative weight exceeds w_{max}, the subtree which starts in that state need not be searched, and an alternative path is selected. The approach is similar to the *stack* algorithm used in sequential decoding (Haccoun and Begin, 1989). Table 7.5 shows code spectra for Figure 7.20(a), where the CCs are defined in Figure 7.16(a) (Ambroze *et al.*, 1998a). The table shows the best result (best d_{free}) for a set of five random interleavers for each value of N. Also, it applies to blockwise decoding, so w_d in the table is now measured over the full length of the codeword and could span a number of error events. The exact values of a_d, w_d and d_{free} depend upon the particular interleaver used, although the table does highlight several important points. First, as N is increased there is a general 'thinning' of the spectrum, which results in a high performance code. Secondly, the d_{free} for a PCCC can be quite low when a random interleaver is used (typically as low as 10). The corresponding union (or

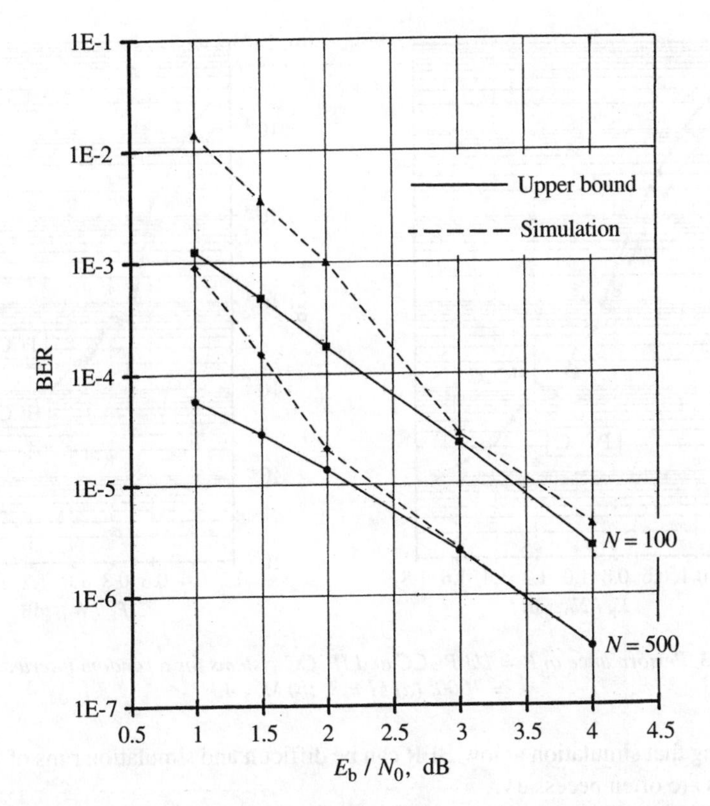

Figure 7.22 *Upper bounds and simulated performance for the Turbo encoder in Figure 7.20(a).*

upper) bound, computed from (7.33) and all available weights in Table 7.5, is shown in Figure 7.22 (using more terms will increase the BER). There is close agreement with simulation for higher E_b/N_0, while the lack of spectral weights and poor convergence of the PCCC simulation contribute to the discrepancy at low E_b/N_0. Another algorithm for finding d_{free} is given by Seghers (1995), and a 'convergence' method is discussed in Ambroze *et al.* (2000).

7.5.3 Performance of PCCCs (Turbo codes)

Given a block of N received vectors ($\mathbf{r}_1$ to $\mathbf{r}_N$), a PCCC (or SCCC) decoder performs forward and backward recursions for each information bit u_i, $i = 1, ..., N$ (Section 5.3.4). This gives an estimate of $\Lambda(u_i)$, and the corresponding extrinsic probabilities. The decoder then repeats this whole process for a specified number of iterations (typically 5–20), and $\Lambda(u_i)$ from one of the decoders is used to estimate u_i (in fact, many blocks of data need to be processed for reliable simulation of BER). *It is the progressive refining of probabilities via iteration that eventually gives a reliable value for* $\Lambda(u_i)$ *and corresponding excellent performance compared to non-iterative decoding,* even though the iterative decoding algorithm is suboptimal. A typical simulation for a rate 1/3 PCCC system having $M = 4$ and $N = 10\,384$ gave BER $\approx 10^{-5}$ at $E_b/N_0 \approx 0.2$ dB after 18 iterations (Anderson, 1995). It is

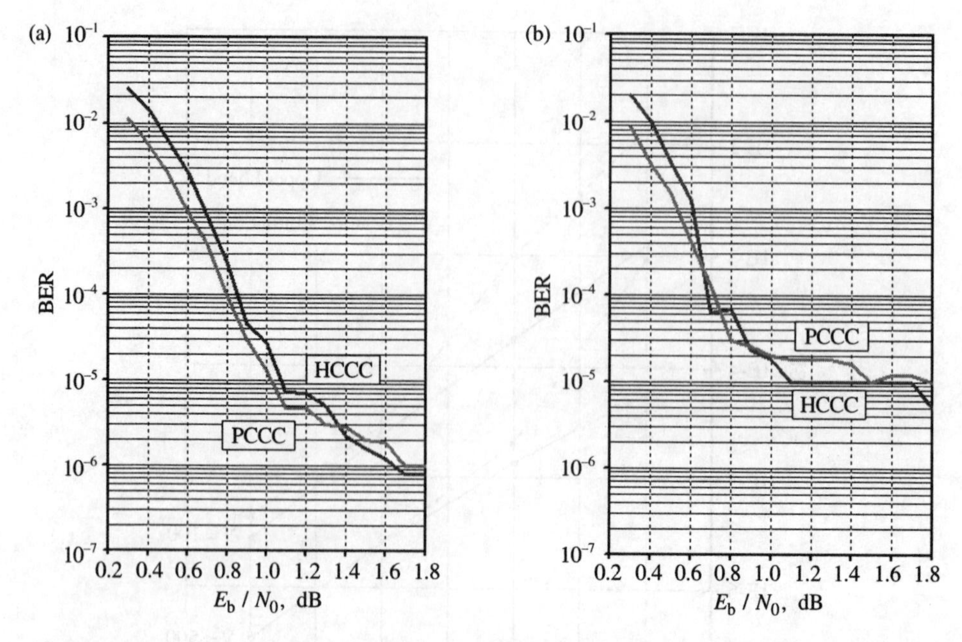

Figure 7.23 *Performance of R = 1/3 PCCC and HCCC systems for a random interleaver and N = 1000: (a) M = 3; (b) M = 4.*

worth noting that simulation at low BER can be difficult and simulation runs of 10^8 information bits are often necessary.

Figure 7.23 illustrates the performance of memory-3 and memory-4 rate 1/3 PCCC systems after 50 iterations. A random interleaver of size $N = 1000$ is used in each case and the corresponding generator matrices of the CCs are

$$M = 3: \quad \mathbf{G}(D) = \left[1, \frac{1+D+D^3}{1+D^2+D^3}\right]$$

$$M = 4: \quad \mathbf{G}(D) = \left[1, \frac{1+D^4}{1+D+D^2+D^3+D^4}\right]$$

The CC for $M = 4$ is in fact a relatively weak code, as reflected in Fig. 7.23(b). Figure 7.24 shows that improved performance can be obtained by using an *S*-interleaver rather than a random interleaver. PCCC schemes tend to exhibit a distinct 'error floor' in the sense that the BER cannot be substantially improved by increasing E_b/N_0 (see Fig. 7.24(b)). This is characteristic of practical single-interleaver PCCC schemes (those using a realistic value of N). The reason for the floor can be seen from (7.33). For moderate to high E_b/N_0 the d_{free} term in (7.33) dominates the BER, in which case

$$P_b(\varepsilon) \approx \frac{1}{2}\frac{w_{d_{free}}}{N} \operatorname{erfc}\left[\sqrt{R\frac{E_b}{N_0}d_{free}}\right] \tag{7.36}$$

where $w_{d_{free}}$ is the total information weight of all codewords of weight d_{free}. A plot of (7.36) against E_b/N_0 is called the *free distance asymptote* (Perez *et al.*, 1996), the negative

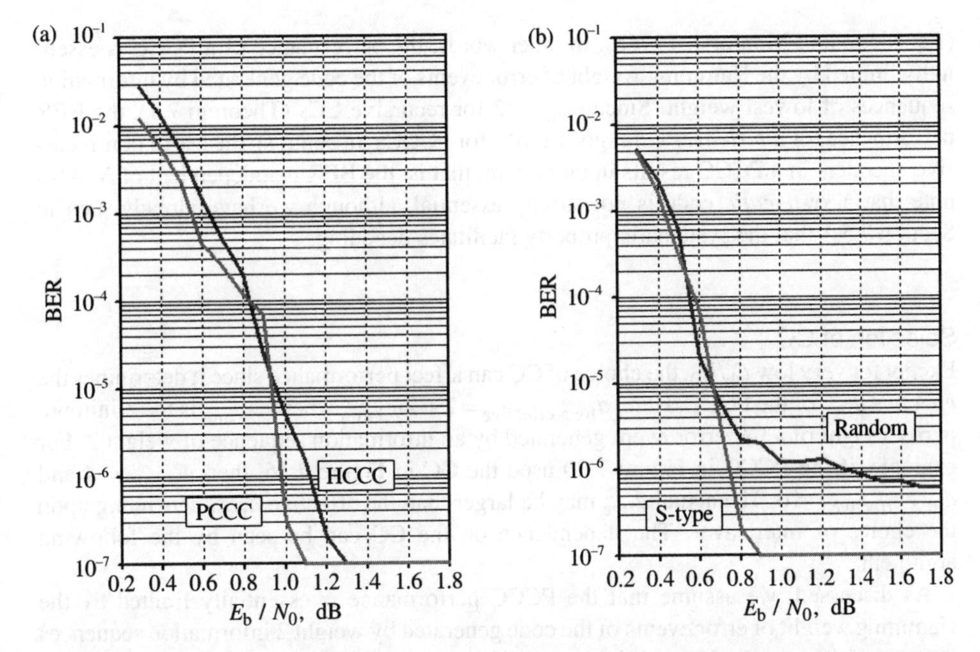

Figure 7.24 *Performance of R = 1/3, M = 3 systems using an S-interleaver: (a) PCCC and HCCC systems, N = 1000; (b) PCCC system, N = 2000, comparison of S and random interleavers.*

slope of which increases as d_{free} is increased. Unfortunately, the d_{free} of a Turbo code can be quite low (see Table 7.5), which means the asymptote falls off only slowly with increase in E_b/N_0. The small-slope asymptote is then seen as an 'error floor' in Turbo decoding. Equation (7.36) suggests that the floor can be lowered by increasing d_{free} or N, or both. Fortunately, careful selection of the interleaver can significantly improve d_{free}, and for Turbo codes this tends to be true when a random interleaver is replaced with an S-interleaver.

The effect of increasing N is more subtle than (7.36) suggests. For a fixed value of d, Table 7.5 shows that a_d tends to decrease as N increases, and this reaches an asymptotic value as $N \to \infty$ (Perez *et al.*, 1996). Clearly, if the number of paths with distance d decreases with N, then the total associated information weight, w_d, of all paths distance d will also tend to decrease (as suggested in Table 7.5). The result is a rather thin and sparse code spectrum in the sense that a_d and w_d grow much less rapidly with d than in a conventional non-recursive convolutional code. Application of the union bound in (7.33) then results in excellent performance compared to a conventional code.

In Section 7.4.3 it was pointed out that, at low E_b/N_0, the whole code weight distribution w_d and not simply d_{free} becomes important when determining the BER. In particular, we noted that punctured (high rate) recursive CCs (RSCs) generally have a better weight distribution than non-recursive CCs, and so seem to be more attractive for use in PCCCs since these operate at low E_b/N_0. In fact, analysis shows that *the CCs in a PCCC must be recursive if the interleaver is to be effective*, that is, the encoders must be recursive. This follows from the fact that the BER in a PCCC system decreases approximately as $N^{1-w_{min}}$, where w_{min} is the minimum information weight associated with an error event of a CC

(Benedetto and Montorsi, 1996b). In other words, the performance of a PCCC is essentially limited by the Hamming weight of error events of the code generated by information sequences of lowest weight. Since $w_{min} = 2$ for recursive CCs (Theorem 7.1), the BER performance (or *interleaver gain*) goes as N^{-1} for PCCCs. In contrast, the use of non-recursive encoders in a PCCC results in unity gain, that is, the BER is independent of N. Also note that a *systematic* code is not strictly essential, although we have already seen in Section 7.5.1 that the systematic property facilitates decoding.

Selection of CC

Except for very low E_b/N_0, the choice of CC can affect performance since it determines the *effective* d_{free} of the PCCC. Here, $d_{free-effective} = 2 + 2w_{p\,min}$ where $w_{p\,min}$ is the minimum parity weight of a CC error event generated by an information sequence of weight 2. For example, if the PCCC in Figure 7.20 used the CC in Figure 7.16, then $w_{p\,min} = 4$ and $d_{free-effective} = 10$. The actual d_{free} may be larger than the effective d_{free}, depending upon the choice of interleaver. The dependence on the CC can be seen by the following argument.

As discussed, we assume that the PCCC performance is essentially limited by the Hamming weight of error events of the code generated by weight 2 information sequences (we note for example that weight 2 error events are associated with low values of d in Table 7.5). Given this assumption, it is clearly helpful to maximise the Hamming weight for weight 2 information sequences, since the decoder will then be less likely to select these paths in preference to the all-zeros path. In terms of Figure 7.12, this means that the weight of $v2(D)$ should be maximised, since $v1(D)$ contributes a fixed weight of 2 for a weight 2 sequence. Assuming an arbitrary weight 2 input sequence we then have

$$v2(D) = (1 + D^i)\frac{n(D)}{p(D)}$$

and $v2(D)$ will be a finite weight sequence (corresponding to an error event) for some value of i since $1 + D^i$ will become divisible by $p(D)$. Intuitively, the larger the value of i the longer the error event and the more code 1s will be generated during the error event. It is therefore necessary to maximise i. Now, since $p(D)$ is the tap polynomial of a linear feedback shift register (Figure 7.12), the register will have a period i for which $p(D)$ divides $1 + D^i$ (Theorem 7.1 proof). According to m-sequence theory (Section 4.4.1) this period is maximised by making $p(D)$ a *primitive* polynomial. In other words, making $p(D)$ primitive maximises the Hamming weight of weight 2 sequences. Finally, since the d_{free} codewords of a PCCC are caused by weight 2 information sequences (providing N is large enough), it follows that selecting $p(D)$ to be primitive tends to maximise d_{free}. In this respect, Figure 7.13 could use either $p(D) = 1 + D + D^3$ or $p(D) = 1 + D^2 + D^3$ since both are primitive. It is interesting to observe that Figure 7.23(a) shows generally better performance than Figure 7.23(b), and the corresponding polynomials, $p(D)$, are primitive and non-primitive respectively.

In summary, some basic guidelines for PCCC (Turbo code) design are:

1. Use only recursive CCs, that is, the constituent encoders must be recursive (also, a systematic CC simplifies decoding).

Table 7.6 *Optimal systematic, rate 1/2 CCs.*

M	$n(D)$	$p(D)$	d_{free}
2	5	7	5
3	17	13	6
4	37	23	6
5	45	67	8
6	101	147	7

2. Select $p(D)$ in each CC to be primitive and of degree M.
3. Select a polynomial $n(D)$ of degree M which gives the best weight spectrum (maximise $w_{p\ \min}$).
4. Use a large value of N, consistent with permissible interleaver delay.
5. Use an S-interleaver rather than a random interleaver.

Tables of optimal recursive systematic CCs having various code rates are available in the literature (Benedetto *et al.*, 1998a). Table 7.6 shows a subset of these tables for rate 1/2 CCs. Here, $n(D)$ and $p(D)$ are in octal notation with the LSB on the right, that is, $13_8 \equiv 1011$, corresponding to $p(D) = 1 + D + D^3$.

7.5.4 Alternative iterative systems

Apart from optimising the interleaver, one possible approach to removing the error floor might be to use an outer code to absorb the residual errors. Another, more fundamental approach, is to use an SCCC rather than a PCCC since the error floor is absent, or at least is at a much lower level. The SCCC tends to force a larger d_{free} which then improves the free distance asymptote (equation (7.36)). However, the distance spectrum becomes somewhat denser and consequently SCCCs are superior at moderate and high E_b/N_0, whereas PCCCs are superior at low E_b/N_0. The SCCC also has considerably better interleaver gain.

Yet another approach to lowering the error rate floor is to extend the Turbo code or PCCC by adding one or more interleaver/encoder pairs. The resulting structure is called a *multiple* PCCC or MPCCC. An encoder for a rate 1/4 3PCCC is shown in Figure 7.25(a) and the corresponding decoder is given in Figure 7.25(b). Clearly, a higher code rate could be achieved by puncturing the four encoder outputs. Note that each MAP decoder in Figure 7.25(b) operates in *parallel* with the other two. In contrast, the Turbo decoder in Figure 7.18 operates in *serial* mode in the sense that MAP1 processes data before MAP2 starts processing. The philosophy behind adding another interleaver/encoder pair is that it provides a second opportunity for 'successful' scrambling of the input data (at a cost of lowering the code rate). In other words, it is less likely that all three encoder inputs will simultaneously give low weight parity sequences, compared to the probability of two encoder inputs simultaneously giving low weight parity sequences (as in a PCCC). This is reflected in improved performance, namely, a marked reduction in the error rate floor, as for SCCCs.

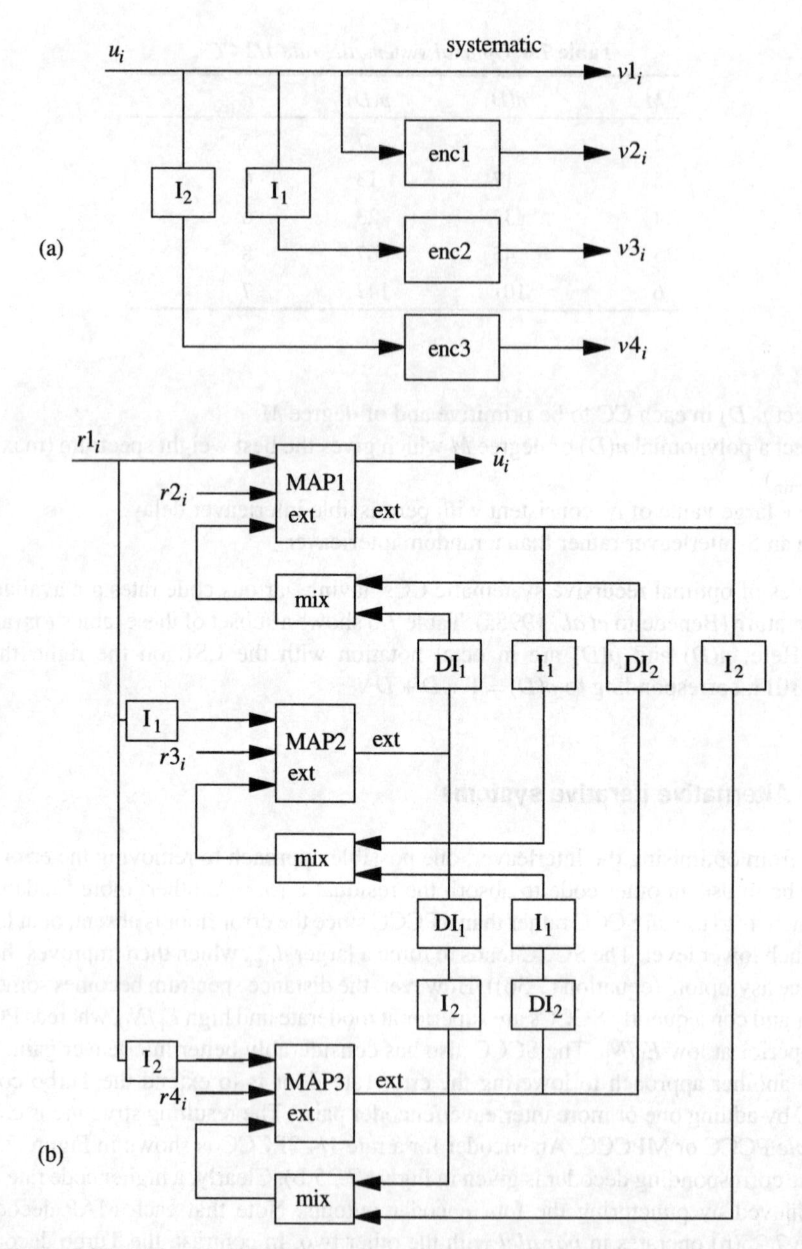

Figure 7.25 *(a) 3-code MPCCC encoder; (b) parallel mode MPCCC decoder for (a).*

7.5.4 Turbo code analysis

In this section we bring together MAP decoder theory and Turbo code concepts in order to generate basic equations for iterative decoding (Section 7.6 then repeats the analysis in more detail in the context of SCCCs). The objective is to derive equations for the output of

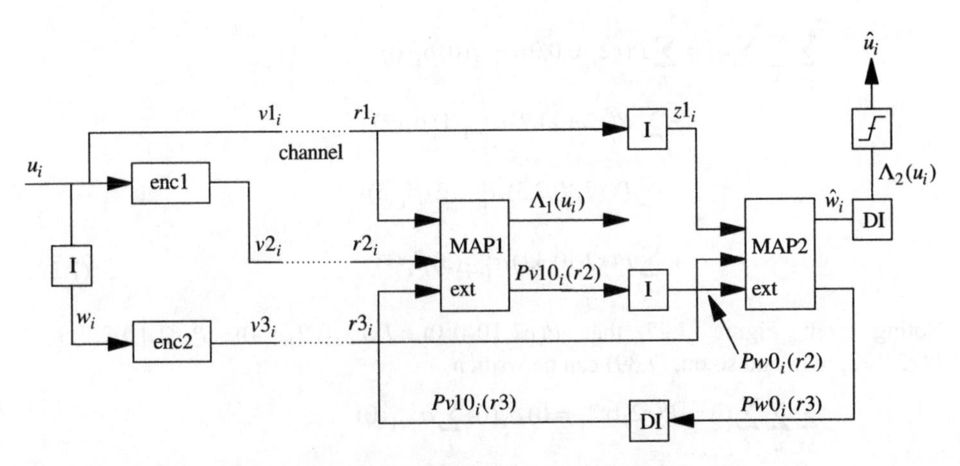

Figure 7.26 *Basic rate 1/3 PCCC system.*

a MAP decoder in an iterative environment, and as such they will apply for MAP decoders in both PCCCs and MPCCCs. In particular, we seek an expression for the *data* output and an expression for the *extrinsic* output (additional information which could be used by another MAP decoder). For simplicity we will restrict consideration to the basic (unpunctured) rate 1/3 Turbo code shown in Figure 7.26, and use the memory-2 (4-state) encoder in Figure 7.27 for each CC.

We start by considering the log-likelihood ratio $\Lambda_1(u_i)$ for MAP1, where $\Lambda_1(u_i)$ is defined in (5.20). Ignoring for the moment the extrinsic input to MAP1, (5.32) gives

$$P(u_i = 0) = P(v1_i = 0) = P(r1_i | u_i = 0) \cdot \sum_m \sum_n \sum_{l=0}^{1} \gamma_0(r2_i, n, m) \alpha_{i-1}^l(n) \beta_i(m) \tag{7.37}$$

where probabilities α and β are defined in (5.22) and (5.23), and (5.31) gives

$$\gamma_0(r2_i, n, m) = P(r2_i | u_i = 0,\ S_{i-1} = n,\ S_i = m) \tag{7.38}$$

The summation in (7.37) can be expanded for the encoder in Figure 7.27 by considering the trellis diagram. In particular, we sum all terms associated with $u_i = v1_i = 0$:

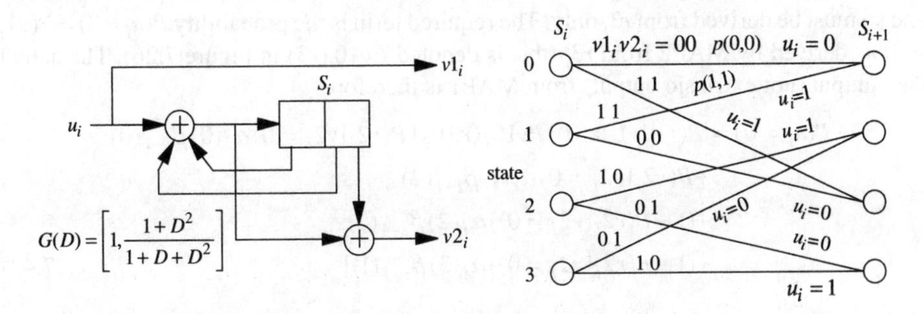

Figure 7.27 *Memory-2 encoder and corresponding trellis.*

$$\sum_m \sum_n \sum_l (\cdot) = \sum_l P(r2_i|0,0,0)\alpha_{i-1}^l(0)\beta_i(0) + \sum_l P(r2_i|0,1,2)\alpha_{i-1}^l(1)\beta_i(2) + \sum_l P(r2_i|0,2,3)\alpha_{i-1}^l(2)\beta_i(3) + \sum_l P(r2_i|0,3,1)\alpha_{i-1}^l(3)\beta_i(1) \quad (7.39)$$

Noting from Figure 7.27 that $P(r2_i|0,0,0) \equiv P(r2_i|v2_i = 0)$, $P(r2_i|0,2,3) \equiv P(r2_i|v2_i = 1)$, and so on, (7.39) can be written

$$\sum_m \sum_n \sum_l (\cdot) = P(r2_i|v2_i = 0)\beta_i(0)\sum_l \alpha_{i-1}^l(0) + P(r2_i|v2_i = 0)\beta_i(2)\sum_l \alpha_{i-1}^l(1) + P(r2_i|v2_i = 1)\beta_i(3)\sum_l \alpha_{i-1}^l(2) + P(r2_i|v2_i = 1)\beta_i(1)\sum_l \alpha_{i-1}^l(3) \quad (7.40)$$

To simplify notation we let

$$\sum_l \alpha_{i-1}^l(\cdot) = \alpha_{i-1}(\cdot) \quad (7.41)$$

Also, (7.37) is based upon the state diagram in Figure 5.8(b), and to make this compatible with the trellis in Figure 7.27 we must increment the state index by one in (7.40). Equation (7.37) then becomes

$$P(u_i = 0) = P(v1_i = 0) = P(r1_i|v1_i = 0)[P(r2_i|v2_i = 0)\alpha_i(0)\beta_{i+1}(0) + P(r2_i|v2_i = 0)\alpha_i(1)\beta_{i+1}(2) + P(r2_i|v2_i = 1)\alpha_i(2)\beta_{i+1}(3) + P(r2_i|v2_i = 1)\alpha_i(3)\beta_{i+1}(1)] \quad (7.42)$$

We must now incorporate the extrinsic information about $u_i = 0$ provided by MAP2 into MAP1 output. Referring to Figure 7.26 we see that this must be independent of $r1_i$ and $r2_i$, and so must be derived from $r3_i$ only. The required term is the probability $P(u_i = 0) = P(v1_i = 0)$ as derived by MAP2 from $r3_i$ (this is denoted $Pv10_i(r3)$ in Figure 7.26). The actual data output (not extrinsic output) from MAP1 is therefore

$$P(u_i = 0) = P(r1_i|v1_i = 0) \cdot Pv10_i(r3) \times [P(r2_i|v2_i = 0)\alpha_i(0)\beta_{i+1}(0) + P(r2_i|v2_i = 0)\alpha_i(1)\beta_{i+1}(2) + (1 - P(r2_i|v2_i = 0))\alpha_i(2)\beta_{i+1}(3) + (1 - P(r2_i|v2_i = 0))\alpha_i(3)\beta_{i+1}(1)] \quad (7.43)$$

MAP1 also generates extrinsic information for MAP2 and reference to Figure 7.26 shows that this must be independent of $r1_i$ and $r3_i$. It must therefore be additional information about $P(w_i = 0)$ which is derived from $r2_i$ (denoted $Pw0_i(r2)$ in Figure 7.26). Since w_i is the interleaved form of $v1_i$ we therefore require the term $Pv10_i(\mathrm{r2})$, and this is readily extracted from (7.42) as follows:

$$
\begin{aligned}
Pv10_i(r2) = {} & P(r2_i | v2_i = 0)\alpha_i(0)\beta_{i+1}(0) \\
& + P(r2_i | v2_i = 0)\alpha_i(1)\beta_{i+1}(2) \\
& + (1 - P(r2_i | v2_i = 0)) \;{}_i(2) \;{}_{i+1}(3) \\
& + (1 - P(r2_i | v2_i = 0) \;{}_i(3) \;{}_{i+1}(1)]
\end{aligned}
\tag{7.44}
$$

Equations (7.43) and (7.44) are the MAP1 decoder outputs assuming the CC in Figure 7.27, and they are easily adapted to the other decoder(s) in a PCCC (MPCCC). Probabilities α and β are readily derived from the state trellis. For instance, according to (7.41) and (5.22), $\alpha_{i+1}(0)$ is the total probability of being in state zero at time $i + 1$, given that the encoder input could be 0 or 1. Referring to the trellis in Figure 7.27, we see that the required recursion is the sum of just two terms:

$$
\alpha_{i+1}(0) \equiv P(S_{i+1} = 0) = p(0,0)\,P(S_i = 0) + p(1,1)\,P(S_i = 1) \tag{7.45}
$$

where $p(0,0)$ and $p(1,1)$ are *transition probabilities,* and $\mathbf{r}$ is understood. The use of the trellis to deduce α and β is discussed in more detail in Section 7.6.

7.6 Serial concatenated convolutional codes

As discussed in Section 7.5, for large interleaver size N, the BER for a PCCC goes approximately as $N^{1-w_{\min}}$, where $w_{\min}$ is the minimum information weight in the finite weight error events of the individual CC (assuming both CCs identical). Since $w_{\min} = 2$ for a recursive code, then the BER for a PCCC goes as N^{-1}. A much stronger dependence on N can be achieved using serial concatenation (SCCC systems). Consider the concatenated system in Figure 7.28. Strictly speaking this is a hybrid between serial and parallel concatenation schemes since only the parity sequence of enc1 is passed (via an interleaver) to enc2. We will therefore refer to it as a hybrid system (HCCC) and a conventional SCCC system (where enc2 encodes *all* outputs of enc1) will be discussed later.

For simplicity, we might assume that the BER of the HCCC scheme is largely determined by error events generated by minimum weight sequences entering enc2. *In this case, the serial structure has the potential for increasing* $w_{\min}$ *for enc2 since its input is a coded sequence from enc1, rather than the information sequence as in a PCCC.* This means that the interleaver gain could be significantly increased (typically to N^{-3}) and the performance enhanced. Unfortunately, this goal is not achieved by the hybrid scheme in Figure 7.28. Suppose for example that we compared the input and output information sequences of the codec in Figure 7.28 in order to obtain a finite error sequence. Due to the linearity of the code we could assume that the all-zeros information sequence is applied to the codec input. A non-zero sequence at the codec output then corresponds to an error event in the decoder, and the same error event would be generated by applying the error sequence to the encoder input. The significant point is that the finite weight error sequence at the encoder input will then very likely generate a weight-2 sequence at the input to enc2.

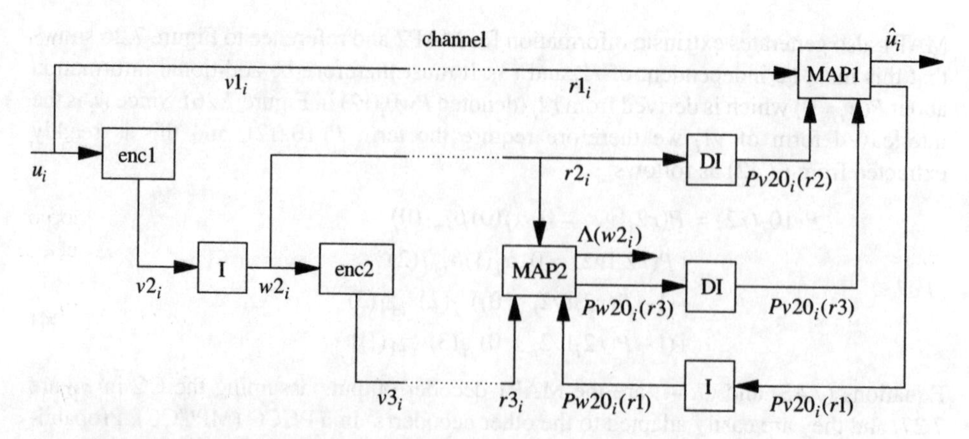

Figure 7.28 *System for a R = 1/3 hybrid concatenated convolutional code (HCCC).*

This is easily verified by simulation. *In other words, error events in the codec are very likely to be generated by weight-2 sequences entering enc2.* We might tentatively conclude that for the HCCC in Figure 7.28, the BER still goes as N^{-1}. On the other hand, Figure 7.28 is relatively simple and serves as a good introduction to SCCC analysis. Simple modifications will then achieve the required enhanced performance associated with an SCCC system.

Before deriving the design equations for Figure 7.28, it is worth highlighting several points. First, the outer MAP decoder (MAP1) has the normal inputs plus an extrinsic input (additional information about symbol $v2_i$). As usual, the extrinsic information is in the form of a probability $P(v2_i = 0)$, denoted $Pv20_i$. Ideally, it is derived only from $r3_i$ and the structure of the inner code, and so is denoted $Pv20_i(r3)$. In this way we avoid duplicating information to MAP1, or feeding back information originally derived from MAP1.

Secondly, MAP1 generates $P(u_i = 0)$ together with an estimate of symbol $v2_i$ in the form of a probability $Pv20_i$. After interleaving this becomes $P(w2_i = 0)$, denoted $Pw20_i$, for symbol $w2_i$. The important point here is that $Pw20_i$ has ideally been derived via $r1_i$ and the structure of the outer code, and is independent of $r2_i$ and $r3_i$ due to the presence of the interleaver (at least for the initial iterations). Feedback probability $Pw20_i(r1)$ therefore acts as an extrinsic input to MAP2, and provides an iterative mechanism.

7.6.1 MAP decoding in an HCCC system

As we have seen in the case of Turbo decoding, the MAP decoder theory in Section 5.3.4 is readily adapted to an iterative system. Consider the inner decoder (MAP2) in Figure 7.28. Ignoring for the moment the extrinsic input, the log-likelihood ratio for the output of MAP2 is obtained from (5.32) by a simple change of variables:

$$\Lambda(w2_i) = \log\left[\frac{P(r2_i|w2_i = 1)\cdot\Sigma_m\Sigma_n\Sigma_{l=0}^{1}\gamma_1(r3_i,n,m)\alpha_{i-1}^{l}(n)\beta_i(m)}{P(r2_i|w2_i = 0)\cdot\Sigma_m\Sigma_n\Sigma_{l=0}^{1}\gamma_0(r3_i,n,m)\alpha_{i-1}^{l}(n)\beta_i(m)}\right] \quad (7.46)$$

As discussed, MAP2 ideally delivers a probability $P(w2_i = 0)$ based upon $r3_i$ only (denoted $Pw20_i(r3)$ in Figure 7.28). This is readily obtained from (7.46) as

$$P(w2_i = 0) \equiv Pw20_i(r3) = \sum_m \sum_n \sum_{l=0}^{1} \gamma_0(r3_i, n, m)\alpha_{i-1}^l(n)\beta_i(m) \tag{7.47}$$

After de-interleaving this becomes the extrinsic term $Pv20_i(r3)$ which must be incorporated into the log-likelihood ratio $\Lambda_1(u_i)$ for MAP1 (see (5.32)). As might be expected, $Pv20_i(r3)$ must also be included in the forward–backward probability computations for MAP1. In practice, instead of computing $\Lambda_1(u_i)$ it suffices to compute the probability

$$P(u_i = 0) = P(r1_i | u_i = 0) \cdot \sum_m \sum_n \sum_{l=0}^{1} \gamma_0(r2_i, n, m)\alpha_{i-1}^l(n)\beta_i(m) \tag{7.48}$$

providing it is modified to account for the extrinsic term. Besides generating $P(u_i = 0)$, MAP1 must also generate a term $P(v2_i = 0) \equiv Pv20_i(r1)$ for iteration. When this is interleaved it becomes extrinsic information for MAP2 since it is independent of $r2_i$ and $r3_i$ (at least for the initial iterations). By replacing u_i with $v2_i$ in the analysis in Section 5.3.4, it can be shown that

$$Pv20_i(r1) = \sum_m \sum_n \sum_{l=0}^{1} \gamma_0(r1_i, n, m)\alpha_{i-1}^l(n)\beta_i(m) \tag{7.49}$$

where γ_0 reduces to either zero or a single probability in $r1_i$, see (5.30). As above, the resulting extrinsic term $Pw20_i(r1)$ is incorporated into the forward–backward computations.

Each MAP decoder in Figure 7.28 performs forward–backward recursions for the complete received sequence, and then the process is repeated for a specified number of iterations. For the first iteration, the extrinsic input $Pw20_i(r1)$ would initially be set to 0.5 (assuming MAP2 decoding is performed first) since the initial probability is unknown. As iteration proceeds, the extrinsic input to each MAP decoder becomes more correlated with the main MAP decoder inputs, and so becomes less effective as extrinsic information. The correlation between the MAP extrinsic input and the MAP extrinsic output also increases with iteration. The net effect is that improvement through iteration becomes marginal.

7.6.2 MAP decoder implementation

The foregoing theoretical ideas are now interpreted as practical equations suitable for realising an HCCC system. For simplicity we assume that the memory-2 code in Figure 7.27 is used for both CCs in Figure 7.28. Consider first a single MAP decoder for Figure 7.27. The forward recursion in (5.26) computes state probability $\alpha_i^k(m)$ from previous probabilities $\alpha_{i-1}^l(n)$, and $\gamma_k(\cdot)$ behaves as a transition probability. Essentially, $\alpha_i^k(m)$ is the probability of going to state $S_i = m$ given a current state and a particular input $u_i = k$ (see (5.22) and Figure 5.8(b)). In fact, it was shown in Section 7.5.5 that we need the *total* state probability, $P(S_i = m)$, as defined in (7.41). The total probability accounts for transitions from different states given different encoder inputs, and is readily deduced from the trellis in Figure 7.27. Following on from (7.45) the forward recursions are

$$P(S_{i+1}=0)=p(0,0)P(S_i=0)+p(1,1)P(S_i=1) \qquad (7.50)$$

$$P(S_{i+1}=1)=p(1,0)P(S_i=2)+p(0,1)P(S_i=3) \qquad (7.51)$$

$$P(S_{i+1}=2)=p(1,1)P(S_i=0)+p(0,0)P(S_i=1) \qquad (7.52)$$

$$P(S_{i+1}=3)=p(0,1)P(S_i=2)+p(1,0)P(S_i=3) \qquad (7.53)$$

where the (unnormalised) transition probabilities are given by

$$p(v1,v2)=\exp[-[(r1_i-(2v1-1))^2+(r2_i-(2v2-1))^2]/2\sigma^2] \qquad (7.54)$$

Equation (7.54) follows from the assumption that the output of the DMC (see Section 5.3.4) has a large alphabet, so that the conditional probabilities (as in (5.27)) tend to Gaussian density functions. Each of the four state probabilities for time $i+1$ must then be normalised by dividing by their sum. Similarly, the backward recursion in (5.28) can be expressed

$$Pb(S_i=0)=p(0,0)Pb(S_{i+1}=0)+p(1,1)Pb(S_{i+1}=2) \qquad (7.55)$$

$$Pb(S_i=1)=p(1,1)Pb(S_{i+1}=0)+p(0,0)Pb(S_{i+1}=2) \qquad (7.56)$$

$$Pb(S_i=2)=p(1,0)Pb(S_{i+1}=1)+p(0,1)Pb(S_{i+1}=3) \qquad (7.57)$$

$$Pb(S_i=3)=p(0,1)Pb(S_{i+1}=1)+p(1,0)Pb(S_{i+1}=3) \qquad (7.58)$$

which again must be normalised. The backward recursion can be initialised by assigning $Pb(S_N=m)=P(S_N=m)$, where m denotes a particular state.

7.6.3 Outer decoder, MAP1

Now consider the MAP1 decoder in Figure 7.28. It is now helpful to write $p(v1,v2) = p(v1)p(v2)$ since $p(v2)$ is obtained from a separate, de-interleaving process. For example, after decrementing the state index for compatibility with Figure 5.8(b), equations (7.50) and (7.51) can be written

$$\begin{aligned}P(S_i=0)=&\ P(r1_i \mid v1_i=0)\cdot Pv20_i(r2)\\ &\times P(S_{i-1}=0)+P(r1_i \mid v1_i=1)\cdot(1-Pv20_i(r2))\cdot P(S_{i-1}=1)\end{aligned} \qquad (7.59)$$

$$\begin{aligned}P(S_i=1)=&\ P(r1_i \mid v1_i=1)\cdot Pv20_i(r2)\\ &\times P(S_{i-1}=2)+P(r1_i \mid v1_i=0)\cdot(1-Pv20_i(r2))\cdot P(S_{i-1}=3)\end{aligned} \qquad (7.60)$$

where

$$P(r1_i \mid v1_i=0)=\exp(-(r1_i+1)^2/\sigma^2) \qquad (7.61)$$

$$P(r1_i \mid v1_i=1)=\exp(-(r1_i-1)^2/\sigma^2) \qquad (7.62)$$

Accounting for the extrinsic information supplied by MAP2, (7.59) and (7.60) become

$$\begin{aligned}P(S_i=0)=&\ P(r1_i \mid v1_i=0)\cdot Pv20_i(r2)\cdot Pv20_i(r3)\cdot P(S_{i-1}=0)\\ &+P(r1_i \mid v1_i=1)\cdot(1-Pv20_i(r2))\cdot(1-Pv20_i(r3))\cdot P(S_{i-1}=1)\end{aligned} \qquad (7.63)$$

$$\begin{aligned}P(S_i=1)=&\ P(r1_i \mid v1_i=1)\cdot Pv20_i(r2)\cdot Pv20_i(r3)\cdot P(S_{i-1}=2)\\ &+P(r1_i \mid v1_i=0)\cdot(1-Pv20_i(r2))\cdot(1-Pv20_i(r3))\cdot P(S_{i-1}=3)\end{aligned} \qquad (7.64)$$

Once all four probabilities have been found they must be normalised. The backward recursion for MAP1 is found in a similar way through modification of the basic recursions in (7.55)–(7.58). For example, accounting for the extrinsic information from MAP2, the recursion to state 0 from states 0 and 2 is, from (7.55)

$$\begin{aligned} Pb(S_i = 0) = {} & P(r1_{i+1} | v1_{i+1} = 0) \cdot Pv20_{i+1}(r2) \cdot Pv20_{i+1}(r3) \cdot Pb(S_{i+1} = 0) \\ & + P(r1_{i+1} | v1_{i+1} = 1) \cdot (1 - Pv20_{i+1}(r2)) \cdot (1 - Pv20_{i+1}(r3)) \cdot Pb(S_{i+1} = 2) \end{aligned} \tag{7.65}$$

The data output from MAP1 is given by an enhanced version of (7.48), that is, we must include the extrinsic input $Pv20_i(r3)$. The first term in (7.48) is already given by (7.61), while the summation terms in (7.48) can be obtained from the trellis in Figure 7.27 by noting the transitions corresponding to $u_i = v1_i = 0$ (there are four such transitions). Each of the corresponding products must be scaled by the appropriate extrinsic term, giving, before normalisation

$$\begin{aligned} P(u_i = 0) = P(r1_i | u_i = 0)\, [& Pv20_i(r2) \cdot Pv20_i(r3) \cdot P(S_{i-1} = 0) \cdot Pb(S_i = 0) \\ & + Pv20_i(r2) \cdot Pv20_i(r3) \cdot P(S_{i-1} = 1) \cdot Pb(S_i = 2) \\ & + (1 - Pv20_i(r2)) \cdot (1 - Pv20_i(r3) \cdot P(S_{i-1} = 2) \cdot Pb(S_i = 3) \\ & + (1 - Pv20_i(r2)) \cdot (1 - Pv20_i(r3) \cdot P(S_{i-1} = 3) \cdot Pb(S_i = 1)] \end{aligned} \tag{7.66}$$

The interpretation of (7.48) as (7.66) warrants a little more explanation. The $\gamma_0(\cdot)$ term in (7.48) is based upon (5.31) which essentially reduces to $P(r2_i | v2_i = 0)$. However, since we are now dealing with a concatenated scheme rather than a single MAP decoder, it is necessary to compute $P(r2_i | w2_i = 0)$ (using an expression similar to (7.61)) and then de-interleave to obtain $Pv20_i(r2)$. Also note that it is necessary to compute $P(u_i = 1)$ in order to normalise $P(u_i = 0)$.

The feedback output from MAP1 is based upon (7.49) and so can be deduced from the trellis by summing all terms associated with $v2_i = 0$. Also note that γ_0 must only be associated with $r1_i$ in order to generate true extrinsic information for MAP2. Expanding (7.49) we have

$$\begin{aligned} P(v2_i = 0) \equiv Pv20_i(r1) = {} & P(r1_i | v1_i = 0) \cdot P(S_{i-1} = 0) \cdot Pb(S_i = 0) \\ & + P(r1_i \mid v1_i = 0) \cdot P(S_{i-1} = 1) \cdot Pb(S_i = 2) \\ & + P(r1_i \mid v1_i = 1) \cdot P(S_{i-1} = 2) \cdot Pb(S_i = 1) \\ & + P(r1_i \mid v1_i = 1) \cdot P(S_{i-1} = 3) \cdot Pb(S_i = 3) \end{aligned} \tag{7.67}$$

7.6.4 Inner decoder, MAP2

Since this uses the same CC, the forward and backward recursions are similar to those for MAP1. For instance, the forward recursion to state 0 in (7.63) becomes

$$\begin{aligned} P(S_i = 0) = {} & P(r2_i | w2_i = 0) \cdot Pv30_i(r3) \cdot Pw20_i(r1) \cdot P(S_{i-1} = 0) \\ & + P(r2_i \mid w2_i = 1) \cdot (1 - Pv30_i(r3)) \cdot (1 - Pw20_i(r1)) \cdot P(S_{i-1} = 1) \end{aligned} \tag{7.68}$$

where, prior to normalisation,

$$Pv30_i(r3) = \exp(-(r3_i + 1)^2 / \sigma^2) \tag{7.69}$$

Term $Pw20_i(r1)$ is the interleaved form of (7.67) and represents the extrinsic input to MAP2. As discussed, MAP2 should generate an output which is independent of both $Pw20_i(r1)$ and $r2_i$, as in (7.47). Again, (7.47) can be deduced from the trellis, giving

$$\begin{aligned} P(w2_i = 0) \equiv Pw20_i(r3) = P(r3_i \mid v3_i = 0) \cdot P(S_{i-1} = 0) \cdot Pb(S_i = 0) \\ + P(r3_i \mid v3_i = 0) \cdot P(S_{i-1} = 1) \cdot Pb(S_i = 2) \\ + P(r3_i \mid v3_i = 1) \cdot P(S_{i-1} = 2) \cdot Pb(S_i = 3) \\ + P(r3_i \mid v3_i = 1) \cdot P(S_{i-1} = 3) \cdot Pb(S_i = 1) \end{aligned} \quad (7.70)$$

which must then be normalised. In a software implementation of the above equations, the encoder state will usually be time-synchronised with the information sequence, that is, it will be in state S_i for input u_i, as in Figure 7.27. Since the state in Figure 5.8(b) is time-slipped, it will usually be necessary to increment the state index by one in (7.59)–(7.70). Also note that numerical problems can lead to inferior results, for example, when the extrinsic probabilities are close to 0 or 1. One solution is to provide numerical limits to the extrinsic terms.

The performance of HCCC systems is illustrated in Figure 7.23 and Figure 7.24(a) and is similar to the corresponding PCCC scheme. This might be expected from the discussion at the start of this section, since the full output of enc1 is not applied to enc2. Figure 7.29 shows the effect of interleaver size, N, in HCCC systems; the memory-2 system uses the CC in Figure 7.27 and shows that the BER goes approximately as N^{-1}, and the memory-3 system uses the memory-3 CC discussed in Section 7.5.3. Appendix D gives a simulation of the rate 1/3 HCCC system in Figure 7.28, where both CCs are defined by Figure 7.27.

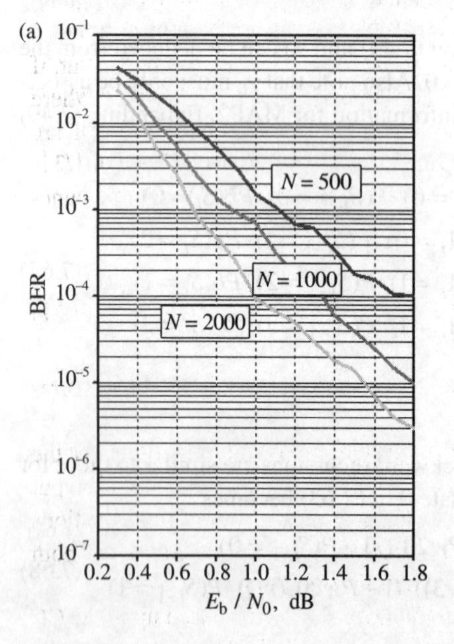

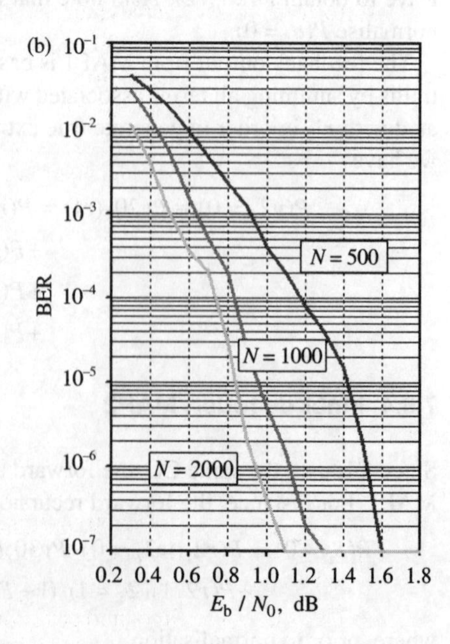

Figure 7.29 *Simulation showing the effect of N for HCCC systems (S-interleaver): (a) memory-2; (b) memory-3.*

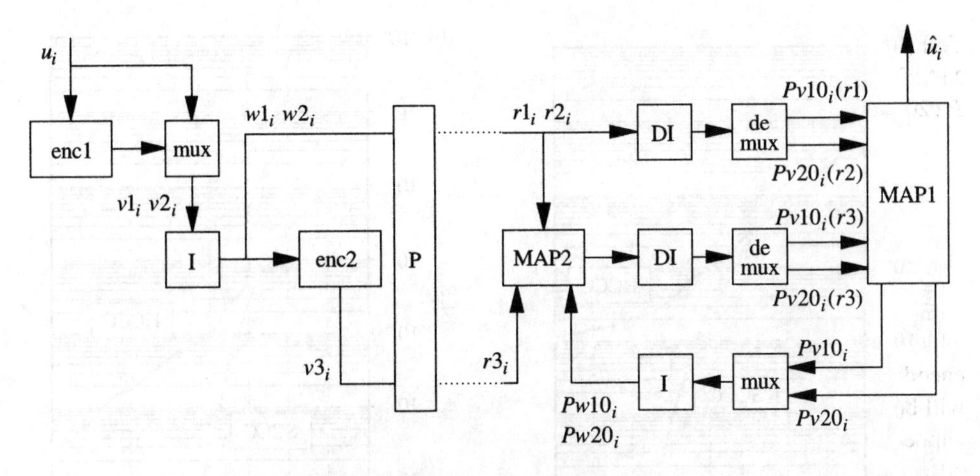

Figure 7.30 *An R = 1/3 SCCC system.*

7.6.5 SCCC scheme

As discussed, the interleaver gain of a PCCC with large interleaver size goes as $N^{1-w_{\min}}$. If it is assumed that the performance of an SCCC scheme is largely determined by error events of the inner code generated by minimum weight sequences entering enc2, it follows that $w_{\min}$ at the input of enc2 should be increased in order to obtain higher interleaver gain. A straightforward way of achieving this (with only small modifications to the foregoing analysis) is shown in Figure 7.30 (Ambroze *et al.*, 1998b). Encoder 2 input now has a minimum weight corresponding to the d_{free} of code 1, that is $w_{\min} = d_{\text{free}}$; in particular, if d_{free} is the free distance of the outer code, the BER for the SCCC now goes as N^{-m} where $m = \lfloor (d_{\text{free}} + 1)/2 \rfloor$ (Benedetto *et al.*, 1996c). For example, if both CCs in Figure 7.30 are defined as in Figure 7.27, then $d_{\text{free}} = 5$ and the BER goes as N^{-3}. If an overall rate of 1/3 is required, this can be achieved by puncturing, as indicated in Figure 7.30. A suitable puncturing matrix is

$$\mathbf{P} = \begin{bmatrix} 1111 \\ 1001 \end{bmatrix} \tag{7.71}$$

which denotes the puncturing of alternate parity bits for every two input bits (corresponding to a rate 2/3 inner code).

Implementation of Figure 7.30 requires small modifications of the equations for the HCCC system. For instance, when decoding u_i, MAP1 must simultaneously use all input information relating to u_i, so (7.66) must be modified accordingly. Similar modifications are required for the extrinsic outputs from MAP1. For example, MAP1 can now use both inputs associated with $v1_i$ when deriving $Pv20_i$ (a small modification of (7.67)).

Figure 7.31 shows the simulated performance of memory-2 systems based upon the CC defined in Figure 7.27. Note that the actual interleaver size in the SCCC system must be double the interleaver delay, N, due to multiplexing. It is apparent from Figure 7.31 that,

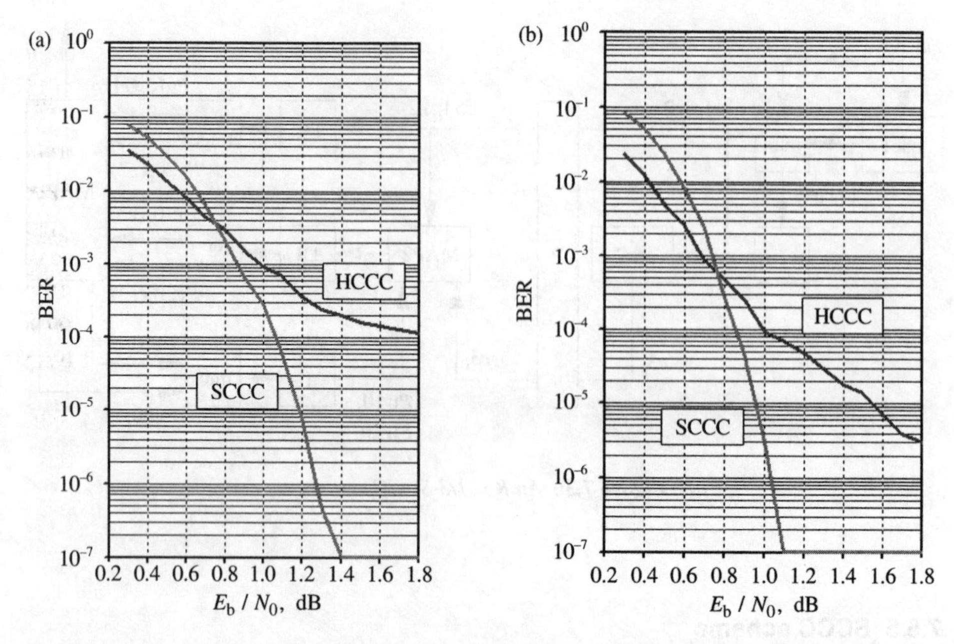

Figure 7.31 *Simulation of rate 1/3 HCCC and rate 1/4 SCCC systems: (a) random interleaver, N = 1000; (b) S-interleaver, N = 2000.*

compared with a PCCC (or HCCC) system, the SCCC effectively eliminates the 'error floor'.

Finally, in contrast to PCCCs, it is not essential for *both* CCs in an SCCC to be recursive. As far as the outer code in an SCCC is concerned, it is important to (a) maximise its d_{free} and (b) minimise both the number of input sequences generating free distance error events, and the information weight associated with these error events (Benedetto *et al.*, 1996). It therefore follows that it can be beneficial to use a *non-recursive* outer code in an SCCC, since, for example, $w_{min} = 1$ (Section 7.4.4). However, it is still essential for the *inner* code to be a recursive convolutional code in order to achieve an interleaver gain.

7.7 Convergence of iterative decoding systems

The finite decoded BER in an iterative decoding system arises from the lack of convergence, and the type of convergence. Consider the PCCC system in Figure 7.32. Each MAP decoder can be considered as a function acting on a probability vector $\mathbf{P_E} = (P_{E1}, P_{E2}, ..., P_{EN})$ where N is the interleaver size (block length) and $P_{Ek} \equiv P_E\{u_k = 1\}, k = 1, ..., N$. That is, P_{Ek} is the probability of information bit u_k being 1 as computed from the *extrinsic* output of the MAP decoder. Starting from an arbitrary point, $\mathbf{P_E}$ may or may not converge to a solution $\mathbf{P_{ES}}$, depending upon whether or not the initial vector falls within a 'contraction region'. In particular, we say the decoder has converged to a fixed point if both extrinsic vectors in Figure 7.32 have values close to 0 or 1 (and are identical), or if they are non-saturated but stable. For each case the vector could still have errors even though it

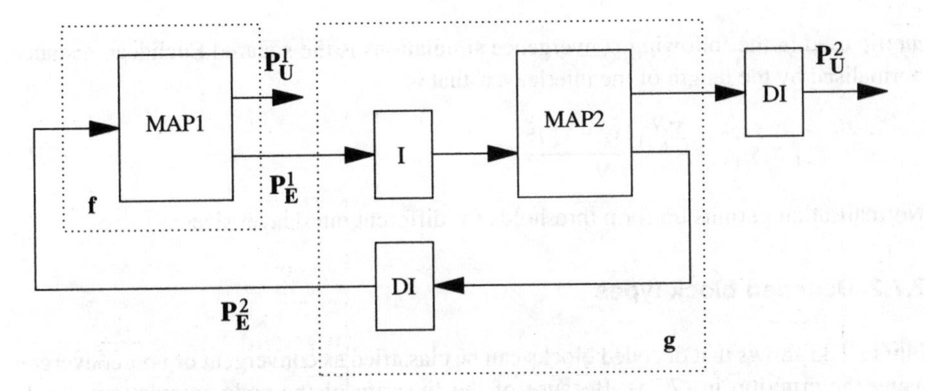

Figure 7.32 *Extrinsic information in a PCCC (Turbo) decoder.*

represents a fixed point. Mathematically, the iterative decoding algorithm can be described as the problem of iteratively solving the equations

$$\begin{aligned}\mathbf{P_E^1} &= \mathbf{f}(\mathbf{P_E^2}) \\ \mathbf{P_E^2} &= \mathbf{g}(\mathbf{P_E^1})\end{aligned} \qquad (7.72)$$

where **f** and **g** represent the two *N*-dimensional MAP functions and **g** is considered to include the interleaving/de-interleaving process. This problem is equivalent to finding a solution for the equation

$$\mathbf{P_E^1} = \mathbf{f}(\mathbf{g}(\mathbf{P_E^1})) = \mathbf{h}(\mathbf{P_E^1}) \qquad (7.73)$$

A vector that satisfies (7.73) is called a fixed point for function **h** and an iterative algorithm will converge to a solution $\mathbf{P_{ES}}$ for (7.73) under certain conditions (Ambroze *et al.*, 2000).

7.7.1 The Cauchy criterion

A practical approach to convergence for a realistic value of *N* is to consider the decoding process as an array of vectors indexed by iteration number, that is, $\mathbf{P_E^1}(1), \mathbf{P_E^2}(2), ..., \mathbf{P_E^1}(n), \ldots$ where

$$\mathbf{P_E^2}(n) = \mathbf{g}(\mathbf{P_E^1}(n)) \qquad (7.74)$$

We then apply the Cauchy criterion (Sawyer, 1978) to determine whether or not the arrays are convergent, and to stop iteration. Essentially the criterion states that an array converges if and only if the amplitude of changes (as measured by a defined distance metric) tends to zero as the number of iterations increases. We therefore establish a small threshold δ (typically 10^{-3}) and iterate until

$$\left\| \mathbf{P_E^1}(n+1), \mathbf{P_E^1}(n) \right\| < \delta \qquad (7.75)$$

Blocks failing to satisfy (7.75) for a given maximum number of iterations (50 is used in Section 7.7.3) are deemed non-convergent. Blocks that satisfy (7.75) are further checked with lower thresholds, the lower limit of δ being determined by machine precision. The metric used in the following convergence simulations is the squared Euclidean distance, normalised by the length of the interleaver, that is

$$\| \mathbf{x}, \mathbf{y} \|^2 = \frac{\Sigma_{k=1}^{N} (x_k - y_k)^2}{N} \tag{7.76}$$

Normalisation permits uniform thresholds for different interleaver sizes.

7.7.2 Decoded block types

Figure 7.33 shows that decoded blocks can be classified as convergent or non-convergent using the criterion in (7.75). Because of the linearity of the code, simulations can be performed by transmitting the all-zeros information sequence, which means that $P_{Uk} = 1$ at the decoder output represents a bit error. For any erroneous block, the information weight (number of data errors/block) and the code weight can be calculated, the latter being obtained by re-encoding the decoded data sequence (as discussed in Section 7.6). This associates a code weight with the information weight; in fact, the identification of low code weight blocks can also be used to estimate the d_{free} of the code (see later). Figure 7.33 also suggests that there are two types of convergent block and these have the following characteristics:

- Type 1: blocks for which vectors $\mathbf{P}_{\mathbf{ES}}^1$ and $\mathbf{P}_{\mathbf{ES}}^2$ have values close to 0 or 1 (saturation). In this case it can be shown that they are identical.
- Type 2: blocks for which the two limit vectors are non-saturated but stable, as in (7.75). In this case they are generally different.

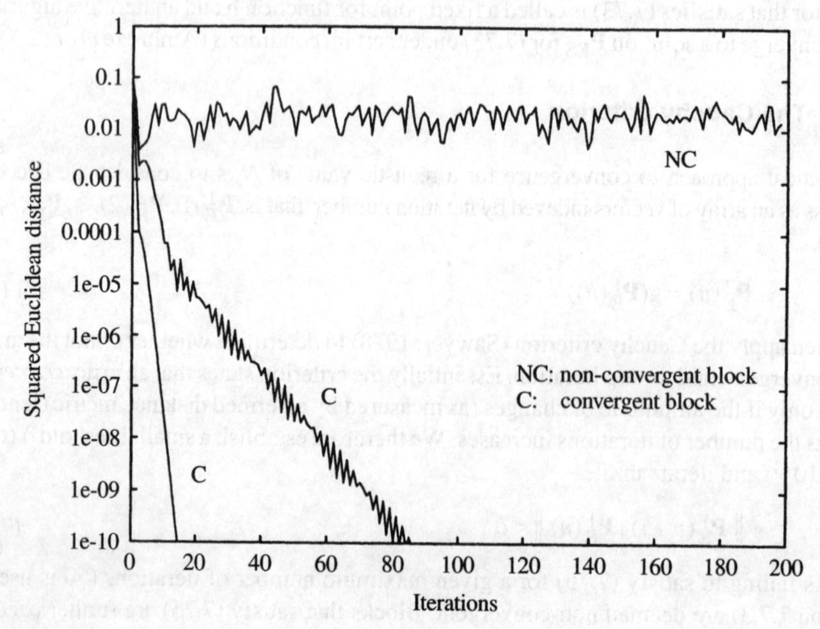

Figure 7.33 *Typical convergence in a PCCC system for three different types of block.*

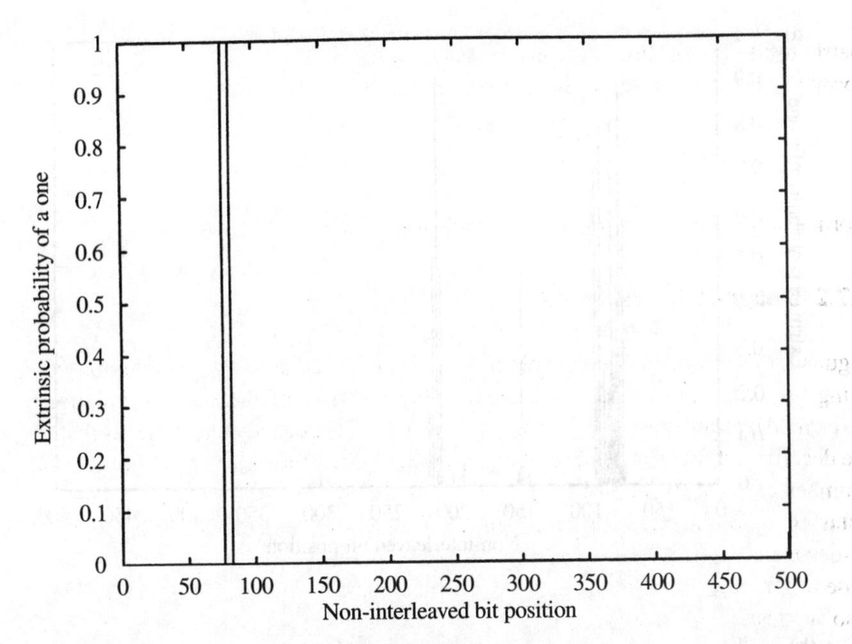

Figure 7.34 *Extrinsic information vector limit for MAP1 and MAP2 (Type 1 decoded block, N = 500).*

An example of a Type 1 block is shown in Figure 7.34 and it represents the limit of the extrinsic information vectors $\mathbf{P}_{\mathbf{E}}^{1}(n)$ and $\mathbf{P}_{\mathbf{E}}^{2}(n)$ for a particular value of δ. Simulation shows that this type of block generally has low information weight and low code weight. The example shown corresponds to an erroneous block with information weight 2 and code weight 18, the latter corresponding to the d_{free} of the turbo code. A special case of this type of decoded block is one that decodes with zero error. An example of a Type 2 decoded block is shown in Figure 7.35 and, clearly, the probability vectors are not saturated. This particular example corresponds to a block with decoded information weight of 3 and a code weight of 292.

Two types of behaviour can now be identified for the extrinsic information vector $\mathbf{P}_{\mathbf{E}}$. For Type 1 blocks, the number of decoded bit errors coincides with the number of ones in $\mathbf{P}_{\mathbf{E}}$, whereas for Type 2 blocks there are only 3 bit errors for a relatively erroneous extrinsic vector. For Type 1 blocks, $\mathbf{P}_{\mathbf{E}}$ is decided with high probability and so dominates the decoding process in the last iterations. For Type 2 blocks, $\mathbf{P}_{\mathbf{E}}^{1}(n)$ and $\mathbf{P}_{\mathbf{E}}^{2}(n)$ are not saturated and so decoding is a compromise between channel values and extrinsic information values.

7.7.3 Simulated convergence

This section illustrates typical convergence performance for the rate 1/3 PCCC system in Figure 7.26 (each CC being defined by Figure 7.27). Simulations are presented for interleaver sizes of 500 and 2000 and use a total of 200,000 blocks for each value of E_b/N_0. Tables 7.7 and 7.8 show only those convergent blocks that decoded in error, where

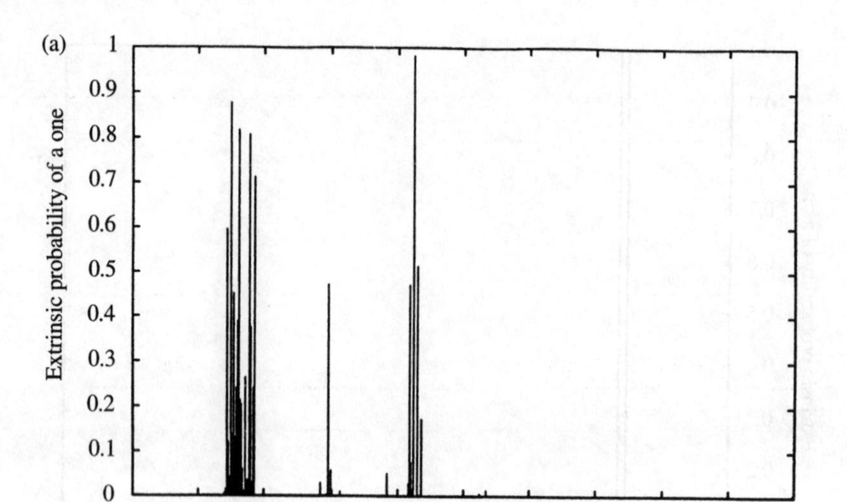

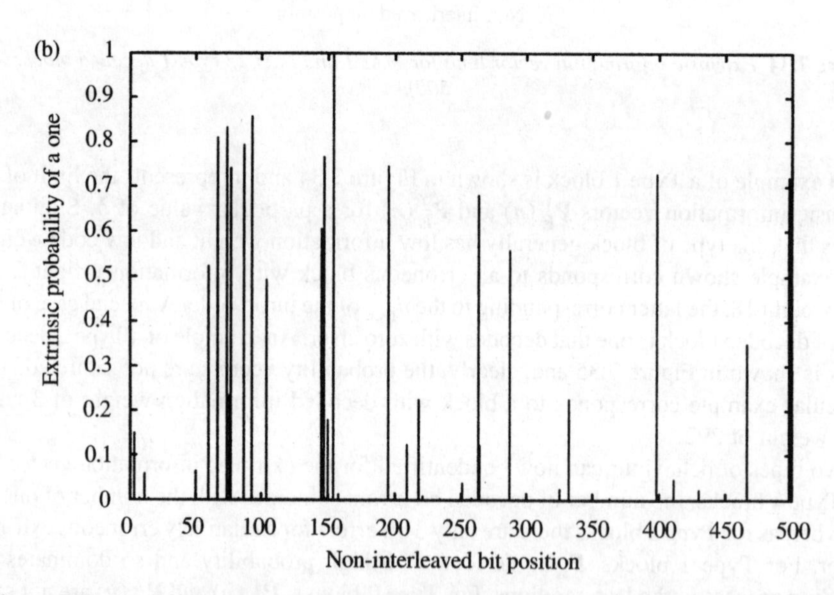

Figure 7.35 *Extrinsic information vector limit for (a) MAP1 and (b) MAP2 (Type 2 decoded block, N = 500).*

convergence satisfied (7.75); the remaining blocks converged without error. Simulation used either a random interleaver or an *S*-interleaver, and leads to several general observations:

- Non-convergence dominates the block error rate at low E_b/N_0. As E_b/N_0 increases, non-convergence decreases and the convergent error events dominate the block error rate.

Table 7.7 *Convergent/non-convergent blocks for N = 500.*

Interleaver	E_b/N_0 dB	1	1.3	1.5	2	3
Random	Convergent	3783	1909	1323	477	112
	Non-convergent	4329	1002	438	53	2
$S = 14$	Convergent	1037	355	223	52	14
	Non-convergent	2008	321	92	4	0

Table 7.8 *Convergent/non-convergent blocks for N = 2000.*

Interleaver	E_b/N_0 dB	0.5	0.7	1	1.3
Random	Convergent	8600	5700	3608	2212
	Non-convergent	8020	1360	284	88
$S = 27$	Convergent	2140	920	392	140
	Non-convergent	4700	680	32	12

- The interleaver can be designed to significantly improve convergence.
- Convergence improves with increasing N.

A simple approach is to stop the iterative decoding process when a maximum number of iterations is reached. On the other hand, simulation shows that different blocks need a different number of iterations in order to converge (see Figure 7.33), and the average decoding time can be reduced by terminating the iteration when no improvement is observed. As discussed, a more adaptive approach is to use the Cauchy criterion in (7.75) to terminate iteration. Clearly, too large a value for δ will increase the BER due to premature termination, that is, before the actual extrinsic limit has been reached, whereas a lower threshold will increase the average number of iterations. Typical average iteration values and corresponding BER statistics for different thresholds are shown in Table 7.9. Here the PCCC has $N = 500$, $S = 14$ and uses the CC defined in Figure 7.27. It is apparent that, providing $\delta \leq 10^{-3}$, there will be only relatively small variation in BER and iteration number.

Finally, it is interesting to observe that the convergence criterion can be used to estimate d_{free} even for large values of N (that is, beyond the limit of the 'code tree' method in Section 7.5.2). As mentioned in Section 7.7.2, this can be done by searching for a block with minimum code weight. Using this approach, the $N = 2000$, $S = 27$ PCCC in Table 7.8

Table 7.9 *Typical convergence performance for a rate 1/3 PCCC decoder.*

	Average number of iterations			BER ($\times 10^{-5}$)		
E_b/N_0 dB	$\delta = 10^{-2}$	$\delta = 10^{-3}$	$\delta = 10^{-6}$	$\delta = 10^{-2}$	$\delta = 10^{-3}$	$\delta = 10^{-6}$
1	4.4	5.5	6.5	67.7	57.3	56.9
1.5	3.1	3.7	4.5	3.1	1.7	1.6
2	2.5	3.1	3.6	0.590	0.161	0.158

has been shown to have $d_{\text{free}} = 20$, but this reduced to $d_{\text{free}} = 10$ when a random interleaver was used. The technique has also been applied to a 3PCCC system and gave $d_{\text{free}} \approx 38$ (Ambroze *et al.*, 2000).

Bibliography

Ambroze, A., Wade, G. and Tomlinson, M. (1998a) Turbo code tree and code performance, *IEE Electronics Lett.*, **34**(4), 353–354.

Ambroze, A., Wade, G. and Tomlinson, M. (1998b) Iterative MAP decoding for serial concatenated convolutional codes, *IEE Proc.-Communications*, **145**(2), 53–59.

Ambroze, A., Wade, G. and Tomlinson, M. (2000) Practical aspects of iterative decoding, *IEE Proc. Communications*, **147**(2), 69–74.

Anderson, J. (1995) Turbo coding for deep space applications, *IEEE Int. Symposium on Information Theory*, Whistler, Canada, p. 36.

Benedetto, S. and Montorsi, G. (1996a) Unveiling Turbo codes: some results on parallel concatenated coding schemes, *IEEE Trans. on Information Theory*, **42**(2), 409–428.

Benedetto, S. and Montorsi, G. (1996b) Design of parallel concatenated convolutional codes, *IEEE Trans. on Communications,* **44**(5), 591–600.

Benedetto, S., Montorsi, G., Divsalar, D. and Pollara, F. (1996) Serial concatenation of interleaved codes: performance analysis, design, and iterative decoding, *TDA Progress Report*, 42-126.

Benedetto, S. and Montorsi, G. (1996c) Iterative decoding of serially concatenated convolutional codes, *IEE Electronics Letts.*, **32**(13), 1186–1187.

Benedetto, S. and Montorsi, G. (1996d) Serial concatenation of interleaved codes: analytical performance bounds, *IEEE GLOBECOM '96*, **1**, London, 106–110.

Benedetto, S., Garello, R. and Montorsi, G. (1998a) A search for good convolutional codes to be used in the construction of Turbo codes, *IEEE Trans. on Communications*, **46**(9), 1101–1105.

Benedetto, S., Divsalar, D., Montorsi, G. and Pollara, F. (1998b) Serial concatenation of interleaved codes: performance analysis, design and iterative decoding, *IEEE Trans. on Information Theory*, **44**(3), 909–926.

Berrou, C., Thitimajshima, P. and Glavieux, A. (1992) *Recursive Systematic Convolutional Codes and Application to Parallel Concatenation*, Ecole Nationale Superieure des Telecomm. De Bretagne, Telecom, Bretagne.

Berrou, C., Glavieux, A. and Thitimajshima, P. (1993) Near Shannon limit error-correcting coding and decoding: Turbo-codes, *Proc. ICC'93*, Geneva, Switzerland, 1064–1070.

Cedervall, M. and Johannesson, R. (1970) A fast algorithm for computing distance spectrum of convolutional codes, *IEEE Trans. Information Theory*, **IT-35**, 720–738.

Divsalar, D. and Pollara, F. (1995) Multiple Turbo codes for deep-space communications, *TDA Progress Report* 42–121.

Forney, G. (1970) Convolutional codes 1: Algebraic structure, *IEEE Trans. Information Theory*, **IT-16**(6), 720–738.

Haccoun, D. and Begin, G. (1989) High-rate punctured convolutional codes for Viterbi and sequential decoding, *IEEE Trans. on Communications*, **37**(11), 1113–1125.

Hagenauer, J. and Hoeher, P. (1989) A Viterbi algorithm with soft-decision outputs and its applications, *Proc. of GLOBECOM '89*, Dallas, Texas, 47.1.1–47.1.7.

Larsen, K. (1973) Short convolutional codes with maximum free distance for rates 1/2, 1/3, and 1/4, *IEEE Trans. Information Theory*, **IT-19**, 371–372.

Lee, P. (1988) Constructions of rate $(n\text{-}1)/n$ punctured convolutional codes with minimum required SNR criterion, *IEEE Trans. on Communications*, **36**(10), 1171–1174.

Paaske, E. (1974) Short binary convolutional codes with maximal free distance for rates 2/3 and 3/4, *IEEE Trans. on Information Theory*, **IT-20**, 683–689.

Perez, L., Seghers, J. and Costello, D. (1996) A distance spectrum interpretation of Turbo codes, *IEEE Trans. on Information Theory*, **42**(6), 1698–1709.

Sawyer, W. (1978) *Numerical Functional Analysis*, Oxford University Press.

Seghers, J. (1995) On the free distance of Turbo codes and related product codes, *Diploma Project No. 6613*, Institute for Signal and Information Processing, Swiss Federal Institute of Technology, Zurich, Switzerland.

Viterbi, A. (1971) Convolutional codes and their performance in communications systems, *IEEE Trans. on Communications Technology*, **COM-19**(5), 751–772.

Viterbi, A. and Omura, J. (1979) *Principles of Digital Communication and Coding*, McGraw-Hill, New York.

Yasuda, Y., Kashiki K. and Hirata, Y. (1984) High-rate punctured convolutional codes for soft decision Viterbi decoding, *IEEE Trans. on Communications*, **COM-32**(3), 315–320.

Problems

Chapter 1

P1.1 The CCIR-601 4:2:2 coding standard for component video signals specifies that the colour-difference signals should be sampled at 6.75 MHz (and be uniformly quantised to 8-bit PCM). If the maximum colour-difference frequency is 1.3 MHz and the PCM signal is subsequently passed through a DAC, what is the maximum (sin x)/x loss at the output of the reconstruction filter?

[0.54 dB]

P1.2 A low-noise analogue video signal is to be encoded and decoded by the system in Figure 1.3. The black-level to white-level range at the ADC input is set to one-half the (uniform) quantising range of the ADC. The two lowpass filters cutoff at 5 MHz and the sampling frequency is 12 MHz. If the SNR (defined as black-to-white range/rms noise) at the output of the anti-imaging filter is to be at least 41 dB, what is the minimum digital resolution, n, of the codec?

[$n = 6$]

Chapter 2

P2.1 State the source entropy for a discrete source of 20 equiprobable symbols.

[4.32 bits/symbol]

P2.2 A ternary Huffman code ($c = 3$ code symbols) is required for the following discrete source:

a_j	a_1	a_2	a_3	a_4	a_5	a_6	a_7	a_8	a_9	a_{10}
$\hat{P}(a_j)$	0.025	0.025	0.03	0.16	0.24	0.26	0.14	0.07	0.03	0.02

What is the average length of the code and the average compression ratio when compared with straight ternary coding of the source?

[$\overline{L} = 1.83$; 1.64]

P2.3 A discrete source has an alphabet $A = \{a_1, a_2\}$ where a_1 and a_2 are independent symbols with probabilities $P(a_1) = 0.7$, $P(a_2) = 0.3$. A binary Huffman code is to be generated for this source. Find the coding efficiency when the symbols are encoded (a) in pairs, and (b) in blocks of 3 symbols.

[(a) 97.35%, (b) 96.95%]

P2.4 A discrete source has an alphabet $\{a_1a_2a_3a_4\}$ and is compressed using Arithmetic Coding. A fixed model for the source is

a_i	$\hat{P}(a_j)$	$\hat{P}_c(a_j)$
a_1	0.01	0
a_2	0.1	0.01
a_3	0.0001	0.11
a_4	0.0011	0.1101

where $\hat{P}_c(a_j)$ is the cumulative probability. A sequence of four symbols is encoded and the transmitted string (corresponding to the fourth code point) is {0.110110010011}. Deduce the decoded sequence.

$[a_4\ a_1\ a_3\ a_2]$

P2.5 A discrete source has alphabet A = {1,2,3,4} and generates the ASCII character sequence

{ 1 1 4 3 2 2 2 2 1 1 4 3 2 1 4 3 3 3 3 1 }

Each ASCII character is coded to 8-bits. The sequence is to be compressed using the LZW algorithm and a fixed codeword length of n bits. Find the maximum value of n if the algorithm is to give compression for this particular sequence (neglect the coding of the last character).

$[n = 12]$

P2.6 An LZW compressor initially loads a 9-bit code table with 256 single 8-bit ASCII characters (locations (0–255)). The remainder of the table is for character strings. The coder transmits 9-bit codewords, for example codeword v_a for character 'a' and codeword v_b for character 'b'. The decoder receives the code sequence

$\{v_b, v_a, v_a, v_b, v_c, 256, 258, 259, v_b, 261\}$

Use the pseudocode in Figure 2.24 to decode the sequence and hence deduce the compression ratio.

[decoded sequence: { b a a b c ba ab bc b b a a }, $CR = 1.33$]

Chapter 3

P3.1 Assume that the discrete-time sequence $\{x_n\}$ at the input to the DPCM coder in Figure 3.4 has zero mean and unit variance, and that its autocorrelation function R_i has the following values:

$R_0 = 1, R_1 = 0.97, R_2 = 0.91, R_3 = 0.86$

Deduce the coefficients of a one-dimensional second-order predictor. If the input signal statistics now change such that $R_i = e^{-\alpha i}$, where α is a constant and $R_1 = 0.94$, what is the prediction gain?

[1.477, –0.523; 9.3 dB]

P3.2 A basic CELP coder uses the signal processing scheme in Figure 3.38. The sampling rate is 8 kHz and the pitch predictor has the form $P_d(z) = \beta z^{-M}$ where M is an integer. A particular speech frame is 'voiced' with a strong fundamental frequency close to 1.5 kHz. Deduce the most likely value of M.

[$M = 5$]

Chapter 4

P4.1 Sketch simple shift register generators (SSRGs) corresponding to the characteristic polynomials $\phi(x) = 1+x+x^2+x^3+x^4$ and $\phi(x) = 1+x^3+x^6$, where $\phi(x)$ is defined by (4.20). Show algebraically that these have cycle lengths of 5 and 9 respectively.

P4.2 Consider the simple shift register generator (SSRG) in Figure 4.23(a) defined by the primitive polynomial $\phi(x)$ in (4.20). If $\phi(x) = 1+x+x^3$ and the initial condition is $\{u_{-1}\ u_{-2}\ u_{-3}\} = \{0\ 0\ 1\}$, use an algebraic approach to determine the random sequence $\{u\} = \{u_0\ u_1 \ldots u_6\}$. Verify your answer using a clock-by-clock analysis.

[$\{u\} = \{1110100\}$]

P4.3 A modular shift register generator (Figure 4.23(b)) is defined by $\phi_M(x) = 1+x+x^4$ using (4.20). Sketch the circuit of the SSRG which generates the same sequence, and deduce the sequence.

[$\{u\} = \{100110101111000\}$]

P4.4 A particular self-synchronising scrambler system is similar to Figure 4.28(a), but uses a 7-stage shift register with taps at stages 3 and 7. The de-scrambler input has three errors represented by the error polynomial $e(x) = 1 + x + x^4$. Determine the error vector at the descrambler output. Hint: refer to the convolutional coder, Figure 7.2.

[**e** = (110100001001)]

Chapter 5

P5.1 A communications system consists of a 1/4-rate FEC coder followed by a QPSK modulator. If a much better coding scheme could be devised with the same bandwidth efficiency, what would be the minimum possible value of E_b/N_0 for a negligible decoded error rate?

[–0.82 dB]

P5.2 The input to a decoder for the BCH(15,7) code can be modelled as a binary symmetric channel (BSC) with transition probability, $p = 10^{-2}$. Estimate an upper bound for the word (block) error probability at the decoder output.

[4.55×10^{-4}]

Chapter 6

P6.1 An odd-parity extended Hamming code is defined by the check matrix

$$\mathbf{H} = \begin{bmatrix} 1 & 1 & 1 & 1 & 0 & 0 & 0 & 0 \\ 1 & 1 & 0 & 0 & 1 & 1 & 0 & 0 \\ 1 & 0 & 1 & 0 & 1 & 0 & 1 & 0 \\ 1 & 1 & 1 & 1 & 1 & 1 & 1 & 1 \end{bmatrix}$$

Three codewords are used to protect decimal numbers in the range 0–999 and for one transmission the received codewords are

01001011, 11101010, 01001100
hundreds tens units
time ⇒

where the units of the number are transmitted last. The first digit transmitted in each codeword is the MSB of the message and the message bits are transmitted in descending significance. Noting that the code can detect and reject double errors (see (6.23)), attempt to find the transmitted number.

[probably 4X5, where X is unknown]

P6.2 Sketch the shift register circuit for multiplying a polynomial $u(x)$ by

$$g(x) = 1 + x^3 + x^4 + x^5 + x^6$$

Compute the circuit output for $u(x) = 1 + x + x^4$ (the data enters as 1 followed by 0011, and is zero thereafter) by examining the shift register state at each clock pulse. Confirm your answer algebraically.

[(11011000111) where '111' is generated first]

P6.3 A BCH(15,5) coder (see Table 6.1) encodes the message $u(x) = 1 + x^2 + x^3 + x^4$ into a systematic binary codeword, $v(x)$, as in (6.42). Hard-decision detection gives the received codeword as

$$r(x) = 1 + x^4 + x^7 + x^8 + x^9 + x^{10} + x^{14}$$

Locate the errors.

$[e(x) = x^9 + x^{12} + x^{13}]$

P6.4 A systematic (7,4) binary cyclic codeword of the form given by (6.42) is generated by the polynomial $g(x) = 1 + x^2 + x^3$. The message is $u(x) = 1 + x + x^3$, and after detection the effective error polynomial is $e(x) = x^4$. Find the first syndrome word $s(x)$ generated by a Meggitt decoder (Figure 6.12) for decoding the first received symbol.

$[s(x) = 1 + x + x^2]$

P6.5 Delete the last three columns in the parity-check matrix in (6.56) in order to generate a (4,1) cyclic code. Sketch a circuit of the decoder for this shortened code

($r(x)$ entering on the left). If the codeword $\mathbf{v} = (1101)$ is received as $\mathbf{r} = (1111)$, show that the decoder corrects the error on the fifth clock pulse.

P6.6 Burst errors in a noisy channel are to be detected using a binary BCH code of block length $n = 127$. If the probability of an undetected burst is to be less than 10^{-9}, determine the highest rate block code that can be used.

[BCH(127, 92, 5)]

P6.7 A BCH(15,7) coder encodes the message $u(x) = 1 + x^2 + x^3$ into the systematic form of (6.42). The codeword is of the form

$$v(x) = v_0 + v_1 x + v_2 x^2 + \ldots + v_{n-1} x^{n-1}$$

If symbols v_3 and v_{10} are received in error and the decoder uses algebraic decoding, deduce the syndrome vector $s = (s_1, s_2, s_3, s_4)$.

[$\mathbf{s} = (\alpha^{12}, \alpha^9, \alpha^7, \alpha^3)$]

P6.8 A BCH(15,7) codeword is received in polynomial form as

$$r(x) = x + x^7 + x^8 + x^{11} + x^{13}$$

and it is to be decoded using Peterson's algebraic method. Using Table 6.5, deduce the relevant syndrome symbols, the error location polynomial, and the decoded codeword.

[$s_1 = \alpha^{12}$, $s_3 = \alpha^6$, $\sigma(x) = x^2 + \alpha^{12} x$, $\mathbf{v} = (010000011001110)$]

P6.9 Find the RS(7,3) codeword $\mathbf{v}$ (in systematic form) for the message $u(x) = \alpha^2 + x + \alpha^5 x^2$.

[$\mathbf{v} = (1\ \alpha\ \alpha\ \alpha^5\ \alpha^2\ 1\ \alpha^5)$]

P6.10 Find the RS(7,5) codeword $\mathbf{v}$ (in systematic form) for the message $u(x) = 1 + x^2 + x^3$.

[$g(x) = x^2 + \alpha^4 x + \alpha^3$, $\mathbf{v} = (\alpha\ \alpha^4\ 1\ 0\ 1\ 1\ 0)$]

P6.11 Using the practical notation in Example 6.34, an RS(7,3) codeword is received as $\mathbf{r} = (6\ 5\ 0\ 6\ 5\ 3\ 6)$. Assuming the decoder uses the approach in Figure 6.17(a), deduce the syndrome vector, $\mathbf{s}$, and hence the corrected codeword. (Use the identities for GF(2^3) given in Example 6.35).

[$\mathbf{s} = (6\ 6\ 3\ 4)$; $\mathbf{v} = (6\ 5\ 0\ 1\ 5\ 1\ 6)$]

P6.12 A decoder for the RS(7,3) code uses the frequency-domain approach in Figure 6.17(b) and the decoder input is

$$\mathbf{r} = (1\ \alpha^4\ \alpha^3\ \alpha^5\ \alpha^2\ 1\ \alpha^5)$$

The elements of $\mathbf{r}$ are drawn from the Galois field defined in Table 6.4. Deduce the frequency domain error vector $\mathbf{E}$, and hence estimate the transmitted codevector.

[$\mathbf{E} = (\alpha^6\ \alpha^5\ 0\ \alpha\ \alpha^5\ \alpha\ \alpha^6)$; $\hat{\mathbf{v}} = (1\ \alpha\ \alpha\ \alpha^5\ \alpha^2\ 1\ \alpha^5)$]

P6.13 Repeat problem P6.12 assuming $\mathbf{r} = (\alpha^5\ 1\ \alpha^6\ \alpha^4\ \alpha^6\ \alpha\ 1)$.

$[\mathbf{E} = (\alpha^3\ \alpha^6\ \alpha^3\ \alpha\ \alpha^2\ 0\ \alpha^3);\ \hat{\mathbf{v}} = (\alpha^5\ 1\ \alpha^2\ \alpha^4\ \alpha^6\ \alpha\ \alpha^3)]$

Chapter 7

P7.1 A rate 1/2 memory-2 convolutional coder has the generator polynomial matrix

$\mathbf{G}(x) = [1+x,\ 1+x+x^2\ ]$

and the first code bit for a single data bit is defined by the $1 + x$ polynomial. Sketch the trellis diagram and hence determine the d_{free} for the code.

$[d_{free} = 4]$

P7.2 If the input sequence to the coder in P7.1 is $\mathbf{u} = \{101100100...\}$, where the leftmost digit enters the coder first, deduce the code sequence.

[$\mathbf{v}$ = (111110001001111101...)]

P7.3 A hard-decision Viterbi decoder for the code in Example P7.1 receives the input sequence $\mathbf{r}$ = (111110101001...). Estimate the data sequence at the coder input.

[$\mathbf{u}$ = {1 0 1 1 0 0 ...}]

P7.4 A convolutional code sequence is generated by the coder in Figure 7.5 and is transmitted over a noisy channel. The corresponding received sequence (codeword) at the input to a hard-decision Viterbi decoder is $\mathbf{r}$ = (111011100100100000...). Perform Viterbi decoding on this sequence up to and including the fifth branch and hence estimate the data sequence at the coder input. Assume the decoder starts from the all-zeros state.

[$\mathbf{u}$ = 1 0 1 1 0 ...]

P7.5 Use an algebraic approach to deduce the binary sequence corresponding to $v2(D)$ in Figure 7.13, if the input sequence is defined by

$u(D) = 1 + D + D^4 + D^5 + D^8$

(Assume u_0 enters first).

[$\mathbf{v}$2 = (100101)]

P7.6 Determine the code sequence $\mathbf{v} = \{v1, v2\}$ for Figure 7.16(a) that corresponds to an error event which has been generated by an input sequence of minimum weight and length > 4 bits. Assume the shortest possible input sequence.

[$\mathbf{v}$ = (11010100010111)]

APPENDIX A

Table of the *Q*-function

The *Q*-function, or normal probability integral, defines the area under the standardized Gaussian tail and is given by

$$Q(x) \triangleq \frac{1}{\sqrt{(2\pi)}} \int_x^{\infty} \exp\left(-\frac{y^2}{2}\right) dy$$

The complementary error function is defined as

$$erfc(x) \triangleq \frac{2}{\sqrt{\pi}} \int_x^{\infty} \exp(-y^2)\, dy = 2Q(\sqrt{2}\, x)$$

x		0	1	2	3	4	5	6	7	8	9
0.0	0.	50000	49601	49202	48803	48405	48006	47608	47210	46812	46414
0.1		46017	45620	45224	44828	44433	44038	43644	43251	42858	42465
0.2		42074	41683	41294	40905	40517	40129	39743	39358	38974	38591
0.3		38209	37828	37448	37070	36693	36317	35942	35569	35197	34827
0.4		34458	34090	33724	33360	32997	32636	32276	31918	31561	31207
0.5		30854	30503	30153	29806	29460	29116	28774	28434	28096	27760
0.6		27425	27093	26763	26435	26109	25785	25463	25143	24825	24510
0.7		24196	23885	23576	23270	22965	22663	22363	22065	21770	21476
0.8		21186	20897	20611	20327	20045	19766	19489	19215	18943	18673
0.9		18406	18141	17879	17619	17361	17106	16853	16602	16354	16109
1.0		15866	15625	15386	15151	14917	14686	14457	14231	14007	13786
1.1		13567	13350	13136	12924	12714	12507	12302	12100	11900	11702
1.2		11507	11314	11123	10935	10749	10565	10383	10204	10027	98525
1.3	0.0	96800	95098	93418	91759	90123	88508	86915	85343	83793	82264
1.4		80757	79270	77804	76359	74934	73529	72145	70781	69437	68112
1.5		66807	65522	64255	63008	61780	60571	59380	58208	57053	55917
1.6		54799	53699	52616	51551	50503	49471	48457	47460	46479	45514
1.7		44565	43633	42716	41815	40930	40059	39204	38364	37538	36727
1.8		35930	35148	34380	33625	32884	32157	31443	30742	30054	29379
1.9		28717	28067	27429	26803	26190	25588	24998	24419	23852	23295
2.0		22750	22216	21692	21178	20675	20182	19699	19226	18763	18309
2.1		17864	17429	17003	16586	16177	15778	15386	15003	14629	14262
2.2		13903	13553	13209	12874	12545	12224	11911	11604	11304	11011
2.3		10724	10444	10170	99031	96419	93867	91375	88940	86563	84242
2.4	0.0^2	81975	79763	77603	75494	73436	71428	69469	67557	65691	63872
2.5		62097	60366	58677	57031	55426	53861	52336	50849	49400	47988
2.6		46612	45271	43965	42692	41453	40246	39070	37926	36811	35726
2.7		34670	33642	32641	31667	30720	29798	28901	28028	27179	26354
2.8		25551	24771	24012	23274	22557	21860	21182	20524	19884	19262
2.9		18658	18071	17502	16948	16411	15889	15382	14890	14412	13949
3.0		13499	13062	12639	12228	11829	11442	11067	10703	10350	10008
3.1	0.0^3	96760	93544	90426	87403	84474	81635	78885	76219	73638	71136
3.2		68714	66367	64095	61895	59765	57703	55706	53774	51904	50094
3.3		48342	46648	45009	43423	41889	40406	38971	37584	36243	34946
3.4		33693	32481	31311	30179	29086	28029	27009	26023	25071	24151
3.5		23263	22405	21577	20778	20006	19262	18543	17849	17180	16534
3.6		15911	15310	14730	14171	13632	13112	12611	12128	11662	11213
3.7		10780	10363	99611	95740	92010	88417	84957	81624	78414	75324
3.8	0.0^4	72348	69483	66726	64072	61517	59059	56694	54418	52228	50122
3.9		48096	46148	44274	42473	40741	39076	37475	35936	34458	33037

x		0	1	2	3	4	5	6	7	8	9
4.0		31671	30359	29099	27888	26726	25609	24536	23507	22518	21569
4.1		20658	19783	18944	18138	17365	16624	15912	15230	14575	13948
4.2		13346	12769	12215	11685	11176	10689	10221	97736	93447	89337
4.3	0.0^5	85399	81627	78015	74555	71241	68069	65031	62123	59340	56675
4.4		54125	51685	49350	47117	44979	42935	40980	39110	37322	35612
4.5		33977	32414	30920	29492	28127	26823	25577	24386	23249	22162
4.6		21125	20133	19187	18283	17420	16597	15810	15060	14344	13660
4.7		13008	12386	11792	11226	10686	10171	96796	92113	87648	83391
4.8	0.0^6	79333	75465	71779	68267	64920	61731	58693	55799	53043	50418
4.9		47918	45538	43272	41115	39061	37107	35247	33476	31792	30190

APPENDIX B

Simulation of Meggitt decoding

```
/* Meggitt.C
 * Simulation of Meggitt decoding
 * Based upon the double-error correcting BCH(15,7)code
 * User supplied information vector and received vector enables random
 * and burst errors to be simulated
 * Shows clock-by-clock state of the syndrome register
 */

#include <stdlib.h>
#include <stdio.h>
#include <math.h>

void clockSR(int,int);
int  s0[17],s1[17],s2[17],s3[17],s4[17],s5[17],s6[17],s7[17];

main()
{
    int  n,k,i,j,m,corr,err;
    int  r[15],e[15],d[7],v[15],vo[15];
    int  p0[15],p1[15],p2[15],p3[15],p4[15],p5[15],p6[15],p7[15];

    n=15; k=7;
    printf("\nBCH(15,7) CODEC SIMULATION \n");
    printf("\nSimulation of encoder and Meggitt decoder\n");

    /* program PROM with syndrome patterns */
    printf("\nsyndrome patterns to be detected by PROM:\n");
    printf("   s0  s1  s2  s3  s4  s5  s6  s7\n");
    e[n-1] = 1;
    for(j=0; j<n; j++) {
      for(i=0; i<n-1; i++) e[i]=0;
        e[n-j-1] = 1; /* error vector set */
        s0[0]=s1[0]=s2[0]=s3[0]=s4[0]=s5[0]=s6[0]=s7[0]=0;

        /* find syndrome for jth error vector */
        for(i=1; i<n+1; i++) clockSR(i,e[n-i]);
        /* SR now holds syndrome for jth error vector */

        printf("   %d   %d   %d   %d   %d   %d   %d   %d\n",
                 s0[n],s1[n],s2[n],s3[n],s4[n],s5[n],s6[n],s7[n]);
        p0[j]=s0[n];p1[j]=s1[n];p2[j]=s2[n];p3[j]=s3[n];
        p4[j]=s4[n];p5[j]=s5[n];p6[j]=s6[n];p7[j]=s7[n];
    }
```

```
/* PROM coding complete */

for(j=0;;j++) {
err=corr=0;
printf("\nenter 7 information bits, i(k-1) last:\n");
printf(".......\n");
for(i=0; i<k; i++) {
  d[i] = getchar();
  if(d[i]==48) d[i]=0;
  if(d[i]==49) d[i]=1;
  v[n-k+i]=d[i];
}
getchar();
s0[0]=s1[0]=s2[0]=s3[0]=s4[0]=s5[0]=s6[0]=s7[0]=0;  /* reset SR */

/* generate code vector */
for(i=1; i<k+1; i++) {
  s7[i]=s6[i-1]+s7[i-1]+d[k-i];if(s7[i]==3) s7[i]=1;if(s7[i]==2) s7[i]=0;
  s6[i]=s5[i-1]+s7[i-1]+d[k-i];if(s6[i]==3) s6[i]=1;if(s6[i]==2) s6[i]=0;
  s5[i]=s4[i-1];
  s4[i]=s3[i-1]+s7[i-1]+d[k-i];if(s4[i]==3) s4[i]=1;if(s4[i]==2) s4[i]=0;
  s3[i]=s2[i-1];
  s2[i]=s1[i-1];
  s1[i]=s0[i-1];
  s0[i]=s7[i-1]+d[k-i];if(s0[i]==2) s0[i]=0;

}
v[0]=s0[k]; v[1]=s1[k]; v[2]=s2[k]; v[3]=s3[k]; v[4]=s4[k]; v[5]=s5[k];
v[6]=s6[k]; v[7]=s7[k];
for(i=0; i<n; i++) { /* integer to char conversion */
  if(v[i]==0) v[i]=48; else v[i] = 49;
}
printf("\nsystematic code vector :\n");
for(i=0; i<n; i++)  putchar((char)v[i]);  printf("\n");
printf("\nenter received vector, r(n-1) last :\n");
for(i=0; i<n; i++) {
  r[i]=getchar();
  if(r[i]==48) r[i]=0;
  if(r[i]==49) r[i]=1;
}
getchar();  printf("\n");

/* Meggitt decoder simulation */

s0[0]=s1[0]=s2[0]=s3[0]=s4[0]=s5[0]=s6[0]=s7[0]=0;  /* reset SR */

printf("state of decoder syndrome register :\n");
printf("\nclock  s0 s1 s2 s3 s4 s5 s6 s7\n");
  for(i=1; i<n+1; i++) {
    clockSR(i,r[n-i]);
    printf(" %2d    %d  %d  %d  %d  %d  %d  %d  %d\n",
              i,s0[i],s1[i],s2[i],s3[i],s4[i],s5[i],s6[i],s7[i]);
  }
/* received vector now in buffer, initial syndrome computed */
printf("received vector now in buffer: commence error correction\n");
```

```
    s0[1]=s0[n]; s1[1]=s1[n]; s2[1]=s2[n]; s3[1]=s3[n];
    s4[1]=s4[n]; s5[1]=s5[n]; s6[1]=s6[n]; s7[1]=s7[n];

    for(i=1; i<n+1; i++) {
      getchar();
      m=15+i;  e[n-i]=0;
      if(s0[i]==1 || s1[i]==1 || s2[i]==1 || s3[i]==1
      || s4[i]==1 || s5[i]==1 || s6[i]==1 || s7[i]==1) err=1;
      for(j=0; j<n; j++) {
      if(s0[i]==p0[j] && s1[i]==p1[j] && s2[i]==p2[j] && s3[i]==p3[j]
      && s4[i]==p4[j] && s5[i]==p5[j] && s6[i]==p6[j] && s7[i]==p7[j])
        { e[n-i]=1; corr=1; }
      }
      vo[n-i]=e[n-i]+r[n-i]; /* error correction */
      if(e[n-i]==1) printf("buffer output complemented\n");
      if(vo[n-i]==2) vo[n-i]=0;
      clockSR(i+1,0);  /* no feedback */

      printf(" %2d    %d  %d  %d  %d  %d  %d  %d  %d\n",
      m,s0[i+1],s1[i+1],s2[i+1],s3[i+1],s4[i+1],s5[i+1],s6[i+1],s7[i+1]);
    }

    for(i=0; i<n; i++) {
      if(vo[i]==0) vo[i]=48; else vo[i]=49;
    }
    if(err==1 && corr==0)
    printf("\nuncorrectable error pattern detected\n"); /* eg burst up to n-k */
    if(corr==1)
    printf("\nreceived vector modified\n"); /* not necessarily corrected */
    printf("\ndecoded vector :     ");
    for(i=0; i<n; i++) putchar((char)vo[i]);
    printf("\ntransmitted vector : "); for(i=0; i<n; i++) putchar((char)v[i]);
    printf("\n");
    }
}

void
clockSR(u,x)
int u,x;
{
  extern int s0[],s1[],s2[],s3[],s4[],s5[],s6[],s7[];

  s7[u]=s7[u-1]+s6[u-1]; if(s7[u]==2) s7[u]=0;
  s6[u]=s7[u-1]+s5[u-1]; if(s6[u]==2) s6[u]=0;
  s5[u]=s4[u-1];
  s4[u]=s7[u-1]+s3[u-1]; if(s4[u]==2) s4[u]=0;
  s3[u]=s2[u-1];
  s2[u]=s1[u-1];
  s1[u]=s0[u-1];
  s0[u]=s7[u-1]+x; if(s0[u]==2) s0[u]=0;
}
```

APPENDIX C

Soft-decision Viterbi decoding

```
/* SOFT-DECISION VITERBI DECODING
 * Simulates R=1/2 encoder, AWGN channel, and soft-decision Viterbi decoder.
 * Constraint length K = 7, M = 6, g1 = 171, g2 = 133
 * Decoding window can be varied.
 * Decodes u[K-1], u[K],..... u[imax-W] inclusive (imax-W-K+2  bits).
 * Array u holds encoder input data, array uh holds decoded data.
 * Array v holds code sequence.
 * Array z holds received analogue data from demodulator.
 */

#include <stdlib.h>
#include <stdio.h>
#include <math.h>

#define  RAND_MAX  32767;

/* noise function, Box-Muller method */
float gnoise(float sigma)
{
 static char n_nr=0;
 static float yn2;
 float yn1,rsq,fac;

 if(n_nr){n_nr = 0; return yn2;}
 do {
   yn1=2*(double)rand()/RAND_MAX-1;
   yn2=2*(double)rand()/RAND_MAX-1;
   rsq=yn1*yn1+yn2*yn2;
 }while(rsq >= 1 || rsq == 0);
 fac = sigma*sqrt(-2*log(rsq)/rsq);
 yn1*=fac;yn2*=fac;n_nr = 1;
 return yn1;
}

long  int   iter,nbits,rbits,tbits,errcnt;
int         cb[1010][64];
float       pm[1010][64];

main()
{
    int    i,imax,j,p,q,r,t,k,loop,seed;
    int    K,W,L1,L2,frame,n,m,s,s0,s1,s2,s3,s4,s5;
    int    a[6],S[64],ns[64],u[1010],uh[1010],opts[1010],v[2020],R[60];
```

```
float  B,DB,BER,sigma,minmet,mp,mq;
float  z[2020];

printf("\nSOFT-DECISION VITERBI DECODING  R=1/2  K=7 :\n");
printf("\nspecify decoding window W (in frames eg 5K) : ");
  scanf("%d",&W); getchar();
printf("\nspecify number of iterations                 : ");
  scanf("%ld",&iter); getchar();
printf("\nspecify channel Eb/No (dB)                   : ");
  scanf("%f",&DB); getchar();
printf("\nspecify random generator seed (1,2,...)      : ");
  scanf("%d",&seed); getchar();

 /* initialization */
 K = 7; /* constraint length */
 imax = 1000;                /* gross data vector size/iteration */
 nbits = imax - W - K + 2;   /* actual number of bits decoded/iteration */
 tbits = iter * nbits;       /* total number data bits decoded */
 n = (1 < (K-1)); m = n/2;
 B = pow(10,(0.1*DB));
 sigma = 1.0/(sqrt(B)); printf("\nsigma = %4.2f",sigma);
 errcnt = 0;
 srand(seed);
 for(j=0; j<60; j++) R[j]=0;  R[0]=1;  /* set 60-stage prbs generator */

for(loop=0; loop<iter; loop++) {
  /* encoder */
  s0=s1=s2=s3=s4=s5=0;
  u[0] = 1; v[0] = 1; v[1] = 1; s5=s4; s4=s3; s3=s2; s2=s1; s1=s0; s0=u[0];
  for(i=1; i<imax; i++) {
    u[i] = (R[0] + R[59]) & 1; for(j=59; j>0; j--) R[j] = R[j-1]; R[0]=u[i];
    v[2*i]   = (u[i] + s0 + s1 + s2 + s5) & 1; if (v[2*i] == 0) v[2*i] = -1;
    v[2*i+1] = (u[i] + s1 + s2 + s4 + s5) & 1; if (v[2*i+1] == 0) v[2*i+1] = -1;
    s5=s4; s4=s3; s3=s2; s2=s1; s1=s0; s0=u[i];
  }

  /* generate demodulator analogue output samples */
  for (i=0; i<(2*imax); i++) z[i] = v[i] + gnoise(sigma);

  /* initialization */
  for (r=0; r<64; r++) { S[r] = 0; ns[r] = 0; }
  /* frame i=0 */
    L1 = -1; L2 = -1; /* u[0] = 0 */
    pm[1][0] = (z[0] - L1)*(z[0] - L1) + (z[1] - L2)*(z[1] - L2);
    cb[1][0] = 0;
    L1 = 1; L2 = 1;   /* u[0] = 1 */
    pm[1][m] = (z[0] - L1)*(z[0] - L1) + (z[1] - L2)*(z[1] - L2);
    cb[1][m] = 0;
    ns[m] = m; S[m] = ns[m];

  /* grow trellis for K-2 frames, to i = K-2 incl. S[r] holds nonzero states */
  for (i=1; i<(K-1); i++) {
    L1 = -1; L2 = -1;  /* u[i] = 0 */
    pm[i+1][0] = pm[i][0]+(z[2*i]-L1)*(z[2*i]-L1)+(z[2*i+1]-L2)*(z[2*i+1]-L2);
    cb[i+1][0] = 0;
```

```
      L1 = 1; L2 = 1;     /* u[i] = 1 */
      pm[i+1][m] = pm[i][0]+(z[2*i]-L1)*(z[2*i]-L1)+(z[2*i+1]-L2)*(z[2*i+1]-L2);
      cb[i+1][m] = 0;
      for (r=1; r<n; r++) {
       if (S[r]>0) {
         p = (int) (r/2);  ns[p]=p;  s = r;            /* u[i] = 0 */
         for(k=5; k>-1; k--) {a[k]=s&1; s=s>1;}
         L1 = (a[0] + a[1] + a[2] + a[5])&1;
         L2 = (a[1] + a[2] + a[4] + a[5])&1;
         if (L1==0)  L1=-1;  if (L2==0) L2=-1;
         pm[i+1][p]=pm[i][r]+(z[2*i]-L1)*(z[2*i]-L1)+(z[2*i+1]-L2)*(z[2*i+
           1]-L2);
         cb[i+1][p] = r;
         q = (int) (r/2) + m; ns[q] = q;                /* u[i] = 1 */
         L1 = (1 + a[0] + a[1] + a[2] + a[5])&1;
         L2 = (1 + a[1] + a[2] + a[4] + a[5])&1;
         if (L1==0)  L1 = -1;  if (L2==0)  L2 = -1;
         pm[i+1][q]=pm[i][r]+(z[2*i]-L1)*(z[2*i]-L1)+(z[2*i+1]-L2)*(z[2*i+
           1]-L2);
         cb[i+1][q] = r;
       }
     }
     for (r=1; r<n; r++)  S[r] = ns[r];  /* add new non-zero states */
    }
    j = 0;
    for(i=K-1; i<imax; i++) {
     for (r=0; r<m; r++) {  /* u[i] = 0 */
       p = 2*r; q = p + 1;  s = p;
       for(k=5; k>-1; k--) {a[k]=s&1; s=s>1;}
       L1 = (a[0] + a[1] + a[2] + a[5])&1;
       L2 = (a[1] + a[2] + a[4] + a[5])&1;
       if (L1==0) L1=-1;  if (L2==0) L2=-1;
       mp=pm[i][p] + (z[2*i] - L1)*(z[2*i] - L1) + (z[2*i+1] -
            L2)*(z[2*i+1] - L2);
       s = q;
       for(k=5; k>-1; k--) {a[k]=s&1; s=s>1;}
       L1 = (a[0] + a[1] + a[2] + a[5])&1;
       L2 = (a[1] + a[2] + a[4] + a[5])&1;
       if (L1==0)  L1=-1;  if (L2==0) L2=-1;
       mq=pm[i][q] + (z[2*i] - L1)*(z[2*i] - L1) + (z[2*i+1] -
            L2)*(z[2*i+1] - L2);
       if (mp<mq) { pm[i+1][r] = mp;  cb[i+1][r] = p; }
       else       { pm[i+1][r] = mq;  cb[i+1][r] = q; }
     }
     for (r=m; r<n; r++) {  /* u[i] = 1 */
       p = 2*(r-m); q = p + 1;  s = p;
       for(k=5; k>-1; k--) {a[k]=s&1; s=s>1;}
       L1 = (1+ a[0] + a[1] + a[2] + a[5])&1;
       L2 = (1+ a[1] + a[2] + a[4] + a[5])&1;
       if (L1==0)  L1=-1;  if (L2==0) L2=-1;
       mp=pm[i][p] + (z[2*i] - L1)*(z[2*i] - L1) + (z[2*i+1] -
            L2)*(z[2*i+1] - L2);
       s = q;
       for(k=5; k>-1; k--) {a[k]=s&1; s=s>1;}
       L1 = (1+ a[0] + a[1] + a[2] + a[5])&1;
```

```
      L2 = (1+ a[1] + a[2] + a[4] + a[5])&1;
      if (L1==0)  L1=-1;  if (L2==0) L2=-1;
      mq=pm[i][q] + (z[2*i] - L1)*(z[2*i] - L1) + (z[2*i+1] - L2)*(z[2*i+1] -
          L2);
      if (mp<mq) { pm[i+1][r] = mp;  cb[i+1][r] = p; }
      else       { pm[i+1][r] = mq;  cb[i+1][r] = q; }
    }
    if (i==(W+5+j))  {  /* decode oldest bit */
      opts[i+1] = 0;   minmet = pm[i+1][0];
      for(r=1; r<n; r++) {
        if(pm[i+1][r] < minmet) {
          minmet = pm[i+1][r];
          opts[i+1] = r;
        }
      } /* sensible starting point r found for chain back */
      for(t=(W+5+j); t>(5+j); t--) {  /* chainback W frames */
        opts[t] = cb[t+1][opts[t+1]];  frame = t;
      }
      if (opts[frame+1] < opts[frame]) uh[j]=0; else uh[j]=1; /* decode
        oldest bit */
      if ((opts[frame+1]==0) && (opts[frame]==0))  uh[j] = 0;
      if ((opts[frame+1]==(n-1)) && (opts[frame]==(n-1)))  uh[j] = 1;
      if (u[(K-1+j)] != uh[j]) errcnt += 1;
      j += 1;
    }
  }
  rbits = (loop+1)*nbits;
  BER = 1.0*errcnt/(1.0*rbits);
  printf("\niteration  %3d   errcnt = %3ld   BER = %10.8f",loop,errcnt,BER);
} /* end loop */
  BER = 1.0*errcnt/(1.0*tbits);
  printf("\nBER = %10.8f Eb/No (dB) = %6.3f window = %d data bits =
    %ld",BER,DB,W,tbits);
}
```

APPENDIX D

Iterative decoding

```
/* HCCC SIMULATION
 * Unpunctured, rate 1/3 HCCC
 * Random interleaver (but easily modified to S-interleaver)
 * Maximum interleaver size N=1000
 * Both constituent codes: K=3, R=1/2, G=[1,(1+D^2)/(1+D+D^2)]
 * The number of decoder iterations can be varied
 */

#include <stdlib.h>
#include <stdio.h>
#include <math.h>
#define  NOBITS   4
#define  X12      4096
#define  X9       512
#define  X3       8
#define  X2       4
#define  X0       1
#define  STRT     3
#define  ELIM     0.00001

int shrand(int dim)
{
  static int sreg=STRT;
  int ib=0,b9,b3,b2,b0;

  do
  {
    if(X9&sreg)  b9=1;
    else         b9=0;
    if(X3&sreg)  b3=1;
    else         b3=0;
    if(X2&sreg)  b2=1;
    else         b2=0;
    if(X0&sreg)  b0=1;
    else         b0=0;
    ib=b9^b3^b2^b0;
    if(ib) sreg|=X12;
    sreg>>=1;
  }while(sreg>dim);

  return sreg-1;
}
```

```
/* noise function, Box-Muller method */
float gnoise(float sigma)
{
  static char n_nr=0;
  static float yn2;
  float yn1,rsq,fac;

  if(n_nr){n_nr = 0; return yn2;}
  do {
    yn1=2*(double)rand()/RAND_MAX-1;
    yn2=2*(double)rand()/RAND_MAX-1;
    rsq=yn1*yn1+yn2*yn2;
  }while(rsq >= 1 || rsq == 0);
  fac = sigma*sqrt(-2*log(rsq)/rsq);
  yn1*=fac;yn2*=fac;n_nr = 1;
  return yn1;
}

int    imax,L;
long int iter,loop,nbits,rbits;
int    rnum[1005],u[1005],uh[1005],v1[1005],v2[1005],v3[1005],w2[1005],w2h[1005];
double P[4][1005];
double Pb[4][1005];

main()
{
  int    i,j,m,w,M,K,s0,s1,errcnt,declcnt,iter2,logic,dseed,state;
  float  r1[1005],r2[1005],r3[1005],Pu0[1005],Pw20[1005],Pv20[1005],Px20[1005],g[2];
  float  ranum,Pu1,Pv21,Pw21,pw20,pw21,pv30,pv31,sigma,sig2,sum,R,B,DB,BER;

  printf("\nspecify channel Eb/No (dB)          :"); scanf("%f",&DB);
    getchar();
  printf("\nspecify data generator seed         :"); scanf("%d",&dseed);
    getchar();
  printf("\nspecify number of data blocks       :"); scanf("%ld",&iter);
    getchar();
  printf("\nspecify number of decoder iterations:"); scanf("%d",&iter2);
    getchar();

  imax = 1000; /* maximum data block size */
  L = imax;
  K = 3;       /* constraint length of CCs */
  nbits = imax - 20;
  M = 4;       /* number of encoder states */
  R = 1.0/3.0; /* HCCC rate */
  B = pow(10,(0.1*DB));
  sigma = 1.0/(sqrt(2.0*R*B)); printf("\nsigma = %4.2f",sigma);
  sig2 = 2.0 * sigma * sigma;
  errcnt = 0;

  /*shift register interleaver*/
  for(j=0;j<L;j++) rnum[j]=shrand(L);

  srand(dseed);  /* set data seed */
```

```
for(loop=1; loop<iter+1; loop++) {
  /* enc1 */
  s0=s1=0; /* initial state */
  for(i=0; i<imax; i++) {
    ranum = 1.0 * rand()/RAND_MAX; if(ranum > 0.5)  u[i] = 1; else u[i] = 0;
    if(i > (imax-K))  u[i] = (s0 + s1)&1;   /* use data tail */

    v1[i] = u[i];   if(v1[i]==0)  v1[i] = -1;
    logic = u[i] + s0 + s1; /* feedback */
    w = logic & 1;
    logic = w + s1;
    v2[i] = logic & 1;
    s1=s0; s0=w;
}
/* interleaver */
for(i=0;i<imax;i++) w2[rnum[i]]=v2[i];

/* enc2 */
s0=s1=0; /* initial state */
for(i=0; i<imax; i++) {
  logic = w2[i] + s0 + s1; /* feedback */
  w = logic & 1;
  logic = w + s1;
  v3[i] = logic & 1; if(v3[i]==0)  v3[i] = -1;
  if(w2[i]==0)  w2[i] = -1;
  s1=s0; s0=w;
}

for(i=0; i<imax; i++) { /* generate received data - add noise */
  r1[i] = v1[i] + gnoise(sigma);
  r2[i] = w2[i] + gnoise(sigma);
  r3[i] = v3[i] + gnoise(sigma);
}
/* r1[i] - r3[i] generated */

/* initial extrinsic value */
for(i=0; i<imax; i++)   Pw20[i] = 0.5;  /* initial extrinsic info */
m = 0; if(w2[imax-1]==-1)  m=1;  /* end information for iteration */

for(j=1; j<iter2+1; j++)  {   /* iterative decoding loop */

/* dec2 forward recursion */
for(state=0; state<M; state++)  P[state][0] = 0;
P[0][0] = 1.0;    /* known start */
for(i=0; i<imax; i++) {
  pw20 = exp(-((r2[i]+1)*(r2[i]+1))/sig2);
  pw21 = exp(-((r2[i]-1)*(r2[i]-1))/sig2);
  pv30 = exp(-((r3[i]+1)*(r3[i]+1))/sig2);
  pv31 = exp(-((r3[i]-1)*(r3[i]-1))/sig2);
  pv30 = pv30/(pv30+pv31);  pv31 = pv31/(pv30+pv31);
  pw20 = pw20/(pw20+pw21);  pw21 = pw21/(pw20+pw21);
  Px20[i] = pw20;

  P[0][i+1] = Pw20[i]*pw20*pv30*P[0][i] + (1-Pw20[i])*pw21*pv31*P[1][i];
  P[1][i+1] = (1-Pw20[i])*pw21*pv30*P[2][i] + Pw20[i]*pw20*pv31*P[3][i];
```

```
    P[2][i+1] = Pw20[i]*pw20*pv30*P[1][i] + (1-Pw20[i])*pw21*pv31*P[0][i];
    P[3][i+1] = (1-Pw20[i])*pw21*pv30*P[3][i] + Pw20[i]*pw20*pv31*P[2][i];
    /* Pw20[i] is from dec1 via interleaver */
    sum = 0;
    for(state=0; state<M; state++)  sum += P[state][i+1];
    for(state=0; state<M; state++)  P[state][i+1] = P[state][i+1]/sum;
  }  /* end i */

  /* backward recursion */
  for(state=0; state<M; state++)
    Pb[state][imax] = P[state][imax];  /* best estimate */
  for(i=imax-1; i>(-1); i--) {
    pw20 = exp(-((r2[i]+1)*(r2[i]+1))/sig2);
    pw21 = exp(-((r2[i]-1)*(r2[i]-1))/sig2);
    pv30 = exp(-((r3[i]+1)*(r3[i]+1))/sig2);
    pv31 = exp(-((r3[i]-1)*(r3[i]-1))/sig2);
    pv30 = pv30/(pv30+pv31);  pv31 = pv31/(pv30+pv31);
    pw20 = pw20/(pw20+pw21);  pw21 = pw21/(pw20+pw21);

    Pb[0][i] = Pw20[i]*pw20*pv30*Pb[0][i+1] + (1-Pw20[i])*pw21*pv31*Pb[2][i+1];
    Pb[1][i] = Pw20[i]*pw20*pv30*Pb[2][i+1] + (1-Pw20[i])*pw21*pv31*Pb[0][i+1];
    Pb[2][i] = Pw20[i]*pw20*pv31*Pb[3][i+1] + (1-Pw20[i])*pw21*pv30*Pb[1][i+1];
    Pb[3][i] = Pw20[i]*pw20*pv31*Pb[1][i+1] + (1-Pw20[i])*pw21*pv30*Pb[3][i+1];
    sum = 0;
    for(state=0; state<M; state++)  sum += Pb[state][i];
    for(state=0; state<M; state++)  Pb[state][i] = Pb[state][i]/sum;

    Pw20[i]=P[0][i]*Pb[0][i+1]*pv30+P[1][i]*Pb[2][i+1]*pv30+
            P[2][i]*Pb[3][i+1]*(1-pv30)+P[3][i]*Pb[1][i+1]*(1-pv30);
    Pw21=P[0][i]*Pb[2][i+1]*(1-pv30)+P[1][i]*Pb[0][i+1]*(1-pv30)+
         P[2][i]*Pb[1][i+1]*pv30+P[3][i]*Pb[3][i+1]*pv30;
    Pw20[i]=Pw20[i]/(Pw21+Pw20[i]);
    if(Pw20[i]<ELIM) Pw20[i]=ELIM;
    if(Pw20[i]>1-ELIM) Pw20[i]=1-ELIM;
  }  /* end i */

  /* de-interleave r2[i] */
  for(i=0; i<imax; i++)  Pv20[i] = Px20[rnum[i]]; /* parity i/p to dec1 */
  /* de-interleave Pw20[i] */
  for(i=0; i<imax; i++)  Px20[i] = Pw20[rnum[i]]; /* extrinsic i/p to dec1 */

  /* dec1 forward recursion */
  for(state=0; state<M; state++)   P[state][0] = 0;
  P[0][0] = 1.0;
  for(i=0; i<imax-1; i++) {
    g[0] = exp(-((r1[i]+1)*(r1[i]+1))/sig2);
    g[1] = exp(-((r1[i]-1)*(r1[i]-1))/sig2);
    P[0][i+1] = g[0]*Px20[i]*Pv20[i]*P[0][i] + g[1]*(1-Px20[i])*(1-Pv20[i])*P[1][i];
    P[1][i+1] = g[0]*(1-Px20[i])*(1-Pv20[i])*P[3][i] + g[1]*Px20[i]*Pv20[i]*P[2][i];
    P[2][i+1] = g[0]*Px20[i]*Pv20[i]*P[1][i] + g[1]*(1-Px20[i])*(1-Pv20[i])*P[0][i];
    P[3][i+1] = g[0]*(1-Px20[i])*(1-Pv20[i])*P[2][i] + g[1]*Px20[i]*Pv20[i]*P[3][i];
    sum = 0;
    for(state=0; state<M; state++)  sum += P[state][i+1];
    for(state=0; state<M; state++)  P[state][i+1] = P[state][i+1]/sum;
  }  /* end i */
```

```
    /* backward recursion */
    for(state=0; state<M; state++) Pb[state][imax] = 0;
    Pb[0][imax]=1;
    for(i=imax-1; i>(-1); i--) {
      g[0] = exp(-((r1[i]+1)*(r1[i]+1))/sig2);
      g[1] = exp(-((r1[i]-1)*(r1[i]-1))/sig2);

      Pb[0][i] = g[0]*Px20[i]*Pv20[i]*Pb[0][i+1] + g[1]*(1-Px20[i])*(1-
                 Pv20[i])*Pb[2][i+1];
      Pb[1][i] = g[0]*Px20[i]*Pv20[i]*Pb[2][i+1] + g[1]*(1-Px20[i])*(1-
                 Pv20[i])*Pb[0][i+1];
      Pb[2][i] = g[0]*(1-Px20[i])*(1-Pv20[i])*Pb[3][i+1] +
                 g[1]*Px20[i]*Pv20[i]*Pb[1][i+1];
      Pb[3][i] = g[0]*(1-Px20[i])*(1-Pv20[i])*Pb[1][i+1] +
                 g[1]*Px20[i]*Pv20[i]*Pb[3][i+1];
      sum = 0;
      for(state=0; state<M; state++)  sum += Pb[state][i];
      for(state=0; state<M; state++)  Pb[state][i] = Pb[state][i]/sum;

      /* MAP1 output */
      Pu0[i]=g[0]*(P[0][i]*Pb[0][i+1]*Px20[i]*Pv20[i]
             +P[1][i]*Pb[2][i+1]*Px20[i]*Pv20[i]);
      Pu0[i]+=g[0]*(P[2][i]*Pb[3][i+1]*(1-Px20[i])*(1-Pv20[i])+
              P[3][i]*Pb[1][i+1]*(1-Px20[i])*(1-Pv20[i]));
      Pu1   =g[1]*(P[0][i]*Pb[2][i+1]*(1-Px20[i])*(1-Pv20[i])+
             P[1][i]*Pb[0][i+1]*(1-Px20[i])*(1-Pv20[i]));
      Pu1+=  g[1]*(P[2][i]*Pb[1][i+1]*Px20[i]*Pv20[i]+
             P[3][i]*Pb[3][i+1]*Px20[i]*Pv20[i]);
      Pu0[i]=Pu0[i]/(Pu1+Pu0[i]);

      /* extrinsic information from MAP1 */
      Pv20[i]=P[0][i]*Pb[0][i+1]*g[0]+P[1][i]*Pb[2][i+1]*g[0]+
              P[2][i]*Pb[1][i+1]*g[1]+P[3][i]*Pb[3][i+1]*g[1];
      Pv21=P[0][i]*Pb[2][i+1]*g[1]+P[1][i]*Pb[0][i+1]*g[1]+
           P[2][i]*Pb[3][i+1]*g[0]+P[3][i]*Pb[1][i+1]*g[0];
      Pv20[i]=Pv20[i]/(Pv21+Pv20[i]);

      if(Pv20[i]<ELIM) Pv20[i]=ELIM;
      if(Pv20[i]>1-ELIM) Pv20[i]=1-ELIM;

      }  /* end i */

      /* interim dec1 decoding */
      dec1cnt = 0;
      for(i=11; i<imax-9; i++)  {
        uh[i] = 0;
        if(Pu0[i] < 0.5)  uh[i] = 1;
        if(uh[i] != u[i]) dec1cnt += 1;
      }
      if(dec1cnt==0)  break;
      BER = 1.0*dec1cnt/(1.0*nbits);
  // printf("\nblock : %ld  iteration: %d   error count = %d",loop,j,dec1cnt);

      /* interleaver */
```

```
        for(i=0;i<imax;i++) Pw20[rnum[i]]=Pv20[i];
    }   /* end j, end iterative decoding */

        /* final dec1 decoding */
        for(i=11; i<imax-9; i++)  {
          uh[i] = 0;
          if(Pu0[i] < 0.5)  uh[i] = 1;
          if(uh[i] != u[i]) {
            errcnt += 1;
            //printf("\ni = %4d   errcnt = %3d",i,errcnt);getchar();
          }
        }
       rbits = loop*nbits;
       BER = 1.0*errcnt/(1.0*rbits);
       printf("\nblock : %ld Eb/No = %4.2f BER = %9.7f decoded bits = %ld
         dec.itns = %d", loop,DB,BER,rbits,iter2); //getchar();

      }  /* end loop */    getchar();
    }
```

Fast 8×8 discrete cosine transform

```
/* FAST 8x8 DISCRETE COSINE TRANSFORM
*  Method              : row-column, radix 4 FFT
*  Function prototype: void ddct8x8s(int isgn, double **a)
*/

/*
  [definition]
    <case1> Normalized 8x8 IDCT
      C[k1][k2] = (1/4) * sum_j1=0^7 sum_j2=0^7
                  a[j1][j2] * s[j1] * s[j2] *
                  cos(pi*j1*(k1+1/2)/8) *
                  cos(pi*j2*(k2+1/2)/8), 0<=k1<8, 0<=k2<8
                  (s[0] = 1/sqrt(2), s[j] = 1, j > 0)
    <case2> Normalized 8x8 DCT
      C[k1][k2] = (1/4) * s[k1] * s[k2] * sum_j1=0^7 sum_j2=0^7
                  a[j1][j2] *
                  cos(pi*(j1+1/2)*k1/8) *
                  cos(pi*(j2+1/2)*k2/8), 0<=k1<8, 0<=k2<8
                  (s[0] = 1/sqrt(2), s[j] = 1, j > 0)
  [usage]
    <case1>
      ddct8x8s(1, a);
    <case2>
      ddct8x8s(-1, a);
  [parameters]
    a[0...7][0...7] :input/output data (double **)
                     output data
                       a[k1][k2] = C[k1][k2], 0<=k1<8, 0<=k2<8
*/

/* Cn_kR = sqrt(2.0/n) * cos(pi/2*k/n) */
/* Cn_kI = sqrt(2.0/n) * sin(pi/2*k/n) */
/* Wn_kR = cos(pi/2*k/n) */
/* Wn_kI = sin(pi/2*k/n) */
#define C8_1R    0.49039264020161522456
#define C8_1I    0.09754516100806413392
#define C8_2R    0.46193976625564337806
#define C8_2I    0.19134171618254488586
#define C8_3R    0.41573480615127261854
#define C8_3I    0.27778511650980111237
#define C8_4R    0.35355339059327376220
```

```
#define W8_4R   0.70710678118654752440

void ddct8x8s(int isgn, double **a)
{
  int j;
  double x0r, x0i, x1r, x1i, x2r, x2i, x3r, x3i;
  double xr, xi;

  if (isgn < 0) {
    for (j = 0; j <= 7; j++) {
      x0r = a[0][j] + a[7][j];
      x1r = a[0][j] - a[7][j];
      x0i = a[2][j] + a[5][j];
      x1i = a[2][j] - a[5][j];
      x2r = a[4][j] + a[3][j];
      x3r = a[4][j] - a[3][j];
      x2i = a[6][j] + a[1][j];
      x3i = a[6][j] - a[1][j];
      xr = x0r + x2r;
      xi = x0i + x2i;
      a[0][j] = C8_4R * (xr + xi);
      a[4][j] = C8_4R * (xr - xi);
      xr = x0r - x2r;
      xi = x0i - x2i;
      a[2][j] = C8_2R * xr - C8_2I * xi;
      a[6][j] = C8_2R * xi + C8_2I * xr;
      xr = W8_4R * (x1i - x3i);
      x1i = W8_4R * (x1i + x3i);
      x3i = x1i - x3r;
      x1i += x3r;
      x3r = x1r - xr;
      x1r += xr;
      a[1][j] = C8_1R * x1r - C8_1I * x1i;
      a[7][j] = C8_1R * x1i + C8_1I * x1r;
      a[3][j] = C8_3R * x3r - C8_3I * x3i;
      a[5][j] = C8_3R * x3i + C8_3I * x3r;
    }
    for (j = 0; j <= 7; j++) {
      x0r = a[j][0] + a[j][7];
      x1r = a[j][0] - a[j][7];
      x0i = a[j][2] + a[j][5];
      x1i = a[j][2] - a[j][5];
      x2r = a[j][4] + a[j][3];
      x3r = a[j][4] - a[j][3];
      x2i = a[j][6] + a[j][1];
      x3i = a[j][6] - a[j][1];
      xr = x0r + x2r;
      xi = x0i + x2i;
      a[j][0] = C8_4R * (xr + xi);
      a[j][4] = C8_4R * (xr - xi);
      xr = x0r - x2r;
      xi = x0i - x2i;
      a[j][2] = C8_2R * xr - C8_2I * xi;
      a[j][6] = C8_2R * xi + C8_2I * xr;
      xr = W8_4R * (x1i - x3i);
```

```
            x1i = W8_4R * (x1i + x3i);
            x3i = x1i - x3r;
            x1i += x3r;
            x3r = x1r - xr;
            x1r += xr;
            a[j][1] = C8_1R * x1r - C8_1I * x1i;
            a[j][7] = C8_1R * x1i + C8_1I * x1r;
            a[j][3] = C8_3R * x3r - C8_3I * x3i;
            a[j][5] = C8_3R * x3i + C8_3I * x3r;
        }
    } else {
        for (j = 0; j <= 7; j++) {
            x1r = C8_1R * a[1][j] + C8_1I * a[7][j];
            x1i = C8_1R * a[7][j] - C8_1I * a[1][j];
            x3r = C8_3R * a[3][j] + C8_3I * a[5][j];
            x3i = C8_3R * a[5][j] - C8_3I * a[3][j];
            xr = x1r - x3r;
            xi = x1i + x3i;
            x1r += x3r;
            x3i -= x1i;
            x1i = W8_4R * (xr + xi);
            x3r = W8_4R * (xr - xi);
            xr = C8_2R * a[2][j] + C8_2I * a[6][j];
            xi = C8_2R * a[6][j] - C8_2I * a[2][j];
            x0r = C8_4R * (a[0][j] + a[4][j]);
            x0i = C8_4R * (a[0][j] - a[4][j]);
            x2r = x0r - xr;
            x2i = x0i - xi;
            x0r += xr;
            x0i += xi;
            a[0][j] = x0r + x1r;
            a[7][j] = x0r - x1r;
            a[2][j] = x0i + x1i;
            a[5][j] = x0i - x1i;
            a[4][j] = x2r - x3i;
            a[3][j] = x2r + x3i;
            a[6][j] = x2i - x3r;
            a[1][j] = x2i + x3r;
        }
        for (j = 0; j <= 7; j++) {
            x1r = C8_1R * a[j][1] + C8_1I * a[j][7];
            x1i = C8_1R * a[j][7] - C8_1I * a[j][1];
            x3r = C8_3R * a[j][3] + C8_3I * a[j][5];
            x3i = C8_3R * a[j][5] - C8_3I * a[j][3];
            xr = x1r - x3r;
            xi = x1i + x3i;
            x1r += x3r;
            x3i -= x1i;
            x1i = W8_4R * (xr + xi);
            x3r = W8_4R * (xr - xi);
            xr = C8_2R * a[j][2] + C8_2I * a[j][6];
            xi = C8_2R * a[j][6] - C8_2I * a[j][2];
            x0r = C8_4R * (a[j][0] + a[j][4]);
            x0i = C8_4R * (a[j][0] - a[j][4]);
            x2r = x0r - xr;
```

```
      x2i = x0i - xi;
      x0r += xr;
      x0i += xi;
      a[j][0] = x0r + x1r;
      a[j][7] = x0r - x1r;
      a[j][2] = x0i + x1i;
      a[j][5] = x0i - x1i;
      a[j][4] = x2r - x3i;
      a[j][3] = x2r + x3i;
      a[j][6] = x2i - x3r;
      a[j][1] = x2i + x3r;
    }
  }
}
```

Index